Fluidization Engineering

SECOND EDITION

Butterworth–Heinemann

Series in Chemical Engineering

SERIES EDITOR

Howard Brenner
Massachusetts Institute of Technology

ADVISORY EDITORS

Andreas Acrivos
The City College of CUNY

James E. Bailey
California Institute of Technology

Manfred Morari
California Institute of Technology

E. Bruce Nauman
Rensselaer Polytechnic Institute

J.R.A. Pearson
Schlumberger Cambridge Research

Robert K. Prud'homme
Princeton University

SERIES TITLES

Chemical Process Equipment: Selection and Design
Stanley M. Walas

Chemical Process Structures and Information Flows
Richard S.H. Mah

Computational Methods for Process Simulations
W. Fred Ramirez

Constitutive Equations for Polymer Melts and Solutions
Ronald G. Larson

Fluidization Engineering, Second Edition
Daizo Kunii and Octave Levenspiel

Fundamental Process Control
David M. Prett and Carlos E. García

Gas-Liquid-Solid Fluidization Engineering
Liang-Shin Fan

Gas Separation by Adsorption Processes
Ralph T. Yang

Granular Filtration of Aerosols and Hydrosols
Chi Tien

Heterogeneous Reactor Design
Hong H. Lee

Molecular Thermodynamics of Nonideal Fluids
Lloyd L. Lee

Phase Equilibria in Chemical Engineering
Stanley M. Walas

Physicochemical Hydrodynamics: An Introduction
Ronald F. Probstein

Transport Processes in Chemically Reacting Flow Systems
Daniel E. Rosner

Viscous Flow: The Practical Use of Theory
Stuart W. Churchill

Fluidization Engineering

SECOND EDITION

Daizo Kunii
Fukui Institute of Technology
Fukui City, Japan

Octave Levenspiel
Chemical Engineering Department
Oregon State University
Corvallis, Oregon

Butterworth−Heinemann
An Imprint of Elsevier
Boston London Singapore Sydney Toronto Wellington

Library of Congress Cataloging-in-Publication Data

Kunii, Daizo, 1923–
    Fluidization engineering / Daizo Kunii, Octave Levenspiel.
—2nd ed.
        p.   cm.—(Butterworth–Heinemann series in chemical engineering)
      ISBN 0-409-90233-0
      1. Fluidization. I. Levenspiel, Octave. II. Title.
      III. Series.
      TP156.F65K8 1991
      660'.284292—dc20   90-31800

British Library Cataloguing in Publication Data

Kunii, Daizo
    Fluidization engineering.
    1. Chemical engineering. Fluidisation
    I. Title II. Levenspiel, Octave
    660.2842

    ISBN 0-409-90233-0

Butterworth–Heinemann
313 Washington Street
Newton, MA 02158-1626

10 9 8 7

Transferred to digital printing 2006

Contents

CHAPTER 3   **Fluidization and Mapping of Regimes**   **61**

CHAPTER 12   **Conversion of Gas in Catalytic Reactions**   **277**

Preface

TO THE

SECOND EDITION

In the early 1960s Davidson's explanation of the movement of gas around a rising gas bubble became the seminal concept that guided research and advanced understanding of dense bubbling fluidized beds. More than anything else our appreciation of the potentialities of this remarkable analysis was what led us to write *Fluidization Engineering*.

In that book we developed a physical-based model to represent the behavior of fine particle systems based on the Davidson bubble and Rowe's finding on bubble wakes. We showed how this model could make sense of a variety of phenomena in dense bubbling fluidized beds. It was the first of a new class of models, the hydrodynamic model, and since its introduction many extensions and variations have been proposed.

Since writing that book, much that is new and exciting has occurred in the field of fluidization—new insights, new understandings, and new predictive methods. First we have the Geldart classification of solids, which divides the behavior of dense beds into four distinct classes. We see the systematic studies of the freeboard region above the dense bed, and of high-velocity fluidization with its significant carryover of particles, which requires the replenishment of bed solids. This regime of operations leads to what is called the *circulating fluidized bed* and *fast fluidization*.

In another direction, the interest in fluidized coal combustion and other large particle systems has spawned many studies of this regime of gas-solid contacting. As a result of these developments, today the term *fluidization* takes on a broader meaning; consequently new predictive methods are being developed to cover this wider range of gas-solid contacting. This has led us to conclude that it is time for a new edition of our book.

In this second edition we expand our original scope to encompass these new areas, and we also introduce reactor models specifically for these contacting regimes. With all these changes, this is largely a new book. Again we generously sprinkle this book with illustrative examples, over 60, plus problems to challenge the student. We hope that these exercises will help cement the ideas developed in the text.

This book does not cover all that is happening in fluidization. Our aim is to distill from these thousands of studies those developments that are pertinent to the engineer concerned with predictive methods, for the designer, and for the user and potential user of fluidized beds. In this sense, ours is an engineering book. We hope that the researcher and practitioner will find it useful.

Grateful thanks to Misses Yoshimi Kawamata and Peggy Offutt for their marathon typing efforts, to Bekki Levien who prepared the illustrations for this volume, to our colleagues at YNU and OSU for their implicit support, and to our wives, Yoneko and Mary Jo, for their quiet encouragement of this project.

Daizo Kunii
Fall 1990   *Octave Levenspiel*

Preface

TO THE

FIRST EDITION

Fluidization has had a rather turbulent history. It hit the industrial scene in a big way in 1942 with catalytic cracking, and has since moved into many other areas. Its proud successes and its spectacular flops spurred research efforts so that there are now thousands of reported studies on the subject. Unfortunately there is still much confusion and contradiction in the reported literature, countless recommended correlations, but little in the way of unifying theory. Most of the research is done in small-scale equipment, even though the designer is well aware that small and large beds behave so differently that extrapolation up to the commercial scale can be quite unreliable. This is particularly true with reactor applications. Consequently, industrial design places much emphasis on previous practice or on careful scale-up coupled with a liberal sprinkling of safety factors. Thus the practice of the art dominates, design from first principles is rarely attempted, and the numerous research findings do not seem to be very pertinent in this effort. Taken together this represents a rather unsatisfactory state of affairs.

In 1963 the two of us started corresponding about fluidization. We both felt that practical design should more closely rely on basic investigations, and we soon agreed that to bridge this gap what was first needed was a reasonable representation of gas-solid contacting in the bed. We tried a number of approaches, in all cases testing these with the reported findings in the literature, and finally settled with a rather simple description that we call *the bubbling bed model*. This model is able to explain a variety of observed kinetic and flow phenomena, its equations are suitable for scale-up and design purposes, and its predictions have since been tested in commercial applications by Kunii. We decided to write this book in early 1964, and this model plays a large part in the book.

As authors, one of our prime responsibilities has been to decide what to include, but more important still, what to leave out. Our book is not intended to be encyclopedic, and certain topics receiving much attention in the literature are barely touched on or are completely ignored. The overriding consideration governing our choice of material is its relevancy to possible use. This strong emphasis on utility is the reason we use the word *Engineering* in the title.

In our presentation we have used theory whenever possible to try to bring order to the chaos of isolated fact and correlation, to help organize the information, and to facilitate understanding. As examples of this program we have a unified representation for bed-wall heat transfer to bring together the seemingly contradictory theories proposed to date, a model accounting for all aspects of elutriation and carryover, and numerous kinetic models to describe the rates of physical and chemical changes, growth and shrinkage of solids, and deactivation and regeneration of catalysts.

To clarify the text we have given 68 illustrative examples. Problems are also included. They extend the ideas in the text, they may be used as an aid to teaching, and of course they may serve as 152 distinct torture devices for students.

This book has not been written with any particular audience in mind. Different readers may find different parts of it of interest. First, we expect that the engineer engaged in design and development of processes requiring gas-solid contact should find the latter part of the book particularly useful. Next, the researcher should be interested in probing the conceptual developments that are presented. These may suggest pertinent questions for further study. In particular, extensions, modifications, and refinements of the bubbling bed model should be well worth exploring. This book could also be used as a text for courses in gas-solid contacting in chemical, metallurgical, and mining engineering programs. Unfortunately, we expect this type of course to become established in American chemical engineering curricula only when its educators become as concerned with the complex and difficult-to-treat multiphase problems as they are now concerned with the classical problems of fluid mechanics.

We would like to express our appreciation to the following people who in various ways helped us in our project:

Drs. Kunio Yoshida, Stephen Szepe, Norman Weinstein, and Thomas Fitzgerald for their critical readings of various parts of the manuscript and for their helpful comments.

The many graduate students at the University of Tokyo and at Illinois Institute of Technology who unknowingly influenced the writing of the book by their questions, discussions, or blank stares.

Miss Kazuko Tanabe of Tokyo, Mrs. Violet Reus, and Miss Diana Aletto of Chicago, whose nimble typing fingers shuttled up to as many as eight drafts of certain chapters back and forth across the Pacific.

And finally, our wives, Yoneko and Mary Jo, who so good-naturedly accepted and cooperated with this project, helped with the typing and in so many other ways, even though they knew that there were more important things to be done.

Daizo Kunii
Octave Levenspiel

October 1968

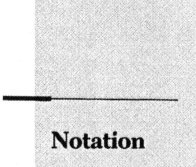

Notation

Symbols and constants that are defined and used locally are not included here. SI units are given to illustrate the dimensions of the various symbols. The equations indicated refer to the location where the symbol is first used or first defined.

a, a'	surface areas of solid per volume of bed and per volume of solid, respectively, m^{-1}; Eq. (3.4)	C_p^*	concentration of vapor in the gas that is in equilibrium with the surface of the particle, mol/m^3; Eq. (16.51)
a, a'	decay constants of clusters in the freeboard, m^{-1}; Eqs. (7.17) and (12.60)	C_{pg}, C_{pl}, C_{ps}	specific heat of gas, liquid, and solid, respectively, $J/kg \cdot K$
a_b	bubble-emulsion interfacial area per bed volume, m^{-1}; Eq. (10.18)	C_s	concentration of tracer solids in the bed, kg/m^3; Eq. (9.4)
$\mathbf{a}$	activity of catalyst, dimensionless; Eq. (15.9)	$\bar{C}_A$	time average concentration of A that a particle encounters, mol/m^3; Eq. (18.43)
A	gaseous reactant	C_A^*	concentration of vapor A in gas that is in equilibrium with bed solids, mol/m^3; Eq. (16.28)
Ar	Archimedes number, dimensionless; Eq. (3.20)		
A_t	cross-sectional area of bed, m^2	C_{Ab}, C_{Ac}, C_{Ae}	concentration of A in the gas bubble, in the cloud-wake region, and in the emulsion phase, respectively, mol/m^3
A_w	area of vessel wall or of heat exchange surface, m^2		
b	stoichiometric coefficient; Eq. (18.1)	C_{Ai}, C_{Ao}, C_{Aex}	concentration of A in the entering gas stream, at the top of the dense bed, and at the exit of the vessel, respectively, mol/m^3
b, B	constants, appear in various places in text		
B	solid reactant		
$C_{d,or}$	orifice coefficient, dimensionless; Eq. (4.12)	C_D	drag coefficient, dimensionless; Eq. (3.28)

C_p	specific heat, J/kg·K
d^*	dimensionless measure of particle diameter; Eq. (3.31)
d_b	effective bubble diameter, m; Eq. (5.2)
d_{bm}	maximum bubble diameter, m; Eq. (6.4)
d_{b0}	bubble diameter just above the distributor, m; Eq. (5.14)
d_{eff}	effective diameter of particles in a bed, m; Eq. (3.3)
d_h	diameter of hole of a multihole tuyere, m; Chap. 4
d_i	outer diameter of tubes inserted into beds, m; Example 6.4
d_{or}	orifice diameter, m
d_p	particle diameter based on screen analysis, m; Eq. (3.5)
d_{sph}	equivalent spherical diameter of a particle, m; Eq. (3.1)
d_t	bed or tube diameter, m
d_{te}	effective diameter of a bed that contains internals, m; Eq. (6.13)
D_{gv}, D_{gh}	vertical and horizontal dispersion coefficients of gas, respectively, m²/s; Eqs. (10.1) and (10.3)
D_{sv}, D_{sh}	vertical and horizontal dispersion coefficients of solids, respectively, m²/s; Eqs. (9.1) and (9.13)
$\mathscr{D}$	molecular diffusion coefficient of gas, m²/s
$\mathscr{D}_e$	effective diffusion coefficient of gas in the emulsion, m²/s; Eq. (10.29)
$\mathscr{D}_m$	diffusion coefficient of moisture, m²/s; Eq. (16.46)
$\mathscr{D}_s$	effective diffusivity of gas through the product blanket, m²/s; Eq. (18.10)
e_s, e_w	emissivity of solids and of wall, respectively, dimensionless; Eq. (13.13)
$\mathbf{E}(t)$	exit age distribution function, s⁻¹; Fig. 10.7

f_c, f_e, f_w	volume fractions of cloud, emulsion, and wake region, respectively, per bubble volume, dimensionless; Eqs. (5.7), (6.32), and (5.11)
f_g, f_s, f_s'	friction factors for gas flow, between dispersed solids and gas, and for gas-solid mixtures, respectively, dimensionless; Eqs. (15.37), (15.36), and (15.35)
F_s	flow rate of solids, kg/s; Eq. (15.15)
F_0, F_1, F_2	feed rate of solids, overflow rate of solids, and carryover rate of solids by entrainment, respectively, kg/s; Eq. (14.1)
$\mathbf{F}(t)$	output tracer concentration versus time for a unit step input of tracer, dimensionless; Fig. 10.7
g	= 9.8 m/s², acceleration of gravity
g_c	$= \dfrac{1\ \text{kg·m}}{\text{N·s}^2} = 32.2\ \dfrac{\text{lb·ft}}{\text{lbf·s}^2}$ $= \dfrac{9.8\ \text{kg·m}}{\text{kg-wt·s}^2}$, conversion factor
G_g	mass flux of gas, kg/m²·s; Eq. (15.23)
G_s	mass flux of solids, kg/m²·s; Eq. (7.1)
G_s^*	saturated mass flux of solids, kg/m²·s; Eq. (7.28)
G_{sd}, G_{su}	downflow and upflow flux of solids, respectively, kg/m²·s; Eq. (7.1)
G_{si}	mass flux of solids from a bed of pure i, kg/m²·s; Eq. (7.3)
G_{s1}, G_{s2}, G_{s3}	mass flux of dispersed solids, of upward-moving clusters and of downward-moving clusters in the freeboard, respectively, kg/m²·s; Eq. (7.20)
h	height, m
h	heat transfer coefficient between wall and bed, W/m²·K; Eq. (13.1)

h^*	heat transfer coefficient at a single sphere falling through a gas, $W/m^2 \cdot K$; Eq. (11.25)	surface reaction, m/s; Eq. (18.6)
h_{bc}	heat transfer coefficient between bubble and cloud, $W/m^2 \cdot K$; Eq. (11.33)	k_d — mass transfer coefficient, m/s
		k_d^* — mass transfer coefficient at a single sphere falling through a gas, m/s; Eq. (11.1)
h_{bed}	apparent heat transfer coefficient between gas and bed based upon total surface area of particles, $W/m^2 \cdot K$; after Eq. (11.27)	$k_{d,bed}$ — apparent mass transfer coefficient between gas and bed, based on the surface area of all the bed particles, m/s; Eq. (11.6)
h_g	gas convection heat transfer coefficient, $W/m^2 \cdot K$; Eq. (13.16)	$k_{d,p}$ — mass transfer coefficient between gas and a single particle, m/s; Eq. (11.3)
h_p	real heat transfer coefficient between gas and single particles, $W/m^2 \cdot K$; Eq. (11.28)	k_r — rate constant for a first-order gas-solid reaction, $m^3/mol \cdot s$; Eq. (18.5)
h_r	radiant heat transfer coefficient, $W/m^2 \cdot K$; Eq. (13.13)	k_e° — effective thermal conductivity of a fixed bed with a stagnant gas, $W/m^2 \cdot K$; (Eq. (13.2)
h_w°	heat transfer coefficient in the wall region of a fixed bed with stagnant gas, $W/m^2 \cdot K$; Eq. (13.5)	k_{ew}° — effective thermal conductivity of a thin layer of bed near the wall surface, $W/m^2 \cdot K$; Eq. (13.4)
H	enthalpy, J/kg; Eq. (15.17)	k_g — thermal conductivity of gas, $W/m^2 \cdot K$; Eq. (11.25)
H_d, H_f	height of lower dense bed and of freeboard, respectively, m; Eq. (8.10)	k_s — thermal conductivity of solid, $W/m^2 \cdot K$; Eq. (13.2)
H_t	$= H_d + H_f$, height of column, m	K_a, K_{a1} — rate constant for the deactivation of catalyst, s^{-1}; Eqs. (15.10) (17.8)
H_{total}	total volumetric heat transfer coefficient between gas and bed, $W/m^2 \cdot K$; Eq. (11.34)	K_{a2} — rate constant for the regeneration of catalyst, s^{-1}; Eq. (17.9)
ΔH_r	heat of reaction, J/kg; Eq. (15.17)	K_{bc} — coefficient of gas interchange between bubble and cloud-wake region, s^{-1}; Eq. (10.13)
$I(R, R_i)$	integral defined in Eq. (14.32)	
I_s	solids flux based on open area of holes, $kg/m^2 \cdot s$; Eq. (9.18)	K_{be} — overall coefficient of gas interchange between bubble and emulsion phase, s^{-1}; Eq. (10.13)
k, k'	rate coefficients of growth or shrinkage of particles, m/s; Eqs. (14.21) and (14.23)	
$\bar{k}$	overall rate coefficient, m/s; Eq. (18.11)	K_{ce} — coefficient of gas interchange between cloud-wake region and emulsion phase, s^{-1}; Eq. (10.13)
k_{bc}, k_{ce}, k_{be}	mass transfer coefficients, m/s; Eqs. (10.26), (10.28), and (10.17)	K_d — interchange coefficient for mass transfer between bubble and emulsion, s^{-1}; Eq. (11.8)
k_c	rate constant for a first-order	

K_f	overall rate constant for a first-order chemical reaction in a fluidized bed, s^{-1}; Eq. (12.13)
K_r	rate constant for a first-order catalytic reaction, s^{-1}; Eq. (12.1)
K_1, K_2, K_3	rate constants for the interchange of solids in freeboard, s^{-1}; Eqs. (7.23–25)
l	length of heat exchange tube, m
l_i	tube pitch or center-to-center distance between adjacent tubes in a tube bundle, m; after Eq. (6.13)
l_{or}	center-to-center distance between neighboring orifices or tuyeres, m; Eq. (5.16)
L	length of solid transporting tube, m; Eq. (15.25)
L_j	jet penetration length, m; before Eq. (4.2)
L_m, L_{mf}, L_f	height of fixed bed, bed at minimum fluidization, and bubbling fluidized bed, respectively, m; Eq. (6.19)
$\mathscr{L}$	latent heat of vaporization, J/kg; Eq. (16.44)
$\mathbf{m}$	equilibrium adsorption constant, dimensionless; Eq. (10.6)
M	solids mixing index, dimensionless; Eq. (9.16)
M	molecular weight, kg/mol
n_b	bubble frequency of an orifice, s^{-1}; Eq. (5.12)
n_i	seed rate of solids based on unit volume of entering solids, m^{-3}; Eq. (14.48)
n_{or}	bubble frequency at an orifice plate distributor, s^{-1}
n_w	bubble frequency in the vicinity of the wall, s^{-1}
N	number of stages in a multistage processing unit
N_A, N_B	number of moles of A and B, respectively
N_t	number of exchanger tubes
Nu_p	$= h_p d_p/k_g$, Nusselt number for gas-particle heat transfer, dimensionless; Fig. 10.6
N_{or}	number of orifices per unit area of distributor, m^{-2}; Eq. (4.13)
p	pressure, Pa
Δp_b	pressure drop across the bed, Pa; Eq. (3.16)
$\Delta p_d, \Delta p_v$	pressure drop across a distributor and across a valve, respectively, Pa; Eqs. (4.3) and (15.4)
$\Delta p_{fr}, \Delta p_f$	frictional pressure drop, Pa; Eqs. (3.6) and (15.35)
$\mathbf{p}, \mathbf{P}$	size distribution functions; Eq. (3.9)
$\mathbf{p}_e$	size distribution of entrained particles, m^{-1}; Eq. (7.5)
$\mathbf{p}_0, \mathbf{p}_1, \mathbf{p}_2, \mathbf{p}_b$	size distribution of feed solids, overflow solids, carryover solids, and solids in the bed, respectively, m^{-1}; Chap. 14
Pr	$= C_{pg}\mu/k_g$, Prandtl number, dimensionless
q	heat transfer rate, W; Eq. (13.1)
q_l	rate of heat loss from equipment to surroundings, W; Eq. (16.24)
Q	moisture fraction of particles, kg of gas or liquid adsorbed/kg of dry solids; Eq. (16.29)
Q_{cr}, Q_f, Q_∞	critical moisture fraction, free moisture fraction, and moisture fraction at infinite time in particles, respectively; Chap. 16
Q_m	moisture fraction at position r in a particle, Eq. (16.46)
r	distance from the center of a particle or a bubble, m
r_c	radius of unreacted core of reactant solid, m; Eq. (18.7)

R	product or intermediate formed by reaction	TDH	transport disengaging height, m; beginning Chap. 7
R, R_i	radius of particle and initial radius of a particle of changing size, respectively, m; Chap. 18	$u, \bar{u}$	velocity and mean velocity of gas, respectively, m/s
R_b, R_c	radius of bubble and radius of cloud surrounding a bubble, respectively, m; Eq. (5.6)	u^*	dimensionless measure of particle velocity; Eq. (3.32)
R_m, R_M	smallest and largest particle size in the feed of a size distribution of solids, respectively; above Eq. (14.17)	u_b	velocity of a bubble rising through a bed, m/s; Eq. (6.8)
		u_{br}	rise velocity of a bubble with respect to the emulsion phase, m/s; Eq. (5.1)
R^{max}	maximum of the size distribution of solids, m; Eq. (14.45)	u_c	gas velocity at which pressure fluctuations in a bubbling bed are maximum, m/s; after Example 3.3
$\bar{R}_s$	surface mean particle radius, m; Eq. (14.41)	u_{ch}, u_{cs}	choking velocity and saltation velocity, respectively, m/s; Chap. 15
Re_p	$= d_p u_o \rho_g/\mu$, particle Reynolds number, dimensionless	u_e	upward superficial velocity of gas through the emulsion phase, m/s; Eq. (6.1)
$\mathbf{R}$	$= 8.314 \text{ J/mol·K}$, ideal gas constant	$u_{e,up}$	upward velocity of emulsion solids, m/s; Eq. (6.10)
$\mathcal{R}$	rate of linear particle growth or shrinkage, m/s; Eq. (14.21)	u_f	$= u_{mf}/\varepsilon_{mf}$, upward velocity of gas at minimum fluidizing conditions, m/s; before Eq. (5.5)
S	a final product in a complex reaction		
S_{be}	surface area of a bubble, m^2; Eq. (10.17)	u_k	superficial gas velocity on entry into the turbulent regime, m/s; after Example 3.3
Sc	$= \mu/\rho_g \mathcal{D}$, Schmidt number, dimensionless		
Sh	$= k_{d,p} d_p y/\mathcal{D}$, Sherwood number, dimensionless	u_{mb}	minimum bubbling velocity, m/s; Eq. (3.27)
$\mathbf{S_R}$	selectivity, moles of desired product R formed per mole of reactant reacted, dimensionless, Chap. 12	u_{mf}	superficial gas velocity at minimum fluidizing conditions, m/s; Eq. (3.18)
t	time, s	u_o	superficial gas velocity (measured on an empty vessel basis) through a bed of solids, m/s
$\bar{t}$	mean residence time of gas or solid in a vessel, s		
T	a final product in a complex reaction	u_{or}	velocity of gas through an orifice, m/s; Eq. (4.12)
T	temperature, K or °C	u_p	slip velocity between gas and solid, m/s; Eq. (8.3)
T_p, T_b, T_g, T_s, T_w	temperature of particle, bed, gas, solid, and wall, respectively, K or °C	u_s	mean downward velocity of solids, m/s

u_t terminal velocity of a falling particle, m/s; Eq. (3.28)

u_1, u_2, u_3 mean velocity of dispersed solids, of upward-moving cluster, and of downward-moving cluster, respectively, m/s; Eqs. (7.23)–(7.25)

v volumetric flow rate of gas, m^3/s

v_{or} volumetric flow rate of gas through an orifice, m^3/s; Eq. (4.1)

V_b volume of a gas bubble, m^3; Eq. (4.1)

V_s volume of solids in a fluidized reactor, m^3; Eq. (12.1)

V_w volume of wake following a rising gas bubble, m^3; Eq. (5.11)

$\dot{w}_{s,actual}$ pumping power, W; Eq. (4.20)

$w_{s,ideal}$ ideal work of compression, J/kg; Eq. (4.17)

$\dot{w}_{s,ideal}$ ideal pumping power, W; Eq. (4.18)

W mass of solids, kg

x_i weight fraction of solids of size i, dimensionless; Eq. (7.3)

X_A, X_B conversion of reactant A and of reactant B, respectively, dimensionless; Eqs. (12.2) and (18.7)

y $= \bar{t}/\tau$, time ratio, dimensionless; Eq. (14.40)

z, z_f distance above the distributor and distance above the mean surface of the fluidized bed, respectively, m; Eq. (7.23)

z_i symmetrical point in the freeboard, m; Fig. 8.6; injection level, m; Eq. (10.2)

Greek Symbols

α measure of kinetic energy of an orifice jet, dimensionless; Eq. (4.15)

α ratio of effective diameter of the wake to diameter of the bubble, dimensionless; Eq. (9.14)

α_w constant representing the mixing of gas in the vicinity of wall, dimensionless; Eq. (13.6)

β weight ratio of product solids referred to the feed solids, dimensionless; Eq. (14.20)

γ $= C_{pg}/C_{vg}$, ratio of specific heats of gas, dimensionless; Eq. (4.18)

$\gamma_b, \gamma_c, \gamma_e$ volume of solids dispersed in bubbles, in the cloud-wake regions, and in the emulsion phase, respectively, divided by the volume of the bubbles, dimensionless; Eq. (6.33)

δ bubble fraction in a fluidized bed, dimensionless; Eq. (6.20)

ε void fraction, dimensionless; Eq. (3.6)

$\varepsilon_e, \varepsilon_f, \varepsilon_m, \varepsilon_{mf}$ void fraction in the emulsion phase of a fluidized bed, in a fluidized bed as a whole, in a fixed bed, and in a bed at minimum fluidizing conditions, respectively

ε_s $= 1 - \varepsilon_f$, volume fraction of solids, dimensionless; Eq. (8.1)

ε_s^* saturated carrying capacity of a gas, or maximum volume fraction of solids that can be pneumatically conveyed by a gas; Eq. (8.6)

ε_{sd} volume fraction of solids in the lower dense region of a fast fluidized bed, dimensionless; Chap. 8

ε_{se} volume fraction of solids at the column exit, dimensionless; Chap. 8

ε_w mean void fraction in the vicinity of wall, dimensionless; Eq. (13.4)

η various measures of efficiency, dimensionless

η_{bed}	conversion efficiency of a fluidized bed; Eq. (12.21)	
η_d	adsorption efficiency factor; Eq. (11.23)	
η_g	efficiency of heat utilization of gas; Eq. (16.14)	
η_g'	efficiency of solute removal from gas; Eq. (16.41)	
η_h	heat transfer efficiency factor; Eq. (11.34)	
η_s	efficiency of heat utilization of solids; Eq. (16.15)	
η_s'	adsorption efficiency of solids; Eq. (16.42)	
$\eta(R)$	cyclone or separator efficiency for particles of size R, dimensionless; above Eq. (14.19)	
θ	angle, degrees	
θ_f, θ_r	angle of internal friction of a mound of solids and angle of repose, respectively, degrees; Eq. (15.21); Fig. 15.7	
κ	elutriation rate constant, s^{-1}; Eq. (7.7)	
κ^*	elutriation rate constant, kg/m^2·s; Eq. (7.6)	
μ	viscosity of gas, kg/m·s	
ρ_B	molar density of solids, mol/m^3	
$\rho_g, \rho_{mf}, \rho_s, \bar{\rho}$	gas density, bulk density of a	

bed at minimum fluidizing conditions, density of solids, and mean density of a gas-solid mixture, respectively, kg/m^3

σ^2 — variance of a residence time distribution curve, s^2; Eq. (10.10)

τ — reactor ability measure, m^3 cat/(m^3 feed/s); Eq. (12.4)

τ — time needed for complete drying or for complete reaction of a feed particle, s; Eqs. (16.56) and (18.9)

ϕ — ratio of heat carrying capacity of a stream of gas and a stream of solids, dimensionless; Eq. (16.13)

ϕ_b — ratio of an equivalent thickness of gas film referring to particle diameter, dimensionless; Eq. (13.3)

ϕ_s — sphericity of a particle, dimensionless; Eq. (3.2)

ϕ_w — ratio of an equivalent thickness of gas film referring to particle diameter in the vicinity of wall, dimensionless; Eq. (13.4)

ψ — ratio of observed bubble flow to that expected from two-phase theory, dimensionless; Eq. (6.3)

Introduction

Fluidization is the operation by which solid particles are transformed into a fluidlike state through suspension in a gas or liquid. This method of contacting has some unusual characteristics, and fluidization engineering puts them to good use.

The Phenomenon of Fluidization

If a fluid is passed upward through a bed of fine particles, as shown in Fig. 1(a), at a low flow rate, the fluid merely percolates through the void spaces between stationary particles. This is a *fixed bed*. With an increase in flow rate, particles move apart and a few vibrate and move in restricted regions. This is the *expanded bed*.

At a still higher velocity, a point is reached where all the particles are just suspended by the upward-flowing gas or liquid. At this point the frictional force between particle and fluid just counterbalances the weight of the particles, the vertical component of the compressive force between adjacent particles disappears, and the pressure drop through any section of the bed about equals the weight of fluid and particles in that section. The bed is considered to be just fluidized and is referred to as an *incipiently fluidized bed* or a bed at *minimum fluidization*; see Fig. 1(b).

In liquid-solid systems, an increase in flow rate above minimum fluidization usually results in a smooth, progressive expansion of the bed. Gross flow instabilities are damped and remain small, and heterogeneity, or large-scale voids of liquid, are not observed under normal conditions. A bed such as this is called a *particulately fluidized bed*, a *homogeneously fluidized bed*, or a *smoothly fluidized bed*; see Fig. 1(c). In gas-solid systems, such beds can be observed only under special conditions of fine light particles with dense gas at high pressure.

Generally, gas-solid systems behave quite differently. With an increase in flow rate beyond minimum fluidization, large instabilities with bubbling and

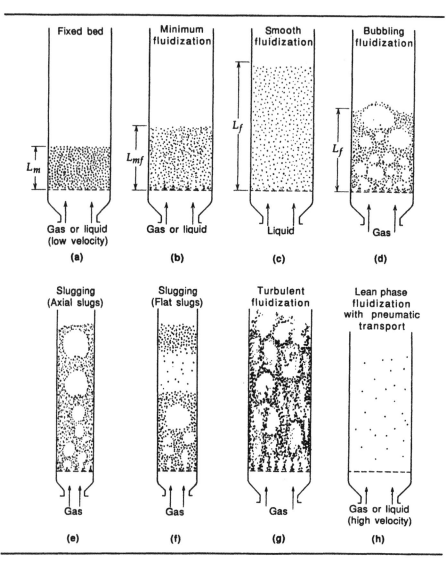

FIGURE 1
Various forms of contacting of a batch of solids by fluid.

channeling of gas are observed. At higher flow rates, agitation becomes more violent and the movement of solids becomes more vigorous. In addition, the bed does not expand much beyond its volume at minimum fluidization. Such a bed is called an *aggregative fluidized bed*, a *heterogeneous fluidized bed*, or a *bubbling fluidized bed*; see Fig. 1(d). In a few rare cases, liquid-solid systems also behave as bubbling beds. This occurs only with very dense solids fluidized by low-density liquids.

Both gas and liquid fluidized beds are considered to be *dense-phase fluidized beds* as long as there is a fairly clearly defined upper limit or surface to the bed.

In gas-solid systems, gas bubbles coalesce and grow as they rise, and in a deep enough bed of small diameter they may eventually become large enough to spread across the vessel. In the case of fine particles, they flow smoothly down by the wall around the rising void of gas. This is called *slugging*, with *axial slugs*, as shown in Fig. 1(e). For coarse particles, the portion of the bed above the bubble is pushed upward, as by a piston. Particles rain down from the slug, which finally disintegrates. At about this time another slug forms, and this unstable oscillatory motion is repeated. This is called a *flat slug*; see Fig. 1(f). Slugging is especially serious in long, narrow fluidized beds.

When fine particles are fluidized at a sufficiently high gas flow rate, the terminal velocity of the solids is exceeded, the upper surface of the bed disappears, entrainment becomes appreciable, and, instead of bubbles, one observes a turbulent motion of solid clusters and voids of gas of various sizes and shapes. This is the *turbulent fluidized bed*, shown in Fig. 1(g). With a further increase in gas velocity, solids are carried out of the bed with the gas. In this state we have a *disperse-*, *dilute-*, or *lean-phase fluidized bed* with *pneumatic transport of solids*; see Fig. 1(h).

In both turbulent and lean-phase fluidization, large amounts of particles are entrained, precluding steady state operations. For steady state operation in these contacting modes, entrained particles have to be collected by cyclones and returned to the beds. In turbulent fluidized beds, inner cyclones can deal with the moderate rate of entrainment, as shown in Fig. 2(a), and this system is

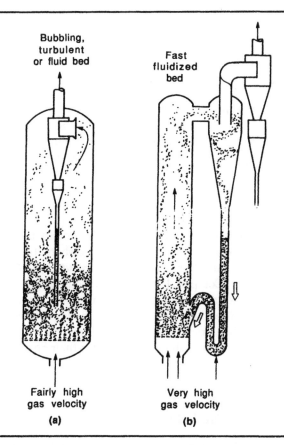

FIGURE 2
Circulating fluidized beds.

Gas

FIGURE 3
Spouted bed.

sometimes called a *fluid bed*. On the other hand, the rate of entrainment is far larger in lean-phase fluidized beds, which usually necessitates the use of big cyclone collectors outside the bed, as shown in Fig. 2(b). This system is called the *fast fluidized bed*.

In fluid beds and fast fluidized beds, smooth and steady recirculation of solids through the dipleg or other solid trapping device is crucial to good operations. These beds are called *circulating fluidized beds*.

The *spouted bed*, sketched in Fig. 3, represents a somewhat related contacting mode wherein comparatively coarse uniformly sized solids are contacted by gas. In this operation, a high-velocity spout of gas punches through the bed of solids, thereby transporting particles to the top of the bed. The rest of the solids move downward slowly around the spout and through gently upward-percolating gas. Behavior somewhere between bubbling and spouting is also seen, and this may be called *spouted fluidized bed* behavior.

Compared to other methods of gas-solid contacting, fluidized beds have some rather unusual and useful properties. This is not the case with liquid-solid fluidized beds. Thus, most of the important industrial applications of fluidization to date are with gas-solid systems, and for this reason this book deals primarily with these systems. It describes their characteristics and shows how they can be used.

Liquidlike Behavior of a Fluidized Bed

A dense-phase gas fluidized bed looks very much like a boiling liquid and in many ways exhibits liquidlike behavior. This is shown in Fig. 4. For example, a large, light object is easily pushed into a bed and, on release, will pop up and

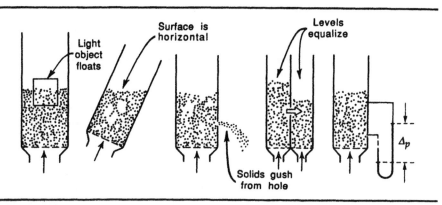

FIGURE 4
Liquidlike behavior of gas fluidized beds.

float on the surface. When the container is tipped, the upper surface of the bed remains horizontal, and when two beds are connected their levels equalize. Also, the difference in pressure between any two points in a bed is roughly equal to the static head of bed between these points. The bed also has liquidlike flow properties. Solids will gush in a jet from a hole in the side of a container and can be made to flow like a liquid from vessel to vessel.

This liquidlike behavior allows various contacting schemes to be devised. As shown in Fig. 5, these schemes include staged countercurrent contacting in a

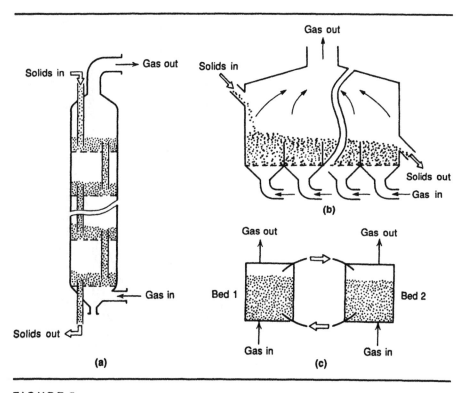

FIGURE 5
Contacting schemes with gas fluidized beds: (a) countercurrent; (b) crosscurrent; (c) solid circulation between two beds.

vessel containing perforated plates and downcomers, crosscurrent contacting in a sectioned bed, and solid circulation between two beds.

To give some insight into the workings of a contacting scheme, consider a *solids circulation system* between two fluidized beds, as shown in Fig. 6. If gas is injected into U-tube C connecting fluidized beds A and B and if the solids everywhere are fluidized, then it can be shown that the difference in static pressure in the two arms of the U-tube will be the driving force causing particles to flow from A to B. A combination of two such U-tubes will then allow complete circulation of solids. The faster the flow, the higher the frictional resistance, and so, as in any hydraulic system of this kind, the rate of circulation is determined by a balance between this frictional resistance and the previously mentioned pressure differences. The circulation is controlled by changing the frictional resistance of the system to flow, say, by slide valves or by varying the average densities of the flowing mixtures in the various portions of the connecting circuit, a procedure that modifies the pressure differences.

For proper operation of circulation and other solids flow systems, the solids must be maintained in dynamic suspension throughout, because any settling of particles can clog the lines and cause a complete shutdown of operations. Thus, special care is needed in the design of such systems: gas injectors must be properly sized, piping liable to settling and clogging should be avoided, and reliable start-up and shutdown procedures must be used.

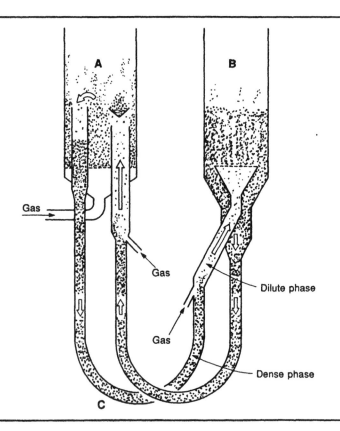

FIGURE 6
Operating principle for stable circulation between two beds.

Circulation systems such as shown in Fig. 6 are used primarily for solid-catalyzed gas-phase reactions. Here, catalyst flows smoothly and continuously between reactor and regenerator. Because of the large specific heat of the solids, their rapid flow between reactor and regenerator can transport vast quantities of heat from one to the other and thus effectively control the temperature of the system. Actually, in highly endothermic or exothermic reactions, the circulation rate of the solids is chosen not only on the basis of the rate of solids deactivation but also as a means of achieving favorable temperature levels in reactor and regenerator. Automatic control of such operations is the rule.

This fluidlike behavior of solids with its rapid, easy transport and its intimate gas contacting is often the most important property recommending fluidization for industrial operations.

Comparison with Other Contacting Methods

Figure 7 sketches the different ways of contacting solids and gas streams, and shows how fluidized beds and pneumatic conveying lines (or fast fluidized beds) compare with the other contacting modes.

In many of the conventional contacting modes, such as fixed beds, moving beds, and rotary cylinders, the gas flow or solid flow closely approximates the ideal of plug flow. Unfortunately, this is not so for single fluidized beds where solids are best represented by well-mixed flow and the gas follows some intermediate and difficult-to-describe flow pattern. Nevertheless, with proper baffling and staging of units and negligible entrainment of solids, contacting in fluidized beds can approach the usually desirable extreme of countercurrent plug flow.

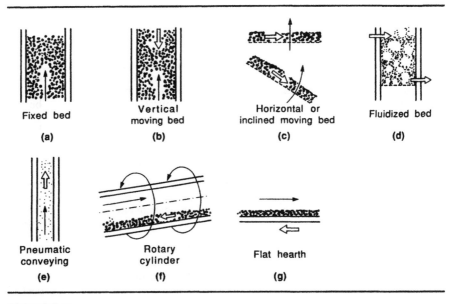

FIGURE 7
Contacting modes for gas-solid reactors.

TABLE 1   Comparison of Types of Contacting for Reacting Gas-Solid Systems.

	Solid-Catalyzed Gas-Phase Reaction	Gas-Solid Reaction	Temperature Distribution in the Bed
Fixed Bed	Only for very slow or nondeactivating catalyst. Serious temperature control problems limit the size of units.	Unsuited for continuous operations, while batch operations yield nonuniform product.	Where much heat is involved large temperature gradients occur.
Moving Bed	For large granular rapidly deactivated catalyst. Fairly large-scale operations possible.	For fairly uniform sized feed with little or no fines. Large-scale operations possible.	Temperature gradients can be controlled by proper gas flow or can be minimized with sufficiently large solid circulation.
Bubbling and Turbulent Fluidized Bed	For small granular or powdery non-friable catalyst. Can handle rapid deactivation of solids. Excellent temperature control allows large-scale operations.	Can use wide range of solids with much fines. Large-scale operations at uniform temperature possible. Excellent for continuous operations, yielding a uniform product.	Temperature is almost constant throughout. This is controlled by heat exchange or by proper continuous feed and removal of solids.
Fast Fluidized Bed and Cocurrent Pneumatic Transport	Suitable for rapid reactions. Attrition of catalyst is serious.	Suitable for rapid reactions. Recirculation of fines is crucial.	Temperature gradients in direction of solids flow can be minimized by sufficient circulation of solid.
Rotary Cylinder (Kiln)	Not used	Widely used, suitable for solids which may sinter or agglomerate.	Temperature gradients in direction of solids flow may be severe and difficult to control.
Flat Hearth	Not used	Suitable for solids liable to sinter or melt.	Temperature gradients are severe and difficult to control.

Particles	Pressure Drop	Heat Exchange and Heat Transport	Conversion
Must be fairly large and uniform. With poor temperature control these may sinter and clog the reactor.	Because of large particle size pressure drop is not a serious problem.	Inefficient exchange, hence large exchanger surface needed. This is often the limiting factor in scale-up.	With plug flow of gas and proper temperature control (which is difficult) close to 100% of the theoretical conversion is possible.
Fairly large and uniform; top size fixed by the kinetics of the solid recirculation system, bottom size by the fluidizing velocity in reactor.	Intermediate between fixed and fluidized beds.	Inefficient exchange but because of high heat capacity of solids, the heat transported by circulating solids can be fairly large.	Flexible and close to ideal countercurrent and cocurrent contacting allows close to 100% of the theoretical conversion.
Wide size distribution and much fines possible. Erosion of vessel and pipelines, attrition of particles and their entrainment may be serious.	For deep beds pressure drop is high, resulting in large power consumption.	Efficient heat exchange and large heat transport by circulating solids so that heat problems are seldom limiting in scale-up.	For continuous operations, mixing of solids and gas bypassing result in poorer performance than other reactor types. For high conversion, staging or other special design is necessary.
Fine solids, top size governed by minimum transport velocity. Severe equipment erosion and particle attrition.	Low for fine particles, but can be considerable for larger particles.	Intermediate between fluidized and moving bed.	Flow of gas and solid both close to cocurrent plug flow, hence high conversion possible.
Any size, from fines to large lumps	Very low	Poor exchange, hence very long cylinders often needed.	Close to countercurrent plug flows, hence conversions can be high.
Both big and small	Very low	Poor exchange	Fair, scrapers or agitators are needed.

For good design, proper contacting of phases is essential. In any case, we should be able to describe the real contacting pattern. In fluidized systems, this can be one of the major problems; consequently, the development of satisfactory methods for predicting contacting patterns is an important consideration in this book.

Advantages and Disadvantages of Fluidized Beds for Industrial Operations

The fluidized bed has desirable and undesirable characteristics. Table 1 compares its behavior as a chemical reactor with other reactors. Its *advantages* are

1. The smooth, liquidlike flow of particles allows continuous automatically controlled operations with easy handling.

2. The rapid mixing of solids leads to close to isothermal conditions throughout the reactor; hence the operation can be controlled simply and reliably.

3. In addition, the whole vessel of well-mixed solids represents a large thermal flywheel that resists rapid temperature changes, responds slowly to abrupt changes in operating conditions, and gives a large margin of safety in avoiding temperature runaways for highly exothermic reactions.

4. The circulation of solids between two fluidized beds makes it possible to remove (or add) the vast quantities of heat produced (or needed) in large reactors.

5. It is suitable for large-scale operations.

6. Heat and mass transfer rates between gas and particles are high when compared with other modes of contacting.

7. The rate of heat transfer between a fluidized bed and an immersed object is high; hence heat exchangers within fluidized beds require relatively small surface areas.

Its *disadvantages* are

1. For bubbling beds of fine particles, the difficult-to-describe flow of gas, with its large deviations from plug flow, represents inefficient contacting. This becomes especially serious when high conversion of gaseous reactant or high selectivity of a reaction intermediate is required.

2. The rapid mixing of solids in the bed leads to nonuniform residence times of solids in the reactor. For continuous treatment of solids, this gives a nonuniform product and poorer performance, especially at high conversion levels. For catalytic reactions, the movement of porous catalyst particles, which continually capture and release reactant gas molecules, contributes to the backmixing of gaseous reactant, thereby reducing yield and performance.

3. Friable solids are pulverized and entrained by the gas and must be replaced.

4. Erosion of pipes and vessels from abrasion by particles can be serious.

5. For noncatalytic operations at high temperature, the agglomeration and sintering of fine particles can require a lowering in temperature of operations, thereby reducing the reaction rate considerably.

The compelling advantage of overall economy of fluidized contacting has been responsible for its successful use in industrial operations. But such success depends on understanding and overcoming its disadvantages. This book, treating as it does the present and possible uses of fluidized beds, considers how such disadvantages can be overcome.

Fluidization Quality

The term *fluidization* has been used in the literature to refer to dense-phase and lean-phase systems, as well as circulation systems involving pneumatic transport or moving beds. The broad field of fluidization engineering thus deals with all these methods of contacting, but the main focus is on dense-phase systems.

The ease with which particles fluidize and the range of operating conditions that sustain fluidization vary greatly among gas-solid systems and numerous factors affect this. First is the *size* and *size distribution of solids*. In general, fine particles tend to clump and agglomerate if they are moist or tacky; thus, the bed must be agitated to maintain satisfactory fluidizing conditions. This can be done with a mechanical stirrer or by operating at relatively high gas velocities and using the kinetic energy of the entering gas jets to agitate the solids. Fine particles of wide size distribution can be fluidized in a wide range of gas flow rates, permitting flexible operations with deep, large beds.

On the contrary, beds of large uniformly sized solids often fluidize poorly, with bumping, spouting, and slugging, which may cause serious structural damage in large beds. The quality of fluidization of these beds can often be spectacularly improved by adding a small amount of fines to act as lubricant. Also, large particles fluidize in a much narrower range of gas flow rates: hence, shallower beds must be used.

A second factor is the *fluid-solid density ratio*. Normally, liquid-solid systems fluidize homogeneously, whereas gas-solids exhibit heterogeneity. However, as mentioned earlier, one may have deviations from the norm with low-density particles in dense gas or high-density particles in low-density liquid.

Numerous other factors may affect the quality of fluidizations, such as vessel geometry, gas inlet arrangement, type of solids used, and whether the solids are free-flowing or liable to agglomerate.

Selection of a Contacting Mode for a Given Application

When a new commercial-scale physical or chemical process is planned, proper selection of a contacting mode is crucial. For catalytic reactions the choice is usually between the fixed bed and the fluidized bed in its various forms. For noncatalytic fluid-solid reactions, the choice is somewhat different, usually between moving beds, shaft kilns, and fluidized, but not fixed, beds. For physical operations such as heating, cooling, and drying of solids or adsorption and desorption of volatiles from solids, one may also want to consider spouted beds. Whatever the final design chosen, one need not use that particular contacting pattern in the initial small-scale experiments.

For example, consider the design for a solid-catalyzed reaction system. Thermodynamic information and laboratory experiments with a batch of solids in a small fixed bed or in a Berty- or Carberry-type mixed flow of gas reactor give the following information:

- The kinetics, whether fast or slow
- The desirable catalyst size range to use
- The extent of heat effects and its expected importance in the large unit
- Whether catalyst deactivation is rapid and regeneration is feasible

If these findings point toward the fluidized bed, additional hydrodynamic tests might be needed to evaluate the friability of the catalyst.

The development strategy differs for different kinds of applications, and later chapters discuss this in more detail.

Some process designers feel that fluidization is an interesting operation but is not for them because it is still too much of an art requiring practical experience and know-how, and because too much uncertainty is involved, particularly in scale-up, at which stage the cost of failure is serious. They point to well-known commercial disasters involving fluidized beds and conclude that it is best to leave the development of such processes to larger companies that have experience with fluidized beds and that, in any case, can absorb the cost of possible failure.

The design of fluidized bed processes is often more complex than other modes of contacting, but not always. It all depends on the operation at hand, whether catalytic reaction, gas-solid reaction, physical operation, or the generation of solids from gas, such as the production of polyethylene. In any case, when technical and economic considerations both point strongly to the fluidized bed, then one must put up with possible difficulties and complications.

In conclusion, we note that our knowledge of what is happening within a fluidized bed, our scale-up methods, and our confidence in these methods have increased significantly in the last 20 years. Fluidization is not such a black art, and if we keep careful tab of our uncertainties and get the needed information for scale-up, we should be able to design fluidized beds successfully.

Overall Plan

Chapter 2 surveys industrial applications of fluidization, showing the many ways that fluidization can be used in various applications.

Chapter 3 follows with some basics and a road map of the various regimes in gas-solid contacting. It is the framework for the rest of the book.

The next group of chapters lays out the state of knowledge in various contacting regimes: pumping power and the distributor zone of a fluidized bed (Chap. 4), bubbles in dense beds (Chap. 5), bubbling beds (Chap. 6), the lean zone above the dense bed (Chap. 7), high-velocity fluidization (Chap. 8), mixing and movement of solids (Chap. 9), and gas dynamics (Chap. 10).

We then consider kinetic phenomena in beds: gas-particle heat and mass transfer (Chap. 11), catalytic reactions in all contacting regimes (Chap. 12), heat transfer at wall surfaces (Chap. 13), and RTD, size distribution of solids, and growth and shrinkage of solids (Chap. 14).

The last section is concerned primarily with design: special problems related to systems of circulating solids (Chap. 15), design for the physical operations of heat transfer, mass transfer and drying (Chap. 16), catalytic reactors (Chap. 17), and noncatalytic gas-solid reactors (Chap. 18).

RELATED READINGS

J.F. Davidson, R. Clift, and D. Harrison, *Fluidization*, 2nd ed., Academic Press, London, 1985.

J.F. Davidson and D. Harrison, *Fluidized Particles*, Cambridge University Press, New York, 1963.

D. Geldart, *Gas Fluidization Technology*, Wiley, New York, 1986.

J.R. Grace, "Fluidization," in *Handbook of Multiphase Systems* (G. Hetsroni, ed.), Section 8, McGraw-Hill, New York, 1982.

M. Leva, *Fluidization*, McGraw-Hill, New York, 1959.

I. Muchi, S. Mori, and M. Horio, *Chemical Reaction Engineering with Fluidization* (in Japanese), Baifukan, Tokyo, 1984.

D.F. Othmer, *Fluidization*, Van Nostrand Reinhold, New York, 1956.

T. Shirai, *Fluidized Beds* (in Japanese), Kagaku-Gijut-susha, Kanazawa, 1958.

V. Vanecek, M. Markvart, and R. Drbohlav, *Fluidized Bed Drying*, translated by J. Landau, Leonard Hill, London, 1966.

J.G. Yates, *Fundamentals of Fluidized-Bed Chemical Processes*, Butterworths, London, 1983.

S.S. Zabrodsky, *Hydrodynamics and Heat Transfer in Fluidized Beds*, translated by F.A. Zenz, MIT Press, Cambridge, MA, 1966.

F.A. Zenz and D.F. Othmer, *Fluidization and Fluid-Particle Systems*, Van Nostrand Reinhold, New York, 1960.

CHAPTER

2

Industrial Applications of Fluidized Beds

- Historical Highlights
- Physical Operations
- Synthesis Reactions
- Cracking of Hydrocarbons
- Combustion and Incineration
- Carbonization and Gasification
- Calcination
- Reactions Involving Solids
- Biofluidization

This chapter discusses uses of the fluidized bed, the diverse designs that have been developed, their weak and strong points, and the reason for choosing them over other designs. The presentation is not exhaustive, and additional applications can be found in other texts on the subject (see "Related Readings" in Chap. 1), in the technical literature, and in patent disclosures.

Historical Highlights

Coal Gasification

Winkler's coal gasifier represents the first large-scale, commercially significant use of the fluidized bed. This unit was fed powdered coal, was 13 m high and 12 m^2 in cross section, and went into smooth operation in 1926. The desired reaction, simply represented, is as follows:

$$\text{coal} \xrightarrow{+O_2, \text{ steam}} \underset{\text{synthesis gas}}{CO + H_2}$$

A number of such units were constructed, primarily in Germany and Japan, to supply raw gas for the synthetic chemicals industries. A typical Winkler gas producer (Fig. 1(a)) shows that considerable space is needed for secondary injection of oxygen above the bed. The resulting temperature rise furthers the decomposition of produced methane to the desired CO and H_2.

Compared to modern technology, the Winkler gas producer is inefficient because of its high oxygen consumption and its large (over 20%) carbon loss by entrainment. With the increased use of petroleum throughout the world, Winkler generators have gradually been replaced by generators that use petroleum feedstocks.

Gasoline from Other Petroleum Fractions

With war threatening in Europe and the Far East around 1940, the United States anticipated a need for vast quantities of high-octane aviation gasoline, so it

15

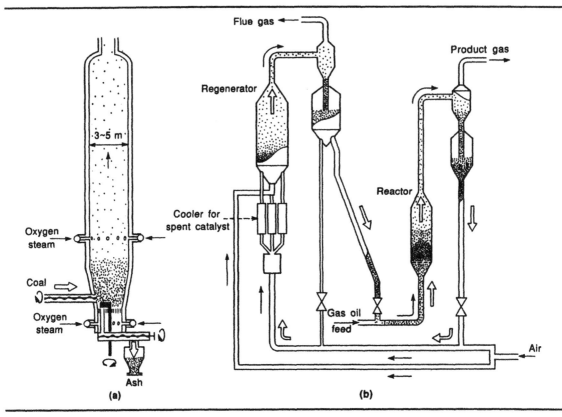

FIGURE 1
Two pioneering fluidized bed reactors: (a) the Winkler gas generator; (b) the first large-scale pilot plant for fluid catalytic cracking.

urged its chemical engineering community to find new ways of transforming kerosene and gas oil into this critical fuel. The Houdry process, in operation since 1937, was already available. However, because it used fixed beds of alumina catalyst requiring intermittent operations to regenerate deactivated catalyst, and because of the complicated arrangements for controlling bed temperatures, this process was unsuited for large-scale production.

One extension of the Houdry process led to the Thermofor catalytic cracking (TCC) process, a reactor-regenerator circuit using two moving beds of relatively large catalyst pellets that are transported from unit to unit by bucket elevator (earlier models) or gas lift (later models). Another variation of the Houdry process was the Hyperforming process, a circuit consisting of a single moving bed of large (4-mm) pellets and a gas lift. The upper part of the bed was used for reaction, the lower part for regeneration.

In parallel with these efforts, research engineers at the Standard Oil Development Company (now Exxon) were trying to develop a pneumatic conveying system for the catalytic cracking of kerosene. However, they were plagued with mechanical problems and problems due to excessive pressure drop in long tubes. At this time, Professors Lewis and Gilliland, on the basis of experiments carried out at the Massachusetts Institute of Technology, confirmed that a completely pneumatic circuit of fluidized beds and transport lines could operate stably, and suggested that one be used. Exxon engineers concentrated on this idea, verified that a standpipe was crucial for smooth circulation, and came up with a large upflow pilot plant (Fig. 1(b)). This was the start of fluid catalytic cracking (FCC). Jahnig et al. [1] and Squires [2] relate these exciting early developments.

In the urgent rush to full-scale commercial operations, Exxon engineers cooperated with engineers at the M.W. Kellogg Company and the Standard Oil Company of Indiana (now Amoco) to overcome difficulties with the collection of catalyst fines entrained by the gases, aeration, erosion of transport lines, attrition of catalyst, and overall instrumentation. This concentrated effort culminated in the first commercial FCC unit, the SOD Model I, being built at Exxon's Baton Rouge refinery. It had a capacity of 13,000 barrels of feed per day, and it went into remarkably smooth operation in 1942, less than two years after the principle of solid circulation was confirmed in the large-scale pilot plant.

To reduce the heavy load on dust collectors, solid upflow beds were soon replaced by downflow fluidized beds, leading to SOD Model II units. More than 30 FCC units of this type were built to produce aviation gasoline during World War II. Successive modifications led to the construction, year after year, of units of improved design and with capabilities increased to about 100,000 barrels/day ($16,000 \text{ m}^3$/day).

At this middle stage of development of FCC units, an amorphous silica-alumina catalyst was being used. However, in 1962, Socony-Mobil Company (now Mobil) developed a new type of catalyst, high-activity zeolite. First tried in FCC units in 1964, zeolite gave higher gasoline yields and better selectivity, and as a result has been widely used in catalytic cracking ever since.

Taking advantage of this remarkable catalyst, reactor designers promptly introduced the riser cracker in which feed is introduced into the upflow pneumatic transport line that carries catalyst from regenerator to reactor. Because of the high activity of the catalyst, close to 90% of the feed is cracked within the transport line itself, resulting in a higher gasoline yield, higher C_3–C_5 olefin content, and less carbon formation.

Because of the giant scale and the worldwide importance of FCC operations, continual efforts are being made to improve these processes.

Gasoline from Natural and Synthesis Gases

In the mid-1940s, vigorous attempts were made in the United States to use the fluidized bed for the Fischer-Tropsch reaction. The driving force for these efforts was the desire to produce high-grade gasoline from cheap and plentiful natural gas. Based on the low cost of natural gas at that time and on the results of laboratory experiments that gave almost 90% conversion, Hydrocarbon Corporation constructed their dense-phase fluidized bed reactor, the Hydrocol unit, at Carthage, Texas.

Unfortunately, conversion in the commercial unit was far below that anticipated, being more in the range of catalytic cracking reactions. Beset with scale-up problems, required modifications, and the rising cost of natural gas, operations on this unit were finally suspended in 1957. Squires [3] tells this sad story in detail and gives a diagram of the reactor.

As an alternative to the Hydrocol process, Kellogg and Sasol (South African Synthetic Oil Limited) jointly developed a synthetic gasoline process based on dilute transfer line contacting—in other words, a fast fluidized bed system. This process was very successful, and has been expanded on a giant scale to meet most of the liquid fuel needs of South Africa. Overall, the gasoline produced comes from synthesis gas, which in turn is produced in Lurgi-designed moving bed gasifiers. Thus,

$$\text{coal} \xrightarrow[\text{moving bed}]{+O_2, \text{ steam}} \underset{\text{synthesis gas}}{CO + H_2} \xrightarrow[\text{fast fluidized bed}]{\text{catalyst,}} \text{gasoline}$$

Synthesis Reactions

The remarkable temperature uniformity of the fluidized bed has strongly recommended it as a vehicle for effecting catalytic reactions, especially highly exothermic and temperature-sensitive reactions. Successful applications in this area include the production of phthalic anhydride by the catalytic oxidation of naphthalene or *ortho*-xylene, the production of alkyl chloride, and the Sohio process for producing acrylonitrile. Although few details are reported in the open literature, one can imagine the enormous effort to develop such processes.

Metallurgical and Other Processes

In 1944, Dorr-Oliver Company acquired rights to Exxon's fluidization know-how for use in fields outside the petroleum industry. Concentrating on noncatalytic gas-solid reactions, they soon developed the FluoSolids system for roasting sulfide ores. The first unit was constructed in 1947 in Ontario, Canada, to roast arsenopyrite and to obtain a cinder suitable for gold production by cyanidation. In 1952 at Berlin, New Hampshire, Dorr-Oliver used the FluoSolids roaster to produce SO_2 from sulfide ores.

Independently, and as early as 1945, the German company Badische Anilin und Soda-Fabrik (BASF) had begun to develop fluidized bed roasters based on experience acquired with the Winkler gas producer. In 1950, their first commercial roaster went on stream at Ludwigshafen with a capacity of 30 tons of ore per day. Scale-up was rapid, and a unit with capacity of 120 tons/day was constructed in 1952.

From the time of their introduction, these remarkable roasters progressively replaced existing technology centering about multihearth roasters and rotary kilns, both in the sulfuric acid industry and for the preparation of a wide variety of solid materials needed in metallurgical industries.

Dorr-Oliver engineers pioneered two additional important uses of fluidized beds: one for drying powdery materials, the other for calcining limestone. Thus, in 1948 the first FluoSolids unit (1.7 m ID) for the drying and sizing of dolomite particles <4 mesh, and having a capacity of 50 tons/day, was put in operation in the Canaan plant of the New England Lime Company. The following year they constructed a large multistage unit for calcining powdery limestone (multistaging was used primarily to reduce the otherwise high fuel consumption for this process).

These initial successes spawned much interest in fluidization, and a variety of new processes have been reported in the literature and in patents. Although we consider examples of these later, we single out three of them now because of their potential influence on three different major industries: polymers, semiconductors, and biotechnology.

First, we have the production of granular polyethylene, today's largest volume plastic, by polymerization of its gaseous monomer in fluidized beds. This process is rapidly replacing liquid-phase technology. Second, the fluidized bed is finding its niche in the expanding semiconductor industry to produce ultrapure silicon and its precursors. Third, it is used in the food and pharmaceutical industries, particularly for efficient cultivation of microorganisms.

On looking back, it seems that the path to commercial success with fluidized processes has been unusually painful and complex, with many stages of scale-up and more than its expected share of embarrassing failures. These difficulties stem largely from the lack of satisfactory answers to the many questions on which design decisions should be based, and this in turn stems from a lack of reliable predictive knowledge about what goes on in these beds. Such design uncertainties, coupled with the large investment involved, have led to a general conservatism and caution in many of these developments. However, the large payoffs that have accompanied successful processes continue to spur research and development efforts on fluidization in numerous applications.

We now look more closely at the use of fluidization in industry.

Physical Operations

Heat Exchange

Fluidized beds have been used extensively for heat exchange because of their unique ability to rapidly transport heat and maintain a uniform temperature. Figure 2(a) illustrates a fluidized bed for the rapid quenching and tempering of hot metalware to a definite temperature so as to obtain the desired properties of an alloy. This kind of operation requires a high heat transfer rate, which is provided by a fluidized bed of fine solids.

An example of a practical noncontacting gas-solid heat exchanger is shown in Fig. 2(b). Here the thermal energy of hot solids is recovered by coolant gas. Figure 2(c) is a sketch of a heat exchanger used for heat recovery and steam generation from hot particles coming from a fluidized bed reactor.

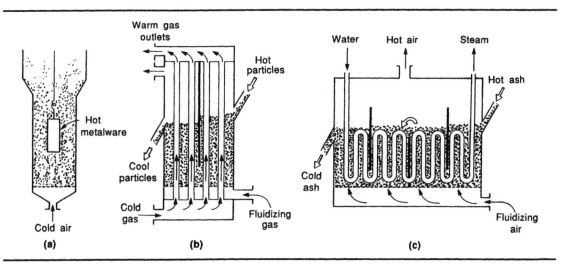

FIGURE 2
Examples of heat exchangers: (a) for rapid quenching of metalware; (b) for indirect heat exchange between coarse particles and gas; (c) for steam generation from hot ash particles.

Solidification of a Melt to Make Granules

To spread urea on fields from the air requires coarse granules in a narrow size range. For this purpose, Mitsui-Toatsu engineers developed the solidification process sketched in Fig. 3(a). Sprayed molten urea falls as droplets through a tall tower while cold air passes upward through the tower, cooling and solidifying the droplets. The few big droplets still needing to be frozen fall into a fluidized bed of urea particles at the base of the tower; they are quickly covered by a layer of smaller solids, move around the bed, and then solidify.

Requirements for better control of particle size then led to the process shown in Fig. 3(b), which combines a shallow fluidized bed with several spouted beds. Molten urea is fed to the nozzle at the bottom of each spout, and air is used for spouting and as the fluidizing gas. Product solid is removed from one end of the unit, and the undersized fines are returned to the other side. This type of operation gives a much narrower size distribution of solids than does the conventional granulation unit.

Coating Metal Objects with Plastic

Looking at Fig. 2(a) again, consider a bed of fine plastic particles fluidized by ambient air. Then metalware, heated to a temperature somewhat higher than the melting point of the plastic, is dipped for a short time (2–12 s) into the bed. Particles impinging on the surface of the metal fuse and adhere to it to form a thin layer. To smooth the coating, the metalware may have to be reheated in hot air. According to Gaynor [4], the thickness of the coating film varies exponentially with time because its deposition rate is proportional to the heat transfer rate, which depends on the film thickness.

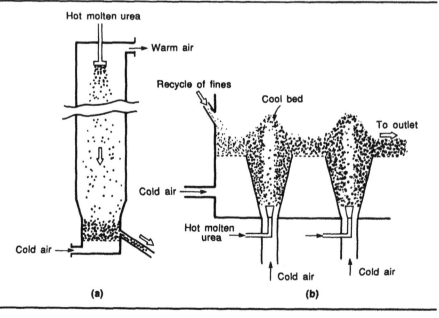

FIGURE 3
Solidification and granulation of molten urea.

This process can be used for objects with uneven or highly concave surfaces, such as metal lawn furniture, and its coating is much thicker than paint. It is economically attractive and widely used in industry because it needs no solvent and utilization of material is complete.

Drying of Solids

The fluidized bed dryer is used extensively in a wide variety of industries because of its large capacity, low construction cost, easy operability, and high thermal efficiency. It is suited to any kind of wet solid as long as the solid can be fluidized by hot gas. Iron and steel companies are using huge driers to dry coal before feeding it to their coke ovens, whereas tiny but efficient driers serve the pharmaceutical and other fine chemical industries. Figure 4 shows several designs of conventional fluidized bed driers.

Inorganic materials, such as dolomite or blast furnace slag, are usually dried in single-bed driers illustrated in Fig. 4(a), because the residence time characteristics of the particles to be dried are not important. Since the water in the particles vaporizes in the bed, the bed temperature need not be high, and 60°–90°C is usually sufficient. Thus, the energy content of hot air or flue gas, often wasted, can be efficiently used in this type of operation.

When the particles require nearly equal drying times, the residence time characteristics of solids in the fluidized beds must be considered. Single-stage operations, as in Fig. 4(a), approximate mixed flow, wherein a large fraction of the solids stay only a short time in the vessel, in effect bypassing it. Multistaging for the flowing solid greatly narrows its residence time distribution and eliminates bypassing. Figures 4(b) and (c) show multistage driers that are formed from vertical partition plates placed in the bed.

Figure 4(d) illustrates a simple design wherein counterflow contacting of gas and solid is achieved. Perforated plates or large screens act as gas redistributors and stage separators, thus eliminating overflow pipes and downcomers.

Very delicate materials, such as some pharmaceuticals, may require identical drying times for all particles. Figure 4(e) is a design for such operations. The distributors rotate on schedule to drop a batch of particles from bed to bed, and this ensures an ideal batch-continuous treatment of the particles.

For certain temperature-sensitive materials, the inlet gas temperature must be kept low. To counter the resulting reduction in thermal efficiency, heat can be recovered from the exiting dry solids. An example of such an operation is shown in the two-stage salt drier of Fig. 4(f).

When the feedstock is very wet, particles are likely to agglomerate and not fluidize at the feed location in the designs of Figs. 4(b), (c), or (d). A possible solution is to first use a backmix dryer, like Fig. 4(a), followed by a plug flow dryer such as 4(b), (c), or (d).

In the designs of Figs. 4(a)–(f), the heat content of the fluidizing gas is the energy source for the drying particles. However, heat can be supplied by heat exchange tubes or plates within the fluidized bed, as shown in Fig. 4(g). With this design, the volume of fluidizing gas needed can be greatly reduced, resulting in smaller pumping cost, less particle attrition, and lower construction cost of the exhaust gas cleaning system.

The design of Fig. 4(g) is suitable for drying very wet feedstock. By operating at high pressure and fluidizing with superheated steam, one can obtain thermal efficiencies far higher than from ordinary dryers. In addition, medium-

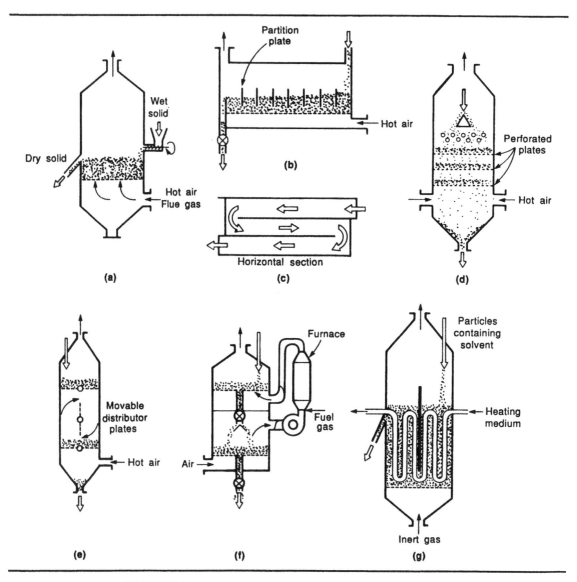

FIGURE 4
Various designs of driers.

or low-pressure steam is produced, which can be used for the next dryer or for some other operation. Alternatively, the fluidizing steam can be recirculated to give a closed system, which may be environmentally attractive if the feedstock gives off undesirable volatiles. The higher the water content, the more advantageous is this drying system; see Jensen [5].

In general, when the wet solids contain considerable amounts of solvent, such as methanol or toluene, one should be alert for possible explosions. One may want to fluidize with an inert gas, or steam, or the vapor of the solvent itself in a completely closed solvent recovery system; see Kjaergaard et al. [6].

Certain materials are not suited to the ordinary fluidized bed drier and need special treatment—for example, cohesive and sticky solids that agglomerate or stick to metal surfaces. For these materials, the vibrofluidized bed may work well. Here, the hot air distributor vibrates in such a way as to convey particles across a shallow bed from entrance to exit without agglomeration. Pesticide granules, ammonium bromide, pharmaceuticals, foodstuffs such as wheat and soy beans, and plastics such as PVC and nylon, are all being dried in such units.

Large uniformly sized particles, such as beans, peas, and other agricultural products, are often awkward to fluidize. For these solids spouted bed driers are sometimes used; see Fig. 1.3.

Finally, comparatively small particles of minerals or salts that are only surface-wetted require very short drying times. Such materials can be effectively dried in lean-phase fluidized beds or in pneumatic transport lines. These units are called *flash driers*.

In some industries, a low-temperature chemical treatment of the solids, such as calcination or roasting, is needed after drying. These situations call for multistage operations where the last stage or last few stages can be used for such heat treatment. Somewhat related to this, the low-temperature roasting of agricultural products, such as coffee beans, has been commercialized in spouted beds; see Sivetz [7].

Two-solid drier-roasters are finding increased use today, particularly in the food industry. Here, small dense particles are fluidized by the hot gas, as shown in Fig. 5. Then large, less dense solids, such as peanuts, are fed in at one end of the unit, float on the surface of the fluidized bed, and leave at the other end. In such operations it is important to select harmless fine particles and to have an efficient means for separating the coarse particles from the fines.

Coating of Objects and Growth of Particles

When a salt solution, such as sodium glutamate, is injected or sprayed into a hot fluidized bed of dry particles, such as sodium chloride, the surfaces of the particles become wet. Subsequent drying of the liquid layer then gives an efficient coating process. Some free-flowing table salt is prepared this way.

This type of operation is also used for growing particles from salt solutions

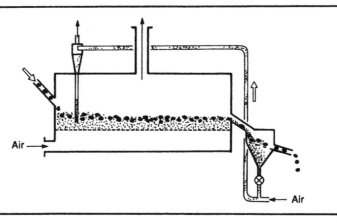

FIGURE 5
Heater for coarse solids.

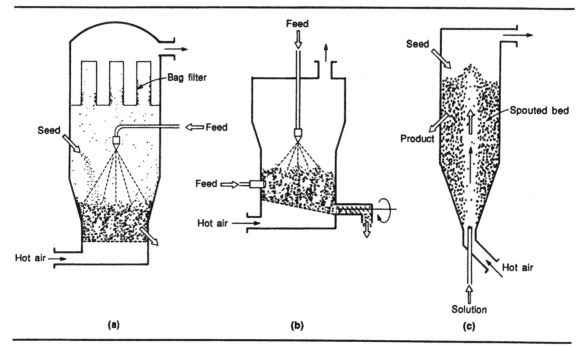

FIGURE 6
Designs for particle coating and/or particle growth.

or from slurries of fine solids. Here, growth proceeds by successive wetting of the fluidized solids with sprayed liquid followed by solidification through drying. The product size and size distribution can be controlled by the size of the seed particles, by adjusting the liquid-to-solid feed ratio and by a proper choice of the ratio of sprayed volume to bed volume. In addition, the feed liquid sometimes has to act as a binder for the fine particles, which then agglomerate to give coarser particles through drying. In these operations, it is important to know the mechanism(s) of agglomeration. Figure 6 illustrates several design features of such processes. Proper location of the spray is essential to avoid unplanned agglomeration of solids and to keep the walls of the vessel from being progressively coated with solid.

A possible related application is the condensation on a sublimable solid of its own vapor, which is present in an inert carrier gas. There is little in the literature on this type of operation, but it may find use for separating the products obtained from high-temperature synthetic gas-phase reactions. According to Ciborowski and Wronski [8], who performed experiments with naphthalene, the efficiency of condensation decreases from 100% to 80–90% with increased gas velocity and increased concentration driving force for the separation.

Adsorption

When very dilute components are to be removed from large flows of carrier gas, then continuous multistage fluidized adsorption processes can become superior to conventional fixed bed processes in which the components are periodically adsorbed onto activated carbon particles and then stripped by steam. This is the case for the separation and concentration of solvents such as carbon disulfide, acetone, methylene chloride, ethanol, and ethyl acetate and for the removal of trace pollutants from flue gas. Figure 7 illustrates some of the designs reported in the open literature.

Figure 7(a), adapted from Avery and Tracey [9], shows the multi-stage process developed by engineers at Courtaulds Ltd. for the recovery of dilute carbon disulfide (~0.1%) from air. To reduce the power consumption needed to handle the very large volume of air to be treated, each stage is very shallow (5–8 cm) and rests on a simple perforated steel plate. Designing the holes in the plate for a pressure drop somewhat smaller than that of the bed itself gives satisfactory fluidization, even in columns up to 16 m ID. The downcomers have no moving parts and are flexible enough to handle a wide

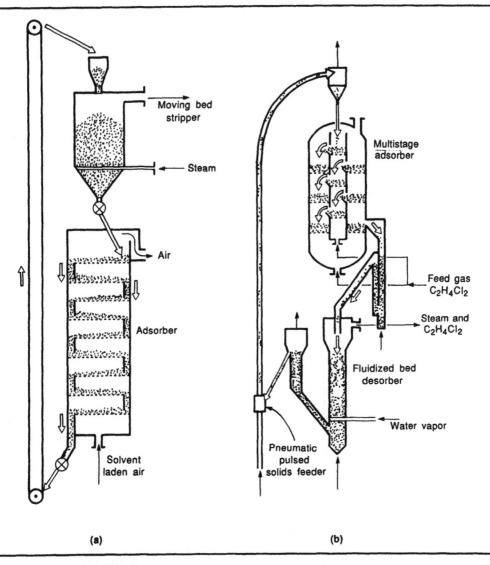

(a)   (b)

FIGURE 7
Adsorption processes, all using activated carbon solids.

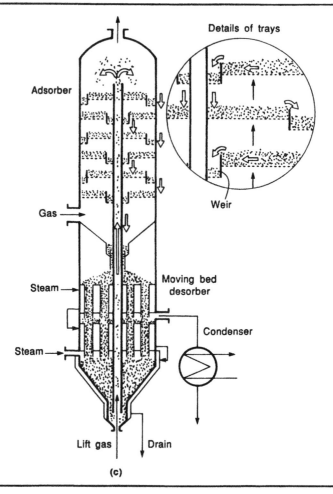

(c)

FIGURE 7 (Contd.)

range of air and solid flow rates. The initial particle size of the carbon is 2–3 mm, but with continuous attrition and makeup the size reduces to 0.1–3 mm after a few months, according to Avery and Tracey [9]. The circulation rate of carbon around the plant is controlled by variable orifice valves at two locations, as shown in Fig. 7(a), and conventional bucket elevators are used to circulate the solids.

The first plant of this type was commissioned in 1959 to recover 1.2 tons/hr of CS_2 from 400,000 m^3/hr of air. Its five-stage adsorber had a bed diameter of 11.6 m, and its upper dual-purpose moving bed stripper-drier was 5.5 m ID. Circulation of carbon was 23 tons/hr, power consumption of the blower was 760 kW, and 90–95% of the entering CS_2 was removed from the air stream, according to Rowson [10]. Following this successful operation, similar plants were constructed to recover acetone, ethanol, and ethyl acetate.

In these multistage units it is important to minimize the attrition of the fragile adsorbent solids, since the cost of makeup solids can well be the dominant operating cost. To prevent mechanical attrition of these solids, it may be preferable to circulate the particles hydraulically. Figure 7(b) is an example of a unit, designed by Chinese engineers and reported by Wang et al. [11], to remove dichloroethane, $C_2H_4Cl_2$, from foul gas. Here, no mechanical device is used to control the circulation rate of solids.

Figure 7(c) shows a process that combines a multistage fluidized adsorber with an indirectly heated desorber. This process was developed by Taiyo Chemical Laboratory to remove solvents and odorous materials from 4000–60,000 m^3/hr of foul air. The adsorbent material is spherical active carbon particles manufactured from petroleum pitch, 0.7 mm in diameter and having a bulk density of 580–650 kg/m^3. The activation process for these solids is shown in Fig. 20(d).

In each stage, the height of the fluidized bed is 2–4 cm (static height ~2 cm), and the solids rest on a flat tray perforated with 3–5 mm holes. This gives a pressure drop for each stage of 0.08–0.15 kPa for superficial gas velocities as high as 1 m/s. Carbon beads flow into the downcomer section whose bed height is somewhat larger than that of the tray itself. By appropriate selection of the open fraction in the tray and downcomer sections, one can obtain stable continuous countercurrent gas-solid contacting in this multistage unit.

Carbon beads are pneumatically and gently transported to the top of the adsorber with very little attrition, such that the adsorbent loss, reported by Amagi et al. [12], is claimed to be an order of magnitude smaller than for other similar processes (0.001–0.002% loss per circulation of solid). Indirect heating in the desorber section recovers solvent at high concentration. This process, named Gastak, has been used commercially to remove perchloroethylene, toluene, carbon tetrachloride, and trichloroethylene from foul gas.

Synthesis Reactions

The main reason for choosing the fluidized bed rather than the fixed bed for these solid-catalyzed gas-phase reactions is the demand for strict temperature control of the reaction zone. There are several possible reasons for this demand: the reaction may be explosive outside a narrow temperature range, the yield of desired product to side products may be sensitive to the temperature level of operations, or hot spots in the catalyst may lead to the rapid deterioration and deactivation of an otherwise stable catalyst that normally does not require regeneration. And to make temperature control difficult, these reactions are generally highly exothermic.

TABLE 1   Examples of Fluidized Bed Catalytic Reactors Commercialized for Chemical Synthesis.

Year	Product or Reaction	Process	Type*
1945	Phthalic anhydride	Sherwin–Williams–Badger	FB
1955	Fischer–Tropsch synthesis	Kellogg, Sasol	FFB
1956	Vinyl acetate	Nihon Gosei	FB
1960	Acrylonitrile	Sohio	FB
1961	Ethylene dichloride	Monsanto	FB
1965	Chloromethane	Asahi Chemical	FB
1970	Maleic anhydride	Mitsubishi Chemical	FB
1977	Polyethylene (low density)	Union Carbide	BB
1984	Polypropylene	Mitsui Petrochemical	BB
1984	o-cresol and 2,6-xylenol	Ashai Chemical	FB

*FB = fluidized bed of fine particles; FFB = fast fluidized bed; BB = bubbling fluidized bed of coarse particles.
In part from Ikeda [13].

Because gases have poor heat transfer characteristics and very low heat capacities compared to their heats of reaction, it is difficult to achieve the necessary positive temperature control in fixed beds. Consequently, extensive heat exchanger surfaces and large dilution of reactant gases are often required. This control is much easier to obtain in fluidized beds because the rapid circulation of solids of relatively high heat capacity efficiently distributes the heat and helps eliminate potential hot spots.

Table 1 shows some noteworthy uses of fluidized beds for synthesis reactions. We discuss their difficulties and the designs that have overcome them.

Phthalic Anhydride

In the presence of a suitable catalyst and excess air, naphthalene is oxidized to produce phthalic anhydride as follows:

$$\text{naphthalene} \xrightarrow{+[O]} \text{phthalic anhydride} \xrightarrow{+[O]} CO_2 + H_2O$$
$$\xrightarrow[+[O] \quad +[O]]{} \text{naphthaquinone}$$

Side reactions produce small quantities of naphthaquinone and maleic anhydride, and no naphthalene appears in the effluent stream.

The problem with this reaction is that it is highly exothermic. Nevertheless, in fluidized operations the bed temperature is very easily controlled within narrow temperature limits, and even with naphthalene or naphthoquinone concentrations well within the flammable region a temperature runaway does not occur because the catalyst bed functions as an extremely efficient heat dispersal medium. The catalyst would prevent an explosion even if the naphthalene were all oxidized to carbon dioxide. In addition, the entrainment of fines into the freeboard is encouraged because these fine particles act as a heat sink to prevent any temperature runaway there.

The naphthalene is not premixed with air but is injected directly into the bed. Thus a high naphthalene-to-air ratio can be used, which would be flammable if premixed. (A low naphthalene-to-air ratio is used in fixed bed operations because the feed has to be premixed. Occasionally explosions still occur in fixed bed reactor inlet chambers; presumably they are caused by deposits of nonvolatile pyrophors from the vaporized naphthalene.)

The intermediate oxidation compound, naphthaquinone, would be minimized if a plug flow reactor (fixed bed) were used. However, the overriding demand for strict temperature control for safe operation with a minimum air usage leads to the use of a fluidized bed.

Figure 8 illustrates one of the reactors used in this process. Here liquid naphthalene is fed through nozzles directly to the bottom of the reactor, which ·is at about 2.7 atm. According to Graham et al. [14], this liquid is immediately vaporized and dispersed in the bed, whose temperature is easily but carefully controlled between 340° and 380°C. The exothermic heat of reaction is removed by direct generation of steam at 7–28 atm in the reactor cooling coils.

To maintain catalyst activity, 1 kg of fresh VO_5 catalyst ($\sim$200 μm) is added to the bed for each 1000 kg of naphthalene treated. The reactor is designed for a contact time of 10–20 s, again carefully controlled, and uses a superficial gas velocity of 30–60 cm/s.

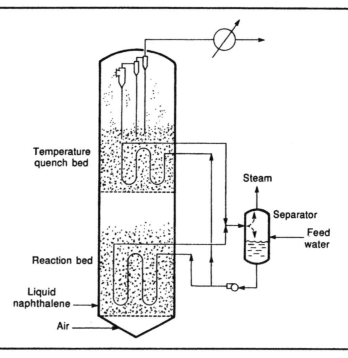

FIGURE 8
Reactor for producing phthalic anhydride from naphthalene (from Graham et al. [14]).

This type of operation was successfully and safely used as early as 1945 by the Sherwin-Williams Company. Other companies have similar operations, and large reactors producing up to 275 tons/day of product have been constructed. Conversion in this process is estimated to be nearly 100%, and yields of anhydride are about 105 kg/100 kg of petroleum naphthalene and about 85 kg/100 kg of coal tar naphthalene.

Even though the reactor operates well within the flammability limits, accumulated operating experience from many plants testifies to the relative safety of the fluidized bed process.

Fischer-Tropsch Synthesis

The synthesis of hydrocarbons from H_2 and CO gases is strongly exothermic and proceeds in a narrow temperature range, around 340°C, as follows:

$$nCO + 2nH_2 \xrightarrow[\text{catalyst}]{\text{iron}} (CH_2)_n + nH_2O, \qquad \text{exothermic}$$

As mentioned earlier in this chapter, Hydrocarbon's ambitious effort in the 1950s to develop a dense-phase fluidized process for producing synthetic gasoline, the Hydrocol process, was not successful. Kellogg took a different route. Their scheme utilized a lean-phase or fast fluidized bed reactor in a solid recirculation system. According to Shingles et al. [15], they carried out pilot-plant studies between 1946 and 1948 in a 14-m-long vertical lean-phase solids upflow reactor connected to a catalyst-disengaging hopper and a 7-cm ID standpipe.

Sasol adopted Kellogg's scheme for commercialization and constructed two such production units for their Synthol process. Not until the early 1960s were reactor operations for their circulating fluidized beds (CFB) firmly established, even though operations started in 1955.

In 1974 Sasol decided to build a second oil-from-coal plant, and selected the Badger Company to assist in the development of their second generation of Synthol CFB reactors. Problems in the first-generation reactors were overcome by the improved design in Fig. 9. In the dilute side of the circuit (voidage 85%), reactant gases, H_2 and CO, carry suspended catalyst upward at 3–12 m/s, and the fluidized bed and standpipe on the other side of the circuit provide the driving force for the smooth circulation of the powdery catalyst. For the removal of reaction heat, tube coolers are positioned in the reactor. This second-generation design has been in operation at Sasol II and, recently, in Sasol III.

Although the CFB has been successfully commercialized in Sasol I, II, and III, Sasol and Badger engineers, in a joint project, turned their attention back to the dense-phase fluidized bed reactor because, if successful, it promised reduced capital, maintenance, and operating costs. Based on experimental findings in cold models (up to 0.64 m ID), they designed a 1-m ID demonstration reactor that incorporated a very close positioning of cooling tubes and internals and used smaller catalyst particles than in the Hydrocol process. Operations of this unit have shown that conversions and selectivities are in line with those obtained in the commercial CFB reactors of Sasol II and III.

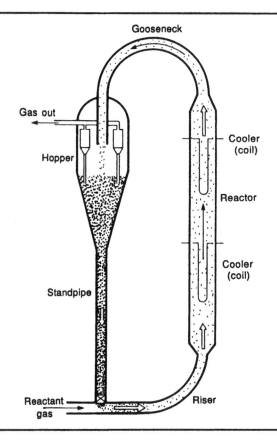

FIGURE 9
Synthol circulating solids reactor (modified from Shingles [15]).

Silverman et al. [16] estimate that on the same production basis the dense-phase fluidized bed reactor should be much smaller, less complex, and cost less than 75% of the CFB unit. Furthermore, the up to 50% lower pressure drop across the unit should result in a significant saving in capital and operating cost of the gas compressors.

At present a commercial dense-phase unit the size of a Sasol I unit is being designed.

Acrylonitrile by the Sohio Process

The Sohio process is considered to be one of the most successful applications of the fluidized bed to synthesis reactions. It produces acrylonitrile by the strongly exothermic catalytic oxidation of propylene and ammonia:

$$CH_2\!:\!CH\cdot CH_3 + NH_3 + \tfrac{3}{2}O_2 \rightarrow CH_2\!:\!CH\cdot CN + 3H_2O, \qquad \Delta H_r = -515\,kJ/mol$$

This process uses a catalyst of high selectivity and weak activity that deactivates slowly but, unfortunately, is easily poisoned by sulfur compounds in the reactant gases. In addition to the above main reaction, side reactions form HCN, acetonitrile, CO, and CO_2. Because of these, the overall heat of reaction can be as high as 670–750 kJ/mol. It is crucial therefore to remove this exothermic heat and keep good temperature control between 400° and 500°C.

After inventing their catalyst, Sohio engineers started research and development for this process in fluidized reactors 7.6 cm and 46 cm ID. The kinetic data needed for designing the commercial unit were obtained in the smaller unit, and catalyst life data and production of small quantities of product for market research were obtained in the larger unit.

The first plant using this process was constructed by Sohio in 1960 with a 20,000-tons/yr capacity. Since then many additional plants using this process have been built worldwide. Presently, close to 90% of the acrylonitrile in the world is being produced by the Sohio process (~2,400,000 tons/yr). The Sohio process reactor is shown in Fig. 10(a). Today's catalysts are multicomponent, composed of Mo, Bi, Fe, and other chemicals impregnated into microspherical silica carrier of the following physical properties:

Size range:	10–200 μm
Mean size:	50–80 μm
Bulk density:	~1 g/cm^3
Fines (−44 μm):	20–40%
Coarse (+88 μm):	10–30%

The reactor diameter ranges from 3 to 8 m, depending on the design capacity. As an example, Nakamura and Ito [17] give a reactor height of 15.2 m, a bed diameter of 3.35 m, and the following range of operating conditions:

Composition of feed:	C_3H_6 : NH_3 : air = 1 : 1–1.2 : 10–12
Temperature and pressure:	400°–500°C, 1.5–3 atm
Superficial gas velocity:	0.4–0.7 m/s
Contact time:	5–20 s

Air is fed uniformly to the fluidized bed through a bottom distributor, whereas the mixture of propylene and ammonia is blown into the bed through a carefully

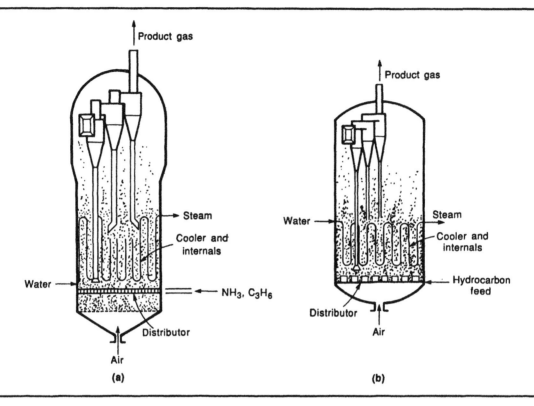

FIGURE 10
Reactors for highly exothermic reactions (adapted and modified from Nakamura and Ito [17]).
(a) Sohio, acrylonitrile production; (b) Mitsubishi Chemical, maleic anhydride production.

designed upper distributor to ensure a uniform distribution of feed gas across the bed. The lower portion of the bed between distributors, being oxygen-rich, serves as a zone for carbon burn-off and catalyst regeneration. To keep the reaction and regeneration zones more distinct and separated, a horizontal perforated plate is sometimes located just below the upper distributor.

In order to maintain good bed fluidity, particles less than 44 μm must be almost completely collected and returned to the bed, emphasizing the importance of proper design of cyclones, not only on collection efficiency but also for easy operation of diplegs.

Water passes through the in-bed cooling tubes to produce high-pressure steam, which is then used to drive the air compressor and to produce process heat for the downstream rectification operations. In addition to the bundle of vertical cooling tubes, vertical internals are located in the reactor to control the fluid dynamic behavior of the bed. Usually, such internals are designed to give an equivalent bed diameter of 1–1.5 m.

In a similar process chemists and engineers of Asahi Chemical Industries aimed at producing metacrylonitrile by ammoxidation of isobutene. Their reaction required more precise control of the reactant feedstream so as to prevent unfavorable carbon deposits on the catalyst. With their new catalyst and a fluidized bed reactor, which prevents hot spots in the reaction zone, they went on stream in 1984 with a 35,000-tons/yr plant.

Maleic Anhydride

Maleic anhydride is normally produced by the catalytic oxidation of benzene in fixed bed reactors. However, Mitsubishi Chemical Industries Co. wanted to use a mixture of butadiene and butene as feedstock because of its availability. A suitable $V_2O_5 \cdot H_2PO_4 \cdot SiO_2$ catalyst was first developed, and in 1967 Mitsubishi started research and development on the reactor.

The kinetics of this reaction can reasonably be represented by [17]

$$C_4H_8, C_4H_6 + O_2 \longrightarrow \underset{\substack{\text{maleic} \\ \text{anhydride}}}{C_4H_2O_3} \longrightarrow H_2O, CO_2, CO, \qquad \Delta H_r = -1420 \text{ kJ/mol}$$

This exothermic heat of reaction is enormous, more than three times that for the combustion of carbon or hydrogen on a molar basis. Considering all factors, the engineers chose the fluidized bed primarily because the reaction could be carried out in the flammability region with a high concentration of reactants.

After bench-scale and pilot-plant studies, a commercial reactor having a capacity of 18,000 tons/yr was constructed and put on stream in 1970. Figure 10(b) shows this reactor, which is 6 m ID and 16 m high. In this design, hydrocarbon feed is vaporized and sent to the specially designed distributor having hundreds of nozzles while the fluidizing air is sent to the bottom of the reactor. Operating conditions are as follows:

Catalyst size:	60–200 μm
Temperature and pressure:	400–500°C, 4 atm
Conversion of hydrocarbon:	>95%
Selectivity of $C_4H_2O_3$:	~60%

The bed contains a bundle of vertical 10-cm cooling tubes to remove the reaction heat. These tubes are also thought to be very effective in hindering the backmixing of gas in the bed and, hence, raising selectivity of the desired intermediate of the reaction.

Other Catalytic Reactions

Vinyl Acetate Monomer. Vinyl acetate monomer is an important starting material for a host of polymeric materials such as vinyl plastic and synthetic leather. It is formed by the exothermic reaction of acetylene with acetic acid, as follows:

$$C_2H_2 + CH_3 \cdot COOH \longrightarrow CH_2:CH \cdot OCOOH_3, \qquad \Delta H_r = -117 \text{ kJ/mol}$$

In the 1950s and 1960s, Nihon Gosei Co. and Denka Co. independently developed fluidized bed processes to make this monomer, Denka going directly from bench scale to the commercial reactor by using an acetylene feed derived from calcium carbide. These reactor designs are similar to those illustrated in Fig. 10, although with simpler distributors and a simpler arrangement of internals and cyclone collectors.

Ethylene Dichloride. Ethylene dichloride $(CH_2Cl)_2$ is made by the oxychlorination of ethylene. Mitsui Toatsu developed an active catalyst for this reaction by spray-drying a gel of mixed $CuCl_2$ and Al_2O_3 and then calcinating it. They then selected the fluidized bed route. Several steps of scale-up were used in developing this process, and the final reactor made extensive use of baffles tc get high contact efficiency. The final reactor was put on stream in 1969 and gave an overall yield of 97%; see Miyauchi et al. [18].

Chlorination of Methane. Chemists and engineers at Asahi Glass started work on methane chlorination in 1962. The main reactions are

$$CH_4 \xrightarrow{+Cl_2} CH_3Cl \xrightarrow{+Cl_2} CH_2Cl_2 \xrightarrow{+Cl_2} CHCl_3 \xrightarrow{+Cl_2} CCl_4$$

with an exothermic heat of reaction of close to 100 kJ/mol for each of these reaction steps.

Data in 1962 from a 7.6-cm ID bench-scale unit and in 1963 from a 49-cm ID pilot-scale unit led to a 10,000-tons/yr commercial plant in 1965, whose capacity was doubled in 1969. Bed temperature was kept $\pm 5°C$, somewhere between 350°–400°C, by cooling jackets about the reactors; a concern about possible explosions led to the decision to operate outside the flammability range (chloromethanes $> 23\%$, HCl $> 29\%$, $N_2 > 44\%$); see Seya [19].

Cresol and 2,6-Xylenol. Asahi Chemical Industries recently developed a fluidized catalytic reactor system to produce cresol and 2,6-xylenol from phenol and methanol:

With their Fe-V catalyst they chose the fluidized bed because of its higher selectivity, slower deactivation, and easy regeneration of catalyst. They designed and constructed the commercial plant directly, without first building a pilot plant, and started commercial production in 1984 [20].

Comments

This discussion on synthesis reactions shows that in most cases the fluidized bed is the reactor of choice whenever the exothermic heat is great, when there is a danger of a temperature runaway or explosion, and thus when strict and reliable temperature control is of paramount importance. Also, because of its large temperature flywheel effect, one can use much higher concentrations of feed in fluidized beds, well within the flammability region, resulting in significant cost savings.

Polymerization of Olefins

Polyethylene, the world's largest-volume plastic today, achieves its preeminent position largely due to a remarkable catalyst in concert with a remarkable fluidized bed process. On coming up with this catalyst, which operates at relatively low pressure and temperature, Union Carbide developed a unique and versatile fluidized bed process, called Unipol, for producing linear low-density polyethylene, which is rapidly replacing conventional processes throughout the world.

In this process (see Fig. 11), reactant gas (ethylene with its comonomers, butene and higher) is fed at a rate of three to six times the minimum fluidizing velocity into a bed of polyethylene particles kept at 75°–100°C and ~20 atm. Extremely small silica-supported catalyst particles are also fed into the bed continuously. Polymerization occurs on the catalyst surface, causing the particles to grow into large granules of 250–1000 μm; see Karol [21] for the mechanism of particle growth. The height of the reactor is reported to be 2.6–4.7 times the bed diameter. One-pass conversion of ethylene is rather low, about 2%, so large recycle flows are needed. Since the reaction is highly exothermic (~3300 kJ/kg of ethylene converted), it is important to avoid hot spots and local accumulation of catalyst at the walls of the reactor [22]. From the engineering point of view this process can be analyzed as a gas–solid reaction with growing solids.

In the Unipol process, two types of catalyst are used: chromium-titanium (or fluorine) compounds on a silica carrier, and Ziegler. These catalysts are so active that more than 10^5 volumes of polymer can be produced by unit mass of active ingredient in the catalyst. Because of the great dilution of catalyst in the granules formed and their large size, the raw product is ready for use without pelletizing it or removing the catalyst. In addition, no solvent is used in the process, and one can make the whole range of product from low- to high-density polymers. All these factors contribute to make this a remarkably efficient, attractive, and economical process.

Following the debut of the Unipol process, fluidized bed polymerization has been extensively investigated by many companies. For example, copolymerization of ethylene with hexene-1 and octene-1 has been developed by Exxon and by Union Carbide; Mitsui Petrochemical and Montedison have developed an ultrahigh performance $MgCl_2/TiCl_4$ catalyst for the gas-phase polymerization of

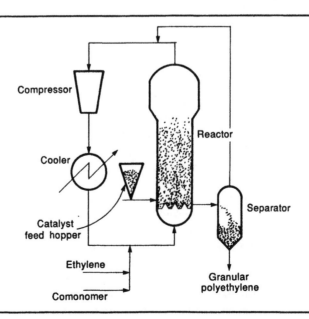

FIGURE 11
Sketch of Unipol process for making polyethylene.

propylene. A commercial plant using this technology went into operation in 1984, according to Koda and Kurisaka [23]. It is also reported that Union Carbide has already developed its own fluidized bed polypropylene process.

Cracking of Hydrocarbons

The catalytic or thermal breakdown of hydrocarbons into lower-molecular-weight materials (cracking reactions) is dominated by two features: the reactions are endothermic and accompanied by carbon deposition on nearby solid surfaces. These features and the large quantities of material to be treated dictate the type of process used industrially for these reactions. Basically, these processes have one location for the absorption of heat, for reaction, and for carbon deposition, and a second location where the deposited carbon is burned off and heat is released. This heat is then returned to the first location to feed the reaction, and the circulating solids are the means for this heat transport. The only way that all of this can be done efficiently is with a solids circulation system employing one or more fluidized beds, and practically all processes today are based on this principle of operation.

Fluid Catalytic Cracking (FCC)

On contact with a suitable catalyst, vaporized heavy hydrocarbons crack into lower-molecular-weight compounds. Numerous compounds are involved, and the key to a successful cracking process is a method for supplying the large amount of heat needed for the endothermic reaction and an effective way to rapidly regenerate tens of tons of catalyst per minute. The FCC process does this efficiently and simply by making the catalyst regeneration step supply the heat for the reaction.

The essential feature of this process is a two-unit assembly: first, a reactor at 480°–540°C, where vaporized petroleum feed is cracked on contact with hot catalyst particles. After a certain residence time, these particles are transported to the regenerator, which is at 570°–590°C, where the carbon deposit is reduced from 1–2% to 0.4–0.8% by burning in air. These heated particles, after a mean stay of 5–10 min, are returned to the reactor. The arrangement of reactor and regenerator, the type and size of catalyst, and the transport lines used vary from process to process; however, the essentials are the same and in all cases involve the use of fluidized beds.

As mentioned at the beginning of this chapter, Exxon's Model II was the first successful FCC unit, and successive improvements and modifications led to advanced designs of high capacity. Figure 12(a) shows Exxon's Model IV, which features a pair of U-tubes for circulating the fine powdery catalyst. Liquid oil is fed to the riser under the reactor, and on vaporization it reduces the bulk density of the upflowing mixture and promotes the circulation of catalyst.

The stacked unit in Fig. 12(b) is an alternative design by Universal Oil Products Company (UOP). It uses a higher pressure in the regenerator than in the reactor, a single riser, and a microspherical catalyst.

Although many variations and sizes of these units have been constructed by UOP, Exxon, Gulf, Texaco, and Kellogg, the reactor is usually 4–12 m ID, 10–20 m high, and constructed of mild steel; for a feedstock with a high sulfur content the inner wall is lined with a resistive alloy. The superficial gas velocity is 31–76 cm/s, and the perforated plate distributor has 3.8–5.1 cm holes. The number of holes is calculated to keep the pressure drop across the plate at

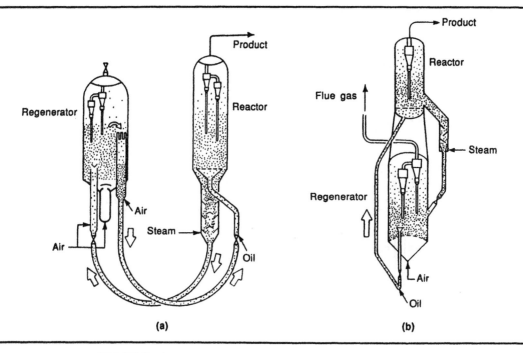

FIGURE 12
FCC units in their middle stage of development: (a) Exxon model IV; (b) UOP stacked unit.

3.5–7.0 kPa; since flat plates buckle easily under normal stresses, concave plates, both upward and downward, are used as distributors.

As mentioned, in the 1960s the highly active zeolite catalyst was created [24], and this gave designers the freedom to develop a new type of FCC, the riser-cracker, in which feed oil is sprayed into the fast upflowing lean-phase stream of regenerated catalyst. Practically all reaction occurs in this upflow riser, plug flow is closely approximated, and selectivity of desired hydrocarbon fractions is thereby markedly improved. Figure 13(a) shows one such design. Catalyst circulates smoothly between the regenerator with its ordinary fluidized bed and the riser reactor with its fast fluidized contacting.

The advantages claimed for the riser-cracker are as follows [25]:

• High conversion in very short contact times.
• Because of closeness to plug flow, overcracking is avoided, resulting in higher yields of gasoline.
• The high activity of the zeolite catalyst can be effectively utilized.
• Formation of liquid products is enhanced and formation of coke is reduced.

Operating conditions for modern riser-crackers are

Reactor:	1.7–3.5 atm, 470°–550°C
Regenerator:	2.0–4.0 atm, 580°–700°C
Coke content in catalyst leaving the reactor:	0.5–1.5 wt%
Coke content in catalyst leaving the regenerator:	0.15–0.35 wt%

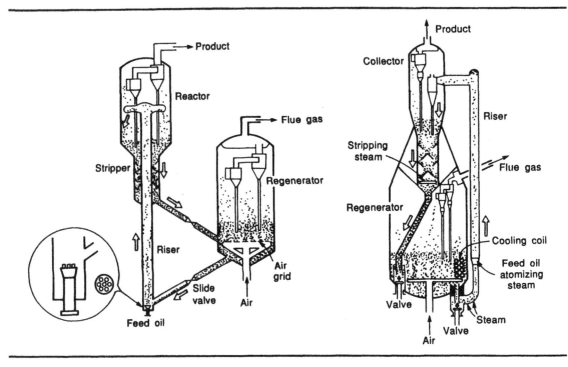

FIGURE 13
Riser cracking FCC units: (a) UOP unit; (b) Kellogg's HOC unit for upgrading heavy oil.

Approximate dimensions for a 30,000-bbl/day unit $(4770 \, \text{m}^3/\text{day})$ are

Reactor:	5 m ID, 13 m high
Regenerator:	8 m ID, 15 m high
Riser:	1.5 m ID
Catalyst circulation rate:	15–30 tons/min

Figure 13(b) shows a riser-reactor FCC unit designed by Kellogg engineers for upgrading heavy oil (atmospheric residue) [26]. Note that the riser is fed both feed oil and steam. Because this feedstock has a high Conradson carbon index (4–9%) and contains much sulfur and heavy metals (S: 0.2–3%, V and Ni: 6–170 ppm), its upgrading in FCC units encounters the following problems:

- The catalyst is rapidly poisoned by vanadium and nickel, lowering the yield of liquid products.
- Increased coke deposition on the catalyst. Its removal involves the release of excess heat that must be removed from the regenerator. .
- Additional flue gas cleaning is required to remove the SO_2 formed.

Because of the first problem, catalyst consumption of this so-called Heavy Oil Cracking (HOC) process is more than 10 times that of ordinary FCC units; to counter the poisoning effect of nickel, an additive liquid containing antimony is mixed with the feed oil. It is claimed that this antimony deposits on the catalyst to effectively remove the nickel.

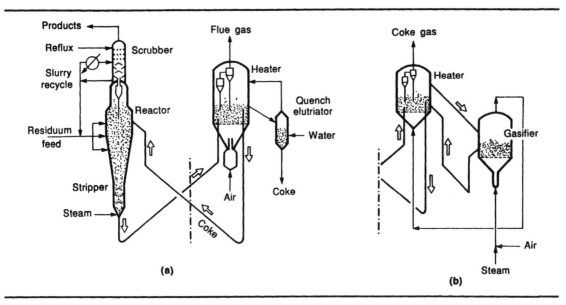

FIGURE 14
Exxon's fluid coker and flexi-coker process (simplified from Matsen [27]).

Fluid Coking and Flexi-Coking

By drawing on their experience with circulating solid systems, Exxon researchers developed a process called Fluid Coking to produce both gas oil and close to spherical coke particles between 20 and 100 mesh from a pitch feed (heavy residuum). Figure 14(a) shows the principle of this process.

In this operation [27] heated pitch is sprayed through nozzles into a coke-containing reactor (480°–570°C) fluidized by steam. Gas oil is formed, coke particles grow, and heat for this endothermic reaction is supplied by a hot coke stream (590°–690°C) coming from a heater. There roughly 5–7% of the feed, or 12–30% of the solids formed is burned to heat the circulating solids. Finally, to control the size distribution of the growing solids, an elutriator is located in the solid circulation stream to remove some of the coarser solids. Since its commercialization in 1954 about 10 such units have been constructed, the largest of these for upgrading bitumen from the Athabasca tar sands in Canada.

For efficient utilization of the by-product coke particles, Exxon combined its fluid coking unit with a giant gasification reactor to develop a process called Flexi-Coking. Figure 14(b) shows the gasifier, which is connected to the heater of the fluid coker to make a giant double-loop circulation system [27, 28]. The first unit was built in Kawasaki, Japan, in 1976, and today this type of unit processes about 3400 tons of vacuum residue per day. Operating conditions are estimated to be as follows:

	Reactor	Heater	Gasifier
pressure, atm	2	3.3	3.9
temperature, °C	510	620	980

Although the inorganic content (V, Ni) of the feed is small, it concentrates in the coke particles, lowering its sintering temperature. This may result in agglomeration of particles in the vicinity of the air distributor where the local temperature can be very high.

Thermal Cracking

In contact with a hot surface, naphtha petroleum fractions crack to produce ethylene and propylene, which are useful starting materials for organic syntheses and polymerizations. The cracking reaction is highly endothermic and proceeds as follows:

$$\underset{\text{naphtha}}{C_5\text{'s}} \longrightarrow H_2, CH_4, C_2\text{'s}, C_3\text{'s}, \ldots$$

In the early 1950s, Lurgi and Fujinagata independently developed thermal cracking processes to produce olefins from naphtha vapor, using a circulation

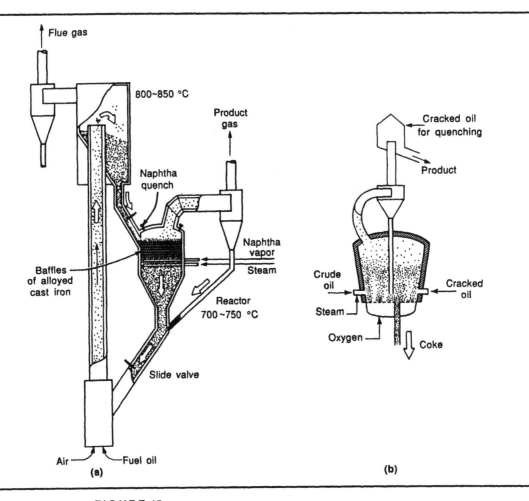

FIGURE 15
Thermal cracking of hydrocarbons to produce olefins: (a) Lurgi sandcracking unit; (b) BASF fluidized coke unit.

system of coarse sand particles (~1 mm) as the heat carrier; see Fig. 15(a). A commercial unit producing 40,000 tons per year of ethylene was built and operated in Japan for about a decade. Other such units have been built in Argentina and China; see Schmalfeld [29].

In order to produce olefins from crude oil, BASF developed a process that used fluidized coke particles, as shown in Fig. 15(b). Two generators of this type were put in operation at Ludwigshafen, producing close to 40,000 tons of ethylene per year, but have since been shut down. In this process, the single fluidized bed serves a dual purpose, to generate heat and crack the sprayed crude oil [30].

BASF engineers have also tried to develop a two-unit solid circulation cracking process using silica-alumina particles as the heat carrier. This process is shown in Fig. 16(a). The separation of the heat generation and cracking functions allows air to be used in place of oxygen for heat generation, and this modification was expected to reduce the overall cost of operations. According to Steinhofer [30], a pilot plant with a capacity of 1.5 tons/hr of crude oil produced 25 wt% ethylene and 11% propylene.

For the rational utilization of very heavy or residual oils with high sulfur content, the Ministry of International Trade and Industry (MITI) in Japan adopted the K-K (Kunii-Kunugi) process as a national project in 1964. A large-scale 120-tons/day pilot plant was constructed and operated successfully until 1981, to give enough information for the commercialization of this process.

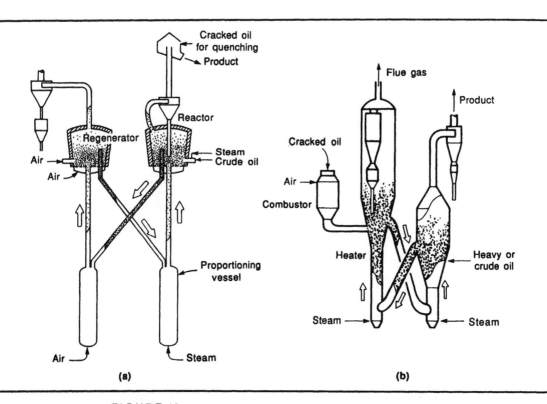

FIGURE 16
Thermal cracking in solid circulation systems to produce olefins: (a) BASF process; (b) K-K process.

In this process (see Fig. 16(b)), coke particles (600–1000 μm) are circulated between reactor and regenerator. Different from the previously mentioned thermal cracking systems, the K-K process chose dense-phase fluidization, even for the upward transport of solids. One reason for taking this route was to prevent any anticipated clogging of the transport lines with large clumps of coke. About 25 wt% ethylene and 11 wt% propylene were produced at 750°C from a feed of 4 tons/hr of paraffinic atmospheric residual oil.

Combustion and Incineration

Fluidized Combustion of Coal

In the hope of finding an alternative combustion system suitable for low-grade coal and oil shale fines, fuels that cannot be burned efficiently in conventional boiler furnaces, researchers in Britain and in China turned to fluidized bed

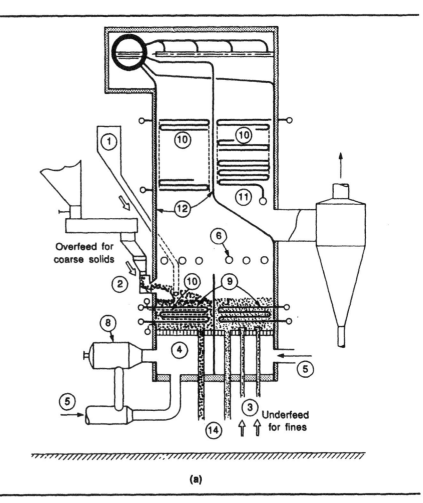

(a)

FIGURE 17
Fluidized bed coal combustors: (a) bubbling bed type; (b) circulating solids type.
1. Limestone chute, 2. spreader feeder, 3. coal-limestone feeder, 4. air distributor, 5. primary air inlet, 6. secondary air nozzle, 7. fluidized air, 8. hot gas generator, 9. evaporator, 10. superheater, 11. economizer (water preheater), 12. water wall, 13. circulator, 14. bed drain pipe (from Fujima [31]).

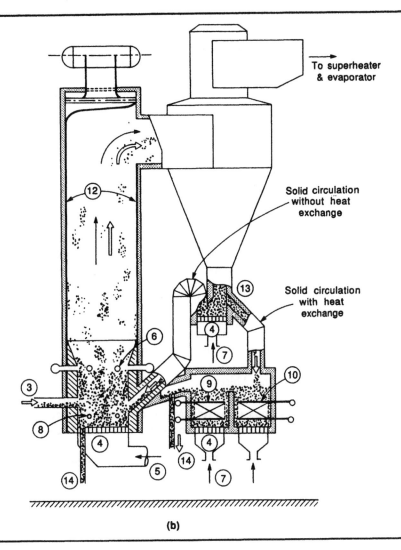

To superheater
& evaporator

Solid circulation
without heat
exchange

Solid circulation
with heat
exchange

(b)

FIGURE 17 (Contd.)

combustion (FBC) in the early 1960s. Spurred by the "oil crisis" in the early 1970s, other technologically advanced countries also focused efforts on FBC. Relatively small compact units were developed commercially and, in certain local circumstances, were found to be economically viable [31].

Figure 17(a) shows typical features of an atmospheric bubbling bed design. First, limestone or dolomite particles are fluidized by primary air entering from below through a distributor, and then small coal particles, 3–6 mm, are pneumatically injected into the bed. These pneumatic feed tubes enter from below in beds of large cross-sectional area or horizontally in smaller beds. Large lumps of coal or filter cake from sedimented fines are thrown onto the bed by a spreader-stoker. Because of the relatively high gas velocities used in these units (on the order of meters per second), considerable elutriation of solids occurs. These fines, which contain unburned carbon, are either trapped and burned in carbon burn-up cells, or else are returned to the fluidized beds from cyclone collectors.

To keep the bed at about 850°C, the temperature at which sulfur compounds are most effectively captured by the CaO and MgO solids, heat exchanger tubes, most often horizontal, are located in the bed, as shown in Figure 17(a). In addition, the walls of the bed itself, as well as the freeboard, are made up of heat exchanger tubes.

The feed rate of sorbent particles is fixed by the required degree of desulfurization. For 80% removal of sulfur compounds, a Ca:S ratio >2 is needed. Pore plugging of the CaO particles is the main reason for the low sulfur capture. In addition to SO_x control, fluidized bed combustion significantly reduces NO_x emission.

In place of heat exchange tubes, the bed temperature can be controlled by the recirculation of bed solids. Figure 17(b) illustrates this concept. Here, a mixed feed of coal and absorbent solids of wide size distribution, but with no coarse material, is fed to the bed in far fewer feed tubes than in the design of Fig. 17(a). With a high gas velocity, particles in the bed are violently fluidized and carried up and out of the combustion section past a heat exchange section to cyclone collectors. Particles are thus cooled and recirculated to the bed to control its temperature.

This fast fluidized bed design results in intense turbulence and a very uniform temperature profile in the combustor. The absence of large absorbent particles combined with a reasonably long contact time for the circulating solids gives close to complete combustion of coal plus very low NO_x and SO_x emissions. Plants using this concept were commercialized in the early 1980s by Lurgi and Ahlstrom.

We mention only two of the numerous alternative designs, for large and small units (2.5–1000 MW), operating at atmospheric and high pressure, that are being developed and commercialized today. The various symposia on FBC held each year, with proceedings running well over a thousand pages, show the interest in this difficult emerging technology. So far no design has dominated and controlled the field.

Incineration of Solid Waste

Incineration of municipal solid waste is inevitable in crowded areas, and chain grate or inclined grate incinerators are being used for this purpose. Countercurrent or crosscurrent modes of contacting, though thermally efficient, are sometimes troublesome because of the noxious odors of the flue gas from these operations. This problem can be avoided with fluidized bed incineration.

Figure 18 illustrates some of today's operating commercial units, from [32]. Garbage is coarsely shredded, iron and steel are removed by magnetic separators, and the garbage is sent to the fluidized bed incinerator. Since the operating temperature is 800°–900°C, organics are decomposed and burned in the bed and freeboard.

To remove surplus heat from the bed, water is poured into the bed in simpler designs, whereas deficient air in the bed followed by secondary combustion in the freeboard is used in more advanced designs. Solids are completely burned, and ash is discharged from the bottom of the bed. Also, a waste heat boiler, carefully designed to handle the dirty corrosive gas stream, is installed to recover heat from the flue gas.

Municipal solid waste often contains large lumps of inorganic materials. Hence, it is important to design the solid feed and discharge systems to

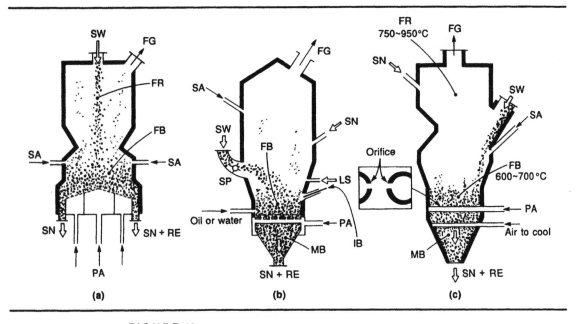

FIGURE 18
Incinerators: (a) Ebara; (b) Mitsui-Raschke; (c) IHI.
FB = fluidized bed, FG = flue gas, FR = freeboard, IB = ignition burner, LS = limestone, MB = moving bed, PA = primary air, RE = residue, SA = secondary air, SN = sand, SP = spreader, SW = solid waste.

accommodate such objects; otherwise more careful and costly pretreatment of the garbage is necessary.

Common to all these units, toxic substances, such as sulfur, nitrogen oxides, chlorine, and vapors of heavy metals (Hg, Pb, Cd), and so forth, should be removed from the flue gas. Since these units are often located in densely populated regions, it is essential that the flue gas cleanup be reliable and efficient.

Carbonization and Gasification

Gasification of Coal and Coke

As mentioned, the Winkler gas generator (Fig. 1(a)) was the first commercial application of the fluidized bed for chemical operations. In this process, powdered coal or coalite <8 mesh is fed into the bed through a screw feeder and is fluidized there by a steam-air-oxygen mixture. These units were the prime source of raw gas for the chemical industry in a few countries until about four decades ago. Then in the 1950s, cheap and abundant petroleum and natural gas became available, mainly from the Near East, and as a result practically all the operating Winkler generators were shut down.

In the 1970s, production of natural gas could not keep up with the ever-rising demand in the United States, so a number of companies started a search for a viable process for producing a high-energy substitute for pipeline gas. The so-called oil crisis of the mid-1970s greatly accelerated the efforts on all fronts to use coal economically, including gasification for the gas turbine–steam turbine binary cycle with which a power plant could generate electricity more efficiently than the conventional steam turbine system.

Today, many different routes and concepts are being pursued for the gasification of coal. They use a variety of contacting methods and ways of supplying heat for the reaction. Figure 19 illustrates those processes that employ fluidized bed gas generators and that have been developed at least until the pilot-plant stage.

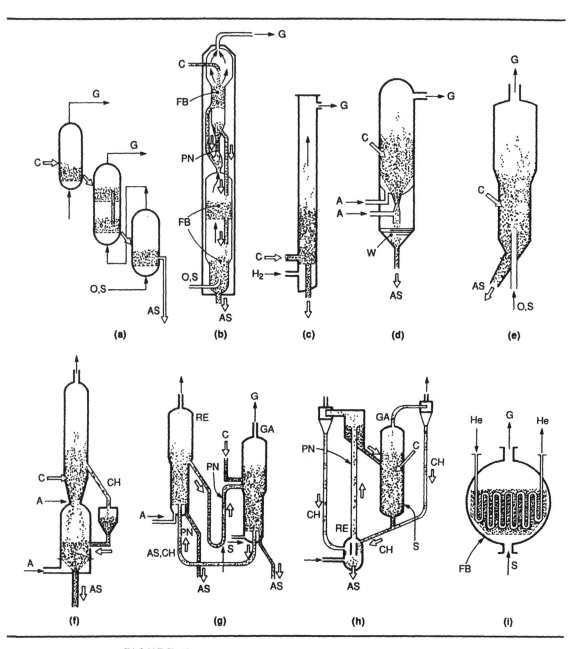

FIGURE 19

Various processes for the gasification of coal: (a) Coed; (b) Hygas; (c) Rheinishe Braunkohlenwerke; (d) U-Gas; (e) Westinghouse; (f) Mitsubishi; (g) Union Carbide; (h) Cogas; (i) Forschungsbau.

A = air, AS = ash, C = coal, CH = char, FB = fluidized bed, GA = gasifier, G = product gas, H_2 = hydrogen, He = helium, O = oxygen, PN = pneumatic conveyer, RE = regenerator, S = steam, W = water.

Figure 19(a) represents the Coal Oil Energy Development (Coed) process, in which coal particles are progressively carbonized at higher temperatures in four fluidized beds (316°, 455°, 538°, 810°C) so as to maximize the yield of hydrocarbon liquid [33]. Figure 19(b) shows the Institute of Gas Technology (IGT) Hygas process for producing pipeline gas. Another version of this process uses the exothermic heat of the hydrogasification reaction to drive the desired gasification reaction, all at pressures as high as 80 atm [34].

Rheinishe Braunkohlenwerke AG has developed a single-stage gasification process, shown in Fig. 19(c), using hydrogen at high pressure (70 atm) to fluidize and combine with coal to produce light hydrocarbons [35]. Figure 19(d) shows the U-gas process developed by IGT in which agglomerated ash is classified and discharged downward from the reactor [36]. Westinghouse has developed another type of gas generator, as shown in Fig. 19(e), which is characterized by a jet nozzle and the discharge of agglomerated ash [37]. Sekitan Giken and Mitsubishi have developed a two-stage gas generator, illustrated in Fig. 19(f), that uses a screw feeder to transport char from the upper bed to the lower [38].

Figure 19(g) sketches a process developed by Union Carbide in which agglomerated ash particles are circulated between the gasification reactor (endothermic) and the regenerator (exothermic) [39]. The Cogas process shown in Fig. 19(h) uses a mixture of char and ash as the heat carrier. The heat needed for the gasification comes from the combustion of fine char collected in the cyclones as well as from the partial combustion of the circulating char [40].

In an attempt to use the energy of high-temperature helium from a gas-cooled nuclear reactor to drive the gasification, Forschungsbau adopted the process shown in Fig. 19(i), in which reactive lignite is fluidized and gasified by steam. High-temperature helium flows through tubes in the bed to provide the heat needed for the gasification, which occurs at about 800°C and 70 atm [41].

In addition to the processes illustrated in Fig. 19, other fluidized bed gasification processes have been developed and tested at the pilot-plant scale, namely the Synthane process by the Pittsburgh Energy Research Center of DOE, the Steam-Iron process by IGT, and the Hitachi process.

Gasification processes using fluidized beds are still in the developmental stage, and their move to the commercial stage is much slower than for processes using other contacting modes, such as entrained flow contactors or moving bed contactors. However, for the large-scale production of fuel gas, such as would be needed for binary cycle power plants, advanced fluidized bed gas generators will likely prevail over other contacting modes.

Activation of Carbon

Charcoal is formed and activated by low-temperature (800°–900°C) endo-thermic gasification with hot combustion gas of wood, peanut shells, and so on. The fluidized bed for this operation is generally a multistage unit, as shown in Fig. 20(a). Multistaging gives a more uniform residence time distribution for the solids and helps to recover heat for the gasification by secondary combustion of CO and H_2 produced from the solids.

During gasification, the density of solids drops to 15–20% of the original value, but the size of the particles remains practically unchanged. Different fluidizing conditions are therefore needed in the various stages of the unit and should be accounted for in design.

Figure 20(b) illustrates a simple alternative design in which the fluidized

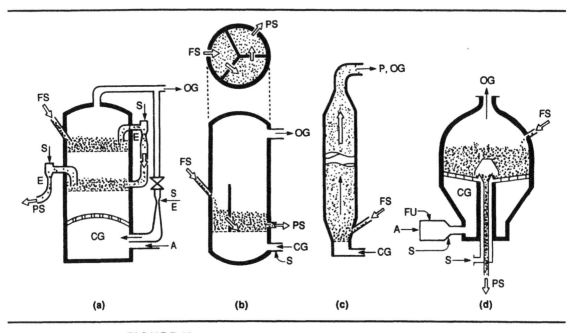

FIGURE 20
Reactors for activation of charcoal from (a) charcoal, (b) charcoal, (c) sawdust, (d) pitch beads.
A = air, CG = hot combustion gas, E = ejector, FS = feed solids, FU = fuel gas, OG = off-gas,
PS = product solid, S = steam.

bed is separated into sections by vertical partitions containing openings. Particles move from stage to stage, giving a better distribution of residence times for the solids than do single-stage operations.

Fast fluidized contacting, shown in Fig. 20(c), can be used for producing charcoal from sawdust. The product of this operation is a good feed for activation in reactors, such as shown in Figs. 20(a) and (b).

To remove pollutants from water, the active carbon used should be strong and inexpensive. Kureha developed a process to meet this requirement with a 3.5-m ID reactor that produced 1000 tons of active carbon per year, as shown in Fig. 20(d). Here, petroleum pitch is fed to a fluidized bed kept at 950°C to form carbon beads (300–800 μm). Because of the high temperature, the distributor plate of this reactor had to be very carefully designed [42].

Gasification of Solid Waste

Municipal solid waste, namely garbage, may contain a variety of toxic and hazardous materials. In ordinary incineration plants costly gas cleaning equipment must be installed to meet the increasingly strict demand for a clean environment. In comparison, the cleanup of combustion gases from gasification plants is much simpler and cheaper because the volume of gas produced is far smaller than that from incinerators.

Extending the concept of the K-K process shown in Fig. 16(b), Tsukishima Co. developed an innovative gasification process for treating municipal solid waste, called the Pyrox process [43]. This process consists of two relatively tall slender fluidized vessels connected by steeply sloping downcomers, with sand as the circulating heat carrier, as shown in Fig. 21(a). Coarsely shredded garbage is

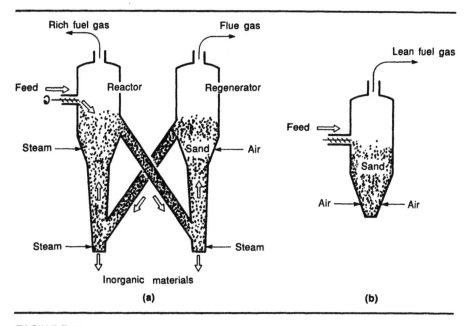

FIGURE 21
Gasifiers for solid waste: (a) Pyrox process; (b) Tsukishima process.

fed continuously to the gasifier to produce rich fuel gas. Char is then carried to the heater by the circulating sand carrier, to be burned there to completion. Hot sand then returns to the gasifier to provide the energy for further gasification.

Three units of this type, each having a capacity of 150 tons/day, were constructed in Funabashi City, near Tokyo, in 1982. They are operating successfully, providing a fuel gas with heating value as high as 21,000–23,000 kJ/N-m^3, and a sterile solid ash containing lumps of iron and steel that come from tin cans, bicycles, and other iron-containing trash.

Tsukishima then came up with a much simpler gasification process, shown in Fig. 21(b), for gasifying plastic rubbish discarded by pulp and paper companies. In this unit, 10–15 tons/day of plastic waste is gasified in a single sand fluidized bed to produce a low-energy fuel gas, about 8500 kJ/N-m^3 which replaces roughly half the fuel oil used in the paper-pulp plant itself. Before installing this unit, the company had to pay to dispose of this waste; now it is generating energy from this waste.

Although gasification plants are more expensive to construct than ordinary incinerators, the growth of populations, society's continued demand for a clean environment, and the inevitable rise in the cost of energy all suggest that gasification processes may eventually prevail, even on an economic basis, over other methods of disposal of solid waste.

Calcination

Particles of limestone and dolomite can be calcined straightforwardly in a fluidized bed by burning fuel directly in the bed:

$$CaCO_3 \xrightarrow{1000°C} CaO + CO_2, \qquad \Delta H_r = +180 \text{ kJ/mole}$$

Since this reaction is highly endothermic and gas and solid both leave at 1000°C, this operation is very wasteful in fuel. To recover much of the heat, multistaging is used, and Fig. 22(a) shows the first commercial unit of this type; see White and Kinsalla [44]. The original unit, designed and built in 1949 for the New England Lime Company, had a diameter of 4 m and a height of 14 m. Raw material 6–65 mesh is fed to the top stage of the unit and flows downward from stage to stage. In the calcination stage, fuel oil is sprayed into the bed through 12 nozzles arranged around the perimeter of the bed, mixed with fluidizing air, and burned.

Mitsubishi later developed the New Suspension Preheating System for cement clinkering, which incorporated a fluidized bed calciner for limestone powder. This was followed by an alternative limestone calcination process that combined a fluidized calciner with suspension preheaters (in effect, cyclone heat exchangers), as shown in Fig. 22(b).

Dorr-Oliver's two-bed process for calcining paper mill lime sludge (<50 μm) operates as follows. Dried and powdered lime mud is fed to the upper bed, which is fluidized by air and is fed fuel that burns in the bed. At a high enough temperature, about 770°C, calcium carbonate calcines to calcium oxide, and trace constituents in the feed, such as sodium carbonate, fuse, act as a binder, and cause agglomeration, eventually resulting in the formation of spherical pellets of lime. Proper control of particle growth is important. The particles then flow to the lower bed, which acts as an air preheater [45].

Fine powdery lime obtained from the calcination of lime sludge is in demand by the steel industry, and Fig. 22(c) shows a reactor designed to produce this material. The fluidized bed contains a carrier of coarse agglomerated lime particles. Fines from the slurry are injected into the bed, stay there for a short time, and are then elutriated, collected, and rapidly cooled to prevent the reverse reaction from proceeding significantly at lower temperature. This rapid cooling is crucial for satisfactory operation.

To upgrade the poor-quality phosphate rock that is plentiful in the western United States, Dorr-Oliver developed a three-stage fluidized calcination system

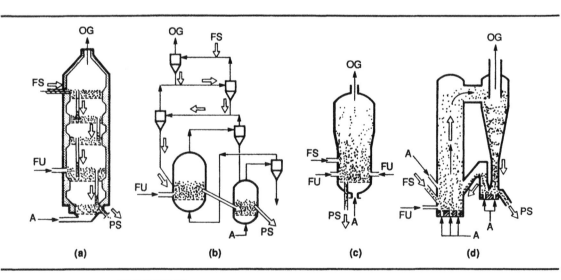

FIGURE 22
Reactors for calcination of (a) particulate limestone, (b) powdery limestone, (c) lime slurry, (d) alumina fines.
A = air, FS = feed solids, FU = fuel, OG = off-gas, PS = product solid.

using 5-m diameter beds. According to Priestley [46], the hydrocarbon content of this phosphate ore (3.5%) provides most of the heat needed for the calcination.

In the late 1960s, Lurgi engineers adopted the fast fluidized bed for the calcination of alumina,

$$Al_2O_3 \cdot 3H_2O \longrightarrow Al_2O_3 \cdot H_2O \longrightarrow \gamma\text{-}Al_2O_3 \longrightarrow \alpha\text{-}Al_2O_3$$

as shown in Fig. 22(d). Stable and smooth circulation of fine particles is required for satisfactory performance of this unit.

Finally, fluidized calciners have been successfully used to defluorinate phosphate rock, to make chicken feed [47], and to reuse mold sand.

Reactions Involving Solids

Roasting Sulfide Ores

Roasting operations are all characterized by a not too exothermic oxidation; hence a single-stage fluidized bed with no outside heating and, if anything, mild cooling is usually satisfactory. These units have a higher capacity than do alternative designs. They also require less excess air, thus giving an off-gas with higher sulfur dioxide concentration. The only problem that occurs is short-circuiting of solids because the solids are well mixed in the single-stage units.

The historical survey given earlier sketched the development of fluidized roasters, and Fig. 23(a) shows the Dorr-Oliver FluoSolids roaster, designed for producing SO_2 from pyrite, zinc blende, and other sulfide ores. According to Noguchi [48], representative operating conditions are as follows:

Reactor:	5.5 m ID, 7.6 m high, atmospheric pressure, 650–700°C
Feed:	−10 mesh, 170–220 tons/day
Gas velocity:	45–50 cm/s
Bed height:	1.2–1.5 m
Residual sulfur:	0.5 wt% in overflow, 1.2 wt% in carryover solids
Product gas:	contains 12% SO_2, entrains 75–80% of the cinder

Reactors of this type, up to 13 m ID, have been constructed to process 700 tons of ore per day.

The uniform temperature of fluidized beds allows sulfide ores containing copper or cobalt to be roasted to the sulfate and then to be separated from the iron oxide cinder by leaching with water or dilute sulfuric acid. Sulfate roasting is usually done at a lower temperature than oxide roasting—for instance, 650°C for copper, 670°C for zinc; consequently, it requires a longer particle residence time. Hence, fluidized beds, such as shown in Fig. 23(a), are more suited to this type of operation than other kinds of reactors having shorter particle residence times.

Another example of a roaster is BASF's design in Fig. 23(b). This has a relatively shallow bed (0.6 m), high gas velocity (1.3–2.3 m/s), large freeboard (~5 m), higher operating temperature (for pyrite roasting, 660°–740°C in the bed and 820°–920°C in the freeboard), and immersed cooling tubes. The distributor consists of flat steel plates with about 1-mm clearance or tuyeres embedded in a refractory plate. Pressure drop across the distributor is 0.5–1.0 kPa; across the whole bed it is 13–15 kPa. The feedstock usually consists of flotation concentrate (−60 mesh), and practically all the solids are carried out

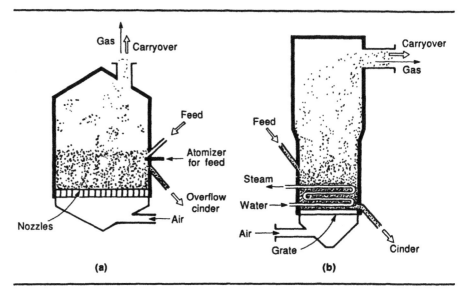

FIGURE 23
Metallurgical roasters for sulfide ores: (a) Dorr-Oliver type; (b) BASF type.

from the bed, continuing their oxidation in the freeboard. To maintain stable and smooth operations under such conditions, coarse solids are separated from the cinder and are returned continuously to the bed. Average conversion from sulfide to oxide is 97%.

Judging from the temperature rise in the freeboard, about 19% of the total reaction occurs there. Manabe [49] measured the residence time distribution of fine particles in an operating hot reactor to be 29–36 s.

This type of roaster is used extensively in the sulfuric acid and mining industries to roast iron pyrite and zinc blende concentrate. Ordinary units treat 80–150 tons/day of ore in beds 2.5–4.2 m in diameter and 8.4–10 m in height. Larger reactors have been constructed to roast as much as 400–1000 tons of sulfide ore daily.

Silicon for the Semiconductor and Solar Cell Industries

There has been an explosive increase in demand for crystalline silicon of exceptional purity for the semiconductor and photovoltaic industries. Many chemical pathways have been explored; those reaching the commercial stage start with metallurgical-grade silicon or liquid silicon tetrachloride, an inexpensive and abundant by-product from chemical vapor deposition (CVD) reactors and from the zirconium and other industries. The various steps to ultrapure silicon are as follows:

$$Si + HCl \xrightarrow{\text{step 1}} SiCl_4(l) \xrightarrow{\text{step 2}} SiHCl_3(g) \xrightarrow{\text{step 3}} SiH_4(g) \xrightarrow[\text{step 5}]{\text{step 4}} Si(s)$$

Steps 1, 2, 4, and 6 involve fluidized beds; steps 3 and 5 involve other types of contacting.

Step 1. Fluidized bed production of $SiHCl_3$ from metallurgical-grade silicon has been practiced by Union Carbide and other companies for over 30 years. In Osaka Titanium's process the reaction takes place at about 300°C. The product gas is then purified in distillation columns and used as a feed for Siemens CVD reactors to produce a high-purity silicon. This process was scaled up directly from bench scale to commercial scale, it yields practically complete utilization of feed, and has been operating successfully since about 1970. Figure 24(a) is a sketch of Texas Instruments' fluidized bed used for this reaction.

Step 2. A joint Union Carbide–MIT effort [51] led to a fluidized bed process for producing $SiHCl_3$ from gaseous $SiCl_4$, as follows:

$$2H_2 + 3SiCl_4 + Si(met) \xrightarrow[500°C]{CuCl_2} 4SiHCl_3$$

As an example of the implementation of this reaction, Osaka Titanium constructed a test reactor of 25 cm ID, corresponding to a production rate of 200 tons/yr of pure silicon. Operating conditions, according to Noda [52], are as follows:

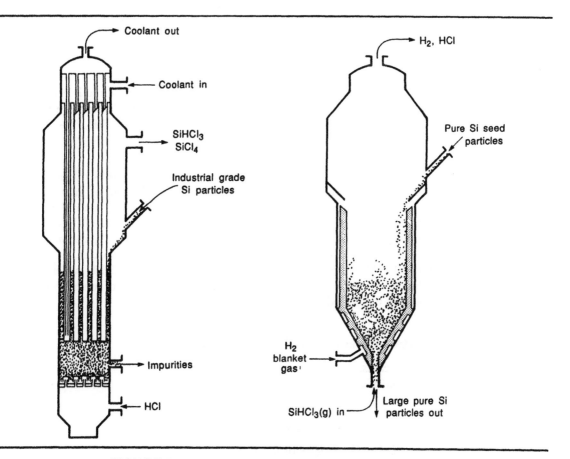

FIGURE 24
Texas Instruments' reactors [50]: (a) for producing $SiHCl_3$ from metallurgical grade Si; (b) for producing very pure silicon particles from $SiHCl_3$.

CuCl catalyst: 500°C, 8.5–9 atm
Feed: $H_2 : SiCl_4 = 1.5–3.0 : 1$
Exit gas: $SiCl_4 = 75–77\%$, $SiHCl_3 = 23–25\%$,
                       $SiH_2Cl_2 = 0.5\%$

Steps 2, 3, and 4 or 5. In the mid-1970s the United States embarked on a national program to reduce the cost of very pure silicon by 90–93%. Industry was invited to participate in this program. Ten processes were selected for follow-up, five of which involved fluidized beds. The process finally selected, proposed by Union Carbide, was a multistep scheme starting with cheap commercially available $SiCl_4$ (steps 2, 3, and 4), in which the first and last steps involved fluidized beds. A 100-tons/yr pilot plant was constructed to produce very pure SiH_4 from cheap starting materials, and in 1985 a 1200-tons/yr commercial plant started operation [53].

For the last step, the conversion of SiH_4 to silicon, Union Carbide, under time pressure, adopted the Komatsu CVD process, which in principle was an extension of the Siemens filament CVD process (step 5). However, they and the Jet Propulsion Laboratory are still pursuing the fluidized bed option for this last step (step 4). We briefly describe this [53, 54].

Since silane becomes unstable when heated, this process simply introduces a cold mixture of SiH_4 and hydrogen directly into a fluidized bed of hot silicon particles:

$$SiH_4(g) \xrightarrow{\sim 700°C} Si(s) + 2H_2$$

Silane decomposes, and the silicon smoke formed fuses directly to the bed material, which then grows. Pilot-plant studies by Union Carbide are trying to minimize dust formation and to find out how to use higher SiH_4/H_2 ratios for the feed.

Step 6. Active development work is also being pursued by other groups in the United States and by NEDO in Japan, all with the aim of growing dense large silicon particles of high purity and low cost. Fluidized beds are the reactors of choice in most of these processes, and Fig. 24(b) shows one of these designs. According to Noda [52], in the Shin-Etsu operation $SiHCl_3$ decomposes at 1000°–1100°C, and fine silicon dust deposits on the bed particles, which grow from 250–500 μm to 800–1500 μm. Silicon yield is about 20%, which is close to the equilibrium for this reaction, and consumption of energy is about 120 MJ/kg Si formed.

Chlorination and Fluorination of Metal Oxides

In some cases, chlorination of the oxide is the only practical path for the production of a pure metal. For example, for titanium the following reactions occur:

$$TiO_2 + 2C + 2Cl_2 \rightarrow TiCl_4 + 2CO$$
$$TiO_2 + C + 2Cl_2 \rightarrow TiCl_4 + CO_2$$

These reactions are carried out at about 1000°C, and for environmental reasons close to complete utilization of chlorine is required. Fluidized bed reactors have been able to meet these strict requirements and are playing an important role in the titanium industry. For the production of zirconium, the process is similar.

Another example is the separation of U-235 from U-238 for the nuclear industry. UO_2 on a carrier of alumina is fluorinated to gaseous UF_6 in fluidized bed reactors operating at about 450°C [55]. This type of reaction may find use for the production of other kinds of valuable metals.

Reduction of Iron Oxide

The fluidized reduction of iron ore (iron oxides) has been extensively studied since about 1960, particularly in the United States, to develop a process for producing iron and steel from fines of high-grade ore or, more importantly, to replace the blast furnace as the basic means for producing iron, if possible. We will describe some of these efforts.

Hydrocarbon Research and Bethlehem Steel jointly developed a process, called the H-Iron process, for the direct reduction of iron ore; see Fig. 25(a). Here, the feed hopper is charged with ore, sealed, purged of air with CO_2, and then pressurized with hydrogen to about 46 atm. Before receiving a batch of fresh ore, the solids in the three stages are successively dumped as follows: first the 98% reduced ore is discharged from the lowest stage; then partly reduced ore (87%) in the middle stage is dropped to the lowest stage; finally, the least reduced ore (47%) in the top stage drops to the middle stage (see Labine [56]).

Using this process, Alan Wood Steel Co. produces 50 tons/day of high-quality pyrophoric iron powder for metallurgical applications. Another 100-tons/day plant was built by Bethlehem Steel. The reactor vessel for the 50-tons/day plant is about 1.7 m ID and 29 m high. Since conversion of hydrogen is low (5%) and the dilute water vapor formed by the reaction

$$\underset{\text{magnetite}}{Fe_3O_4} + 4H_2 \rightarrow Fe + 4H_2O$$

must be separated by cooling, large amounts of hydrogen must be circulated, and heat consumption for the hydrogen preheat is necessarily high. Roughly 1.4 tons of high-grade magnetite ore, 0.051–0.056 ton of hydrogen, and 0.25 ton of oxygen are needed to produce each ton of iron by this process.

Besides the preceding process, numerous pilot-plant ore reducers have been constructed. United States Steel developed its Nu-Iron process for reduction with hydrogen of −10 mesh ore [57]. Other processes for iron ore reduction with hydrogen include the Stelling process for form cementite and the Armco process with its two-bed reactor.

Exxon has developed one such process, using multistage reactors, called the Fluid Iron Ore Direct Reduction (Fior) process [58]. Having demonstrated the technical feasibility of the process with a 5-tons/day pilot plant, they built an experimental 300-tons/day continuous plant at the Imperial Oil refinery in Nova Scotia, which started operations in late 1965. The product of this process consists of a free-flowing powder containing as high as 89% metallic iron with a total iron content as high as 93%. The powder consists of 25–45% through 325 mesh with hardly any material larger than the 4–8 mesh material. A full-sized commercial unit, sketched in Fig. 25(b), was constructed in Venezuela by Arthur G. McKee, which licensed this process from Exxon. This plant has been producing iron briquettes since 1978 at the scheduled production rate of 1000 tons/day since 1980.

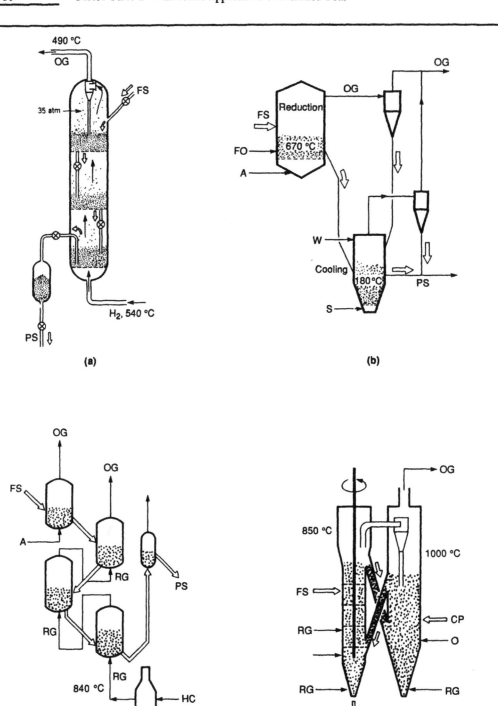

FIGURE 25

Iron ore reduction processes: (a) H-Iron; (b) Fior; (c) FluoSolids; (d) Kawasaki Iron and Steel. A = air, CP = coke particles, FO = feed oil, FS = feed solids, GR = gasifier, H_2 = hydrogen, HC = hydrocarbon, O = oxygen, OG = off-gas, PS = product solid, RG = reducing gas, S = steam, W = water.

Tomasicchio [59] reports on the FluoSolids process in which the reduction is accomplished by direct injection of fuel oil into a hot bed that is fluidized by substoichiometric air; see Fig. 25(c). This process was first used on a commercial scale at the Montecatini plant in Follonica, Italy. It reduces 400 tons/day of hot (530°C) hematitic pyrite cinder to magnetite.

Gradual economic changes in the iron and steel industries are leading them to use smaller, more efficient, direct reduction processes to serve the so-called minimills, which are an order of magnitude smaller in capacity than conventional steel mills. Fluidized contacting is a likely choice of reactor for these processes, and one can expect much development work in this line.

One example is shown in Fig. 25(d), which Kawasaki Iron and Steel developed to pilot scale. Here coarse iron ore particles are fed into the 700°C reduction reactor fitted with perforated disks that rotate to prevent agglomeration of the solids. In addition, fine coke particles are circulated between reactor and a heater, wherein carbon is burned and gasified with oxygen. The off-gas from the heater can be cleaned and used as feed gas to the reactor.

Biofluidization

The cultivation of microorganisms appears to be one of the more interesting applications of fluidization. Kikkoman Co. has pioneered this use, and Fig. 26 illustrates the design developed by them for producing soy sauce. Wheat bran is first treated and pasteurized by superheated steam, sized to −28 mesh, and then fluidized by sterilized air. Water is sprayed onto the bed to keep the moisture content of the solids at about 70% on a dry basis, and seed spores of the microorganism are sent into the bed through an ejector (weight ratio of the seed

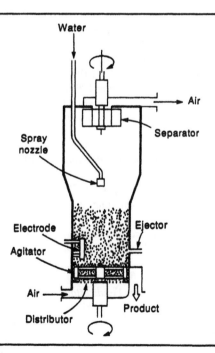

FIGURE 26
Fluidized bed cultivator to produce threadlike fungus (*Aspergillus sojae*) (adapted from Akao and Okamoto [60]).

spore to wheat bran is about 4%). Fluidized cultivation is reported to be superior to the conventional layer cultivation in the following areas:

- Large effective growing surface of microorganisms.
- Easy oxygen transfer results in an active metabolism.
- Heat and carbon dioxide generated by this active metabolism are efficiently removed.
- Temperature, moisture, and pH level are easily and automatically controlled.

Figure 26 shows various features of this bioreactor: the rotary agitator just above the air distributor to prevent defluidization in the lower portion of the bed, the rotating separator in the freeboard to return elutriated particles to the bed, and an electrode to detect the water content of the particles.

Batch cultivation at about 30°C for a few days yields a microorganism of high quality with 5–15 times the activity obtainable by conventional methods. A pilot plant 1.5 m ID below and 2.1 m ID above, with a static bed height of 2 m, was built in Japan to supply the *Aspergillus sojae* needed for Kikkoman's soy sauce production; see Akao and Okamoto [60].

We may expect biofluidized reactors to be increasingly used in the food and pharmaceutical industries.

REFERENCES

1. C.E. Jahnig et al., in *Fluidization III*, J.R. Grace and J.M. Matsen eds., p. 3, Plenum, New York, 1980.

2. A.M. Squires, in *Circulating Fluidized Bed Technology*, P. Basu ed., p. 1, Pergamon, New York, 1986.

3. A.M. Squires, in Proc. CIESC/AIChE Joint Meeting, p. 332, Chem. Ind. Press, Beijing, 1982.

4. J. Gaynor, *Chem. Eng. Prog.*, **56**, 75 (July 1960).

5. A.S. Jensen, in *Fluidization V*, K. Østergaard and A. Sørensen eds., p. 651, Engineering Foundation, New York, 1986.

6. O. Kjaergaard et al., in *Fluidization V*, K. Østergaard and A. Sørensen eds., p. 656, Engineering Foundation, 1986.

7. M. Sivetz, U.S. Patent 3,964,175, 22 June 1976.

8. J. Ciborowski and S. Wronski, *Chem. Eng. Sci.*, **17**, 481 (1962).

9. D.A. Avery and D.H. Tracey, in *Fluidization*, Tripartite Chem. Eng. Conf., Montreal, p. 28, Institution of Chemical Engineers, 1968.

10. H.M. Rowson, *Brit. Chem. Eng.*, **8**, 180 (1963).

11. Y. Wang, J. Yong, and M. Kwauk, in *Fluidization '85, Science and Technology*, M. Kwauk et al. eds., p. 11, Science Press, Beijing, 1985.

12. Y. Amagi et al., *Monthly Rep. Soc. Chem. Ind. Japan*, **29**, 30 (1976).

13. Y. Ikeda, in *Fluidization '85, Science and Technology*, M. Kwauk et al., eds., p. 1, Science Press, Beijing, 1985.

14. J.J. Graham, P.F. Way, and S. Chase, *Chem. Eng. Prog.*, **58(1)**, 96 (1962);
J.J. Graham, *Chem. Eng. Prog.*, **66(9)**, 54 (1970); and private communication.

15. T. Shingles et al., *ChemSA*, **12**, 79 (1986).

16. R.W. Silverman et al., in *Fluidization V*, K. Østergaard and A. Sørensen eds., p. 441, Engineering Foundation, New York, 1986.

17. T. Nakamura and M. Ito, in *Chemical Reactors*, K. Hashimoto ed., pp. 168, 173, Baifukan, Tokyo, 1984 (in Japanese).

18. T. Miyauchi et al., *Adv. Chem. Eng.*, **11**, 275 (1981); *Jap. Petrol. Inst. J.*, **20**, 624 (1977).

19. H. Seya, preprint, Seminar on Development to Commercial Scale, Feb. 1971, Kanto Branch, Soc. Chem. Eng. Japan.

20. T. Katsumata and T. Dozono, *AIChE Symp. Ser.*, **83(255)**, 86 (1987).

21. F.J. Karol, *Catal. Rev.-Sci. Eng.*, **26**, 557 (1984).

22. M. Sugimoto, *Kagaku-Kogaku*, **49**, 341 (1985).

23. H. Koda and T. Kurisaka, in *Fluidization '85, Science and Technology*, M. Kwauk et al. eds., p. 402, Science Press, Beijing, 1985.

24. K. Izumi, *Kagaku Kogaku*, **43**, 218 (1979).

25. H. Nakaishi, in *Chemical Reactors*, K. Hashimoto ed., p. 177, Baifukan, 1984.

26. Y. Ikeda and S. Tashiro, *Fluidization Technology*, SCEJ Kanto Branch, Gakkai, p. 97, 1981 (in Japanese).

27. J.M. Matsen, in *Handbook of Multiphase Systems*,

G. Hetsroni ed., pp. 8–178, McGraw-Hill, New York, 1982.

28. D.E. Blaser and A.M. Edelman, paper, API Refining Dept. 43rd Midyear Meeting, Toronto, May 1978.

29. P. Schmalfeld, *Hydrocarbon Processing and Petroleum Refiner*, **42**, 145 (July 1963).

30. A. Steinhofer, *Hydrocarbon Processing and Petroleum Refiner*, **44**, 134 (August 1965).

31. Y. Fujima, private communication.

32. *Technical Guide Book*, Research Group on Environmental Technology, pp. 1083, 1982 (in Japanese).

33. J.F. Jones et al., in *Proc. Int. Symp. on Fluidization*, A.A.H. Drinkenburg ed., p. 676, Netherlands Univ. Press, Amsterdam, 1967.

34. B.S. Lee et al., paper, AIChE National Meeting, New Orleans, 1967.

35. L. Schrader and G. Felgener, in *Fluidization*, J.F. Davidson and D.L. Keairns eds., p. 221, Cambridge Univ. Press, New York, 1978.

36. A. Rehmat and A. Poyal, in *Fluidization IV*, D. Kunii and R. Toei eds., p. 647, Engineering Foundation, New York, 1983.

37. D.L. Keairns, Lecture in Professional Develop. Seminar, October 1981, Montreal.

38. Report of the Advisory Committee for Coal Gasification, New Energy Development Organization, Japan, 1981.

39. A.L. Conn, *Chem. Eng. Prog.*, **64**, 56 (Dec. 1973).

40. H.R. Hoy and R.T. Eddinger, in *Proc. 7th Synthetic Pipeline Gas Symp.*, p. 87, Chicago, 1975.

41. H. Kubriak et al., in *Proc. 16th Int. Symp. on Heat and Mass Transfer*, Dubrovnik, 1984.

42. H. Murakami and T. Okamoto, Chemical Apparatus (85), p. 38, 1975 (in Japanese).

43. M. Hasegawa, J. Fukuda, and D. Kunii, *Conservation and Recycling*, **3**, 143 (1980); *ACS Symp. Ser.*, **130**, 525 (1980).

44. F.S. White and E.L. Kinsalla, *Mining Eng.*, **4**, 903 (1952).

45. R.B. Thompson, in *Fluidization*, D.F. Othmer ed., p. 212, Van Nostrand Reinhold, New York, 1956.

46. R.J. Priestley, in *Proc. Int. Symp. on Fluidization*, A.A.H. Drinkenburg ed., p. 701, Netherlands Univ. Press, Amsterdam, 1967.

47. M. Azegami, *Kagaku Kogaku*, **34**, 1019 (1970).

48. N. Noguchi, Lecture on Fluidized Bed Driers and Roasters, Japan Science Foundation, Tokyo, 1963.

49. A. Manabe, Lecture on Fluidized Bed Roasters, Japan Science Foundation, Tokyo, 1967.

50. F.A. Padovani et al., U.S. Patent No. 4,092, 446 (1978).

51. DOE/JPL contract 955382-79 (1979).

52. T. Noda, in *Proc. Flat-Plate Solar Array Project Workshop on Low Cost Polysilicon for Terrestrial Photovoltaic Solar Cell Applications*, Jet Propulsion Laboratories, Las Vegas, October 1985.

53. N. Rohatgi and G. Hsu, Report from Jet Propulsion Laboratories, California Institute of Technology, October 1983.

54. *Flat-Plate Solar Array Project, 10 Years of Progress*, E. Christensen ed. Jet Propulsion Laboratories, California Institute of Technology, for DOE and NASA, October 1985.

55. L.S. Anastasia, P.G. Alfredson, and M.J. Steindler, *Ind. Eng. Chem. Process Des. Dev.*, **10**, 150 (1971).

56. R.A. Labine, *Chem. Eng.*, **67**, 96 (Feb. 1960).

57. C.S. Cronan, *Chem. Eng.*, **67**, 64 (April 4, 1960).

58. J.W. Brown et al., *J. Metals*, **18**, 237 (1966).

59. G. Tomasicchio, in *Proc. Int. Symp. on Fluidization*, A.A.H. Drinkenburg ed., p. 725, Netherlands Univ. Press, Amsterdam, 1967.

60. T. Akao and Y. Okamoto, *Kagaku Kogaku*, **49**, 349 (1985).

3

Fluidization and Mapping of Regimes

When a bed of solids is kept suspended by fluid upflow, the bed can behave in various ways—smoothly fluidized, bubbling, slugging, spouting, and so on. This chapter considers the mapping of these flow regimes.

Fixed Beds of Particles

Characterization of Particles

The size of spherical particles can be measured without ambiguity; however, questions arise with nonspherical particles. Here one can define the size in several ways. We adopt a size d_{eff} that is useful for flow and pressure drop purposes.

The size of *larger particles* (> 1 mm) can be found by calipers or micrometer if the particles are regular in shape, or by weighing a certain number of particles if their density is known, or by fluid displacement if the particles are nonporous. From these measurements we first calculate the equivalent spherical diameter, defined as follows:

$$d_{sph} = \left(\begin{array}{c} \text{diameter of sphere having the} \\ \text{same volume as the particle} \end{array} \right) \tag{1}$$

Various measures of nonsphericity are available, and are summarized by Zenz and Othmer [1]. For our purposes we choose the one-parameter measure called the sphericity, ϕ_s, defined as

$$\phi_s = \left(\frac{\text{surface of sphere}}{\text{surface of particle}} \right)_{\text{of same volume}} \tag{2}$$

With this definition $\phi_s = 1$ for spheres and $0 < \phi_s < 1$ for all other particle shapes. Table 1 lists calculated sphericities for different solids.

Next we represent a bed of nonspherical particles by a bed of spheres of

TABLE 1   Sphericity of Particles

Type of Particle	Sphericity ϕ_s	Source
Sphere	1.00	(a)
Cube	0.81	(a)
Cylinder		
$h = d$	0.87	(a)
$h = 5d$	0.70	(a)
$h = 10d$	0.58	(a)
Disks		
$h = d/3$	0.76	(a)
$h = d/6$	0.60	(a)
$h = d/10$	0.47	(a)
Activated carbon and silica gels	0.70–0.90	(b)
Broken solids	0.63	(c)
Coal		
anthracite	0.63	(e)
bituminous	0.63*	(e)
natural dust	0.65	(d)
pulverized	0.73	(d)
Cork	0.69	(d)
Glass, crushed, jagged	0.65	(d)
Magnetite, Fischer-Tropsch catalyst	0.58*	(e)
Mica flakes	0.28	(d)
Sand		
round	0.86*	(e)
sharp	0.66*	(e)
old beach	as high as 0.86	(f)
young river	as low as 0.53	(f)
Tungsten powder	0.89	(d)
Wheat	0.85	

(a) From geometric considerations
(b) From Leva [2]
(c) From Uchida and Fujita [3]
(d) From Carman [4]
(e) From Leva et al. [5]
(f) From Brown et al. [6]
*Photographs available.

diameter d_{eff} such that the two beds have the *same total surface area* and *same fractional voidage* ε_m. This representation should ensure almost the same frictional resistance to flow in these two beds. Then by geometry we can show that

$$d_{eff} = \phi_s d_{sph} \qquad (3)$$

The *specific surface* of particles in either bed is then found to be

$$a' = \left(\frac{\text{surface of a particle}}{\text{volume of a particle}}\right) = \frac{\pi d_{sph}^2/\phi_s}{\pi d_{sph}^3/6} = \frac{6}{\phi_s d_{sph}}, \quad [m^{-1}] \qquad (4a)$$

and for the whole bed

$$a = \left(\frac{\text{surface of all particles}}{\text{total volume of particles in the bed}}\right) = \frac{6(1 - \varepsilon_m)}{\phi_s d_{sph}}, \quad [m^{-1}] \qquad (4b)$$

where ϕ_s is measured directly, estimated from Table 1, or evaluated by the procedure just before Example 1.

TABLE 2 Tyler Standard Screens

Mesh Number[a]	Aperture[b]		Mesh Number	Aperture	
	(in)	(μm)		(in)	(μm)
3	0.263	6680	35	0.0165	417
4	0.185	4699	48	0.0116	295
6	0.131	3327	65	0.0082	208
8	0.093	2362	100	0.0058	147
10	0.065	1651	150	0.0041	104
14	0.046	1168	200	0.0029	74
20	0.0328	833	270	0.0021	53
28	0.0232	589	400	0.0015	38

[a]Number of wires per inch
[b]Opening between adjacent wires

For *intermediate particle sizes* screen analysis is the most convenient way to measure particle size. Numerous calibrated screens are available, and Table 2 shows the size of openings for the Tyler standard screens. Particles passing through a 150 mesh screen but resting on a 200 mesh screen are called $-150 + 200$ mesh particles and have a screen size

$$d_p = \frac{104 + 74}{2} = 89 \ \mu m$$

Since there is no general relationship between d_{eff} and d_p, the best we can say without doing experiments is the following:

- For irregular particles with no seemingly longer or shorter dimension (hence, isotropic in shape),

$$d_{eff} = \phi_s d_{sph} \cong \phi_s d_p \tag{5a}$$

- For irregular particles with one somewhat longer dimension, but with a length ratio not greater then 2:1 (eggs, for example),

$$d_{eff} = \phi_s d_{sph} \cong d_p \tag{5b}$$

- For irregular particles with one somewhat shorter dimension, but with a length ratio not less than 1:2, then roughly,

$$d_{eff} = \phi_s d_{sph} \cong \phi_s^2 d_p \tag{5c}$$

- For very flat or needlelike particles, estimate the relationship between d_p and d_{eff} from the ϕ_s values for the corresponding disks and cylinders.

Most fluidized bed operations treat particles whose sizes are measured with screen analysis. In addition, most of these solids are irregular with no seemingly larger or smaller dimension. Therefore we take the particle size to be given by Eq. (5a). Where the particles are needlelike, flat, or flaky, we might want to use Eq. (5b) or (5c) to relate d_p to d_{sph}. However, this approach is not really practical. In these situations we recommend the experimental procedure outlined just before Example 1. This will give an effective sphericity that can then reliably be used with Eq. (5a).

For *very small particles* ($< 40 \ \mu m$) we cannot use screen analysis, so we rely on

- Scanning of magnified photographs of particles.
- Sedimentation of particles in a known fluid; the terminal velocity of these particles will give the diameter of the equivalent sphere.   '

Fixed Beds—One Size of Particles

According to Brown et al. [6], the *fraction void* ε_m in a packed bed is related to particle sphericity, as shown in Fig. 1; in addition, for vessels of small diameter the wall effect becomes important and influences the bed voidage. Since ε_m is easy to measure, we suggest it be found experimentally.

The *frictional pressure drop*, always positive, through fixed beds of length L containing a single size of isotropic solids of screen size d_p has been correlated by Ergun [7] by the equation

$$\frac{\Delta p_{fr}}{L_m} g_c = 150 \frac{(1 - \varepsilon_m)^2}{\varepsilon_m^3} \frac{\mu u_o}{(\phi_s d_p)^2} + 1.75 \frac{1 - \varepsilon_m}{\varepsilon_m^3} \frac{\rho_g u_o^2}{\phi_s d_p} \tag{6}$$

The measured pressure drop is

$$\Delta p_{measured} = \Delta p_{fr} \pm \frac{\rho_g L_m}{g_c}, \qquad \text{all } \Delta p \text{ positive} \tag{7}$$

where the + sign stands for upflow of fluid. The last term may be appreciable for flowing liquids, but it can safely be ignored for flowing gases unless one is dealing with deep beds at high pressure. Thus, in most cases with gases, we may write

$$\Delta p = \Delta p_{fr} = \Delta p_{measured}, \qquad \text{all } \Delta p \text{ positive} \tag{8}$$

For randomly packed granular materials this expression has been found to represent the data within ±25%; however, it may not be expected to extend to nonrandomly packed beds, to beds of solids of abnormal void content (e.g.,

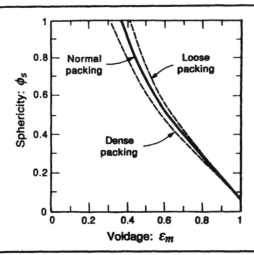

FIGURE 1
Voidage of a randomly packed bed of uniformly sized particles increases as particles become less spherical; from Brown et al. [6].

Raschig rings), or to highly porous beds (e.g., fibrous beds where $\varepsilon_m = 0.6$ to 0.98). At these high porosities the pressure drop can be much greater than that predicted by Eq. (6) (see Carman [4]).

Other procedures and expressions for finding the pressure drop are given by Perry [8], based on the work of Chilton and Colburn [9], by Carman [4], and by Brown et al. [6], based on the work of Brownell and Katz [10]. Brown's procedure accounts for all ranges of ϕ_s and ε_m.

Fixed Beds—Solids with a Distribution of Sizes

Before considering the behavior of beds containing solids of different sizes, we must be able to describe usefully the size distribution of a batch of solid particles. For this, define the size distribution functions **P** and **p** as follows. Let **P** be the volume fraction of particles smaller than size d_p and let $\mathbf{p}\, d(d_p)$ be the volume fraction of particles of size between d_p and $d_p + d(d_p)$. Typical distribution curves and their properties are shown in Fig. 2. From this we see that **p** gives the volume (or weight or numbers) distribution of particles directly and has units of reciprocal length, whereas **P** gives the cumulative distribution of sizes and is dimensionless.

The relationship between **p** and **P** is found by considering particles of any particular size, d_{p1}, for which we have

$$\mathbf{P}_1 = \left(\frac{d\mathbf{P}}{d(d_p)}\right)_1 \qquad \text{or} \qquad \mathbf{P}_1 = \int_0^{d_{p1}} \mathbf{p}\, d(d_p) \tag{9}$$

For a discrete distribution of particles with equal or unequal size intervals, we have the situation of Fig. 3, with the relation between **p** and **P** at any d_{pi} given by

$$\mathbf{p}_i = \left(\frac{\Delta\mathbf{P}}{\Delta d_p}\right)_i \qquad \text{or} \qquad \mathbf{P}_i = \sum_1^i (\mathbf{p}\,\Delta d_p)_i = \sum_1^i x_i \tag{10}$$

where x_i is the fraction of material in size interval i.

We next find the specific surface and mean diameter of a mixture of isotropic particles of different sizes. Many averages or means may be defined; however, for pressure drop in flow-through beds, the surface area is of prime consideration. Consequently, a mean size and shape should be defined to give the same total surface area for the same total bed volume. Thus, using the size distribution, we define the mean specific surface as

$$\bar{a}' = \int_0^{d_{p,max}} a'\mathbf{p}\, d(d_p) = \int_0^{d_{p,max}} \frac{6}{\phi_s d_p} \mathbf{p}\, d(d_p) \tag{11}$$

or, for a discrete distribution,

$$\bar{a}' = \sum^{\text{all } i} a_i'(\mathbf{p}\,\Delta d_p)_i = \frac{6}{\phi_s} \sum^{\text{all } i} \frac{(\mathbf{p}\,\Delta d_p)_i}{d_{pi}} = \frac{6}{\phi_s} \sum^{\text{all } i} \left(\frac{x}{d_p}\right)_i \tag{12}$$

Since the mean specific surface, defined in terms of mean diameter, is

$$\bar{a}' = \frac{6}{\phi_s \bar{d}_p} \tag{13}$$

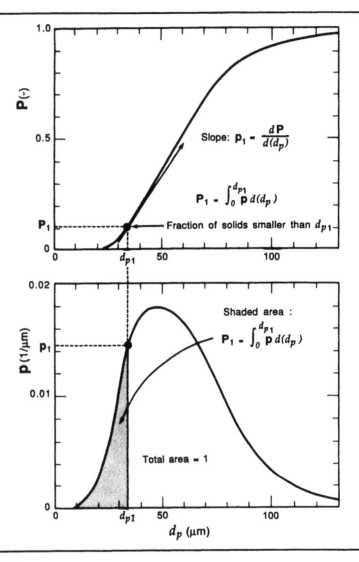

FIGURE 2
Typical size distribution of catalyst designed for fluidized reactors; adapted from Miyauchi et al.
[11].

we have, on combining Eqs. (11) or (12) with (13),

$$\bar{d}_p = \frac{6}{\phi_s \bar{a}'} = \frac{1}{\int_0^{d_{p.max}} (\mathbf{p}/d_p)\, d(d_p)} \tag{14}$$

or

$$\bar{d}_p = \frac{1}{\sum^{\text{all } i} [\mathbf{p}\, \Delta d_p)_i / d_{pi}]} = \frac{1}{\sum^{\text{all } i} (x/d_p)_i} \tag{15}$$

The frictional pressure drop in beds of mixed particles approximately follows the Ergun equation for single size of particles, Eq. (6), but with d_p replaced by $\bar{d}_p$.

The voidage for a mixture of sizes cannot be estimated reliably; factors that must be considered include how solids are introduced into the vessel, size of the

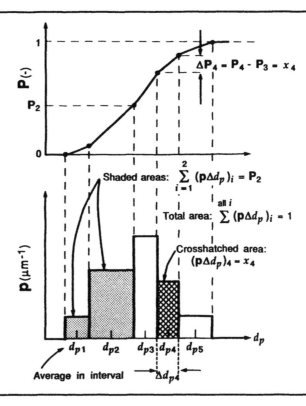

FIGURE 3
Relationship between **p** and **P** for a discrete size distribution of solids; useful for numerical calculations.

solids, and shape of the size distribution curve. For example, if the size variation is large, the fines can fit into the voids between the large particles, thus greatly decreasing voidage. Since the bed voidage is a relatively simple matter to determine experimentally, do so.

Determination of the Effective Sphericity $\phi_{s,\text{eff}}$ from Experiment

Some serious problems arise when using ϕ_s. First, all sorts of particle shapes can have the same sphericity, for example, pencils, doughnuts, and coins. Next, how does one quantify the "egg-shapedness" of an irregular particle or account for particle roughness? And most important, it is very difficult and tedious to evaluate properly the sphericity of irregular particles.

Therefore, we recommend the following experimental procedure for finding an effective value for ϕ_s. Carefully and accurately determine the bed voidage ε_m. Then measure the frictional pressure drop of this bed at several flow velocities. Finally insert d_p, ε_m, Δp_{fr}, and all of the system properties into the Ergun equation and extract the value of ϕ_s that best fits the data. This gives the relationship

$$d_{\text{eff}} = \phi_{s,\text{eff}} \, d_p$$

This value of effective sphericity can be used with the measured screen size d_p to predict frictional losses in beds of this solid of any size and for wide size distribution. In general, this probably is the most reliable measure of particle size for pressure drop purposes.

EXAMPLE 1

Size Measure of Nonuniform Solids

Calculate the mean diameter d_p of material of the following size distribution:

Cumulative weight of a representative 360-g sample ...	...having a diameter smaller than d_p (μm)
0	50
60	75
150	100
270	125
330	150
360	175

SOLUTION With Fig. 3 as a guide, make the following table:

Diameter range (μm)	d_{pi} (μm)	Weight fraction in interval $(p \, \Delta d_p)_i = x_i$	$(x/d_p)_i$
50–75	62.5	$(60 - 0)/360$ = 0.167	0.167/62.5 = 0.00267
75–100	87.5	$(150 - 60)/360$ = 0.250	0.250/87.5 = 0.00286
100–125	112.5	0.333	0.00296
125–150	137.5	0.167	0.00121
150–175	162.5	0.083	0.00051
			$\sum (x/d_p)_i = 0.01021$

From Eq. (15) the mean diameter is

$$\bar{d}_p = \frac{1}{\sum^{\text{all } i} (x/d_p)_i} = \frac{1}{0.01021} = 98 \, \mu m$$

Fluidization without Carryover of Particles

Minimum Fluidizing Velocity, u_{mf}

Consider a bed of particles resting on a distributor designed for uniform upflow of gas—for instance, a porous sintered metal plate. As stated in Chap. 1, the onset of fluidization occurs when

$$\left(\begin{array}{c} \text{drag force by} \\ \text{upward moving gas} \end{array} \right) = \left(\begin{array}{c} \text{weight of} \\ \text{particles} \end{array} \right) \tag{16a}$$

or

$$\left(\begin{array}{c} \text{pressure drop} \\ \text{across bed} \end{array} \right) \left(\begin{array}{c} \text{cross-sectional} \\ \text{area of tube} \end{array} \right) = \left(\begin{array}{c} \text{volume} \\ \text{of bed} \end{array} \right) \left(\begin{array}{c} \text{fraction} \\ \text{consisting} \\ \text{of solids} \end{array} \right) \left(\begin{array}{c} \text{specific} \\ \text{weight} \\ \text{of solids} \end{array} \right) \tag{16b}$$

or, with Δp always positive,

$$\Delta p_b A_t = W = A_t L_{mf}(1 - \varepsilon_{mf}) \left[(\rho_s - \rho_g) \frac{g}{g_c} \right] \tag{16c}$$

By rearranging, we find for minimum fluidizing conditions that

$$\frac{\Delta p_b}{L_{mf}} = (1 - \varepsilon_{mf})(\rho_s - \rho_g)\frac{g}{g_c} \qquad (17)$$

At the onset of fluidization, the voidage is a little larger than in a packed bed, actually corresponding to the loosest state of a packed bed of hardly any weight. Thus, we may estimate ε_{mf} from random packing data, or, better still, we should measure it experimentally, since this is a relatively simple matter. Table 3 records experimental values of ε_{mf}.

The superficial velocity at minimum fluidizing conditions, u_{mf}, is found by combining Eqs. (17) and (6) (a reasonable extrapolation for this packed bed expression). In general, for isotropic-shaped solids this gives a quadratic in u_{mf}:

$$\frac{1.75}{\varepsilon_{mf}^3 \phi_s}\left(\frac{d_p u_{mf}\rho_g}{\mu}\right)^2 + \frac{150(1 - \varepsilon_{mf})}{\varepsilon_{mf}^3 \phi_s^2}\left(\frac{d_p u_{mf}\rho_g}{\mu}\right) = \frac{d_p^3 \rho_g(\rho_s - \rho_g)g}{\mu^2} \qquad (18)$$

or

$$\frac{1.75}{\varepsilon_{mf}^3 \phi_s}Re_{p,mf}^2 + \frac{150(1 - \varepsilon_{mf})}{\varepsilon_{mf}^3 \phi_s^2}Re_{p,mf} = Ar \qquad (19)$$

where the Archimedes number is defined as

$$Ar = \frac{d_p^3 \rho_g(\rho_s - \rho_g)g}{\mu^2} \qquad (20)$$

Some authors call this dimensionless group the Galileo number, Ga.

In the special case of *very small particles*, Eq. (18) simplifies to

$$u_{mf} = \frac{d_p^2(\rho_s - \rho_g)g}{150\mu}\frac{\varepsilon_{mf}^3 \phi_s^2}{1 - \varepsilon_{mf}}, \qquad Re_{p,mf} < 20 \qquad (21)$$

TABLE 3   Voidage at Minimum Fluidizing Conditions ε_{mf}

Particles	Size, d_p (mm)						
	0.02	0.05	0.07	0.10	0.20	0.30	0.40
Sharp sand, $\phi_s = 0.67$	—	0.60	0.59	0.58	0.54	0.50	0.49
Round sand, $\phi_s = 0.86$	—	0.56	0.52	0.48	0.44	0.42	—
Mixed round sand	—	—	0.42	0.42	0.41	—	—
Coal and glass powder	0.72	0.67	0.64	0.62	0.57	0.56	—
Anthracite coal, $\phi_s = 0.63$	—	0.62	0.61	0.60	0.56	0.53	0.51
Absorption carbon	0.74	0.72	0.71	0.69	—	—	—
Fischer-Tropsch catalyst $\phi_s = 0.58$	—	—	—	0.58	0.56	0.55	—
Carborundum	—	0.61	0.59	0.56	0.48	—	—

From Leva [2].

For *very large particles,*

$$u_{mf}^2 = \frac{d_p(\rho_s - \rho_g)g}{1.75\rho_g} \, \varepsilon_{mf}^3\phi_s \,, \qquad Re_{p,mf} > 1000 \tag{22}$$

When ε_{mf} and/or ϕ_s are not known, one can still estimate u_{mf} for a bed of irregular particles with no seemingly longer or shorter dimension as follows. First, rewrite Eq. (19) as

$$K_1 \, Re_{p,mf}^2 + K_2 \, Re_{p,mf} = Ar \tag{23}$$

where

$$K_1 = \frac{1.75}{\varepsilon_{mf}^3\phi_s} \quad \text{and} \quad K_2 = \frac{150(1 - \varepsilon_{mf})}{\varepsilon_{mf}^3\phi_s^2} \tag{24}$$

Wen and Yu [12] were the first to note that K_1 and K_2 stayed nearly constant for different kinds of particles over a wide range of conditions (Re = 0.001 to 4000), thus giving predictions of u_{mf} having a $\pm 34\%$ standard deviation. Since then, other investigators (see Table 4) have reported on K_1 and K_2.

Solving Eq. (23) for minimum fluidizing conditions and using the values for K_1 and K_2 recommended by Chitester et al. [17] *for coarse particles* gives

$$\frac{d_p u_{mf}\rho_g}{\mu} = \left[(28.7)^2 + 0.0494\left(\frac{d_p^3\rho_g(\rho_s - \rho_g)g}{\mu^2}\right)\right]^{1/2} - 28.7 \tag{25a}$$

or

$$Re_{p,mf} = [(28.7)^2 + 0.0494 \, Ar]^{1/2} - 28.7 \tag{25b}$$

For fine particles the values recommended by Wen and Yu [12] give

$$Re_{p,mf} = [(33.7)^2 + 0.0408 \, Ar]^{1/2} - 33.7 \tag{26}$$

Other recommended values of the constants in Eq. (25) are given in Table 4.

This expression is only useful as a rough estimate of u_{mf}. Naturally, if information on ε_{mf} and ϕ_s is available, Eqs. (18), (19), (21), or (22) should be used, since they may be expected to give more reliable predictions of u_{mf}.

TABLE 4   Values of the Two Constants in Eq. (25)

Investigators	First, $K_2/2K_1$	Second, $1/K_1$
Wen and Yu [12] (1966)	33.7	0.0408
284 data points from the literature		
Richardson [13] (1971)	25.7	0.0365
Saxena and Vogel [14] (1977)	25.3	0.0571
Dolomite at high temperature and pressure		
Babu et al. [15] (1978)	25.3	0.0651
Correlation of reported data until 1977		
Grace [16] (1982)	27.2	0.0408
Chitester et al. [17] (1984)	28.7	0.0494
Coal, char, Ballotini; up to 64 bar		

K_1 and K_2 are given by Eq. (24).

Since u_{mf} is the most important measurement needed for design, it has been the focus of a tremendous amount of experimentation under a wide variety of conditions. Numerous correlations have been proposed for its prediction, and they are summarized in useful form by Couderc [18]. For elevated pressure and temperature, see Yang et al. [19].

Pressure Drop-versus-Velocity Diagram

The Δp-versus-u_o diagram is particularly useful as a rough indication of the quality of fluidization, especially when visual observations are not possible.

Not-too-Small Uniformly Sized Particles. Figure 4, with uniformly sized 160-μm sand, is typical of systems of uniformly sized particles that are not too small. For the relatively low flow rates in a fixed bed, the pressure drop is approximately proportional to gas velocity, as indicated by Eq. (6), and usually reaching a maximum, Δp_{max}, slightly higher than the static pressure of the bed. With a further increase in gas velocity, the fixed bed "unlocks"; in other words, the voidage increases from ε_m to ε_{mf}, resulting in a decrease in pressure drop to the static pressure of the bed, as given by Eq. (17). With gas velocities beyond minimum fluidization, the bed expands and gas bubbles are seen to be present, resulting in nonhomogeneity. Despite this rise in gas flow, the pressure drop remains practically unchanged. To explain this constancy in pressure drop, note that the dense gas-solid phase is well aerated and can deform easily without appreciable resistance. In its hydrodynamic behavior, we can liken it to a liquid. If a gas is introduced at the bottom of a tank containing a liquid of low viscosity, we find that the pressure required for injection is roughly the static pressure of the liquid and is independent of the flow rate of gas. The constancy in pressure drop in the two situations, the bubbling liquid and the bubbling fluidized bed, are somewhat analogous.

When gas velocity decreases, the fluidized particles of Fig. 4 settle down to

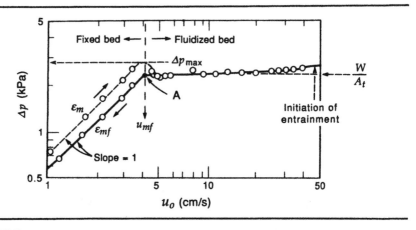

FIGURE 4

Δp versus u_o for uniformly sized sharp sand gives ideal textbook behavior; $d_t = 4.1$ cm, distributor consists of a fixed bed of larger solids; from Shirai [20].

form a loose fixed bed of voidage ε_{mf}. With gas flow eventually turned off, a gentle tapping or vibration of the bed will reduce its voidage to its stable initial value of ε_m. Usually, u_{mf} is taken as the intersection of the Δp-versus-u_o line for the fixed bed of voidage ε_{mf}, with the horizontal line corresponding to W/A_t (point A).

Figure 4, with not too much pressure fluctuation, represents a well-behaved bubbling bed of these not-too-small solids (Geldart **B**, discussed later in the chapter). Large, fairly regular fluctuations suggest that slugging is occurring, as shown in Fig. 1.1(e) or (f). On the other hand, an observed pressure drop lower than W/A_t indicates a partly fluidized bed.

Wide Size Distribution of Particles. When the gas velocity u_o is increased through these beds of solids, the smaller particles are apt to slip into the void spaces between the larger particles and fluidize while the larger particles remain stationary. Then partial fluidization occurs, giving an intermediate Δp.

With increasing gas velocity, Δp approaches W/A_t, showing that all the solids eventually fluidize. Figure 5, typical of such systems, shows that hysteresis is negligible. For mixtures containing rather large particles ($d_p > 1$ mm), segregation and settling of these larger particles may occur, giving a stepwise increase in Δp and hysteresis of the Δp-versus-u_o curve. However, this hysteresis disappears when the bed contains a large enough fraction of fines (see Saxena and Vogel [14]). In mixed particle systems, u_{mf} is defined by convention as the intersection of the fixed bed Δp-versus-u_o line with the W/A_t line (point B in Fig. 5), and this is what is reported in the literature.

One warning: u_{mf} should be determined for the size distribution of solids actually in the bed. This may differ considerably from that of fresh feed due to elutriation of fines, attrition or agglomeration of solids, or the growth or shrinkage of particles resulting from reaction. These matters are taken up in Chaps. 14 and 18.

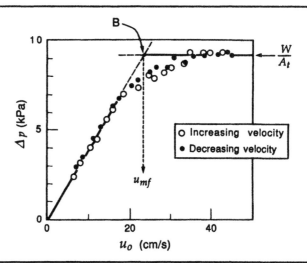

FIGURE 5
For a wide distribution of solids, the onset of fluidization is gradual but is defined as point B; dolomite, $d_p = 180-1400$ μm, $d_t = 15.2$ cm; adapted from Saxena and Vogel [14].

Transition from Smooth to Bubbling Fluidization. The fluidizing velocity at which bubbles are first observed is called the minimum bubbling velocity, u_{mb}. In liquid-solid systems, one usually has particulate or smooth fluidization throughout, so u_{mb} has no meaning. On the other hand, in gas-solid beds of large particles, bubbles appear as soon as the gas velocity exceeds u_{mf}; hence $u_{mb} \cong u_{mf}$.

Now consider gas fluidized beds of small, light, nearly spherical particles of mixed size. For these solids the Δp-versus-u_0 relationship looks more like that of Fig. 4 than of Fig. 5. An FCC catalyst with a size range of 5–100 μm is typical, and Fig. 6 shows the Δp-versus-u_0 curve for this material.

Figure 6 also shows the bed expansion of this material, where L_f is the average fluidized bed height. With increasing gas velocity beyond u_{mf}, the bed expands smoothly with no observed bubbling. However, at a gas velocity of about $3u_{mf}$, bubbles begin to form and bed height begins to decrease. Figure 7 shows how the particle properties affect u_{mb}.

Geldart and Abrahamsen [22] measured u_{mb} for 23 different particles ($\bar{d}_p = 20$–72 μm, $\rho_s = 1.1$–4.6 g/cm^3), using ambient air, helium, argon, carbon dioxide, and Freon-12. They found that u_{mb}/u_{mf} was strongly dependent on the weight fraction of particles smaller than 45 μm, thus $P_{45\mu m}$, and for these systems they gave, in SI units,

$$\frac{u_{mb}}{u_{mf}} = \frac{2300\rho_g^{0.13}\mu^{0.52}\exp(0.72P_{45\mu m})}{\bar{d}_p^{0.8}(\rho_s - \rho_g)^{0.93}} \tag{27}$$

This expression, with Fig. 7, should give a reasonable estimate of u_{mb}.

Finally, the range of particulate fluidization in gas-solid systems can be

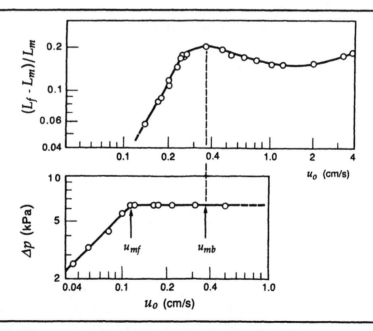

FIGURE 6
For FCC catalyst, the bed expands smoothly and expands progressively above u_{mf}; then bubbling occurs and expansion ceases; $\bar{d}_p = 64.7$ μm, $\rho_{bulk} = 0.5$ g/cm^3, $d_t = 6.6$ cm, $L_m = 130.8$ cm, perforated plate distributor; from Morooka et al. [21] and Miyauchi et al. [11].

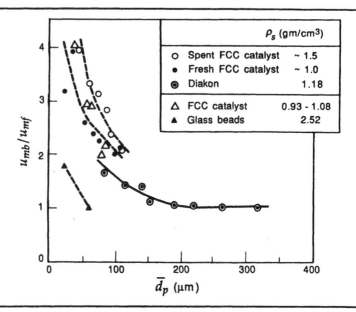

FIGURE 7
Minimum bubbling velocity ratio decreases sharply with increase in particle size; circles from Geldart and Abrahamsen [22], triangles from Morooka et al. [21]. Diakon is a spherical plastic molding powder.

extended considerably by adding a small fraction of fine particles of special characteristics; see Brooks and Fitzgerald [23].

Effect of Pressure and Temperature on Fluidized Behavior

The effect of pressure has been studied by many investigators [14, 17, 24–30] and we summarize these findings for beds of porous carbon powder, coal, char, and uniformly sized glass beads at pressures up to 80 bar as follows:

- ε_{mf} increases slightly (1–4%) with a rise in operating pressure.
- u_{mf} decreases with a rise in operating pressure. However, this decrease is negligible for beds of fine particles ($\bar{d}_p < 100$ μm), but becomes significant (up to 40%) for larger particles ($\bar{d}_p \cong 360$ μm). These experimental findings are consistent with the predictions of Eqs. (21) and (22). In general, u_{mf} can be reasonably predicted at all pressures by Eqs. (18) or (19).
- u_{mb}/u_{mf} for coarse alumina ($\bar{d}_p = 450$ μm) increases up to 30% for a rise in operating pressure. This suggests that an increase in operating pressure widens the range of particulate fluidization in gas-solid systems.

The effect of temperature has also been studied by numerous researchers [31–39], and Saxena and Vogel [14] and Kitano et al. [40] studied the combined effects of high temperature and high pressure. Although there are still contradictions between some of the reported findings, we may tentatively summarize them as follows:

- ε_{mf} increases with temperature for fine particles (up to 8% for temperatures up to 500°C), but seems to be unaffected by temperature for coarse particles.

- u_{mf} can be reasonably predicted by Eqs. (18)–(22) when the correct ε_{mf} value is used.

Sintering and Agglomeration of Particles at High Temperature

A potentially serious problem at high temperature is that of sintering of particles, because when this occurs the behavior of the fluidized bed can change drastically. Gluckman et al. [41] investigated this phenomenon by slowly heating a fluidized bed of copper shot, $-16 + 20$ mesh, at close to u_{mf}. At about 900°C, the bed took on a sluggish appearance and suddenly defluidized. Gas flow then had to be progressively increased with temperature, up to three times the 900°C value, to keep the bed fluidized, as shown in Fig. 8(a). All this happened at temperatures below the melting point of copper or cupric oxide, a material that may be expected to coat the surface of the copper particles at these high temperatures. The melting point of cupric oxide is shown in this figure.

The onset of sintering can be measured by a dilatometer. Here a sample of particles is placed in a cylinder and compressed by a constant force. The cylinder is then heated slowly, and the length of the sample is noted. With copper, the results are as shown in Fig. 8(b). At first the sample expands due to thermal expansion. Then expansion slows, ceases, and the sample begins to contract. At the point where the slope of the expansion curve is zero, the thermal expansion is just balanced by the contraction due to sintering. Gluckman et al. [41] call this the *initial sintering temperature*.

This phenomenon of sintering should be kept in mind for high-temperature operations, especially with industrial materials such as coal ash or metal

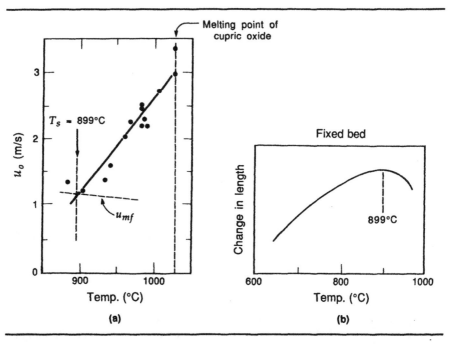

FIGURE 8
Defluidization of $-16 + 20$ mesh copper shot caused by sintering: (a) sintering (above 899°C) causes u_{mf} to increase; (b) fixed bed experiments can determine the onset of sintering; adapted from Gluckman et al. [41].

ores, which contain a variety of impurities. These materials may form low-melting-point eutectics at the surface of particles, resulting in unexpectedly low sintering temperatures. Information on sintering should be obtained early in the development of high-temperature processes.

Although sintering may cause unexpected undesirable behavior for certain high-temperature processes, it is often used to advantage to develop processes for the agglomeration of fine particles. For these operations, one should note that when a cooler bed is fluidized by a hot gas it is the smallest particles that heat up most rapidly and thus become the first to agglomerate. In any case, when planning for high-temperature operations, one should observe carefully the mechanisms of cohesion, sintering, and agglomeration in small-scale operations so as not to be surprised when going to large-scale operations.

EXAMPLE 2

Estimation of u_{mf}

Calculate the minimum fluidizing velocity u_{mf} for the bed of sharp sand particles used by Shirai [20] and reported in Fig. 5.

Data

Bed:	$\varepsilon_{mf} = 0.55$
Fluidizing gas:	ambient air, $\rho_g = 0.0012 \, \text{g/cm}^3$, $\mu = 0.00018 \, \text{g/cm} \cdot \text{s}$
Solids:	sharp irregular sand, not longish or flattish

$$\bar{d}_p = 160 \, \mu\text{m}, \qquad \phi_s = 0.67, \qquad \rho_s = 2.6 \, \text{g/cm}^3$$

SOLUTION

Because the particles are small, we use Eq. (21) to find u_{mf}. Thus

$$u_{mf} = \frac{(0.0160)^2(2.6 - 0.0012)(980)}{(150)(0.00018)} \cdot \frac{(0.55)^3(0.67)^2}{1 - 0.55} = 4.01 \, \text{cm/s}$$

Now check to see whether Eq. (21) is applicable:

$$Re_{p,mf} = \frac{d_p u_{mf} \rho_g}{\mu} = \frac{(0.0160)(4.01)(0.0012)}{0.00018} = 0.43 < 20$$

This justifies the use of the simplified equation, and we conclude that

$$u_{mf} = \underline{4.01 \, \text{cm/s}}.$$

This compares well with the measured value in Fig. 4.

Comment. Suppose that neither ε_{mf} nor ϕ_s is known. Then using Eq. (25) for these not-so-fine particles gives

$$u_{mf} = \frac{0.0018}{(0.0160)(0.0012)} \left[\left\{ (28.7)^2 + 0.0494 \frac{(0.0160)^3(0.0012)(2.6 - 0.0012)(980)}{(0.00018)^2} \right\}^{1/2} - 28.7 \right] = \underline{3.10 \, \text{cm/s}}$$

This value is 22% below the experimentally reported value.

Types of Gas Fluidization without Carryover

So far we have discussed two types of fluidization, *bubbling* (aggregative, heterogeneous) and *nonbubbling* (particulate, homogeneous, smooth). They occur at gas velocities slightly above minimum, or u_o less than about $10u_{mf}$. We restrict our discussion to these flow rates in beds supported by porous, high-pressure drop distributors that give uniform gas flow.

Numerous attempts have been made to devise a criterion to predict the mode of fluidization and the transition from one mode to another. Wilhelm and Kwauk [42] were the first. They considered interparticle forces in the vicinity of bubbles and proposed using the Froude number as the criterion for flow transition. Romero and Johanson [43] extended this idea to four dimensionless groups, which included the Reynolds number and the Froude number. Zenz [44], taking a different approach, presented an empirical plot of bed voidage versus ρ_s/ρ_g with particle size as parameter. Criteria for indicating when bubbles would form, based on stability theory as applied to growth rate of pressure disturbances in the bed, were proposed by three groups [45–47]. Others [48,49], opposed to these criteria, assumed that bubbles were always present in fluidized beds but were not observable below a certain size. Even the existence of shock waves was proposed as a criterion for transition from nonbubbling to bubbling behavior [50].

Geldart [51] approached this question in a different way. He focused on the characteristics of the particles that make them fluidize in one way or another. His approach is simple, has great generalizing power, and is very useful.

The Geldart Classification of Particles

By carefully observing the fluidization of all sorts and sizes of solids, Geldart [51] came up with four clearly recognizable kinds of particle behavior. From smallest to largest particle, they are as follows:

- Group **C**: cohesive, or very fine powders. Normal fluidization is extremely difficult for these solids because interparticle forces are greater than those resulting from the action of gas. Face powder, flour, and starch are typical of these solids.
- Group **A**: aeratable, or materials having a small mean particle size and/or low particle density $(<\sim 1.4\,\text{g/cm}^3)$. These solids fluidize easily, with smooth fluidization at low gas velocities and controlled bubbling with small bubbles at higher gas velocities. FCC catalyst is representative of these solids.
- Group **B**: sandlike, or most particles of size $40\,\mu\text{m} < \bar{d}_p < 500\,\mu\text{m}$ and density $1.4 < \rho_s < 4\,\text{g/cm}^3$. These solids fluidize well with vigorous bubbling action and bubbles that grow large.
- Group **D**: spoutable, or large and/or dense particles. Deep beds of these solids are difficult to fluidize. They behave erratically, giving large exploding bubbles or severe channeling, or spouting behavior if the gas distribution is very uneven. Drying grains and peas, roasting coffee beans, gasifying coals, and some roasting metal ores are such solids, and they are usually processed in shallow beds or in the spouting mode.

Geldart's classification is clear and easy to use and is readily displayed in Fig. 9 for air fluidization at ambient conditions and for u_o less than about $10u_{mf}$. For any solid of known density ρ_s and mean particle size $\bar{d}_p$, this graph shows the type of fluidization to be expected. This grouping of solids is widely used today, with the solids simply called *Geldart A solids*, and so forth. We follow this practice here.

We now discuss the distinctive characteristics of solids in these groupings.

Geldart C Particles. In small-diameter beds Geldart C particles, which are difficult to fluidize, tend to rise as a plug of solids, whereas in larger-

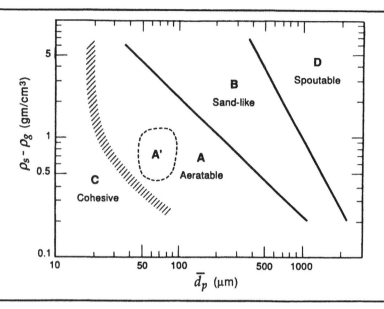

FIGURE 9
The Geldart classification of particles for air at ambient conditions; adapted from Geldart [51].
Region **A'**: Range of properties for well-behaved FCC catalyst; from Miyauchi et al. [11].

diameter beds channels form from distributor to bed surface with no fluidization of solids. These particles have been studied by Geldart [52].

One way of processing these solids is to introduce them into a bed of the same material but of larger size, preferably Geldart **B**. Even though the fines are very small, they are not entrained immediately, but may stay in the bed an average of several minutes. This usually is long enough for a physical or chemical transformation of these solids.

Geldart A Particles. When these solids are fluidized, the bed expands considerably before bubbles appear, as mentioned earlier. At gas velocities higher than u_{mb}, the bed shifts to the bubbling mode, characterized as follows:

- Gas bubbles rise more rapidly than the rest of the gas, which percolates through the emulsion.
- These gas bubbles appear to split and coalesce frequently as they rise through the bed. There is a maximum bubble size, usually less than 10 cm, even in a large bed.
- Internals do not appreciably improve fluidization.
- Gross circulation of solids occurs even when only a few bubbles are present. This circulation is especially pronounced in large beds.
- When bubbles grow to the vessel diameter, they turn into axial slugs (see Fig. 1.1).

Fines act as a lubricant to make it easier to fluidize the bed. Thus, the ratio u_{mb}/u_{mf} increases with added fines, namely $\mathbf{P}_{45\ \mu m}$, as indicated by Eq. (27). The size distribution of typical Geldart **A** solids that fluidize well is shown in Fig. 2.

Geldart B Particles. In beds of Geldart **B** solids, bubbles form as soon as the gas velocity exceeds u_{mf}. Thus, $u_{mb}/u_{mf} \cong 1$, as opposed to Geldart **A** solids. At higher gas velocities, the bed behaves as follows:

- Small bubbles form at the distributor and grow and coalesce as they rise through the bed.
- Bubble size increases roughly linearly with distance above the distributor and excess gas velocity, $u_o - u_{mf}$.
- Bubble size is roughly independent of mean particle size.
- Vigorous bubbling encourages the gross circulation of solids.

The majority of gas-solid reactions, metallurgical and others, are run in this regime because the mean size and size distribution of feed particles are usually determined by the upstream processing of the raw materials.

Geldart D Particles. Fluidized beds of Geldart **D** solids have the following properties:

- Bubbles coalesce rapidly and grow to large size.
- Bubbles rise more slowly than the rest of the gas percolating through the emulsion.
- The dense phase has a low voidage.
- When the bubble size approaches the bed diameter, flat slugs are observed (see Fig. 1.1).
- These solids spout easily, whereas Geldart **B** solids do not.

Large particle beds are usually undesirable for physical or chemical operations. However, in some industries, for instance, in processing agricultural products, in chemical agglomeration, and in the reaction of composite pellets, one cannot avoid this.

An enormous amount of gas is needed to fluidize these solids, often far more than required for the physical or chemical operation. In such situations, one may want to use spouted beds, since they need much less gas.

Bubbling can be made to occur with these solids if the bed is shallow, has sufficient diameter, and the gas velocity is not much more than u_{mf}. To avoid slugging, especially at onset of fluidization, the vessels are sometimes designed with a larger-diameter upper section, as shown in Figs. 2.1(a) and 2.16(b).

As shown in Fig. 2.3, in some operations two modes of fluidization are combined (i.e., spouting and bubbling) to get better contacting.

Extensions of the Geldart Chart. Figure 9 was originally proposed for beds only at ambient conditions. Further studies have led to a number of proposed modifications and refinements. For example, an **AC** classification for particles in the uncertain transition region between Geldart **A** and Geldart **C** solids has been proposed. These solids flow well when fluidized (type **A** influence), but they permanently defluidize on any horizontal surface and thus block or plug horizontal pipes (type **C** influence).

Another area of study seeks to locate more clearly the boundary between regions. This should depend not only on densities and mean solid size (see Fig. 9) but also on u_o/u_{mf}, gas properties, and the size distribution of solids. Grace [53] presents the latest findings in this area.

We consider the whole question of flow regime in a much broader context in the last section of this chapter, where a generalized flow diagram is displayed.

Fluidization with Carryover of Particles

So far we have considered only fluidized beds at moderate gas flows, or $u_o < 10 u_{mf}$. At higher gas flows, more and more particles are projected into the freeboard above the bed, some to return to the bed, others to be carried out of the bed. The bed surface becomes agitated and hazy. This is the *turbulent fluidized bed*. At even higher gas velocities the density of particles in the freeboard rises, the bed blends into freeboard, and carryover of solids from the bed becomes high. This is the *fast fluidized bed*.

Recirculation of solids is needed when carryover is significant, for if this is not done there soon will be no bed left. Hence we use the general term *recirculating fluidized beds* for these conditions. Before we consider these operations, we must know how to estimate the terminal velocity of particles in fluids.

Terminal Velocity of Particles, u_t

When a particle of size d_p falls through a fluid, its terminal free-fall velocity can be estimated from fluid mechanics by the expression

$$u_t = \left[\frac{4 d_p (\rho_s - \rho_g) g}{3 \rho_g C_D} \right]^{1/2} \tag{28}$$

where C_D is an experimentally determined drag coefficient. In general, Haider and Levenspiel [54] find

$$C_D = \frac{24}{\text{Re}_p} [1 + (8.1716 e^{-4.0655 \phi_s}) \text{Re}_p^{0.0964 + 0.5565 \phi_s}]$$

$$+ \frac{73.69 (e^{-5.0748 \phi_s}) \text{Re}_p}{\text{Re}_p + 5.378 e^{6.2122 \phi_s}}, \qquad [-] \tag{29}$$

For spherical particles this expression reduces to

$$C_D = \frac{24}{\text{Re}_p} + 3.3643 \, \text{Re}_p^{0.3471} + \frac{0.4607 \, \text{Re}_p}{\text{Re}_p + 2682.5}, \qquad \text{for } \phi_s = 1 \tag{30}$$

Figure 10 is a graphical representation of these equations, which allows a direct evaluation of u_t, given d_p and the physical properties of the system. This chart introduces a dimensionless particle size d^* and a dimensionless gas velocity u^*. These useful measures are defined as follows:

$$d_p^* = d_p \left[\frac{\rho_g (\rho_s - \rho_g) g}{\mu^2} \right]^{1/3} = \text{Ar}^{1/3} = (\tfrac{3}{4} C_D \, \text{Re}_p^2)^{1/3}, \qquad [-] \tag{31}$$

and

$$u^* = u \left[\frac{\rho_g^2}{\mu (\rho_s - \rho_g) g} \right]^{1/3} = \frac{\text{Re}_p}{\text{Ar}^{1/3}} = \left(\frac{4}{3} \frac{\text{Re}_p}{C_D} \right)^{1/3}, \qquad [-] \tag{32}$$

Haider and Levenspiel [54], using the equation form suggested by Turton and Clark [55], present the following useful approximation for the direct evaluation of the terminal velocity of particles:

$$u_t^* = \left[\frac{18}{(d_p^*)^2} + \frac{2.335 - 1.744 \phi_s}{(d_p^*)^{0.5}} \right]^{-1}, \qquad 0.5 < \phi_s < 1 \tag{33}$$

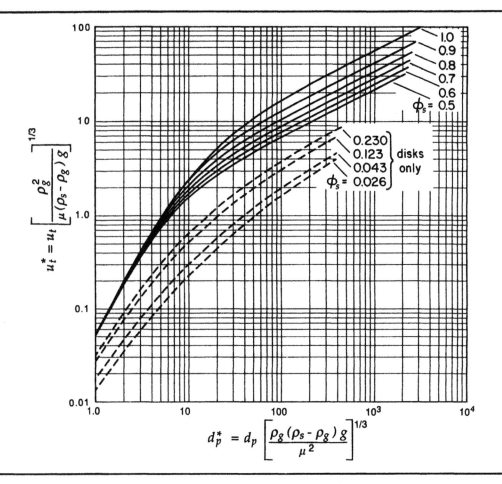

FIGURE 10
Chart for determining the terminal velocity of particles falling through fluids; from Haider and
Levenspiel [54].

For spherical particles this expression reduces to

$$u_t^* = \left[\frac{18}{(d_p^*)^2} + \frac{0.591}{(d_p^*)^{0.5}} \right]^{-1}, \qquad \phi_s = 1 \tag{34}$$

To find the terminal velocity of single free-falling particles, use Fig. 10 or Eqs.
(33) or (34), as illustrated in Example 3.

To avoid or reduce carryover of particles from a fluidized bed, keep the
gas velocity between u_{mf} and u_t. In calculating u_{mf}, use the mean diameter $\bar{d}_p$
for the size distribution actually present in the bed, whereas for u_t use the
smallest size of solids present in appreciable quantities in the bed.

The ratio u_t/u_{mf} strongly depends on particle size. Thus, for spherical
particles of one size and $\varepsilon_{mf} = 0.4$, we find

$$\text{for fine solids:} \qquad \frac{u_t \text{ from Eq. (33)}}{u_{mf} \text{ from Eq. (21)}} = 78 \tag{35}$$

$$\text{for large solids:} \qquad \frac{u_t \text{ from Eq. (33)}}{u_{mf} \text{ from Eq. (22)}} = 9.2 \tag{36}$$

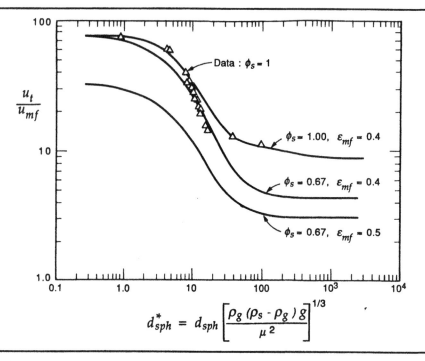

$$d^*_{sph} = d_{sph}\left[\frac{\rho_g(\rho_s-\rho_g)g}{\mu^2}\right]^{1/3}$$

FIGURE 11
Ratio of terminal to minimum fluidization velocity, from Eqs. (35) and (36); data from Pinchbeck and Popper [56].

More generally, Fig. 11 shows how this ratio is affected by particle size and particle sphericity. Roughly,

$$d^*_p < 1, \quad \text{for small particles} \tag{37}$$

$$d^*_p > 100, \quad \text{for large particles} \tag{38}$$

This ratio indicates the flexibility of possible operations in the nonentrained regime and shows that the useful velocity range for large particle systems is much smaller than that for small particle systems.

Fluidized beds, however, can be made to operate at velocities well beyond the terminal velocity of practically all the solids, without excessive carryover of solids. This is possible because a large fraction of the gas flows through the bed as high-speed gas bubbles, bypassing most of the bed solids. If cyclone separators are used to return the entrained solids to the bed, even higher gas velocities can be used.

EXAMPLE 3

Estimate the Terminal Velocity of Falling Particles

Calculate u_t for the sharp irregular sand particles used by Shirai [20] and reported in Fig. 4.

Data

Air: $\rho_g = 1.2 \times 10^{-3}\,\text{g/cm}^3$, $\mu = 1.8 \times 10^{-4}\,\text{g/cm}\cdot\text{s}$
Sand: $\bar{d}_p = 160\,\mu\text{m}$, $\phi_s = 0.67$, $\rho_s = 2.60\,\text{g/cm}^3$

SOLUTION

First, calculate d_p^* from Eq. (31):

$$d_p^* = 0.0160\left[\frac{0.0012(2.6 - 0.0012)980}{(0.00018)^2}\right]^{1/3} = 7.28$$

Next, from Eq. (33) or from Fig. 11, we find

$$u_t^* = \left[\frac{18}{(7.28)^2} + \frac{2.335 - 1.744 \times 0.67}{(7.28)^{0.5}}\right]^{-1} = 1.2954$$

Finally, from Eq. (32),

$$u_t = u_t^*\left[\frac{\mu(\rho_s - \rho_g)g}{\rho_g^2}\right]^{1/3}$$

$$= 1.2954\left[\frac{0.00018(2.6 - 0.0012)980}{(0.0012)^2}\right]^{1/3} = \underline{88\text{ cm/s}}$$

Turbulent and Churning Fluidization

Small particle beds. Consider a bed of fine particles as sketched in Fig. 1.1(d). With increasing gas velocity, the bubbling action becomes increasingly vigorous, and is accompanied by increasing pressure fluctuations, as measured just above the distributor. These fluctuations peak, decrease sharply, and level off, as

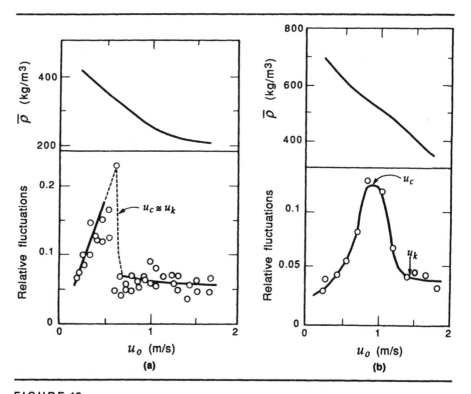

FIGURE 12

Pressure fluctuations and mean density in a 15.2-cm bed for two solids 0–130 μm, $\bar{d}_p$ = 49 μm; adapted from Yerushalmi and Cankurt [63]: (a) FCC catalyst: ρ_s = 1070 kg/m^3, u_t = 7.78 cm/s, u_c = 61 cm/s, u_k = 61 cm/s; (b) silica alumina catalyst: ρ_s = 1450 kg/m^3, u_t = 10.6 cm/s, u_c = 91 cm/s, u_k = 137 cm/s.

shown in Fig. 12. This progression corresponds to the transition from bubbling to turbulent fluidization, and has been extensively investigated [57–65].

Yerushalmi and Cankurt [63] characterized this transition in terms of two velocities, namely u_c, at which the pressure fluctuations peak, and u_k, where the pressure fluctuations begin to level off, as shown in Fig. 12. They call u_k the *onset* of turbulent fluidization. This onset occurs at gas velocities far beyond u_t for the mean size of bed solids. Typically, for fine catalyst, $u_k/u_t = 8$–13. This ratio decreases with increase in pressure.

In the turbulent regime, bubbles (or slugs in narrow columns) no longer appear distinct. Clusters and strands of particles as well as voids of elongated and distorted shapes are seen to move about violently, making it difficult to distinguish continuous and discontinuous phases in the bed.

Bed voidage is large, and at the indistinct bed surface clusters and strands of particles are continually ejected into the freeboard. These clusters disperse into single particles, the smaller of which are carried out of the bed. Consequently, cyclones and diplegs are needed to maintain the bed inventory. However, since the mass flux of these fines that are to be trapped and returned to the bed is not excessive, internal cyclones, those within the vessel, can be used as shown in Fig. 1.2(a).

Figure 12 shows bed voidages and pressure fluctuations of typical fine particle systems.

Large Particle Beds. Visual observations show that large particle beds behave differently than small particle beds. First, in approaching u_c one can very quickly generate very large exploding bubbles, especially with Geldart **D** solids. The transition to turbulent flow occurs at lower relative velocity, u_k/u_t, and even dips below u_t with very large solids.

As an example, in large beds and at various pressures, Canada et al. [61] found that $u_k/u_t = 0.5$–0.6 and 0.3–0.35 for 650-μm and 2600-μm glass beads, respectively.

In contrast to the small clusters and strands of material moving violently about the bed in fine particle systems, one observes large-scale uniform motion, gross circulation of bed material, and severe channeling. We call this coarse particle regime *churning fluidization* to distinguish it from the more general term, *turbulent fluidization*, used primarily for fine particle systems.

Pneumatic Transport of Solids

Consider the upflow of air plus a continuous feed of fine solids to a vertical tube. If the air velocity u_o is high enough and the feed rate of solids is small enough, as sketched in Fig. 13(a), then all the solids will be carried up the tube as separate particles widely dispersed in the gas. The relative velocity between gas and solid is known as the *slip velocity*, $u_p = u_o - u_s$. Up to a point, one may change the flow rates of gas or solid and still maintain a lean dispersed upflowing gas-solid mixture. This regime is called the *pneumatic transport* regime.

Just above the solids feed, the particles are accelerated upward by the flowing gas stream to give a vertical distribution of solids shown as curve *PQR* in Fig. 13(e).

Conventional pneumatic conveying operates in this regime using high gas velocities (roughly $20u_t$ for small particles) in order to prevent the settling (saltation) of particles, particularly on horizontal surfaces of the flow system. The

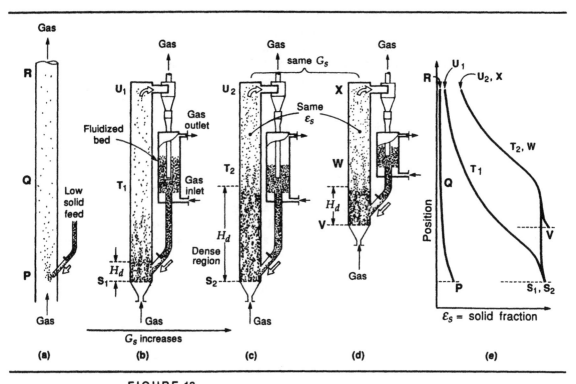

FIGURE 13

Sketches showing how the feed rate of solids (at given gas flow) determines the density distribution of solids in the vessel: (a) very low feed rate of solids, pneumatic transport; (b) low feed rate of solids; (c) high feed rate of solids; (d) same high feed rate of solids but in a shorter bed; (e) vertical distribution of solids fraction.

mass flow ratio of solid to gas is usually 1:20, which represents a very high voidage. For example, for an air-sand system, this corresponds to a voidage of 0.999–0.980. In systems this dilute, one can reasonably assume no interaction between particles; hence, far enough downstream from the particle feed port we can assume that $u_p = u_t$.

When gas velocity is reduced or solid flow rate is increased, a condition is reached where the character of the mixture changes drastically, with clumping, slugging, and solids falling below the solids feed port. This transition is called the *choking condition*, and it represents the limit of the pneumatic transport regime.

We consider pneumatic transport in more detail in Chap. 15.

Fast Fluidization

When the feed rate of solids exceeds the choking condition at a given gas velocity, one should place an appropriate gas distributor in the column, as shown in Fig. 13(b). Here solids are "pushed" into the bottom of the column, and the distribution of solid density adjusts itself to account for this forced solid input. This represents a situation completely different from pneumatic transport. At high gas velocity ($u_o > 20u_t$) with very fine solids, this situation represents *fast fluidization*.

We now discuss the behavior of gas-solid systems in the fast fluidized regime with the help of Fig. 13. At a low solid feed rate, we have the situation of Fig. 13(b) with its corresponding solids fraction curve $S_1 T_1 U_1$. Here we have a denser region at the bottom of the bed (5–15% solid), which gradually blends into a leaner region higher up (1–5% solids). These solid fractions are much lower than for bubbling or turbulent beds (30–60% solids), but much higher than that for pneumatic transport (< 1% solids). The density of exiting solids (point U_1) is governed by the feed rate of solids to the vessel.

For a higher flow rate of solids into the bed, we have the situation of Fig. 13(c) with its corresponding solids density curve $S_2 T_2 U_2$. Note the upward shift in the curve for density versus height.

Figure 13(d) has the same solid flow rate as Fig. 13(c) but in a shorter vessel, and the solid fraction trace is given by curve VWX. Note that the solid fraction at point U_2 is the same as at point X.

Figures 13(b)–(d) show that the density trace moves up or down the vessel to give the correct solid fraction at the vessel exit for the imposed feed rate of solids. The measurements of Li and Kwauk [70] in 9-cm vessels clearly show this shift.

Based on experimental findings in a 15.2-cm column, Yerushalmi et al. [63, 64] characterized the fast fluidized bed as follows:

- Solid concentration somewhere between dense-phase beds and pneumatic transport conditions
- Clusters and strands of particles that break apart and reform in quick succession
- Extensive back mixing of solids
- Slip velocity of particles one order of magnitude larger than u_t

In vessels of larger diameter, a layer of particles is seen to flow down along the wall, whereas dense packets are carried upward in the central core of the vessel. Such findings suggest that severe voidage maldistributions may be expected in large-diameter vessels, and one should be wary of extrapolating findings from small vessels to large ones.

Solid Circulation Systems. A typical circulation system is shown in Fig. 13. Here the solids leaving the fast fluidization vessel are separated from the gas by a cyclone separator and enter a fluidized bed that is deep enough to provide the pressure head needed to feed solids pneumatically into the vessel without recourse to mechanical devices. The circulation of solids is controlled by a valve in the feeder tube.

Suppose that the gas flow to the vessel is kept constant. Then for a very small feed rate of solids, the vessel contains a lean phase of solids in the pneumatic transport regime. This mixture becomes denser as the feed rate of solids is increased. At a sufficiently high feed rate, the dispersed solids clump into clusters and strands with a density variation from the bottom to the top of vessel, as discussed before. Fast fluidization is thus said to have been established.

If one now increases the gas velocity keeping the solid flow rate unchanged, one observes the reverse progression of events. The bed density goes down, the clusters and strands of particles disintegrate, and at a gas velocity u_{pt} pneumatic transport is reestablished. Thus, with a fluidized bed in the circulation system, one can control u_0 and G_s independently, and the particle inventory in the vessel becomes the dependent variable.

Now suppose we remove the fluidized bed receiver of solids from the circulation system. Then u_0 becomes the only operational variable. Thus, if we raise u_0, the circulation of solids increases and can become excessive, thus clogging cyclone and downcomer. On the other hand, if we lower u_0 enough, turbulent bed behavior results. The same changes occur if u_0 is kept constant and if the bed inventory is too large or too small. Note that the transition from fast fluidized bed behavior to turbulent bed behavior is gradual and not clear-cut.

This discussion shows that the presence of a fluidized bed in the circulation system acts as a surge tank to allow more flexibility and better control of the operations. For more details see Chap. 15.

Voidage Diagrams for All Solid Carryover Regimes

Over a wide range of operating conditions, Fig. 14 shows typical distributions of solids with position in a vessel. These curves show that each flow regime has its own distinctive density-height curve.

Avidan and Yerushalmi [64] gave the bed voidage-versus-velocity diagram of Fig. 15. Although the values given in this chart represent a particular gas-solid system, the progression of changes should apply to other gas-solid systems.

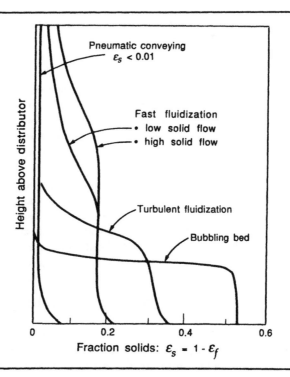

FIGURE 14
Each regime of fluidization has its own distinctive voidage profile in the vessel.

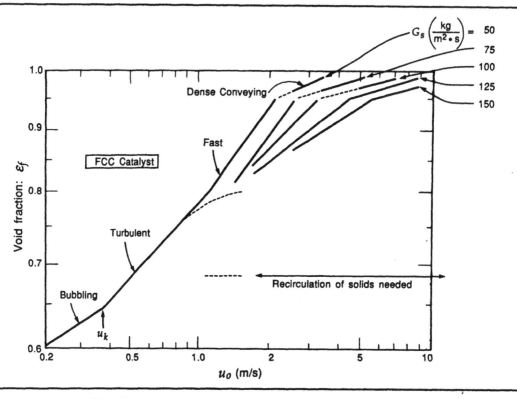

FIGURE 15
Changing the gas velocity changes the void fraction and flow regime for the gas-solid system, for FCC catalyst; adapted from Avidan and Yerushalmi [64].

The Mapping of Fluidization Regimes

Before we can predict the behavior of a specific gas-solid operation, we must know what contacting regime will be encountered. We can then use the appropriate performance expressions for that regime. We can also tell whether solid recirculation, cyclones, and so forth, are needed. This whole question is especially important to the design engineer concerned with practical applications.

Various investigators have constructed charts to map these regimes (see Table 5). Each diagram has its particular use, but the one developed by Grace [53], using coordinates first used by Zenz and Othmer [1], seems to be most useful for engineering applications. Consequently, we adopt it here. The axes of Fig. 16 are labeled with the dimensionless variables d_p^* and u^*, defined in Eqs. (31) and (32), and Fig. 16 then represents the information from Grace's original diagram plus information from other sources.

- They show the onset of fluidization and the terminal velocity in beds of single-size particles.
- They locate the modified boundaries for the Geldart classification of solids. Thus, to account for other than ambient conditions and for gases in addition to air, the **AB** boundary is given by

$$(d_p^*)_{\mathbf{AB}} = 101\left(\frac{\rho_g}{\rho_s - \rho_g}\right)^{0.425}$$

(39)

TABLE 5   Flow Regime Diagrams for Gas-Solid Contacting

Author	Abscissa	Ordinate
Reh [66] (1968, 71)	Re_p	$1/C_D$
Čatipović et al. [67] (1978, 79)	u_o	d_p
Yerushalmi and Cankurt [63] (1978, 79)	$\varepsilon_s = 1 - \varepsilon_f$	slip velocity, $u_p = u_o - u_s$
van Deemter [68] (1980)	u_o	d_p
Werther [69] (1980)	Re_p	$1/C_D$
Li and Kwauk [70] (1980)	u_o	ε_f
Avidan and Yerushalmi [64] (1982)	u_o	ε_f
Matsen [71] (1982, 83)	u_o	u_s
Squires et al. [72] (1985)	u_o	$\varepsilon_s = 1 - \varepsilon_f$
Horio et al. [73] (1986)	Re_p	Ar
Grace [53] (1986)	d_p^*	u^*

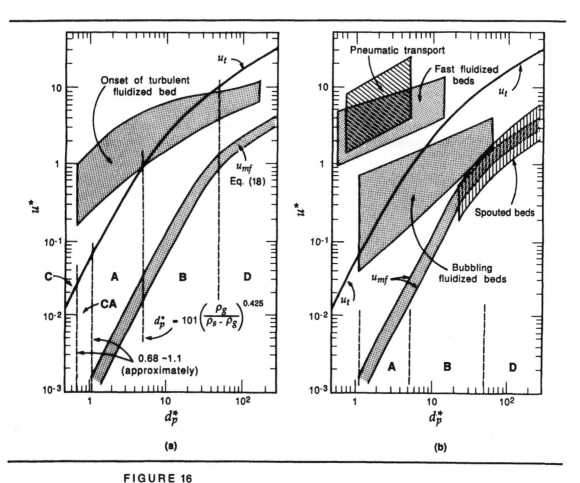

FIGURE 16

General flow regime diagram for the whole range of gas-solid contacting, from percolating packed beds to lean pneumatic transport of solids; letters **C, A, B,** and **D** refer to the Geldart classification of solids; adapted from Grace [53], but also including information from van Deemter [68], Horio et al. [73], and Čatipović et al. [67].

The **CA** boundary is uncertain and is affected by cohesive forces between particles. Thus, stronger surface forces will shift the boundary to the right, and increased humidity of the gases will shift the boundary to the left.

• They show that spouting is characteristic of Geldart **D** solids and can be made to occur at gas velocities even lower than u_{mf}.

• Normal bubbling beds are seen to operate stably over a wide range of conditions and particle size, for Geldart **A** and **B** particles. For larger particles, these beds only operate over a relatively narrow range of gas velocities. For smaller particles, however, bubbling only starts at many multiples of u_{mf} and continues way beyond the terminal velocity of the particles.

• The onset of turbulent flow is gradual, and hence is not clearly shown on this graph, but it can be seen to occur beyond u_t for very small particle systems. For larger particles, it occurs close to u_{mf} (the churning flow regime).

• Fast fluidization is only practical for very small particles and at very high gas velocities, as high as $1000u_{mf}$.

This flow map represents experimental data by many researchers at various conditions as follows:

Gases: air, N_2, CO_2, He, H_2, Freon-12, CCl_4
Temperature: 20°–300°C
Pressure: 1–85 bar

Grace reports that it is generally possible to extend the various operations well beyond the boundaries indicated on Fig. 16; however, most industrial reactors are designed to operate within the regions indicated. This graph is the best we

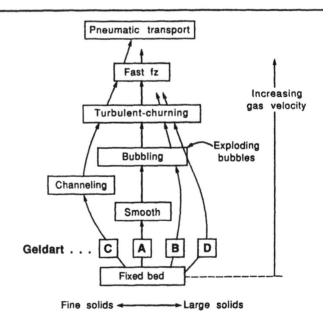

FIGURE 17
Progressive change in gas solid contacting with change in gas velocity.

have today, and, as results are reported in the future, it will be refined and modified accordingly.

Finally, Fig. 17 shows the progression of changes in behavior of a bed of solids as the gas velocity is progressively increased.

EXAMPLE 4

Prediction of

Flow Regime

Predict the mode of fluidization for particles of density $\rho_s = 1.5 \, g/cm^3$ at superficial gas velocities of $u_o = 40$ and $80 \, cm/s$.

(a) $d_p = 60 \, \mu m$, $\rho_g = 1.5 \times 10^{-3} \, g/cm^3$, $\mu = 2 \times 10^{-4} \, g/cm \cdot s$
(b) $d_p = 450 \, \mu m$, $\rho_g = 1 \times 10^{-3} \, g/cm^3$, $\mu = 2.5 \times 10^{-4} \, g/cm \cdot s$

SOLUTION

(a) *The smaller particles.* Equations (31) and (32) give

$$d_p^* = 0.006 \left[\frac{(1.5 \times 10^{-3})(1.5 - 1.5 \times 10^{-3})(980)}{(2 \times 10^{-4})^2} \right]^{1/3} = 2.28$$

$$u_o^* = u_o \left[\frac{(1.5 \times 10^{-3})^2}{(2 \times 10^{-4})(1.5 - 1.5 \times 10^{-3})(980)} \right]^{1/3}$$

$$= 0.07885 \text{ and } 1.577, \quad \text{for } u_o = 40 \text{ and } 80 \, cm/s$$

From Fig. 16, we have

at $u_o = 40 \, cm/s$:   onset of turbulent fluidization in an ordinary bubbling bed

at $u_o = 80 \, cm/s$:   fast fluidization (requires a circulating solid system)

(b) *The larger particles.* Following the same procedure, we find

$$d_p^* = (0.045) \left[\frac{(0.001)(1.5 - 0.001)(980)}{(2.5 \times 10^{-4})^2} \right]^{1/3} = 12.89$$

$$u_o^* = u_o \left[\frac{(0.001)^2}{(2.5 \times 10^{-4})(1.5 - 0.001)(980)} \right]^{1/3}$$

$$= 0.559 \text{ and } 1.12, \quad \text{for } u_o = 40 \text{ and } 80 \, cm/s$$

From Fig. 16, we can expect bubbling fluidization at both gas velocities.

PROBLEMS

1. Calculate the minimum fluidizing velocity u_{mf} for a bed of crushed anthracite coal fluidized by gas.
    (a) Use information on ϕ_s and ε_{mf}.
    (b) Do not use information on ϕ_s and ε_{mf}.

    Data

    Solids: $\rho_s = 2 \, g/cm^3, d_p = 100 \, \mu m, \phi_s = 0.63$
    Gas: $\rho_g = 1.22 \times 10^{-3} \, g/cm^3, \mu = 1.8 \times 10^{-4} \, g/cm \cdot s$

    Use Table 3 to estimate ε_{mf}.

2. Calculate the minimum fluidizing velocity u_{mf} for a bed of microspherical catalyst of wide size distribution ranging between 0 and 260 μm. Compare your value with the experimental value of $u_{mf} = 2.6$ cm/s.

Data

> *Solids:* $\rho_s = 1.83$ g/cm^3, $\varepsilon_{mf} = 0.45$
>
> *Gas:* $\rho_g = 1 \times 10^{-3}$ g/cm^3, $\mu = 1.7 \times 10^{-4}$ g/cm·s

Size frequency distribution

$d_p \times 10^3$ (cm)	5	8	10	12	14	15	16	17	18	20	22	24	26
p (1/cm)	5	13	23	45	95	135	145	115	88	50	22	12	1

3. Calculate the minimum fluidizing velocity for a bed of large particles.
(a) Use the information given for ε_{mf} and ϕ_s.
(b) Do not use information on ε_{mf} and ϕ_s.

Data

> *Solids:* $\rho_s = 2.93$ g/cm^3, $d_p = 1$ mm, $\phi_s = 0.75$, $\varepsilon_{mf} = 0.5$
>
> *Gas:* $\rho_g = 0.01$ g/cm^3, $\mu = 0.0003$ g/cm·s

4. Calculate the terminal velocity of
(a) 10-μm spheres
(b) 1-mm spheres
(c) 10-μm irregular particles, $\phi_s = 0.67$
(d) 1-mm irregular particles, $\phi_s = 0.67$
for $\rho_s = 2.5$ g/cm^3, $\rho_g = 1.2 \times 10^{-3}$ g/cm^3, and $\mu = 1.8 \times 10^{-4}$ g/cm·s.

5. Estimate the mode of fluidization and describe the type of behavior expected of a bed of particles of density $\rho_s = 1.5$ g/cm^3.
(a) $d_p = 100$ μm, $\rho_g = 0.01$ g/cm^3, $\mu = 3 \times 10^{-4}$ g/cm·s, $u_o = 40$ cm/s
(b) $d_p = 1$ mm, $\rho_g = 0.001$ g/cm^3, $\mu = 2 \times 10^{-4}$ g/cm·s, $u_o = 2$ m/s

REFERENCES

1. F.A. Zenz and D.F. Othmer, *Fluidization and Fluid Particle Systems*, Van Nostrand Reinhold, New York, 1960.
2. M. Leva, *Fluidization*, McGraw-Hill, New York, 1959.
3. S. Uchida and S. Fujita, *J. Chem. Soc., Ind. Eng. Section (Japan)* **37**, 1578, 1583, 1589, 1707 (1934).
4. P.C. Carman, *Trans. Inst. Chem. Eng.*, **15**, 150 (1937).
5. M. Leva et al., *Chem. Eng. Prog.*, **44**, 511, 707 (1948); *Ind. Eng. Chem.*, **41**, 1206 (1949).
6. G.G. Brown et al., *Unit Operations*, Wiley, New York, 1950.
7. S. Ergun, *Chem. Eng. Prog.*, **48**, 89 (1952).
8. J.H. Perry, *Chemical Engineers' Handbook*, 3rd ed., McGraw-Hill, New York, 1963.
9. T.H. Chilton and A.P. Colburn, *Trans. Am. Inst. Chem. Eng.*, **26**, 178 (1931).
10. L.E. Brownell and D.L. Katz, *Chem. Eng. Prog.*, **43**, 537 (1947).
11. T. Miyauchi et al., *Adv. Chem. Eng.*, **11**, 275 (1981).
12. C.Y. Wen and Y.H. Yu, *AIChE J.*, **12**, 610 (1966).
13. J.F. Richardson, in *Fluidization*, J.F. Davidson and H. Harrison, eds., p. 26, Academic Press, New York, 1971.
14. S.C. Saxena and G.J. Vogel, *Trans. Inst. Chem. Eng.*, **55**, 184 (1977); *Chem. Eng. J.*, **14**, 59 (1977).

15. S.P. Babu, B. Shah, and A. Talwalkar, *AIChE Symp. Ser.*, **74(176)**, 176 (1978).

16. J.R. Grace, in *Handbook of Multiphase Systems*, G. Hetsroni, ed., p. 8-1, Hemisphere, Washington, D.C., 1982.

17. D.C. Chitester et al., *Chem. Eng. Sci.*, **39**, 253 (1984).

18. J.P. Couderc, in *Fluidization*, 2nd ed., J.F. Davidson et al., eds., Chap. 1, Academic Press, New York, 1985.

19. W.C. Yang et al., *AIChE J.*, **31**, 1085 (1985).

20. T. Shirai, *Fluidized Beds*, Kagaku-Gijutsu-Sha, Kanazawa, 1958.

21. S. Morooka, M. Nishinaka, and Y. Kato, *Kagaku Kogaku*, **37**, 485 (1973).

22. D. Geldart and A.R. Abrahamsen, *Powder Technol.*, **19**, 133 (1978); **26**, 34, 47 (1980); in *Fluidization III*, J.R. Grace and J.M. Matsen, eds., p. 453, Plenum, New York, 1980.

23. E.F. Brooks and T.J. Fitzgerald, "Aggregation and Fluidization Characteristics of a Fibrous Carbon," paper delivered at annual meeting of American Institute of Chemical Engineers, November 1985.

24. T. Varadi and J.R. Grace, in *Fluidization*, J.F. Davidson and D.L. Keairns, eds., p. 50, Cambridge Univ. Press, New York, 1978.

25. J.R.F. Guedes de Carvalho, D.F. King, and D. Harrison, in *Fluidization*, J.F. Davidson and D.L. Keairns, eds., p. 59, Cambridge Univ. Press, New York, 1978; D.F. King and D. Harrison, *Trans. Inst. Chem. Eng.*, **60**, 26 (1982); J.R.F. Guedes de Carvalho, *Chem. Eng. Sci.*, **36**, 413 (1981).

26. M.E. Crowther and J.C. Whitehead, in *Fluidization*, J.F. Davidson and D.L. Keairns, eds., p. 65, Cambridge Univ. Press, New York, 1978.

27. P.N. Rowe et al., in *Fluidization IV*, D. Kunii and R. Toei, eds., p. 53, Engineering Foundation, New York, 1983.

28. L.E.L. Sobreiro and J.L.F. Monteiro, *Powder Technol.*, **33**, 95 (1982).

29. A.W. Weimer and G.J. Quarderer, *AIChE J.* **31**, 1019 (1985); A.W. Weimer, *AIChE J.*, **32**, 877 (1986).

30. W.C. Yang et al., *AIChE J.*, **31**, 1086 (1985).

31. T. Mii, K. Yoshida, and D. Kunii, *J. Chem. Eng. Japan*, **6**, 100 (1973).

32. M. Avedesian and J.F. Davidson, *Trans. Inst. Chem. Eng.*, **51**, 121 (1973).

33. B. Singh, G.R. Rigby, and T.G. Callcott, *Trans. Inst. Chem. Eng.*, **51**, 93 (1973).

34. A. Desai, H. Kikukawa, and A.H. Pulsifer, *Powder Technol.*, **16**, 143 (1977).

35. M.A. Doheim and C.N. Collinge, *Powder Technol.*, **21**, 289 (1978).

36. G.H. Hong et al., *Kagaku Kogaku Ronbunshu*, **6**, 557 (1980).

37. K. Svoboda and M. Hartman, *Ind. Eng. Chem. Process Des. Dev.*, **20**, 319 (1981).

38. J.S.M. Botterill, Y. Teoman, and K.R. Yuregir, *Powder Technol.*, **31**, 101 (1982); J.S.M. Botterill and Y. Toeman, in *Fluidization III*, J.R. Grace and J.M. Matsen, eds., p. 93, Plenum, New York, 1980.

39. R. Yamazaki, G.H. Hong, and G. Jimbo, in *Fluidization IV*, D. Kunii and R. Toei, eds., p. 121, Engineering Foundation, New York, 1983; R. Yamazaki, N. Ueda, and G. Jimbo, *J. Chem. Eng. Japan*, **19**, 251 (1986).

40. K. Kitano et al., *Kagaku Kogaku Ronbunshu*, **12**, 354 (1986).

41. M.J. Gluckman, J. Yerushalmi, and A.M. Squires, in *Fluidization Technology*, D.L. Keairns, ed., vol. 2, p. 395, McGraw-Hill, New York, 1976.

42. R.H. Wilhelm and M. Kwauk, *Chem. Eng. Prog.*, **44**, 201 (1948).

43. J.B. Romero and L.N. Johanson, *Chem. Eng. Prog. Symp. Ser.*, **58(38)**, 28 (1962).

44. F.A. Zenz, *Petrol. Refiner*, **36**, 321 (1957).

45. R. Jackson, *Trans. Inst. Chem. Eng.*, **41**, 13 (1963).

46. O. Molerus, in *Proc. Int. Symp. on Fluidization*, A.A.H. Drinkenburg, ed., p. 134, Netherlands Univ. Press, Amsterdam, 1967.

47. R.L. Pigford and T. Baron, *Ind. Eng. Chem. Fundamentals*, **4**, 81 (1965).

48. D. Harrison, J.F. Davidson, and J.W. de Kock, *Trans. Inst. Chem. Eng.*, **39**, 202 (1961).

49. H.C. Simpson and B.W. Rodger, *Chem. Eng. Sci.*, **16**, 153 (1961).

50. J. Verloop and P.M. Heertjes, *Chem. Eng. Sci.*, **25**, 825 (1970).

51. D. Geldart, *Powder Technol.*, **7**, 285 (1973); **19**, 133 (1978).

52. D. Geldart, *Chem. Eng. Sci.*, **39**, 1481 (1984); **40**, 653 (1985).

53. J.R. Grace, *Can. J. Chem. Eng.*, **64**, 353 (1986).

54. A. Haider and O. Levenspiel, *Powder Technol.*, **58**, 63 (1989).

55. R. Turton and N.N. Clark, *Powder Technol.*, **53**, 27 (1987).

56. P.H. Pinchbeck and F. Popper, *Chem. Eng. Sci.*, **6**, 57 (1956).

57. K.P. Lanneau, *Trans. Inst. Chem. Eng.*, **38**, 125 (1960).

58. P.W.K. Kehoe and J.F. Davidson, *Inst. Chem. Eng. Symp. Ser.*, **33**, 97 (1971).

59. L. Carotenuto, S. Crescitelli, and G. Donsi, *Quad. Ing. Chim. Ital.*, **10**, 185 (1974); S. Crescitelli et al., *CHISA Conference*, 1 (1978).

60. W.J. Thiel and O.E. Potter, *AIChE J.*, **24**, 561 (1978).

61. G.S. Canada, M.H. McLaughlin, and F.W. Staub, *AIChE Symp. Ser.*, **74(176)**, 27 (1978).

62. J. Yerushalmi, D.H. Turner, and A.M. Squires, *Ind. Eng. Chem. Process Des. Dev.*, **17**, 47 (1976).

63. J. Yerushalmi and N.T. Cankurt, *Chemtech*, **8**, 564 (1978); *Powder Technol.*, **24**, 187 (1979).

64. A. Avidan and J. Yerushalmi, *Powder Technol.*, **32**, 223 (1982); in *Fluidization*, 2nd ed., J.F. Davidson et al., eds., p. 225, Academic Press, New York, 1985.

65. D. Geldart and M.J. Rhodes, in *Circulating Fluidized Bed Technology* P. Basu, ed., p. 21, Pergamon, New York, 1986; M.J. Rhodos and D. Geldart, in *Fluidization V*, K. Østergaard and A. Sørensen, eds., p. 281, Engineering Foundation, New York, 1986.

66. L. Reh, *Chem.-Ing.-Techn.*, **40**, 509 (1968); *Chem. Eng. Prog.*, **67**, 58 (1971).

67. N.M. Ćatipović, G.N. Jovanović, and T.J. Fitzgerald, *AIChE J.*, **24**, 543 (1978).

68. J.J. van Deemter, in *Fluidization III*, J.R. Grace and J.M. Matsen, eds., p. 69, Plenum, New York, 1980.

69. J. Werther, *Intn. Chem. Eng.*, **20**, 529 (1980).

70. Y. Li and M. Kwauk, in *Fluidization III*, J.R. Grace and J.M. Matsen, eds., p. 537, Plenum, New York, 1980.

71. J.M. Matsen, *Powder Technol.*, **32**, 21 (1982); in *Fluidization IV*, D. Kunii and R. Toei, eds., p. 225, Engineering Foundation, New York, 1983.

72. A.M. Squires, M. Kwauk, and A. Avidan, *Science*, **230**, 1329 (1985).

73. M. Horio et al., in *Circulating Fluidized Bed Technology*, P. Basu, ed., p. 255, Pergamon, New York, 1986.

4

The Dense Bed: Distributors, Gas Jets, and Pumping Power

This chapter focuses on what happens at the bottom of a dense fluidized bed and on the proper introduction of gas feed. We consider various distributor designs, their accompanying gas jetting problems, and the role of nozzles and their large gas jets as a means of promoting the circulation of bed solids. The chapter ends with a discussion of design procedures for distributors and of the pumping power requirement to keep the bed fluidized.

Distributor Types

Ideal Distributors

Most small-scale studies in fluidization use ceramic or sintered metal porous plate distributors, because they have a sufficiently high flow resistance to give a uniform distribution of gas across the bed. This situation is ideal. Many other materials can do this—for instance, filter cloth, compressed fibers, compacted wire plate, or even a thin bed of small particles. Of course, some of these materials should be reinforced by sandwiching between metal or wire plates with large openings.

Although gas-solid contacting is superior with such distributors, for industrial operations they have several drawbacks:

- High-pressure drop leads to increased pumping power requirements, often a major operating cost factor.
- Low construction strength, hence impractical for large-scale use.
- High cost for some materials.
- Low resistivity against thermal stresses.
- Possible gradual clogging by fine particles or by products of corrosion.

Despite these disadvantages, compacted wire plates or sandwiched beds of small particles are sometimes used.

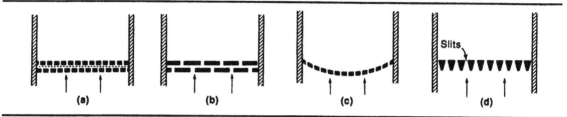

FIGURE 1
Plate and grate distributors are cheap and easy to construct: (a) sandwiching perforated plates; (b) staggered perforated plates; (c) dished perforated plate; (d) grate bars.

Perforated or Multiorifice Plates

Perforated plate distributors are widely used in industry because they are cheap and easy to fabricate. Figure 1 illustrates several variations of a simple perforated plate distributor. Type (a) consists of two perforated plates sandwiching a metal screen that prevents solids from raining through the orifices when the gas flow is stopped. A variation of this, type (b), uses two staggered perforated plates and no screen.

One problem with this design is lack of rigidity. Large perforated plates deflect unpredictably under heavy load; hence, they need reinforcing for support. In addition, during thermal expansion gas leakage at the bed perimeter is possible.

When it is impractical to have a reinforcing structure to support a flat perforated plate against heavy loads, curved plates, such as type (c), are sometimes used. Curved plates will withstand heavy loads and thermal stresses. Because bubbling and channeling tend to occur preferentially near the center of a fluidized bed, design (c) helps to counter this tendency. Distributor plates curved upward achieve good contacting only with more orifices near the perimeter and fewer near the center, a disadvantage for fabrication. Alternatively, parallel grate bars, type (d), may be used. These bars may be considered as two-dimensional versions of perforated plates, and they have only seen limited use; see Fig. 2.23(b).

In some operations, large amounts of solids enter the bed with the inlet gases, for example in Exxon's model IV FCC reactor, or in the multistage fluidized limestone calciner. In these situations perforated plates without screens are recommended.

The diameter of orifices in perforated plate distributors may range from 1 to 2 mm in small experimental beds to as much as 50 mm in large FCC units with their solid-entrained gases.

Tuyeres and Caps

Perforated plate distributors cannot be used under severe operating conditions, such as high temperature or a highly reactive environment. Tuyere designs (Fig. 2) are used in these situations. The multiple porous plate, type (a), gives good gas distribution above each filter, but particles will settle between adjacent tuyeres. Also, special precautions must be taken to ensure that the incoming gas is free of filter-clogging material. Types (b), (c), and (d) are frequently used and prevent solids from falling through the distributor. However, with all these

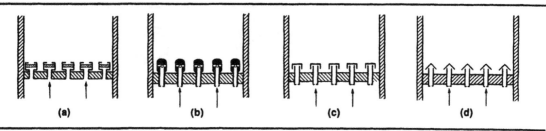

FIGURE 2
Tuyere distributors: (a) porous plate type; (b) nozzle type; (c) bubble cap type; (d) slit nozzle type.

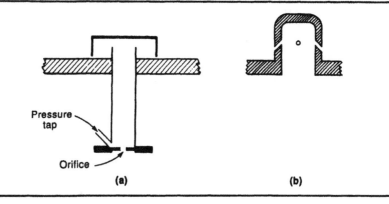

FIGURE 3
Details of tuyeres: (a) tuyere with inflow orifice to control the gas velocity of gas exiting into the bed; (b) cap type.

designs, particles are apt to settle, sinter, and stick on the distributor plate itself. A variety of designs have been proposed and used to minimize this effect.

To ensure equal gas flow through the tuyeres of type (b), (c), or (d), each tuyere may be fitted with a high-resistance orifice at its gas inlet, as shown in Fig. 3(a). The cap-type tuyere of Fig. 3(b) has no orifice at its gas inlet. Instead, the orifices around the cap are designed to create a sufficient pressure drop for uniform fluidization. A disadvantage of this design involves the jetting effect of the high-velocity gas issuing from the orifices. This can cause considerable particle attrition. Conversely, the velocity of the gas issuing from the tuyeres of Fig. 3(a) can be chosen as desired because the rate of gas flow is fixed by the high-resistance inlet orifice. Because of their complicated construction, tuyere-type distributors are much more expensive than perforated plate distributors.

Pipe Grids and Spargers

Experience shows that internals, such as properly placed heat exchanger tubes, substantially improve gas-solid contacting by breaking up growing bubbles and by preventing gulf streaming, or gross circulation of solids. In fact, proper design of internals can improve the quality of fluidization so much that refined high-resistance distributors are not needed. In such cases, a pipe grid or sparger, such as shown in Fig. 4(a), may be all that is needed to introduce reactant gas into beds fluidized by a second carrier gas coming from below. This is the

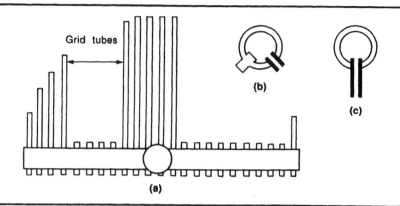

FIGURE 4
Examples of sparger designs.

situation for the Sohio acrylonitrile process shown in Fig. 2.10(a). Downward-pointing nozzles of Figs. 4(b) and (c) prevent clogging of the spargers by particles when gas flow is stopped. In addition, downward gas ejection gives somewhat better bubble formation than upward ejection, which will be discussed later.

The examples discussed illustrate some of the many possibilities in the design of distributors, and only good judgment and experience will tell what combination is best for any application. Distributors should be selected and designed with care, for this is the first step in the development of a fluidized bed process.

Gas Entry Region of a Bed

An enormous amount of attention has focused on the gas-solid contacting just above the distributor, because contacting is very good here, and this in turn can strongly affect the progress of fast heat transfer, mass transfer, and reaction processes. We consider, in turn, the behavior in the vicinity of the different types of distributors.

Above a Porous Plate. For a uniform gas flow, $u_o > u_{mf}$, a highly expanded gas-solid dispersion forms directly above the distributor. This is unstable, and a few millimeters above the plate the dispersion divides into many little bubbles plus an emulsion phase. On rising upward, these bubbles grow very rapidly by coalescence, as shown in Fig. 5(a).

Above a Single Orifice with Background Flow, u_{mf}. With the bed kept at u_{mf}, introduce extra gas at a volumetric flow rate v_{or} m³/s through a single orifice in the porous distributor. Assuming that the bubble formed detaches when its rise velocity exceeds its linear growth rate, Davidson and Schüler [2] calculated that the bubble volume should be

$$V_b = 1.138 \frac{v_{or}^{6/5}}{g^{3/5}}, \qquad [\text{m}^3] \tag{1}$$

Harrison and Leung [3] found that this expression fitted their experimental findings at low gas flow rates. Figure 6(a) shows this behavior.

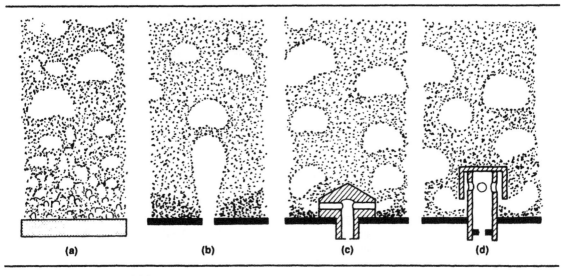

FIGURE 5
Behavior of bubbles just above the distributor; sketched from Werther [1]: (a) porous plate; (b) perforated plate; (c) nozzle-type tuyere; (d) bubble cap tuyere.

Nguyen and Leung [6] then asked what fraction K of this orifice gas actually becomes part of the bubble, and what fraction enters the emulsion region. They found $K = 0.53$. Later, Yates et al. [5] obtained the following values in beds of Geldart **AB** particles

at height of 10 cm $K = 0.36$

at height of 25 cm $K = 0.79$

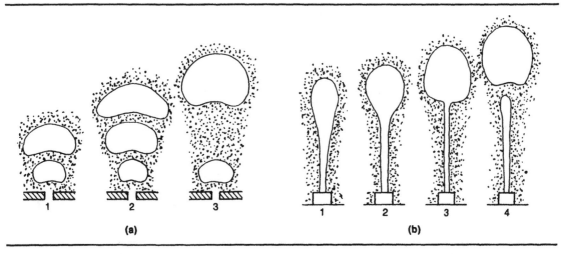

FIGURE 6
Two modes of bubble formation just above a single orifice into an incipiently fluidized bed. (a) For relatively low gas velocity, a chain of bubbles form at the orifice; sketched from Massimilla et al. [4]. (b) For high orifice velocity (here $u_{or} = 120$ m/s), a standing jet forms at the orifice. This breaks into bubbles; sketched from Yates et al. [5].

Thus, 10 cm above the distributor plate only about one-third of the orifice gas appears as rising bubbles with well-defined boundaries, whereas two-thirds of this gas enters the emulsion to increase its voidage above ε_{mf}. As these bubbles rise and coalesce and grow, some of this emulsion gas returns to the bubble so that 25 cm above the distributor over two-thirds of the orifice gas is accounted for by the bubble flow. Findings similar to these have been reported by Sit [7], and we may expect this to represent general behavior at orifice flows that are not too high.

At increasing orifice flow rates, bigger bubbles form, the distance between successive bubbles decreases, and the bubbles coalesce to form a plume or jet. In a typical sequence of events, shown in Fig. 6(b), the jet elongates, the top balloons, a bubble detaches, and the process repeats.

Above a Perforated Plate Distributor. Here jets form, as shown in Fig. 5(b). In addition, particles are seen to settle on the flat surfaces between orifices to form a dead zone. Wen et al. [8] investigated the effect of operating conditions on this dead zone and found that it shrank with increasing gas velocity, increasing particle size, increasing orifice size, and decreased spacing of orifices. They also found that at high enough orifice flow rates and with large enough particle size these dead zones could be completely eliminated.

At a Nozzle Tuyere of Figure 2(b). At low gas flows, bubbles form and detach in orderly procession, as shown in Fig. 5(c). At high velocities, however, these bubbles coalesce into horizontal jets. Wen et al. [8] found that these jets caused large disturbances and gave better mixing of particles than did ordinary perforated plates for the same orifice area and gas flow. Hence, these distributors are favored if dead zones are to be avoided on the distributor plate.

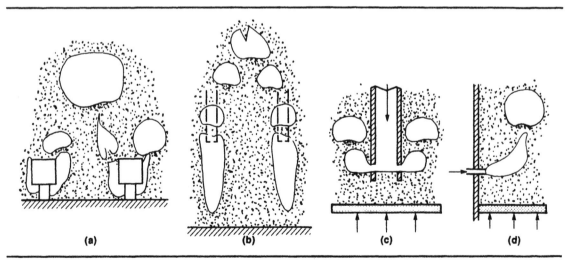

FIGURE 7
Downward and horizontal gas entry into a fluidized bed: (a) from a bubble cap tuyere; sketched from Liu et al. [9]; (b) from a downpointing sparger pipe at high gas velocity; sketched from Jir et al. [10]; (c) from a slotted pipe at low gas velocity; sketched from Massimilla [4]; (d) from the vessel wall, at high velocity; sketched from Massimilla [4].

At a Bubble Cap Tuyere of Figure **2(c).** Bubbles form at bubble cap tuyeres as shown in Figs. 5(d) and 7(a). At higher gas flows, larger bubbles are generated, but without jet action.

At Pipes in Beds and at Bed Walls. In general, for downpointing or horizontal gas inlet pipes, low gas velocities give a succession of bubbles and high velocities give a standing, flickering jet or plume attached to the entry pipe. Figures 7(b)–(d) show what these look like.

Gas Jets in Fluidized Beds

Gas jets form at distributors, from tubes inserted at various angles into beds, and at bed walls. This jet flow and jet penetration length are studied because

- The high jet velocity from orifices (up to 180 m/s in experiments, but 30–40 m/s in commercial distributors) entrains solids, which gives a very energetic sandblasting action and rapid erosion of any impinged surface. Thus, it is important to keep the bed internals, such as heat exchanger tubes, a sufficient distance from the distributor.
- Knowing the jet penetration length will help the designer to design nozzles for feeding large quantities of gas into the bed.
- Attrition of friable solids in beds occurs primarily at these jets. Hence, a knowledge of jet action will help to control the size distribution of solids in the bed.
- In certain processes involving fast physical and chemical changes— combustions, gasifications, flame reactions, granulations, coatings, de-volatilizations—the character and quality of product obtained depends largely on what happens just as the feed enters the bed.

These are some reasons for studying jets and jet action.

But first, we give some definitions of jet penetration length. For an orifice in a plate, the jet penetration length L_j is defined by Filla et al. [11] as the distance between the plate and center of the bubble at the instant when it detaches from the jet; see Fig. 6(b). For upward-pointing tubes in fluidized beds, Knowlton and Hirsan [12] noted that the jet length fluctuated greatly, so they defined $L_{j,min}$ and $L_{j,max}$ as the minimum and maximum jet heights, and L_j as the furthest penetration of the jet bubbles. This is the distance beyond which it would be safe to locate bed internals.

We now extract some findings of general interest from the many studies on jets. For perforated plates, given solids, fixed $u_o = 20$ cm/s, the same fraction of openings in the plates (0.3%), and the same jet velocity ($u_{or} = 67$ m/s), Werther [13] found that

for many small holes, $d_{or} = 2.1$ mm: $L_j = 10\text{–}15$ cm

for a few large holes, $d_{or} = 9.5$ mm: $L_j = 50\text{–}60$ cm

This means that, for a given u_{or}, small holes give shorter jets but are accompanied by a larger pressure drop across the distributor.

For a single orifice plus background flow, Yates et al. [5] found a large effect of background flow upon L_j, and that L_j rises to a maximum when the total flow is close to u_{mf} in the bed. See Fig. 8(a). At higher flows, L_j decreases because of the lateral movement of bed solids caused by the gas jet. They also found that high system pressure and high u_{or} gave longer jets. See Fig. 8(b).

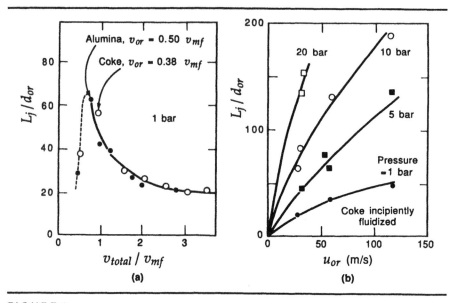

FIGURE 8
Effect of background flow and pressure upon the penetration depth of a vertical jet, d_{or} = 1.55 mm; adapted from Yates et al. [5].

They correlate their findings for jet penetration length as follows:

$$\frac{L_j}{d_{or}} = 21.2\left(\frac{u_{or}^2}{gd_p}\right)^{0.37}\left(\frac{d_{or}u_{or}\rho_g}{\mu}\right)^{0.05}\left(\frac{\rho_g}{\rho_s}\right)^{0.68}\left(\frac{d_p}{d_{or}}\right)^{0.24} \tag{2}$$

Equation (2) is just one of the many correlations proposed for jet length. As with this equation, all other investigations have presented their findings in terms of L_j/d_{or}. Massimilla [4] tabulates and then compares these findings in two diagrams, L_j/d_{or} versus u_o and L_j/d_{or} versus operating pressure, and finds disagreement up to a factor of 100 or more. Since many of the pertinent variables, such as solid properties, size, and orifice diameter, differ from study to study, this is not surprising.

In conclusion, to predict the jet penetration length for a particular application, choose the correlation for conditions that most closely match the system at hand and apply that correlation with caution.

Pressure Drop Requirements across Distributors

Experience shows that distributors should have a sufficient pressure drop Δp_d to achieve equal flows over the entire cross section of the bed. According to Zuiderweg [14], in the early years of fluidization engineering rules of thumb were followed, such as

$$\Delta p_d = (0.2 - 0.4)\, \Delta p_b \tag{3}$$

where Δp_b is the pressure drop across the bed, given by Eq. (3.17). It is also clear that increased Δp_d will ensure a more even distribution of entering gas. However, an excessive Δp_d has its drawbacks.

- Power consumption and construction cost for the blower or compressor increases with the total pressure drop, or $\Delta p_t = \Delta p_b + \Delta p_d$, and Δp_d can represent a significant portion of the total pressure drop.

- For tuyere distributors without inlet orifices (see Fig. 3(b)), the required distributor pressure drop to satisfy Eq. (3) may require the use of excessively high gas velocities at the nozzles. This may result in erosion and breakage of particles and an undesirable shift in size distribution of bed solids.

It is important, therefore, to know the minimum Δp_d that would ensure uniform fluidization in the required range of operations. From orifice theory and fixed bed equations, we can show that

$$\Delta p_d \propto u_o \qquad \text{for porous plates} \tag{4a}$$

$$\Delta p_d \propto u_o^2 \qquad \text{for perforated plates and tuyere distributors} \tag{4b}$$

Several papers have published recommendations for relating Δp_d with Δp_b for satisfactory operations [15–21]. Hiby [15] lowered the gas flow to a portion of a bed and considered the bed stable if this caused $\Delta p_b + \Delta p_d$ to decrease in this zone. From these experiments Hiby recommends that for stable operations, one should have

$$\frac{\Delta p_d}{\Delta p_b} = 0.15 \qquad \text{for } \frac{u_o}{u_{mf}} = 1\text{–}2 \tag{5a}$$

$$\frac{\Delta p_d}{\Delta p_b} = 0.015 \qquad \text{for } \frac{u_o}{u_{mf}} \gg 2 \tag{5b}$$

Considering the channeling that may result from perturbations in a bed at close to u_{mf}, Siegel [16] came up with the following criterion for stable operations:

$$\frac{\Delta p_d}{\Delta p_b} \geqslant 0.14 \tag{6}$$

Extending this channeling model, Shi and Fan [21] conclude that one can guarantee full fluidization if

$$(\Delta p_d + \Delta p_b)_{\text{at any } u_o} \cong (\Delta p_d + \Delta p_b)_{\text{at } u_{mf}} \tag{7}$$

where, at u_{mf},

$$\frac{\Delta p_d}{\Delta p_b} > \begin{cases} 0.14 & \text{for porous plates} \\ 0.07 & \text{for perforated plates} \end{cases} \tag{8a} \tag{8b}$$

Finally, for bubbling beds of fine particles, meaning $u_o > u_{mb}$ for Geldart A particles, supported on porous or perforated plate distributors, Mori and Moriyama [17] considered the consequences of shutting off the flow to part of the bed. With resumption of flow would the slumped solids refluidize? The result of their analysis suggests that for full fluidization one should have

$$\frac{\Delta p_d}{\Delta p_b} \geq \left(\frac{L_f}{L_{mf}} - 1 \right) \frac{1}{1 - (u_{mf}/u_o)^n} \tag{9}$$

where $n = 1$ for porous plates, $n = 2$ for perforated plates. This expression shows that Δp_d must be large when operating close to u_{mf}, but can be lower when the bed operates at high u_o.

At $u_0/u_{mf} > 10$ for porous plates and $u_0/u_{mf} > 3$ for perforated plates, Eq. (9) reduces to

$$\frac{\Delta p_d}{\Delta p_b} \geq \frac{L_f}{L_{mf}} - 1 \tag{10}$$

and if we take $L_f/L_{mf} = 1.2$–1.4, typical of bubbling beds, Eq. (10) then reduces to $\Delta p_d = (0.2$–$0.4)\,\Delta p_b$, which is identical to the rule of thumb given in Eq. (3).

Using large beds (up to 2.4 m square) with tuyere-type distributors, Whitehead et al. [22] carried out extensive experiments with a variety of Geldart **B** sands. On slowly increasing the gas flow to the bed, they observed that a number of tuyeres became active as soon as u_0 exceeded u_{mf}, as shown in Fig. 9(a). The number increased progressively until all tuyeres were in use when u_0 reached u_1. When they reduced the gas flow, all the tuyeres kept operating until a critical velocity u_2 was reached. Then the number of active tuyeres decreased until u_{mf} was reached. Figure 9(a) shows this hysteresis behavior.

Whitehead et al. also found that Δp_d and u_2 were related to L_m, ρ_s, and u_{mf}. Their final correlation is given in Fig. 9(b), and shows that deeper beds and beds close to u_{mf} require a larger Δp_d than do shallow beds at high u_0.

We summarize these findings with the following design recommendations:

a. For even distribution of fluidizing gas to a bed where u_0 is close to u_{mf} choose

$$\frac{\Delta p_d}{\Delta p_b} \geq 0.15$$

b. The required $\Delta p_d/\Delta p_b$ decreases as u_0/u_{mf} increases.
c. $\Delta p_d/\Delta p_b$ is roughly independent of bed height, or Δp_b.
d. For the same bed, same u_0, and same u_{mf}, but different distributors,

$$\left(\frac{\Delta p_d}{\Delta p_b}\right)_{porous\ plate} > \left(\frac{\Delta p_d}{\Delta p_b}\right)_{orifice\ plate} \tag{11}$$

This difference is greater when u_0 is close to u_{mf} and decreases to zero for $u_0 \gg u_{mf}$.

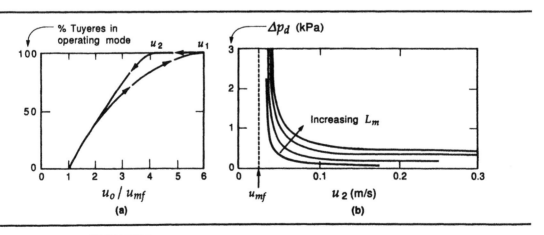

FIGURE 9
Characteristics of a multituyere distributor; adapted from Whitehead et al. [22].

e. The original rule of thumb

$$\Delta p_d = (0.2\text{–}0.4)\,\Delta p_b \tag{3}$$

is verified by various analyses and experiments and represents a reasonable upper bound to the required distributor pressure drop for smooth operations. This value can be made lower in specific cases.

Especially in large-scale commercial-type operations, a severe problem occurs when a distributor is designed for a particular range of operating velocities but is operated in a different range. The problem occurs because Δp_d varies strongly with u_o, as given by Eq. (4), whereas Δp_b remains practically independent of u_o. So suppose that a distributor is designed according to Eq. (5) for $u_o = 2u_{mf}$. Then at $u_o = 20u_{mf}$, with Eq. (4), we have

$$\frac{\Delta p_d}{\Delta p_b} = \begin{cases} (0.15)\dfrac{20}{2} = 1.5 & \text{for porous plates} \\[2ex] (0.15)\left(\dfrac{20}{2}\right)^2 = 15 & \text{for perforated plates and tuyeres} \end{cases}$$

Such excessive pressure drops could well exceed the capacity of the blowers.

Conversely, if a distributor is designed to operate according to Eq. (10) for $u_o = 20u_{mf}$, then, at $u_o = 2u_{mf}$, Δp_d becomes negligible and the distributor cannot be expected to sustain even fluidization. This discussion suggests that the porous plate distributor can operate satisfactorily over a wider range of gas velocities than can other types of distributors.

Finally, with a properly designed distributor, one without excessive jet penetration, the interaction between the distributor and the bed is limited to a narrow bottom zone of bed. Above this zone, gas-solid contacting is governed primarily by the hydrodynamic properties of the bed itself and cannot be altered much by changing Δp_d; see Chap. 5.

Design of Gas Distributors

Perforated plates and most tuyere distributors can be designed directly from orifice theory, and since the orifice pressure drop is only a small fraction of the total pressure drop, we can use the following procedure.

1. Determine the necessary pressure drop across the distributor, Δp_d, on the basis of the previous discussion, or simply by using Eq. (3).

2. Calculate the vessel Reynolds number, $Re_t = d_t u_o \rho_g/\mu$, for the total flow approaching the distributor and select the corresponding value for the orifice coefficient, $C_{d,or}$.

Re_t	100	300	500	1000	2000	>3000
$C_{d,or}$	0.68	0.70	0.68	0.64	0.61	0.60

3. Determine the gas velocity through the orifice, measured at the approach density and temperature:

$$u_{or} = C_{d,or}\left(\frac{2\Delta p_d}{\rho_g}\right)^{1/2} \tag{12}$$

The ratio u_o/u_{or} gives the fraction of open area in the distributor plate. See that this is less than 10%.

4. Decide on N_{or}, the number of orifices per unit area of distributor, and find the corresponding orifice diameter from the equation

$$u_o = \frac{\pi}{4} d_{or}^2 u_{or} N_{or} \tag{13}$$

For a tuyere with an inlet orifice, as in Fig. 3(a), N_{or} should be the number of tuyeres per unit area. On the other hand, for tuyeres as in Fig. 2(b) but without an inlet orifice (see Fig. 3(b)), N_{or} is given by

$$N_{or} = \left(\frac{\text{tuyeres}}{\text{area}}\right)\left(\frac{\text{number of holes}}{\text{tuyere}}\right) \tag{14}$$

Agitating Distributors. For beds of fine solids such as FCC catalyst, a well-designed distributor should also act as a stirrer, promoting the mixing of solids and keeping the bed well fluidized. The factor that measures this stirring effect is α, defined as

$$\alpha = \frac{\rho_g u_{or}^2/2g_c}{\Delta p_b} = \frac{\text{kinetic energy of the orifice jets}}{\text{resistance of the bed}} \tag{15}$$

If $\alpha > 1$, the jets will punch right through the bed, causing severe gas bypassing attrition of particles, and erosion of bed internals. If $\alpha \ll 1$, the jets will not contribute much to bed stirring, and bubbles rising in the bed will have to do this.

What are typical values of α? For a 1-m-high bed of cracking catalyst at ambient conditions, Eq. (3.16) gives, in SI units,

$$\Delta p_b = \frac{(1 - \varepsilon_{mf})(\rho_s - \rho_g)gL_{mf}}{g_c} = \frac{(1 - 0.4)(1000 - 1)(9.8)(1)}{(1)}$$

$$= 6000 \text{ Pa} \quad (\cong 60 \text{ cm H}_2\text{O})$$

For perforated plate distributors, common practice has u_{or} ranging between 30 and 60 m/s. Replacing these values in Eq. (15) gives

$$\alpha = \frac{(1.2 \text{ kg/m}^3)(30\text{–}60 \text{ m/s})^2}{2(1 \text{ kg-m/s}^2\text{N})(6000 \text{ Pa})} = 0.09\text{–}0.36$$

These values of α show that gas jets at orifice plates designed for $u_{or} = 30$ m/s contribute very little to the stirring of bed solids. At $u_{or} = 60$ m/s, quite reasonable stirring by the distributor jets can be expected; nevertheless, for friable particles one may prefer to select the distributor plate with smaller u_{or}.

Avoidance of High Jet Velocities at Distributors. When a high jet velocity cannot be used to provide the needed distributor pressure drop because of particle attrition, then one must use inlet orifices for the tuyeres, as shown in Fig. 3(a). For orifice plates, one can meet these requirements by properly selecting N_{or} and d_{or}.

EXAMPLE 1

Design of a Perforated Plate Distributor

Design a perforated plate distributor for use in a commercial fluidized bed reactor

Data

$d_t = 4$ m	$L_{mf} = 2$ m	$\varepsilon_{mf} = 0.48$
$\rho_s = 1500$ kg/m^3	$\rho_g = 3.6$ kg/m^3	$\mu = 2 \times 10^{-5}$ kg/m·s

Pressure and superficial velocity of inlet gas:

$$p_0 = 3 \text{ bar} \quad \text{(absolute)} \qquad u_0 = 0.4 \text{ m/s}$$

To avoid unnecessary attrition of bed solids:

maximum allowable jet velocity from holes, $u_{or} = 40 \text{ m/s}$

SOLUTION We solve this problem in SI units. In following the recommended procedure, we use three steps.

Step 1. Determine the minimum allowable pressure drop through the distributor. According to Eq. (3.17), in SI units

$$\Delta p_b = \frac{(1 - \varepsilon_{mf})(\rho_s - \rho_g)gL_{mf}}{g_c} = \frac{(1 - 0.48)(1500 - 3.6)(9.8)(2)}{} \qquad (1)$$
$$= 15251 \text{ Pa} \quad (\cong 153 \text{ cm } H_2O)$$

Then from Eq. (3), taking an average value, choose

$$\Delta p_d = 0.3 \, \Delta p_b = 4575 \text{ Pa} \quad (\cong 46 \text{ cm } H_2O)$$

Step 2. Determine the orifice coefficient. For flow approaching the plate

$$\text{Re}_t = \frac{d_t u_0 \rho_g}{\mu} = \frac{(4)(0.4)(3.6)}{2 \times 10^{-5}} = 288,000 > 3000$$

Hence

$$C_{d,or} = 0.6$$

Step 3. Calculate u_{or} by Eq. (12). Thus

$$u_{or} = 0.6 \left[\frac{2(4575)}{3.6} \right]^{1/2} = 30.2 \text{ m/s}$$

This value is satisfactory since it does not exceed the maximum allowable jet velocity.
    The fraction of open area in the perforated plate is then given by

$$\frac{u_0}{u_{or}} = \frac{0.4}{30.2} = 0.01325, \text{ or } 1.3\%$$

The relationship between the number and size of orifices that will meet the above requirement is found from Eq. (13). Solving, we find the following possible combinations:

d_{or} (m)	0.001 (1 mm)	0.002 (2 mm)	0.004 (4 mm)
N_{or} (m^{-2})	16900 (1.7/cm^2)	4200 (0.42/cm^2)	1060 (0.1/cm^2)

Orifices that are too small are liable to clog, whereas those that are too large may cause uneven distribution of gas. In light of these considerations, choose

$$d_{or} = 2 \text{ mm} \quad \text{and} \quad N_{or} = \underline{4200/m^2}$$

This means one orifice in a square of side 1.54 cm.

EXAMPLE 2

Design of a Tuyere Distributor

Design a distributor of the kind shown in Fig. 2(b) for the commercial reactor of Example 1, giving the recommended pitch, l_{or}, the number of holes/tuyere, N_h, and the diameter of these holes, d_h.

Data

Minimum allowable pitch or tuyere spacing:	$l_{or} = 0.1$ m.
Maximum allowable jet velocity from the tuyere:	$u_{or} = 30$ m/s.

SOLUTION   Again we use SI units. Start by assuming the minimum spacing for the tuyeres. Hence, for the tuyere plate having $l_{or} = 0.1$ m,

$$N_{or} = 1/[(0.1)(0.1)] = 100 \text{ tuyeres/m}^2$$

Then, from Eq. (13), the diameter of the inlet orifice to the tuyere, as shown in Figs. 2(b) or 3(a), is given by

$$d_{or} = \left[\frac{4}{\pi} \left(\frac{u_0}{u_{or}} \right) \left(\frac{1}{N_{or}} \right) \right]^{1/2}$$

$$= \left[\frac{4}{\pi} (0.01325) \left(\frac{1}{100} \right) \right]^{1/2} = 0.013 \text{ m} = 13 \text{ mm}$$

With the restriction of 30 m/s on the exit gas flow from the holes of the tuyeres, we have, per tuyere,

$$\left(\begin{array}{c} \text{volumetric} \\ \text{flow rate} \end{array} \right) = (0.1 \text{ m} \times 0.1 \text{ m})(0.4 \text{ m/s}) = N_h \left(\frac{\pi}{4} d_h^2 \right)(30 \text{ m/s})$$

Solving gives

$N_h \left(\dfrac{\text{number of holes}}{\text{tuyere}} \right)$	8	6	4
d_h (m)	0.0046 (4.6 mm)	0.0053 (5.3 mm)	0.0065 (6.5 mm)

Because a rectangular pitch was chosen for the tuyeres, choose the six-hole tuyere. This should discourage dead zones between the tuyeres. You may want fewer tuyeres, but this only encourages the formation of dead zones on the distributor plate, so stay with the minimum pitch design.

Thus the design chosen is as follows:

- Tuyeres are as shown in Fig. 2(b) on a 10-cm rectangular pitch.
- Six 5.3-mm holes per tuyere.
- An incoming high-pressure-drop orifice 13 mm ID for each tuyere.

Final Note. We check the pressure drop for each of the six holes in the tuyere. If the drop is large enough, we may have to modify the high-pressure-drop inlet orifice to each tuyere.

Consider each of the six holes to be an orifice. The pressure drop through each hole is given by Eq. (12), or

$$\Delta p_{\text{hole}} = \frac{\rho_g}{2} \left(\frac{u_0}{C_d} \right)^2 = \frac{3.6}{2} \left(\frac{30}{0.6} \right)^2 = 4500 \text{ Pa}$$

But, from Example 1, this is all that is needed to give a sufficiently high distributor pressure drop. Thus, we can dispense with the inlet orifice to each tuyere, and our modified and final design is as follows:

- Tuyeres are as shown in Fig. 2(b) on a 10-cm rectangular pitch, but with no incoming orifice.
- Six holes 5.3 mm ID for each tuyere.

Power Consumption

Power consumption is a significant cost factor in any process using fluidized beds, and occasionally it can be so high that it cancels the advantages of this type of operation. Therefore, roughly estimate the power requirement in the early design stages—certainly before making a detailed design or deciding to pilot plant.

Suppose a stream of gas is to be compressed from an initial pressure p_1 to a higher pressure p_2 to run a fluidized bed unit. Then

$$p_2 - p_1 = \Delta p_b + \Delta p_d + \Delta p_{\text{cyclones and filters}} \tag{16}$$

For adiabatic reversible operations with negligible kinetic and potential energy effects, the ideal shaft work to compress each kilogram of gas is given by

$$-w_{s,\text{ideal}} = \int_{p_1}^{p_2} \frac{dp}{\rho_g}, \quad \text{[J/kg]} \tag{17}$$

With the additional reasonable assumption of ideal gas behavior, or $pv = \dot{n}RT$, the ideal pumping requirement then becomes

$$-\dot{w}_{s,\text{ideal}} = \frac{\gamma}{\gamma - 1} \, p_1 v_1 \left[\left(\frac{p_2}{p_1} \right)^{(\gamma - 1)/\gamma} - 1 \right], \quad \text{[W]} \tag{18a}$$

$$= \frac{\gamma}{\gamma - 1} \, p_2 v_2 \left[1 - \left(\frac{p_1}{p_2} \right)^{(\gamma - 1)/\gamma} \right], \quad \text{[W]} \tag{18b}$$

where v is the volumetric flow rate of gas (in m^3/s) and $\gamma = C_{pg}/C_{vg}$.

Adiabatic reversible compression of a gas from p_1 to p_2 raises its temperature, and from thermodynamics we have

$$T_2 = T_1 \left(\frac{p_2}{p_1} \right)^{(\gamma - 1)/\gamma} \tag{19}$$

where $\gamma \cong 1.67$, 1.40, and 1.33 for monatomic, diatomic, and triatomic gases, respectively.

For real operations with its frictional losses, the actual shaft work required is always greater than the ideal and is given by

$$-\dot{w}_{s,\text{actual}} = \frac{-\dot{w}_{s,\text{ideal}}}{\eta} \tag{20}$$

where η is the compressor efficiency, roughly given by

$$\eta = 0.55\text{--}0.75 \quad \text{for a turboblower}$$
$$= 0.6\text{--}0.8 \quad \text{for a Roots blower}$$
$$= 0.8\text{--}0.9 \quad \text{for an axial blower or a two-stage reciprocating compressor}$$

The actual temperature of gas leaving a well-insulated (adiabatic), but not 100% efficient compressor, is then

$$T_{2,\text{actual}} = T_1 + \frac{T_1}{\eta} \left[\left(\frac{p_2}{p_1} \right)^{(\gamma - 1)/\gamma} - 1 \right] \tag{21}$$

This temperature is higher than the ideal, given by Eq. (19).

EXAMPLE 3

Power Require-ment for a Fluidized Coal Combustor (FBC)

Calculate the compressor power requirement to run an atmospheric fluidized bed coal combustor under the following conditions:

(a) Distributor pressure drop: $\Delta p_d = 3\,\text{kPa}$
(b) Distributor pressure drop: $\Delta p_d = 10\,\text{kPa}$
(c) 50% of the required air bypasses the bed and is introduced into the freeboard to burn the volatile gases released in the bed by the coal. Take $\Delta p_d = 3\,\text{kPa}$.

Data

Entering air:	$p_0 = 101\,\text{kPa}$, $T_0 = 20°\text{C}$, $\gamma = 1.4$
Across the bed:	$\Delta p_b = p_2 - p_3 = 10\,\text{kPa}$
At the bed exit:	$p_3 = 103\,\text{kPa}$
Coal:	feed rate = 8 tons/hr
	gross heating value = 25 MJ/kg
	air at standard conditions needed = 10 nm^3/kg (at 15% excess)
Efficiency:	of compressor $\eta = 0.75$
	of power plant (electric power/heat in coal) = 36%

We use the nomenclature of Fig. E3.

SOLUTION

(a) For $\Delta p_d = 3\,\text{kPa}$ or $\Delta p_d = 0.3\,\Delta p_b$: From Eq. (3), this represents a reasonable distributor pressure drop for good fluidization. We next evaluate pressures and inlet flow rates

$$p_0 = 101\,\text{kPa}$$
$$p_3 = 101 + 2 = 103\,\text{kPa}$$
$$p_2 = 103 + 10 = 113\,\text{kPa}$$
$$p_1 = 113 + 3 = 116\,\text{kPa}$$

and

$$v_0 = (8000\,\text{kg/hr})(10\,\text{m}^3/\text{kg})\left(\frac{293}{273}\right)(\text{hr}/3600\,\text{s}) = 23.85\,\text{m}^3/\text{s}$$

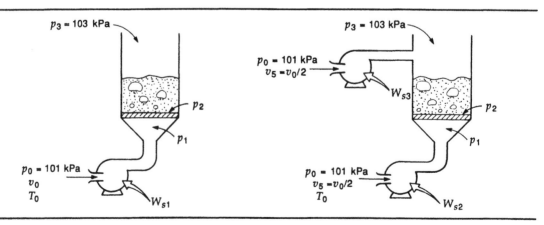

FIGURE E3
Different air feed arrangements for a fluidized coal combustor.

Thus the compressor power for this FBC is given by Eqs. (18) and (20) as

$$-\dot{W}_{s,1} = \frac{1.4}{1.4-1}\,(101\text{ kPa})(23.85\text{ m}^3/\text{s})\left[\left(\frac{116}{101}\right)^{(1.4-1)/1.4} - 1\right]\frac{1}{0.75}$$

$$= \underline{455\text{ kW}}\text{ (or }610\text{ hp)}$$

(b) For $\Delta p_d = 10$ kPa, this represents a distributor plate with excessive pressure drop, since $\Delta p_d = \Delta p_b$. Evaluating pressures gives

$$p_0 = 101\text{ kPa}$$
$$p = 103\text{ kPa}$$
$$p_2 = 113\text{ kPa}$$
$$p = 123\text{ kPa}$$

Following the same procedure as in part (a), we find

$$-\dot{W}_{s,\text{actual}} = \underline{651\text{ kW}}\text{ (or }873\text{ hp)}$$

Thus, the power requirement increases by almost 200 kW.

(c) For bypass into the freeboard of 50% of the air and $\Delta p_d = 3$ kPa, we have no change in pressure drops from part (a). Thus

$$p_0 = 101\text{ kPa}$$
$$p_3 = 103\text{ kPa} \qquad u_5 = \frac{23.85}{2} = 11.925\text{ nm}^3/\text{s}$$
$$p_2 = 113\text{ kPa}$$
$$p_1 = 116\text{ kPa}$$

So for the primary air, from part (a),

$$-\dot{W}_{s,\text{actual}} = \frac{455}{2} = 227.5\text{ kW}$$

For the air bypassed into the freeboard we need another blower. Its power requirement is again given by Eq. (18). Thus

$$-\dot{W}_{s,\text{actual}} = \frac{1.4}{1.4-1}\,(101)(11.925)\left[\left(\frac{103}{101}\right)^{(1.4-1)/1.4} - 1\right]\frac{1}{0.75}$$

$$= 31.6\text{ kW}$$

So the total power requirement for the two blowers is

$$-\dot{W}_{s,\text{actual,total}} = 227.5 + 31.6 = \underline{259\text{ kW}}$$

This design gives a 43% savings in pumping power over the design of part (a).

PROBLEMS

1. A perforated plate distributor is to be designed for a fluidized bed. Determine the fraction of open area needed and the relationship between orifice diameter and number of orifices per area.

    Data

    Solids: $\rho_s = 2\text{ g/cm}^3$, $\varepsilon_{mf} = 0.48$, $L_{mf} = 3\text{ m}$
    Gas: $\rho_g = 2 \times 10^{-3}\text{ g/cm}^3$, $\mu = 2 \times 10^{-4}\text{ g/cm}\cdot\text{s}$, $u_o = 60\text{ cm/s}$

    Take $d_t = 6\text{ m}$, $\Delta p_d = 0.3\,\Delta p_b$.

2. For the bed of Prob. 1, design an eight-hole tuyere distributor of the type shown in Fig. 3(b) without an inlet orifice. The tuyeres are to be on a square arrangement at least 10 cm apart.

3. For the bed of Prob. 1, design an eight-hole tuyere distributor of the type shown in Fig. 2(b). The tuyeres are to be on a square arrangement at least 10 cm apart. Also, since the bed solids are friable, the gas velocity issuing from the holes in the tuyere is restricted to 20 m/s.

4. A fixed bed of refractory spheres, $\varepsilon_m = 0.4$, sandwiched between two perforated plates is to serve as a distributor for a fluidized bed. Determine the diameter of spheres to be used if the thickness of the fixed bed is to be 20 cm and $\Delta p_d = 4$ kPa. To simplify the calculations, assume that the two perforated plates have large enough open areas so that they do not contribute to the pressure drop across the distributor.

5. Estimate the compressor power needed to send reactant gas into the bottom of a fluidized bed reactor. Also determine what fraction of this compressor power is used to overcome the frictional loss of the distributor.

Data

Solid:	$\rho_s = 1.5$ g/cm^3, $\varepsilon_{mf} = 0.5$, $L_{mf} = 4$ m
Gas entering the	
compressor:	20°C, 101 kPa, 20,000 m^3/hr
Gas at reactor exit:	1000 kPa

Use $\gamma = 1.4$, $\eta = 0.8$, $\Delta p_d = 0.3\,\Delta p_b$.

6. Calculate the temperature rise resulting from the compression of inlet gases of Example 2 and Prob. 5.

REFERENCES

1. J. Werther, _German Chem. Eng._, **1**, 166 (1978).
2. J.F. Davidson and B.O.G. Schüler, _Trans. Inst. Chem. Eng._, **38**, 335 (1960).
3. D. Harrison and L.S. Leung, _Trans. Inst. Chem. Eng._, **39**, 409 (1961).
4. L. Massimilla, in _Fluidization_, J.F. Davidson et al., eds., p. 133, Academic Press, New York, 1985; M. Filla, L. Massimilla et al., in _Fluidization V_, K. Østergaard and A. Sørensen, eds., p. 71, Engineering Foundation, New York, 1986.
5. J.G. Yates, P.N. Rowe, and D.J. Cheesman, _AIChE J._, **30**, 890 (1984); J.G. Yates, V. Bejek, and D.J. Cheesman, in _Fluidization V_, K. Østergaard and A. Sørensen, eds., p. 79, Engineering Foundation, New York, 1986.
6. X.T. Nguyen and L.S. Leung, _Chem. Eng. Sci._, **27**, 1748 (1972).
7. S.P. Sit, in _Fluidization V_, K. Østergaard and A.

Sørensen, eds., p. 39, Engineering Foundation, New York, 1986.
8. C.Y. Wen et al., in _Fluidization_, J.F. Davidson and D.L. Keairns, eds., p. 32, Cambridge Univ. Press, New York, 1978.
9. J. Liu et al., _Kagaku Kogaku Ronbunshu_, **9**, 102 (1983).
10. Y. Jin et al., in _Fluidization, Science and Technology_, M. Kwauk and D. Kunii, eds., p. 224, Science Press, Beijing, 1982.
11. M. Filla, M. Massimilla, and S. Vaccaro, _Int. J. Multiphase Flow_, **9**, 259 (1983).
12. I. Hirsan, C. Sishtla, and T.M. Knowlton, paper delivered at AIChE annual meeting, November 1981; T.M. Knowlton and I. Hirsan, in _Fluidization III_, J.R. Grace and J.M. Matsen, eds., p. 315, Plenum, New York, 1980.
13. J. Werther, in _Fluidization_, J.F. Davidson and

D.L. Keairns, eds., p. 7, Cambridge Univ. Press, New York, 1978.

14. F.J. Zuiderweg, in *Proc. Int. Symp. on Fluidization*, A.A.H. Drinkenburg, ed., p. 739, Netherlands Univ. Press, Amsterdam, 1967.

15. J.W. Hiby, *Chem.-Ing.-Techn.*, **36**, 228 (1964).

16. R. Siegel, *AIChE J.*, **22**, 590 (1976).

17. S. Mori and A. Moriyama, *Int. Chem. Eng.*, **18**, 245 (1978).

18. D. Sathiyamoorthy and C.S. Rao, *Powder Technol.*, **24**, 215 (1979).

19. A.E. Qureshi and D.E. Creasy, *Powder Technol.*, **22**, 113 (1979).

20. G.L. Sirotkin, *Chem. Petrol. Eng.*, **15**, 113 (1979).

21. Y.F. Shi and L.T. Fan, *AIChE J.*, **30**, 860 (1984).

22. A.B. Whitehead et al., in *Proc. Int. Symp. on Fluidization*, A.A.H. Drinkenburg, ed., pp. 284, 802, Netherlands Univ. Press, Amsterdam, 1967; in *Fluidization*, J.F. Davidson and D. Harrison, eds., p. 781, Academic Press, New York, 1971.

5

Bubbles in
Dense Beds

Chapter 3 presented the essentials of fluidized contacting and mapped out the expected flow regimes in these systems. Chapter 4 then treated in detail the gas-solid interaction in the gas entry region of the bed. This chapter and the next three consider gas-solid contacting in the main portion of the bed. Bubbling and slugging are taken up in this chapter and the next, and other types of contacting are considered in Chaps. 7 and 8.

A dense bubbling fluidized bed has regions of low solid density, sometimes called gas pockets or voids. We call these regions *gas bubbles* or, simply, *bubbles*. The region of higher density we call the *emulsion* or *dense phase*. This chapter treats, in turn, single bubbles rising in fluidized beds, the interaction of pairs and streams of bubbles, and bubble formation, and ends with a consideration of slugging.

Single Rising Bubbles

Rise Rate of Bubbles

In a surprising number of ways a bubbling bed behaves like a bubbling liquid of low viscosity.

1. The shapes of bubbles are somewhat alike, close to spherical when small, flattened and distorted when larger, and spherical cap-shaped when large.

2. For both systems, small bubbles rise slowly and large bubbles rise rapidly.

3. For both systems, a train of bubbles may coalesce to give larger bubbles. The interaction of a train gives a different rise velocity, and the direction of this change is the same in both cases.

4. Wall effects act in the same direction on the bubbles' rise velocity.

5. The bubbles' rise velocity depends on the same factors and is described by similar expressions in the two systems. Thus, in a liquid, the rate of rise of a large spherical cap bubble is well described by the theoretical expression of

Davies and Taylor [1],

$$u_{br} = \tfrac{2}{3}(gR_n)^{1/2} \tag{1}$$

where R_n is the radius of curvature at the nose of the bubble. More convenient- ly, the experimental rate of rise, as measured by Davidson et al. [2], Harrison and Leung [3], Reuter [4], Rowe et al. [5], Toei et al. [6], and summarized by Clift and Grace [7], can be expressed as

$$u_{br} = 0.711(gd_b)^{1/2} \tag{2}$$

where d_b is the diameter of sphere having the same volume as the spherical cap bubble and where wall effects do not intrude.

For calculations in this and the following chapters, we will take the velocity of rise of single bubbles in fluidized beds to be

$$u_{br} = 0.711(gd_b)^{1/2}, \qquad \frac{d_b}{d_t} < 0.125 \tag{3}$$

Wall effects retard the rise of bubbles when $d_b/d_t > 0.125$. Somewhat similar to Wallis's [8] suggestion, we take

$$u_{br} = [0.711(gd_b)^{1/2}]1.2\exp\left(-1.49\,\frac{d_b}{d_t}\right), \qquad 0.125 < \frac{d_b}{d_t} < 0.6 \tag{4}$$

For $d_b/d_t > 0.6$, the bed should be considered not to be bubbling, but slugging, as shown in Fig. 15. This regime is considered at the end of this chapter.

6. Further experiments in bubbling beds indicate roughly that all gas in excess of that needed to just fluidize the bed passes through the bed as bubbles, and the emulsion phase remains close to minimum fluidizing conditions.

7. As opposed to gas-liquid systems, there is an interchange of gas between the bubble and dense phase in fluidized beds.

These findings show that a bed at minimum fluidizing conditions can be treated as a liquid of low or negligible viscosity. At higher velocity the excess gas goes through the bed as bubbles, which rise as in an ordinary liquid of low viscosity. The voidage of a bed, not counting bubbles, remains close to ε_{mf}. At minimum fluidizing conditions, the solids are relatively quiescent. At higher gas velocities the rising bubbles cause the observed churning, mixing, and flow of solids.

The Davidson Model for Gas Flow at Bubbles

The first significant breakthrough was made by Davidson, whose elegant model successfully accounted for the movement of both gas and solids and the pressure distribution about rising bubbles. Extensions and alternative analyses have since been proposed, and these will be mentioned later; however, because of its simplicity and essential correctness, we shall concentrate on this model, a complete exposition of which is given by Davidson and Harrison [9]. This model was developed for two- and three-dimensional beds (a two-dimensional bed is one formed between two closely spaced parallel plates) and is based on the following postulates.

Postulate 1. A gas bubble is solid-free and circular in shape, and thus is spherical in the three-dimensional case and cylindrical in the two-dimensional case.

Postulate 2. As a bubble rises, particles move aside, as would an incompressible inviscid fluid of bulk density $\rho_s(1 - \varepsilon_{\mathrm{mf}})$.

Postulate 3. The gas flows in the emulsion phase as an incompressible viscous fluid; hence, the relative velocity between gas and solid must satisfy Darcy's law. Thus for any direction x,

$$(u_{\mathrm{gas}} - u_{\mathrm{solid}})_x = -K \frac{\partial p}{\partial x}$$

The following reasonable boundary conditions are also used in the development:

- Far from the bubble the undisturbed pressure gradient exists. This is given by Eq. (3.17).
- The pressure in the bubble is constant.

These postulates and boundary conditions are sufficient to give the flow pattern for solids and for gas, as well as the pressure distribution, all in the vicinity of the rising bubble. Thus, Postulate 2 allows us to find the motion of solids directly from potential flow theory. This is shown in Fig. 1.

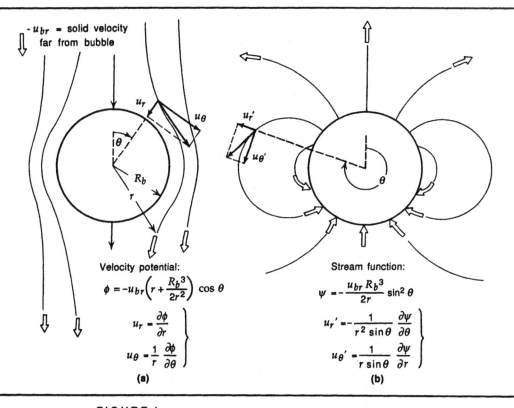

Velocity potential:

$$\phi = -u_{br}\left(r + \frac{R_b{}^3}{2r^2}\right) \cos\theta$$

$$\left. \begin{aligned} u_r &= \frac{\partial \phi}{\partial r} \\[4pt] u_\theta &= \frac{1}{r}\frac{\partial \phi}{\partial \theta} \end{aligned} \right\}$$

(a)

Stream function:

$$\psi = -\frac{u_{br} R_b{}^3}{2r} \sin^2\theta$$

$$\left. \begin{aligned} u_r{}' &= -\frac{1}{r^2 \sin\theta}\frac{\partial \psi}{\partial \theta} \\[4pt] u_\theta{}' &= \frac{1}{r \sin\theta}\frac{\partial \psi}{\partial r} \end{aligned} \right\}$$

(b)

FIGURE 1
The motion of solids in the vicinity of a rising three-dimensional bubble, from potential flow theory: (a) as viewed by an observer moving with the bubble; (b) as viewed by a stationary observer; from the Davidson model.

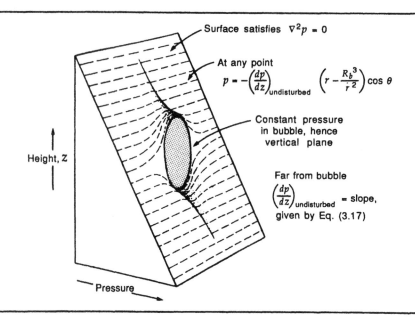

Surface satisfies $\nabla^2 p = 0$

At any point
$$p = -\left(\frac{dp}{dz}\right)_{\text{undisturbed}} \left(r - \frac{R_b{}^3}{r^2}\right) \cos \theta$$

Constant pressure in bubble, hence vertical plane

Far from bubble
$\left(\dfrac{dp}{dz}\right)_{\text{undisturbed}} =$ slope, given by Eq. (3.17)

Height, Z

Pressure

FIGURE 2
Representation of the pressure distribution in the vicinity of a three-dimensional bubble; from the Davidson model.

Postulate 3 shows that the pressure distribution around a bubble must satisfy the Laplace equation. Postulate 3 along with the boundary conditions and other postulates give the complete pressure distribution. This distribution is sketched in Fig. 2. There we see that the pressure in the lower part of the bubble is lower than that in the surrounding bed, whereas in the upper part it is higher. Thus, gas flows into the bubble from below and leaves at the top. This is verified when the velocity distribution of gas is calculated as shown in Fig. 3. The resulting flow pattern is solely dependent on the relative velocity of bubble u_{br} with emulsion gas, $u_f = u_{\text{mf}}/\varepsilon_{\text{mf}}$. Figure 3 also shows a distinct difference in the gas flow pattern, depending on whether the bubble rises faster or slower than the emulsion gas. Consider the following flow patterns.

The Cloudless or Slow Bubble: $u_{\text{br}} < u_f$. Here, the emulsion gas rises faster than the bubble; hence it uses the bubble as a convenient shortcut on its way through the bed. It enters the bottom of the bubble and leaves at the top. However, an annular ring of gas does circulate within the bubble, moving upward with it. The amount of this accompanying gas increases as the bubble velocity slows to the rise velocity of emulsion gas.

The Clouded or Fast Bubble: $u_{\text{br}} > u_f$. As with the slow bubble, emulsion gas enters the lower part of the bubble and leaves at the top. However, the bubble is rising faster than the emulsion gas; consequently, the gas leaving the top of the bubble is swept around and returns to the base of the bubble. The region around the bubble penetrated by this circulating gas is called the cloud. The rest of the gas in the bed does not mix with the recirculating gas but moves aside as the fast bubble with its cloud passes by.

Figure 3 shows that the transition from slow to fast bubble is smooth. The

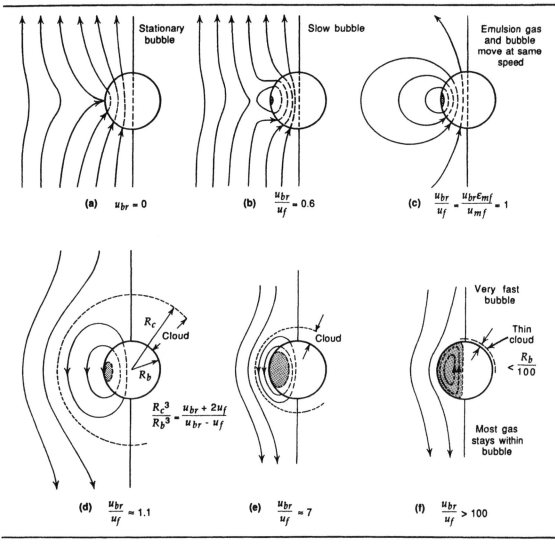

FIGURE 3
Gas streamlines near a single rising bubble. (a) Upper sketches show the slow cloudless bubble; (b) lower sketches show the fast clouded bubble. Only flow on the left side is shown; the right side is symmetric; from the Davidson model.

cloud is infinite in thickness at $u_{br} = u_f$, but thins with increasing bubble velocity. Its size is given by

$$\frac{R_c^2}{R_b^2} = \frac{u_{br} + u_f}{u_{br} - u_f} \qquad \text{for a two-dimensional bed} \qquad (5)$$

or

$$\frac{R_c^3}{R_b^3} = \frac{u_{br} + 2u_f}{u_{br} - u_f} \qquad \text{for a three-dimensional bed} \qquad (6)$$

Hence the ratio of cloud to bubble volume is

$$f_c = \frac{2u_f}{u_{br} - u_f} = \frac{2u_{mf}/\varepsilon_{mf}}{u_{br} - u_{mf}/\varepsilon_{mf}} \qquad \text{for two-dimensional bubbles} \qquad (7)$$

and

$$f_c = \frac{3u_f}{u_{br} - u_f} = \frac{3u_{mf}/\varepsilon_{mf}}{u_{br} - u_{mf}/\varepsilon_{mf}} \qquad \text{for three-dimensional bubbles} \qquad (8)$$

Flow Rate of Gas into and out of a Bubble. This theory also shows that the upward flow of gas into and out of the bubble is

$$v = 4u_{mf}R_bL = 4u_f\varepsilon_{mf}R_bL \qquad \text{for a two-dimensional bed of thickness } L \quad (9)$$

or

$$v = 3u_{mf}\pi R_b^2 = 3u_f\varepsilon_{mf}\pi R_b^2 \qquad \text{for a three-dimensional bed} \qquad (10)$$

This means that at its maximum cross section, the upward flow of gas through a two-dimensional bubble is $2u_{mf}$, and through a three-dimensional bubble it is $3u_{mf}$. The bubble is thus processing two or three times the amount of gas processed by the equivalent section of emulsion phase in the same time interval. For a stationary bubble this gas is all fresh; for a slow cloudless bubble it is partly recirculating gas, and for a fast clouded bubble all the gas being processed recirculates through the bubble and cloud.

Other Models for Gas Flow at Bubbles

Collins [10] and Stewart [11] chose postulates similar to Davidson's but used kidney-shaped bubbles with indented bases. Jackson [12] retained the spherical bubble but allowed the voidage of the emulsion phase to vary. Murray [13] developed a somewhat similar model, and Stewart [11], in a useful table, compared the fundamental postulates of all of these models.

Comparison of Models with Experiment

Clouded and Cloudless Bubbles. When $u_{br} > u_f = u_{mf}/\varepsilon_{mf}$ (fast clouded bubble), Davidson predicts that a cloud of circulating gas should envelop the bubble as it rises up the bed (see Fig. 3(d)–(f)). Rowe et al. [14] show photographs of single bubbles formed by injecting nitrogen dioxide, a visible brownish gas, into a two-dimensional bed to create bubbles. In a typical fast bubble, shown in Fig. 4(a), one clearly sees the cloud region that surrounds the bubble as it rises through the bed. The shape of the cloud deviates somewhat from the Davidson predictions at the back of the bubble; nevertheless, the presence of the cloud clearly verifies the essential predictions of this model.

When $u_{br} < u_f = u_{mf}/\varepsilon_{mf}$ (slow bubble), Davidson predicts that the bubble should be cloudless and that emulsion gas would pass through the bubble just once as it overtakes the bubble (see Fig. 3(a)–(c)). Again, photographs by Rowe et al. [14] taken under these conditions verify the predictions. Figure 4(b), a sketch of one of these photographs, shows another important feature of bubbling beds, namely that the flow of gas through the emulsion far from any bubble is essentially laminar.

Cloud Thickness. Lignola et al. [15] and Hatano and Ishida [16] compared measured cloud thicknesses at the nose of bubbles with the predictions of the various models (Fig. 5). The dispersion of the data does not allow us to say which model best fits the data.

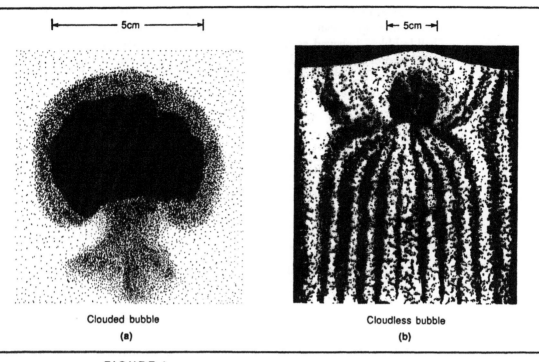

Clouded bubble
(a)

Cloudless bubble
(b)

FIGURE 4
Photographs by Rowe et al. [14] show the flow pattern of gas around rising bubbles. (a) For $u_{br} = 2.4 u_f$, from the sketch one clearly sees the cloud surrounding the fast-rising bubble. (b) For $u_{br} < u_f$, streaklines of tracer gas show the emulsion gas overtaking the slow-rising bubble.

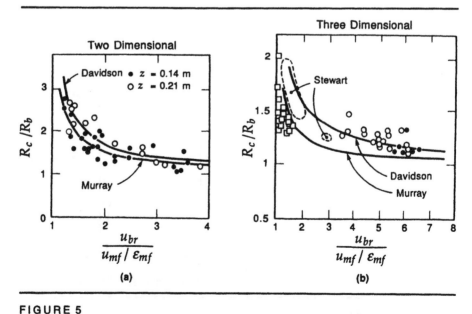

FIGURE 5
Cloud radius at the nose of three-dimensional bubbles, theory versus experiment; (a) adapted from Lignola et al. [15]; (b) from Hatano and Ishida [16].

Pressure Distribution. Reuter [4] measured gas pressures around bubbles sliding up a wall of a rectangular bed. Stewart [11] argued that these "half-bubbles" behaved as three-dimensional bubbles and, on this basis, compared Reuter's findings with the various models, as shown in Fig. 6.

Above the bubble, only Davidson's model fits the data well. In addition, only Davidson's model can account for pressure recovery in the emulsion below the bubble.

Gas Throughflow in Bubbles. Hilligardt and Werther [17] measured the upflow of gas through cloudless bubbles in freely bubbling beds and compared it with the predictions of the Davidson model. They found the following upflow velocity through the bubble's maximum cross section.

	Experiment	*Theory* (see Eqs. (9) and (10))
Two-dimensional bubbles	$1.84u_{mf}$	$2u_{mf}$
Three-dimensional bubbles	$2.70u_{mf}$	$3u_{mf}$

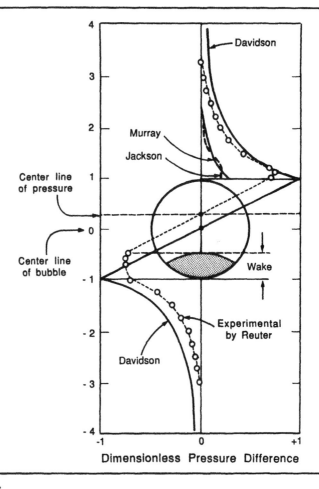

FIGURE 6
Pressure distribution in the vicinity of a rising bubble, comparison of measurements by Reuter [4] with the three theories; adapted from Stewart [11].

This represents a difference of about 10%, a reasonable fit considering the many uncertainties of freely bubbling beds.

Evaluation of Models for Gas Flow at Bubbles

The models discussed above explain and make sense of the main features of gas bubbles in dense fluidized beds—why they even exist, in that it is the upflow of gas that keeps the roof of the bubble from collapsing. These models also explain why one observes two distinctly different kinds of bubbles, namely the slow cloudless bubble that simply acts as a shortcut for the throughflow of emulsion gas and the fast clouded bubble with its captive vortex ring of rapidly recirculating gas.

In comparing models, we see that in some aspects the Davidson model does not fit the data as well as some of the other models do. First, its spherical bubble is less representative of reality than the kidney-shaped bubbles of some of the other models. As a consequence, it does not fit the back end of the bubble and its overall cloud thickness is greater than that observed and predicted by the other models. In other aspects, this model fits the data as well or better than the other models do.

The Davidson model has one overwhelming advantage over all other models: it is much simpler in every way and therefore much easier to use. Thus, we find it to be the best suited for analyzing the complicated phenomena that we will encounter in fluidizing engineering. We use this model in one way or other throughout this book.

The Wake Region and the Movement of Solids at Bubbles

Solids move out of the way as a bubble rises. At the front half of the bubble, this movement is well represented by the Davidson model predictions, shown in Fig. 1. However, typical bubbles are not spherical but have a flattish, or even concave, base, as shown in Fig. 4. The region just below the bubble is the *wake region* and it most likely forms because the pressure in the lower part of the bubble is less than in the nearby emulsion. Thus, gas is drawn into the bubble, causing instability, partial collapse of the bubble, and turbulent mixing. For fast clouded bubbles, this is the reason for the observed leakage of circulating bubble gas into the wake, as shown in Fig. 4(a). This turbulence also results in solids being drawn up behind the bubble and forming the wake region.

Figure 7, drawn from a sequence of photographs, clearly shows that a rising bubble drags a wake of solids up the bed behind it, and that the wake sheds and leaks solids as it rises. This means that there is a continuous, although not necessarily large, interchange of solids between wake and emulsion.

Using x-ray photography, Rowe and Partridge [18] observed the wake angle, θ_w, as well as the wake volume, V_w, defined as the volume occupied by the wake within the sphere that circumscribes the bubble, as shown in Fig. 8. The wake fraction, defined as

$$f_w = \frac{V_w}{V_b} \tag{11}$$

is shown for various solids in Fig. 8. Note that the wake angle decreases with

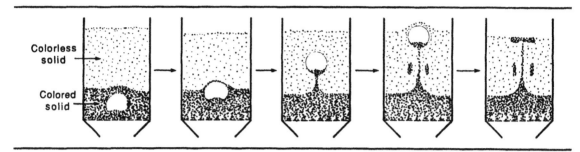

FIGURE 7

Sketches of photographs by Rowe and Partridge [18] showing the entrainment of solids by a rising bubble.

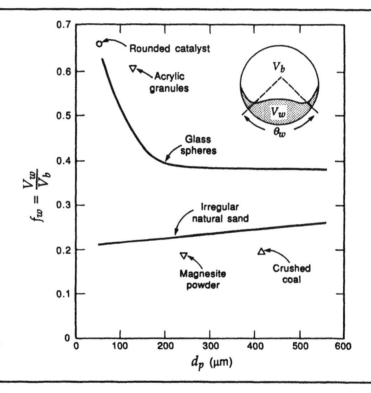

FIGURE 8

Wake angle θ_w and wake fraction of three-dimensional bubbles at ambient conditions; evaluated from x-ray photographs by Rowe and Partridge [18].

smaller particles, meaning that bubbles are flatter in small particle beds. Kawabata et al. [19] reports a similar direction of change as the pressure on the system is increased, up to 8 bar.

Solids within Bubbles

So far, nothing has been said about the possibility that bubbles contain solids. However, three groups of investigators, using different techniques, independently found that solids are present in bubbles. Toei et al. [6] photographed bubbles, using a lens with extremely shallow depth of field. Hiraki et al. [20] used the Tyndall effect of dispersed particles illuminated by a thin beam of light. Kobayashi et al. [21] measured the bulk density of rising bubbles with a sensitive microphototransistor. All three groups found that bubbles contain 0.2–1.0% solids by volume.

Aoyagi and Kunii [22] used a rapid combustion technique in their study of dispersed particles, as follows. They injected bubbles of air into a very hot bed of carbon particles fluidized at u_{mf} by air or nitrogen, reasoning that any particle finding itself in the air bubble would ignite and become white hot and visible. In such experiments, ignited particles are clearly seen, as shown in Fig. 9. In addition, an analysis of ciné sequences shows that these bright particles did not fall from the roof of the bubbles but flew upward from the base of the bubbles.

The existence of particles dispersed in rising bubbles has been ignored in just about all kinetic models. Even though their volume fraction might be as small as 0.1%, they could enormously influence practical operations in which rapid kinetic operations occur. For example, for highly exothermic catalytic reactions, which are often effected in fluidized beds, catalyst particles may

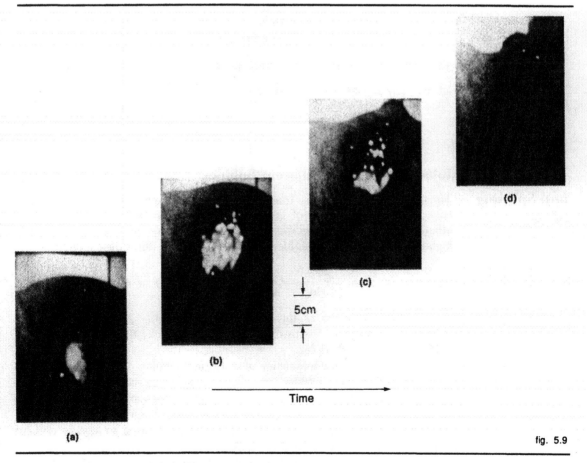

fig. 5.9

FIGURE 9
Glowing particles dispersed in an air bubble rising in a nitrogen-fluidized bed of carbon at 707°C; from Aoyagi and Kunii [22].

"ignite" in bubbles of fresh reactant. This may result in changed selectivity or a slow progressive deterioration of the catalyst, particle by particle. We consider this phenomena in Chap. 17.

EXAMPLE 1

Characteristics

of a Single

Bubble

A shot of gas is injected into a 60-cm ID incipiently fluidized bed of 300-μm sand for which $u_{mf} = 3$ cm/s and $\varepsilon_{mf} = 0.5$. A 5-cm bubble forms. For this bubble find

 (a) The rise velocity, u_{br}.
 (b) The cloud thickness, $R_c - R_b$.
 (c) The volume of wake to volume of bubble, f_w.

SOLUTION

(a) Since $d_b/d_t = 5/60 < 0.125$, Eq. (3) gives the rise velocity of the bubble. Thus

$$u_{br} = 0.711(980 \times 5)^{1/2} = \underline{49.8\ cm/s}$$

(b) From Eq. (6), noting that $R_b = 2.5$ cm and $u_f = u_{mf}/\varepsilon_{mf} = 3/0.5 = 6$ cm/s,

$$\frac{R_c}{R_b} = \left[\frac{49.8 + 2(6)}{49.8 - 6}\right]^{1/3} = 1.12$$

$$\therefore R_c = 1.12(2.5) = 2.80\ cm$$

$$\therefore R_c - R_b = 2.80 - 2.5 = \underline{0.30\ cm}$$

This means that the cloud surrounding the bubble is about five particles thick.

(c) From Fig. 8, the wake fraction is

$$f_w = \frac{V_w}{V_b} = \underline{0.24}$$

Coalescence and Splitting of Bubbles

Interaction of Two Adjacent Bubbles

The formation of bubbles from an orifice was discussed in Chap. 4. Now consider pairs and chains of bubbles issuing from an orifice in a bed that is otherwise at minimum fluidizing conditions. Assuming that Eq. (4.1) holds at higher orifice flows, the bubble frequency just above the orifice should be

$$n_b = \frac{v_{or}}{V_b} = \frac{g^{3/5}}{1.138 v_{or}^{1/5}} = \frac{54.8}{v_{or}^{1/5}}, \qquad [s^{-1}] \quad \text{with } v \text{ in cm}^3/s \tag{12}$$

For $v_{or} = 200$–2000 cm^3/s, this equation gives $n_b = 19 - 12$ s^{-1}, as opposed to the observed $n_b \cong 7$ s^{-1}. This difference is explained by the rapid formation of bubble doublets and triplets.

Next, consider the interaction of two rising bubbles, one trailing the other. In water, when the bubbles are close enough together, the trailing one accelerates and is drawn into the leader. This phenomenon may be explained by supposing that the trailing bubble accelerates when it enters the wake of the leading bubble. The same kind of reasoning can be used to explain vertical coalescence of bubbles in fluidized beds.

Label the leading and trailing bubbles 1 and 2, respectively. Then, Fig. 10 shows that perceptible interaction starts when the distance between the nose of bubble 2 and the center of bubble 1 is less than three times the radius of the

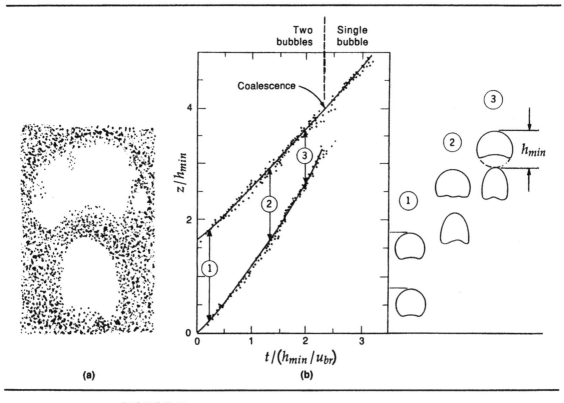

FIGURE 10

Coalescence of bubbles; from Toei et al. [6]: (a) sketch from an x-ray photograph; (b) dimensionless correlation for coalescence.

circumscribed circle around bubble 1. At this point, bubble 2 starts to lengthen as it reaches into the wake of bubble 1, which in turn starts to flatten.

When not in vertical alignment, the lower bubble first drifts sideways behind the upper bubble and then rises into the upper bubble. In a similar manner, a large bubble rising past many smaller ones will sweep them up, always by absorption of the smaller bubbles through the base of the larger bubble.

Clift and Grace [7] refer to numerous studies of the many other aspects of coalescing bubbles, including the modeling of gas flow around bubble pairs and the coalescence of bubble chains.

Coalescence, Bubble Size, and Bubble Frequency

Experiments show that bubble size in fluidized beds increases with gas velocity and with height above the distributor, and varies widely from system to system, as shown in Fig. 11. This is to be expected, since an important variable, the excess gas flow rate, measured by $u_o - u_{mf}$, is not accounted for in this figure.

Bubble frequency should also be related to the excess gas flow rate; nevertheless, Fig. 12 shows that the bubble frequency versus height forms a narrow band of data. This suggests that a somewhat similar mechanism of

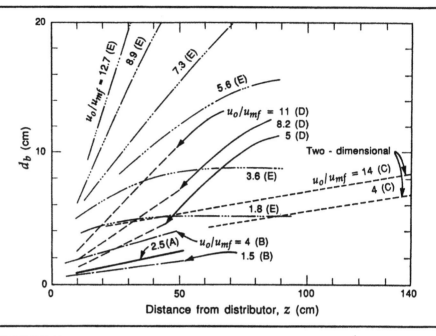

FIGURE 11
Size of bubbles at different levels in beds of Geldart **A** and **B** particles; from Kunii and Levenspiel [23]: (A) Yasui and Johanson [24]; (B) Toei et al. [6]; (C) Hiraki et al. [20], two-dimensional; (D) Kobayashi et al. [21]; (E) Hiraki et al. [20].

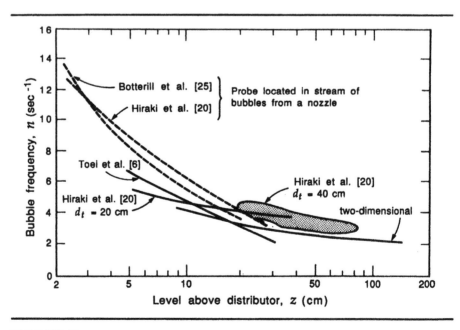

FIGURE 12
Frequency of bubbles passing a point in a fluidized bed at various gas flow rates; from Kunii and Levenspiel [23].

bubble coalescence acts in all systems and that it is important to find the initial bubble frequency near the bottom of the bed.

Various models have been proposed to account for the coalescence of bubbles in freely bubbling beds. Some only consider the coalescence of leading and following bubbles in a bubble chain; others only consider lateral coalescence between side-by-side bubbles; still others try to account for both forms of coalescence with computer simulation models.

Actually, one would expect all forms of coalescence to act. For example, in the lower portion of beds supported by perforated plate distributors, vertical coalescence of the many tiny bubbles of a bubble chain should predominate. Higher up the bed, as the bubbles grow larger, lateral coalescence should become important. Bubble splitting should also occur, and we consider this next. Finally, in large beds, circulation of solids may affect the bubble size distribution.

The overall characteristics of bubbling beds and a comparison of theory with experiment will be taken up in Chap. 6.

Splitting of Bubbles and Maximum Bubble Size

Rowe [14] observed that the roof of a bubble sometimes develops a downward cusp, which then frequently grows rapidly to cause the bubble to split vertically. When this knifing action slices off a small daughter bubble, it is almost immediately reabsorbed in the mother bubble. When the two bubbles formed are nearly equal, the larger one first grows at the expense of the smaller, which is then absorbed by the larger bubble, as shown in Fig. 13. In some cases, the faster-rising larger bubble is able to pull away from the smaller bubble, leaving two bubbles. In beds of fine particles, recoalescence is less frequent than in large particle beds.

Toei et al. [6] measured the frequency of bubble splitting in two-dimensional beds of uniform particles ranging from 210 to 360 μm ($u_{mf} = 3$–10 cm/s) and concluded that the frequency is inversely proportional to u_{mf},

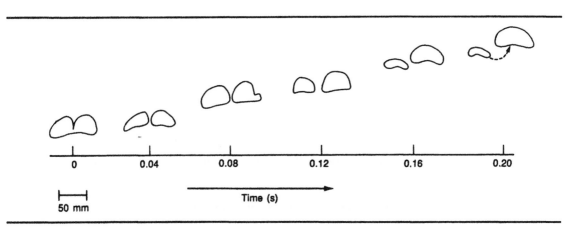

0   0.04   0.08   0.12   0.16   0.20

Time (s)

50 mm

FIGURE 13
Splitting of single bubbles caused by knifing, 60-μm Ballotini. X-ray sequence sketched from Rowe [14].

ranges from 3 to 20 s^{-1}, and is almost independent of bubble size, which ranged from 3 to 13 cm.

The consequence of coalescence and splitting is that a maximum bubble size may be present in the bed. Since bubble splitting is more frequent in Geldart **A** solids and less frequent in beds of larger particles, the maximum bubble size is large in beds of coarse solids but small in beds of fine particles. This should be kept in mind when we examine the overall bed properties in the next chapter.

Bubble Formation above a Distributor

Low Gas Flow Rate. If the number of orifices per unit area is N_{or} [cm^{-2}] and all the gas in excess of u_{mf} forms bubbles of equal size, the volumetric flow rate of gas from each of the orifices, v_{or}, is found from the expression

$$u_o - u_{mf} = v_{or} N_{or} \qquad . \qquad (13)$$

For a low enough flow rate so that the initial bubbles from adjacent orifices are not big enough to touch each other, or $d_{b0} < l_{or}$, the size of bubble that just forms is given by the single orifice expression of Eq. (4.1), which can be rewritten as

$$d_{b0} = 1.30 \frac{v_{or}^{0.4}}{g^{0.2}}, \qquad [cm] \qquad (14)$$

Combining Eqs. (13) and (14) gives the initial bubble size as

$$d_{b0} = \frac{1.30}{g^{0.2}} \left[\frac{u_o - u_{mf}}{N_{or}} \right]^{0.4}, \qquad d_{b0} \le l_{or}, \quad [cm] \qquad (15)$$

If l_{or} is the spacing between adjacent holes, then

$$N_{or} = \frac{1}{l_{or}^2} \qquad \text{for a square array of holes} \qquad (16)$$

$$N_{or} = \frac{2}{\sqrt{3} l_{or}^2} \qquad \text{for an equilateral triangle array of holes} \qquad (17)$$

Chiba and Kobayashi [26], using a different approach, came up with similar results.

High Gas Flow Rate. When the initial bubbles are so big that they touch and overlap when formed, then Eq. (15) cannot be used. In this case, we must take the initial bubble size to be that which just accommodates the imposed gas flow and views the neighboring bubbles as just touching. For an equilateral triangle array of touching bubbles, this condition is given by

$$d_{b0}^2 = \frac{2}{\sqrt{3} N_{or}'} \qquad (18)$$

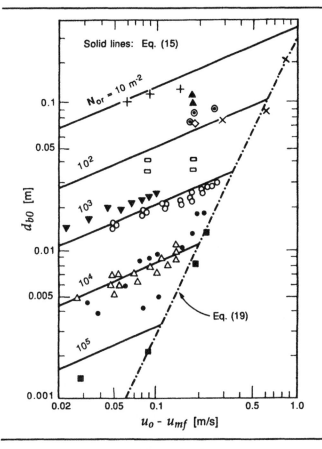

FIGURE 14
Comparison of theory with measurements of initial bubble size generated at perforated plate distributors having a triangular array of orifices. Data taken from Miwa et al. [27], Mori and Wen [28], and Glicksman et al. [30].

where N'_{or} represents the fictitious orifice spacing that corresponds to these touching bubbles.

Replacing N_{or} in Eq. (15) by N'_{or} gives the initial bubble size at these higher gas flow rates as

$$d_{b0} = \frac{2.78}{g}\,(u_o - u_{mf})^2, \qquad \text{with } d_{b0} > l_{or}, \quad [\text{cm}] \tag{19}$$

This analysis parallels that of Miwa et al. [27]. Figure 14 shows that the analysis fits the data well; in particular, note the transition between Eqs. (15) and (19).

Porous Plate Distributor. If we view a porous plate distributor as a perforated plate with numerous tiny triangular arranged holes, then Eq. (19) applies equally well to it.

EXAMPLE 2

Initial Bubble Size at a Distributor

Estimate the size of bubbles forming above the following distributors in a bed where $u_0 = 15\,cm/s$ and $u_{mf} = 1\,cm/s$.

(a) porous plate
(b) perforated plate with a triangular array of orifices having a 2-cm pitch

SOLUTION

(a) For a porous plate, Eq. (19) gives

$$d_{b0} = \frac{2.78}{980}(15-1)^2 = \underline{0.56\ cm}$$

(b) For a perforated plate with a triangular spacing of holes, Eq. (17) gives

$$\frac{1}{N_{or}} = \frac{\sqrt{3}}{2}(2)^2 = 3.46\ cm^2$$

First guessing that the initial bubble size will be smaller than the hole spacing, we use Eq. (15) to find

$$d_{b0} = 1.30(980)^{-0.2}[(15-1)(3.46)]^{0.4} = \underline{1.54\ cm}$$

Since the initial bubble size is smaller than the hole spacing, our guess was right and we accept this result.

Slug Flow

In fluidizing a tall, narrow bed of solids, bubbles formed at the distributor may grow to the bed diameter to form *slugs*. For beds of fine particles of good fluidity, particles will rain at the bed wall to match the rise rate of these slugs, as shown in Fig. 15(a). These are called *axial slugs*. At higher gas velocity and with either angular particles or rough vessel wall, the rising slugs tend to adhere to and slide up the wall, as shown in Fig. 15(b), according to Clift and Grace [7]. These are called *wall slugs*.

With larger Geldart **D** solids, another mode of slugging is seen, as shown in Fig. 15(c). Here the bed separates into slices of emulsion separated by gas. These slices of gas and of emulsion rise up the bed, matched by a continuous raining of solids from slice to slice. The topmost slice of emulsion, unreplenished by raining solids, eventually disappears, while new slices and slugs form at the bottom of the bed. These are called *flat slugs*.

In liquid-solid systems, Stewart and Davidson [33] analyzed the forces holding a slug in place against a downflow of water flowing at u_{br} and found, theoretically, that

$$u_{br} = 0.361(gd_t)^{1/2} \tag{20}$$

Experiments with various liquids in columns of different diameters show good agreement with this equation. For larger-diameter beds and low-viscosity liquids, a constant of 0.35 in this equation probably best fits the data.

Equation (20), developed for liquid-solid systems, was then tested on fluidized beds (see Stewart and Davidson [33] for details) by injecting *single slugs* into incipiently fluidized beds and measuring their rise velocities. Generally, the data agreed remarkably well with the equation for gas-liquid systems. Thus, for slugs in fluidized beds, we take

$$u_{br} = 0.35(gd_t)^{1/2} \tag{21}$$

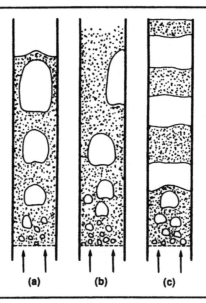

FIGURE 15
Types of slugs formed in fluidized beds: (a) axial slugs—fine smooth particles; (b) wall slugs—fine rough particles, rough walls, high velocity; (c) flat slugs—large particles (Geldart **D**).

For a *continuous introduction of gas* into the narrow bed, Stewart and Davidson reasoned that the excess gas beyond u_{mf}, thus $u_o - u_{mf}$, would push the slugs or slices of solids up the bed at a rise velocity larger than u_{br}, and be given by

$$u_b = \text{const}(u_o - u_{mf}) + 0.35(gd_t)^{1/2}, \tag{22}$$

Ormiston et al. [34] tested this equation with experiment and found that the constant was about 1.

Stewart and Davidson [33] also stated that below the following bubble-rise velocity slugging should not take place:

$$u_{b,ms} = u_{mf} + 0.07(gd_t)^{1/2} \tag{23}$$

Baeyens and Geldart [35] carried out experiments in air fluidized beds, using four column diameters ($d_t = 5$–30 cm), various solids ($\rho_s = 0.85$–2.8 g/cm^3), and a wide range of mean particle size ($\bar{d}_p = 55$–3380 μm). They found that, except for the smallest column diameter, neither particle size nor size distribution had any effect on slugging. They gave the height in the bed at which complete slugging sets in as

$$z_s = 60d_t^{0.175}, \quad [\text{cm}] \tag{24}$$

We conclude, then, that slugging should be the mode of contacting in tall beds if the superficial gas velocity is in excess of u_{ms}, given by Eq. (23), and should set in at a height z_s above the distributor, given by Eq. (24). Beds shallower than z_s should show no slugging.

PROBLEMS

1. A shot of gas is injected into a 30-cm ID incipiently fluidized bed of crushed solid ($d_p \cong 500 \, \mu$m) for which $u_{mf} = 25$ cm/s and $\varepsilon_{mf} = 0.44$. A 3-cm bubble forms. Determine
    (a) The rise velocity of the bubble.
    (b) The cloud thickness about the bubble.
    (c) The volume ratio of wake to bubble.

2. Repeat Prob. 1 for a larger shot of gas equivalent to a 12-cm bubble.

3. Repeat Prob. 1 for an even larger shot of gas equivalent to a 20-cm bubble.

4. Calculate the initial size of bubbles forming above a porous distributor plate for a bed wherein $u_0 = 35$ cm/s and $u_{mf} = 3$ cm/s.

5. Calculate the initial size of bubbles forming in the bed of Prob. 4 if the porous plate distributor is replaced by a perforated plate with a square array of 3-mm holes spaced 1 cm apart.

6. The upflow of gas is progressively increased through a 50-cm-deep bed of particles in a 30-cm ID vessel. At $u_0 = 3$ cm/s, the bed just fluidizes.
    (a) At what superficial gas velocity will the bed begin to slug?
    (b) At what height in the bed will slugs just appear?

7. Repeat Prob. 6 for a bed 3 m deep.

8. For a gas velocity of $30u_{mf}$ in the bed of Prob. 7, find the rise velocity of the slugs or bubbles near the top surface of the bed.

REFERENCES

1. R.M. Davies and G.I. Taylor, *Proc. Roy. Soc.*, **A200**, 375 (1950).

2. J.F. Davidson et al., *Trans. Inst. Chem. Eng.*, **37**, 323 (1959); *Ann. Rev. Fluid Mech.*, **9**, 55 (1977).

3. D. Harrison and L.S. Leung, *Trans. Inst. Chem. Eng.*, **39**, 409 (1961); **40**, 146 (1962).

4. H. Reuter, *Chem.-Ing.-Techn.*, **35**, 98, 219 (1963).

5. P.N. Rowe and B.A. Partridge, *Trans. Inst. Chem. Eng.*, **43**, T157 (1965); P.N. Rowe and R. Matsuno, *Chem. Eng. Sci.*, **26**, 923 (1971); P.N. Rowe and H. Masson, *Trans. Inst. Chem. Eng.*, **59**, 177 (1981).

6. R. Toei et al., *Kagaku Kogaku*, **29**, 851 (1965); *Mem. Fac. Eng. Kyoto Univ.*, **27**, 475 (1965); in *Proc. Int. Symp. on Fluidization*, A.A.H. Drinkenburg, ed., p. 271, Netherlands Univ. Press, Amsterdam, 1967.

7. R. Clift and J.R. Grace, in *Fluidization*, 2nd ed., J.F. Davidson et al., eds., p. 73, Academic Press, New York, 1985.

8. G.B. Wallis, *One-Dimensional Two-Phase Flow*, McGraw-Hill, New York, 1969.

9. J.F. Davidson and D. Harrison, *Fluidized Particles*, Cambridge Univ. Press, New York, 1963.

10. R. Collins, *Chem. Eng. Sci.*, **20**, 788 (1965).

11. P.S.B. Stewart, *Trans. Inst. Chem. Eng.*, **46**, T60 (1968).

12. R. Jackson, *Trans. Inst. Chem. Eng.*, **41**, 22 (1963); in *Fluidization*, J.F. Davidson and D. Harrison, eds., p. 65, Academic Press, New York, 1971.

13. J.D. Murray, *J. Fluid Mech.*, **22**, 57 (1965).

14. P.N. Rowe, in *Fluidization*, J.F. Davidson and D. Harrison, eds., p. 121, Academic Press, New York, 1971; U.K. Atomic Energy Agency Research

Group Report AERE-R 4383 (1963); P.N. Rowe, B.A. Partridge, and E. Lyall, *Chem. Eng. Sci.*, **19**, 973 (1964).

15. P.G. Lignola, G. Donsi, and L. Massimilla, *AIChE Symp. Ser.* **79**(**222**), 19 (1983).

16. H. Hatano and M. Ishida, *Kagaku Kogaku Ronbunshu*, **8**, 219 (1982); **10**, 184 (1984).

17. K. Hilligardt and J. Werther, in *Proc. 3rd World Cong. Chem. Eng.*, Tokyo, 1986.

18. P.N. Rowe and B.A. Partridge, *Proc. Symp. on Interaction between Fluids and Particles*, p. 135, *Inst. Chem. Eng.*, 1962; *Trans. Inst. Chem. Eng.*, **43**, 157 (1965).

19. J. Kawabata et al., *J. Chem. Eng. Japan*, **14**, 85 (1981).

20. I. Hiraki et al., *Kagaku Kogaku*, **29**, 846 (1965); **33**, 680 (1969).

21. H. Kobayashi, F. Arai, and T. Chiba, *Kagaku Kogaku*, **29**, 858 (1965).

22. M. Aoyagi and D. Kunii, *Chem. Eng. Comm.*, **1**, 191 (1974); K. Yoshida, T. Mii, and D. Kunii, in *Fluidization and Its Applications*, p. 512, Cepadues, Toulouse, 1973; K. Yoshida and D. Kunii, *J. Chem. Eng. Japan*, **7**, 34 (1974).

23. D. Kunii and O. Levenspiel, *Fluidization Engineering*, p. 129, Krieger, Melbourne FL, 1978.

24. G. Yasui and L.N. Johanson, *AIChE J.*, **4**, 458 (1958).

25. J.S.M. Botterill, J.S. George, and H. Besford, *Chem. Eng. Prog. Symp. Ser.*, **62**(**62**), 7 (1966).

26. T. Chiba and H. Kobayashi, *Chem. Eng. Sci.*, **27**, 965 (1972).

27. K. Miwa et al., *Int. Chem. Eng.*, **12**, 187 (1972); *Kagaku Kogaku*, **35**, 770 (1971).

28. S. Mori and C.Y. Wen, *AIChE J.*, **21**, 109 (1975).

29. C. Fryer, Ph.D. thesis, Monash Univ., Australia, 1974; in S. Mori and C.Y. Wen [28].

30. L.R. Glicksman, W.K. Lord, and M. Sakagami, *Chem. Eng. Sci.*, **42**, 479 (1987).

31. A.B. Whitehead and A.D. Young, in *Proc. Int. Symp. on Fluidization*, A.A.H. Drinkenburg, ed., p. 284, Netherlands Univ. Press, Amsterdam, 1967.

32. J. Werther, in *Fluidization*, J.F. Davidson and D.L. Keairns, eds., p. 7, Cambridge Univ. Press, New York, 1978.

33. P.S.B. Stewart and J.F. Davidson, *Powder Technol.*, **1**, 60 (1967); P.S.B. Stewart, Ph.D. dissertation, Cambridge Univ., 1965.

34. R.M. Ormiston, F.R.G. Mitchell, and J.F. Davidson, *Trans. Inst. Chem. Eng.*, **43**, T209 (1965).

35. J. Baeyens and D. Geldart, *Chem. Eng. Sci.*, **29**, 255 (1974).

6 Bubbling Fluidized Beds

— Experimental Findings
— Estimation of Bed Properties
— Physical Models: Scale-up and Scale-down
— Flow Models for Bubbling Beds

Chapter 5 dealt with the single rising bubble and its interaction with its neighbors. This chapter deals with the behavior of the bubbling bed as a whole. In many applications the performance of fluidized beds depends on this bubbling behavior, in which case the control and improvement of performance can only come after this gas-solid contacting is understood.

The earliest view of the bubbling bed was that all gas in excess of u_{mf}, thus $u_o - u_{mf}$, passed through the bed as bubbles while the emulsion remained at minimum fluidizing conditions, stationary except when moving aside to let bubbles through. This is called the *simple two-phase model*. Numerous experimental investigations in the last 30 years have shown that things are somewhat more complex than as viewed by this model.

Experimental Findings

Emulsion Movement for Small (Geldart B) and Fine (Geldart A) Particles

The many studies in larger beds (>30 cm) indicate that the assumptions of the simple two-phase theory are not well met in that

- The bubble gas is not given by $u_o - u_{mf}$.
- The emulsion voidage ε_e does not stay at ε_{mf} as gas velocity is raised above u_{mf}.
- The emulsion is not essentially stagnant but develops distinct flow patterns, called gulf streaming, induced by the uneven rise or channeling of gas bubbles.

These experimental results are not all consistent, possibly because the different types, sizes, and size distributions of solids used may lead to quite different bed behavior. However, we will try to make some generalizations.

We first look at some of the reported findings on emulsion movement.

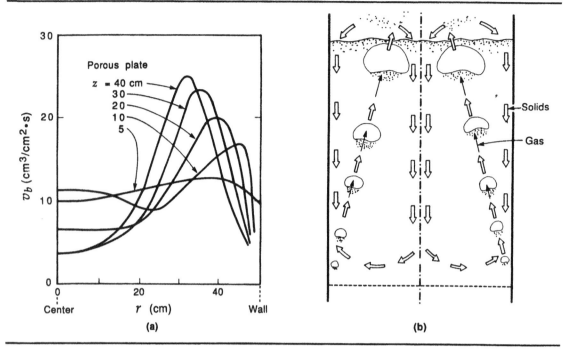

FIGURE 1
Flow of visible bubbles in a shallow bed of Geldart **B** solids, $d_t = 1$ m, quartz sand $\bar{d}_p = 103\ \mu$m, $u_{mf} = 1.35$ cm/s, $u_o = 20$ cm/s; (a) at various heights above a porous plate; (b) general pattern of movement of emulsion; adapted from Werther and Molerus [1].

Werther and Molerus [1] used a high-pressure-drop porous plate distributor which should ensure an even distribution of gas across the bed. Instead, as shown in Fig. 1, they found a strong upflow of emulsion solids close to the vessel walls and starting close to the bottom of the bed. Higher up the bed this upflow region shifted toward the center of the bed.

Using Geldart **B** solids, Whitehead [2], Yamazaki et al. [3], and Lin et al. [4] found somewhat similar solid circulation patterns, with downflow at the center of shallow beds at low gas flow rates, but a reversal to upflow at higher flow rates. With more detailed measurements in a bed of fine Geldart **A** solids Tsutsui et al. [5] found a more complex emulsion flow pattern, with both upflow and downflow at the bed axis, as shown in Fig. 2. From these and other related studies, we tentatively make the following generalizations regarding the emulsion flow in fluidized beds of Geldart **B** solids:

- At low fluidizing velocity in beds of aspect ratio (height/diameter) close to, but less than, unity, the emulsion solids circulate as a vortex ring with upflow at the wall and downflow at the bed axis; see Fig. 3(a). However, at high gas flow rates this flow pattern may reverse because of the large rising bubbles in the bed; see Fig. 3(b).
- As the bed aspect ratio approaches unity, emulsion solids begin to move down the wall near the bed surface, as shown in Fig. 3(c).
- In deeper beds (aspect ratio >1), a second vortex ring forms above the original vortex ring, with upflow at the centerline of the bed; see Fig. 3(d). At higher gas flows, the solid circulation in the upper vortex ring becomes more vigorous and dominates the overall movement of the emulsion.

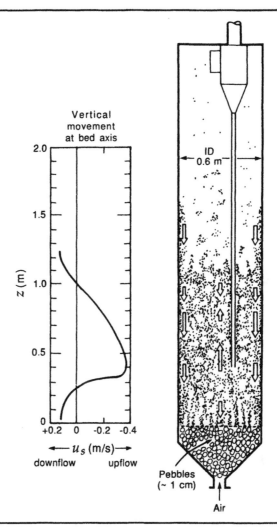

FIGURE 2
Movement of bubbles and emulsion in a Geldart **A** bed of FCC catalyst at $u_0 = 0.33$ m/s; from Tsutsui et al. [5].

- In very shallow beds (aspect ratio < 0.5) supported on uniform distributors, vortex rings of aspect ratio $\cong 1$ may develop; see Fig. 3(e); but with high-pressure-drop tuyeres, these distributors may determine the circulation pattern of the emulsion; see Fig. 3(f).
- In beds of Geldart **A** FCC catalyst, the transition to upflow of emulsion occurs much closer to u_{mf} than in beds of Geldart **B** solids.

This emulsion flow reflects the rise pattern of gas bubbles in the bed. The upflow emulsion region should be rich in bubbles, and the downflow regions should have few, if any, rising bubbles.

Emulsion Movement for Large (Geldart D) Particles

With the increasing interest in the use of large particle beds, researchers have examined them. Their findings are reviewed by Fitzgerald [6]. Figure 4 shows

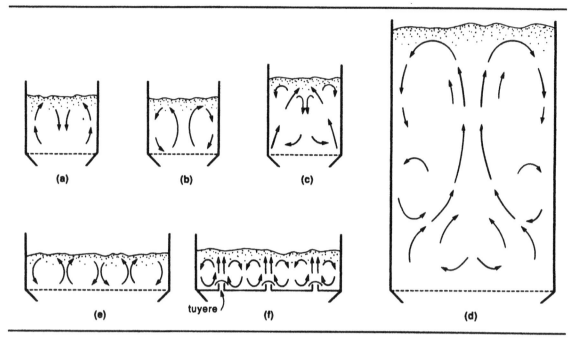

FIGURE 3
Movement of solids in bubbling fluidized beds: (a) $z/d_t \cong 1$, low u_o; (b) $z/d_t \cong 1$, high u_o; (c) $z/d_t \cong 2$, high u_o; (d) general pattern in deep beds; (e) shallow bed, uniform distributor; (f) shallow bed, with tuyeres.

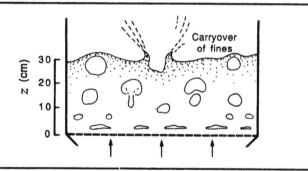

FIGURE 4
Sketch of typical bubbling condition in fluidized beds of coarse (Geldart **D**) solids; sketched from Geldart and Cranfield [7].

typical conditions in large Geldart **D** beds. One sees long lenticular cavities close to the perforated plate distributor. These cavities move slowly upward to transform into nearly spherical bubbles higher up the bed. These bubbles grow rapidly and do not follow any preferred path. Also, note that this nearly spherical bubble shape suggests a very small wake, in contrast to what one finds with small particle systems. Actually, the wake fraction is $f_w \approx 0.1$.

Canada et al. [8], using higher velocities and higher pressures, probably in the slugging and turbulent regimes, found bubbles coalescing into large voids, which produced large bed oscillations and cyclic heaving of the bed surface. Geldart et al. [9] and Miller et al. [10] also studied this flow regime.

Glicksman et al. [11] found no significant variation in gas flow across the bed cross section and no distinct circulation of emulsion solids, which differs from small particle systems. Also, they found that the primary bubble-bubble interaction in shallow beds was caused not by vertical coalescence but by the lateral absorption of smaller bubbles by their larger neighbors. Horizontal tube banks were found to reduce bubble coalescence; consequently, bubble diameter and rise velocity did not increase appreciably with increased gas flow. With a triangular array of tubes that occupied 8% of the bed volume, the bubble size was roughly 1.5 times the tube spacing.

Emulsion Gas Flow and Voidage

Simple two-phase theory assumes that the voidage ε_e and superficial gas velocity u_e through the emulsion remain at ε_{mf} and u_{mf} at all gas flow rates u_o. Abrahamson and Geldart [12] found that u_e and ε_e did change with u_o. In addition, these changes were interrelated and could be reasonably represented for small Geldart **A** and **AB** solids by

$$\left(\frac{\varepsilon_e}{\varepsilon_{mf}}\right)^3\left(\frac{1-\varepsilon_{mf}}{1-\varepsilon_e}\right) = \left(\frac{u_e}{u_{mf}}\right)^{0.7} \tag{1}$$

Thus, knowing any three quantities, say u_{mf}, ε_{mf}, and u_e, gives the fourth quantity. The applicability of this expression to larger particle systems has not yet been tested.

For larger Geldart **B** and **D** particles, Hilligardt and Werther [13] found that u_e did not change appreciably with height in the bed, but that u_e was significantly greater than u_{mf} and dependent on u_o as follows:

$$\frac{u_e - u_{mf}}{u_o - u_{mf}} = \begin{cases} \frac{1}{3} & \text{for three-dimensional beds} \\ \frac{1}{8} & \text{for two-dimensional beds} \end{cases} \tag{2}$$

This means that some of the gas that is expected to go through the bed as bubbles does not. Glicksman et al. [11] estimate that in Geldart **D** beds about 45% of the expected bubble gas actually passes through the emulsion.

Effect of Pressure on Bed Properties

Many commercial fluidized beds operate at high pressure so that more feed can be processed without a corresponding increase in bed cross section. We know that u_{mf} decreases with pressure for Geldart **A** and **B** solids and that bed properties can change drastically with change in pressure. Several researchers have examined the effect of pressure on the behavior of bubbling beds, and a comprehensive summary of these studies is given by Hoffmann and Yates [14]. Some of these findings are given in Chap. 3. More completely, we generalize as follows for an increase in pressure:

Void fraction in the emulsion, ε_e
    Geldart **A**: from 1 to 70 bar, 20–40% increase
    Geldart **AB**: no change

Bubble shape and size
    Geldart **A**: flatter, smaller, less stable
    Geldart **B**: no change

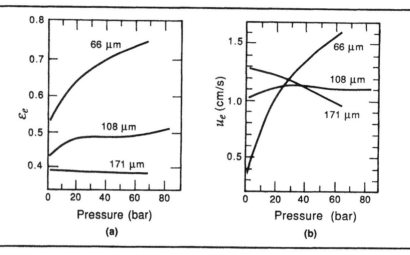

FIGURE 5
Changes in emulsion voidage and gas flow of Geldart **A** and **AB** beds with change in pressure; adapted from Weimer and Quarderer [15].

Bubble splitting
> Geldart **A**: from below, more frequent
> Geldart **B**: from the roof, not more frequent

General bed behavior
> Geldart **B** changes to Geldart **A**, smoother fluidization, less slugging, sharp increase in entrainment

Emulsion voidage and flow velocity
> Equation (1) reasonably relates u_e and ε_e with u_{mf} and ε_{mf} at all pressures for Geldart **A** and **AB** particles.

Figure 5 illustrates some of these findings. Note that the 66-μm particles are Geldart **A** solids, whereas the 171-μm particles are Geldart **AB** solids. The relationship between u_e and ε_e in these figures is consistent with Eq. (1).

Effect of Temperature on Bed Properties

Chemical reactions are often carried out hot, and this effect on u_{mf} has been discussed in Chap 3. As for the other bed properties, we may generalize the experimental findings on changes in bed behavior for a rise in temperature as follows:

*Geldart **A** solids*
> Increase in bubble frequency, significant decrease in bubble size, much smoother fluidization

*Geldart **B** solids*
> Constant or somewhat smaller bubble size, enlarged region of good fluidization

*Geldart **D** solids*

Constant or larger bubble size.

The onset of sintering can be a most important concern in high-temperature operations. This has been discussed in Chap. 3; however, the onset temperature may be much lower in fine particle systems, with their very large specific surfaces, than in large particle systems where the large kinetic energy of the individual particles can significantly raise the safe operational temperature of these systems.

Estimation of Bed Properties

Gas Flow in the Emulsion Phase

Gas flow through the emulsion of small Geldart **A** and **AB** solids can be estimated by using Eq. (1) with the curves of Fig. 5. For larger Geldart **B** and **D** particles, Eq. (2) may give reasonable estimates of the flow of emulsion gas.

Bubble Gas Flow

As has been mentioned, not all the excess gas, $u_o - u_{mf}$, passes through the bed as observable bubbles. Define

$$\psi = \left(\frac{\text{observed bubble flow}}{\text{excess flow, from two-phase theory}} \right) = \frac{v_b}{(u_o - u_{mf})A_t} \tag{3}$$

For small and large particles, Hilligardt and Werther [16] found that ψ changed with height z in the bed, as shown in Fig. 6. Thus, up to $z/d_t \cong 1$, we have approximately $\psi = 0.8$, 0.65, and 0.26 for Geldart **A**, **B**, and **D** particles,

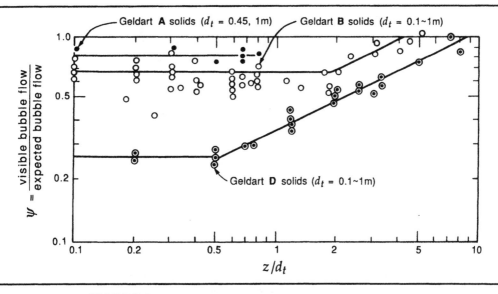

FIGURE 6

Flow of bubble gas increases with height in the bed for all kinds of solids: solids **A**: FCC catalyst, $u_{mf} = 0.2$ cm/s; solids **B**: sand, $d_p = 100\ \mu$m, $u_{mf} = 1.35$ cm/s; solids **D**: sand, $u_{mf} = 18$ cm/s; from Hilligardt and Werther [16].

respectively. Above $z/d_t = 1$, the rising bubbles seem to progressively deaerate the emulsion.

Bubble Size and Bubble Growth

As mentioned in Chap. 5, bubbles in a bubbling bed can be quite irregular in shape and may vary greatly in size. This makes it difficult to characterize a mean bubble size, but such a measure is needed. So for application purposes, we define the mean as a spherical bubble of diameter d_b that represents the bubbles in the bed, usually a mean volumetric size. However, for certain extremely fast kinetic processes, one should more strongly weight the smaller bubbles since most of the transfer or reaction occurs near the bottom of the bed, where the bubbles are small. This point will be discussed in Chap. 11.

Geldart A Particles. In beds of fine particles, such as FCC catalyst, bubbles quickly grow to a few centimeters in size and stay at that size as a result of the equilibrium between coalescence and splitting; see Fig. 7. Occasionally, a

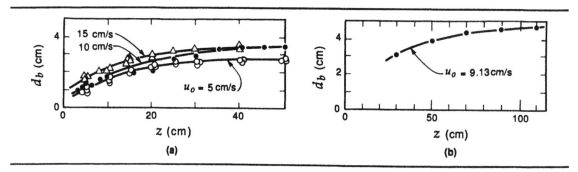

FIGURE 7
Bubble growth in about 0.5-m ID beds of fine (Geldart **A**) particles: (a) cracking catalyst, $u_{mf} = 0.23$ cm/s, porous plate distributor; from Werther [13]; (b) spent cracking catalyst, $u_{mf} = 0.13$ cm/s, $\bar{d}_p = 63$ μm, perforated plate distributor; from Yamazaki et al. [17].

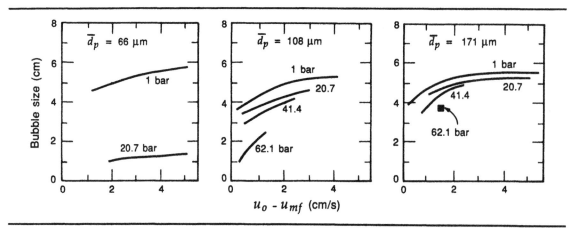

FIGURE 8
Effect of pressure on bubble size in beds of Geldart **A** and **B** solids, inferred from experimental data of bubble rise velocity; adapted from Weimer and Quarderer [15].

larger bubble (8–11 cm) may be observed. Figure 8 shows that at higher pressure bubbles shrink drastically in small particle (Geldart **A**) beds but hardly at all in larger particle (Geldart **B**) beds.

Geldart AB, B, and D Particles. Figures 9 and 10(a) show that bubbles grow steadily with height in the bed and reach tens of centimeters in size. Also,

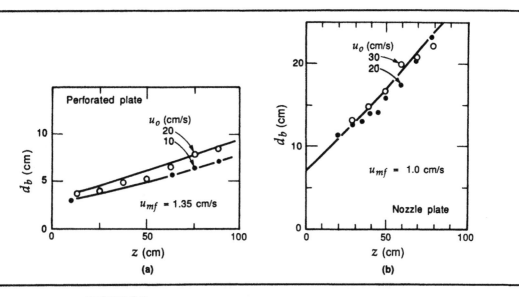

FIGURE 9
Effect of distributor on bubble growth in beds of Geldart **AB** quartz sand, $d_p \cong 100 \ \mu$m: (a) perforated plate, 2.1-mm holes, $l_{or} = 5.2$ cm, triangular arrangement, $d_t = 0.45$ m; (b) nozzles, 3.5-cm OD, $l_{or} = 23.5$ cm, triangular arrangement, six 8.65-mm horizontal ports per nozzle located 1 cm above the plate, $d_t = 1.0$ m; adapted from Werther [18].

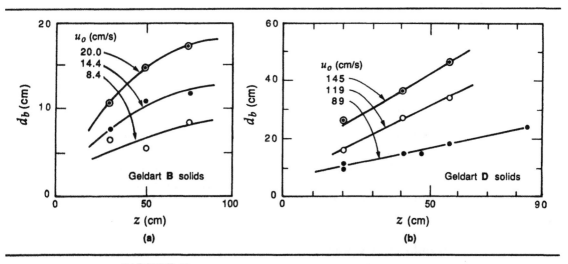

FIGURE 10
Bubble growth in large beds (1 × 1 m, and larger) supported by perforated plate distributors: (a) Geldart **B** solids, $\bar{d}_p = 184 \ \mu$m, $u_{mf} = 3.5$ cm/s; from GOLFERS [19]; (b) Geldart **D** solids, $\bar{d}_p = 1$ mm, $u_{mf} = 58$ cm/s; from Glicksman et al. [11].

nozzle or tuyere distributors give bigger bubbles than perforated distributors do at similar fluidizing conditions. Figure 10(b) shows that bubbles seem to grow quite large in these large particle Geldart **D** systems.

Bubble Size Correlations. Overall, bubbles reach a small limiting size in fine particle systems, are larger in larger particle systems, and seem to grow without limit in very large particle systems.

Several correlations to estimate bubble growth in fluidized beds have been developed from experiments, mainly in small-diameter beds of Geldart **B** solids. Still, according to GOLFERS [19], these correlations reasonably apply to large-diameter beds. We give three of these correlations.

Mori and Wen [20], for Geldart **B** and **D** solids, proposed that the bubble size d_b at any height z in the bed be given as

$$\frac{d_{bm} - d_b}{d_{bm} - d_{b0}} = e^{-0.3z/d_t} \tag{4}$$

where d_{b0} is the initial bubble size formed near the bottom of the bed, given by Eqs. (5.14), (5.16), or (5.19), and d_{bm} is the limiting size of bubble expected in a very deep bed. This maximum is given as

$$d_{bm} = 0.65\left[\frac{\pi}{4}\, d_t^2(u_o - u_{mf})\right]^{0.4}, \qquad [\text{cm}] \tag{5}$$

Thus, calculating d_{b0} and d_{bm} and inserting their values into Eq. (4) gives the mean bubble size d_b at any level z in a bed of diameter d_t. The range of conditions from which this correlation was obtained is

$$d_t \leq 1.3\,\text{m} \qquad\qquad 0.5 \leq u_{mf} \leq 20\,\text{cm/s}$$
$$60 \leq d_p \leq 450\,\mu\text{m} \qquad u_o - u_{mf} \leq 48\,\text{cm/s}$$

In another approach, Werther [18] gives the following expression for bubble size at any height z in a bed of Geldart **B** solids supported by a porous plate distributor:

$$d_b = 0.853[1 + 0.272(u_o - u_{mf})]^{1/3}(1 + 0.0684z)^{1.21}, \qquad [\text{cm}] \tag{6}$$

with the following applicable range of operating conditions:

$$d_t > 20\,\text{cm} \qquad\qquad 1 \leq u_{mf} \leq 8\,\text{cm/s}$$
$$100 \leq d_p < 350\,\mu\text{m} \qquad 5 \leq u_o - u_{mf} \leq 30\,\text{cm/s}$$

In beds having other than porous plate distributors, the d_b-versus-z curve should be shifted accordingly to fit the initial bubble size d_{b0} (from Chap. 5) at initial height of bubble formation z_0 (from Chap. 4 for tuyeres). Example 2 shows how to do this. Horio and Nonaka [21] modified Eq. (4) so that it can be used over a much wider range of solid sizes, from Geldart **A** through Geldart **D**.

Bubble Rise Velocity

On the basis of simple two-phase theory, Davidson and Harrison [22] proposed the following rise velocities:

For single bubbles:

$$u_{br} = 0.711(gd_b)^{1/2}$$

(5.2) or (7)

For bubbles in bubbling beds:

$$u_b = u_o - u_{mf} + u_{br}$$

(8)

Figures 11(a)–(e) present experimental findings on bubble rise velocity in Geldart **A**, **B**, and **D** beds, and show how well Eqs. (7) and (8) fit the data.

Figure 11(a), for Geldart **A** beds, shows that bubbles rise two or three times as fast as predicted when $u_o \gg u_{mf}$. Figure 11(b), for smaller Geldart **B** solids, again shows faster rise velocities when $u_o \gg u_{mf}$. However, for all sizes of Geldart **B** solids at u_o close to u_{mf} and for Geldart **BD** solids, Figs. 11(b) and (c) show that the bubble rise data are bounded by Eqs. (7) and (8). Figure 11(d), for tube-filled beds of Geldart **D** solids, shows that the bubble rise velocity is again reasonably represented by Eqs. (7) and (8). Finally, Fig. 11(e) shows, for large and small particles, that as the bubble size approaches roughly 50% of the bed diameter then the rising bubbles transform into slugs with rise velocity given by Eq. (5.23).

Whenever the rise velocity can be represented by either Eq. (7) or Eq. (8), we will use Eq. (8) because it represents the more conservative estimate for design purposes.

We now suggest some reasons for these findings. In fine particle beds, both large and small, bubbles are likely to be accompanied by relatively large wakes (see Fig. 5.8), and this is probably the reason for the observed emulsion circulation mentioned earlier. In addition to this phenomenon, large beds of fine particles develop gulf stream circulation of solids. In this state the upflow regions of the bed become richer in bubbles and have a lower bulk density, and this creates the driving force for the maintenance of this vortex stream. With increasing bed size, the emulsion becomes more fluid, which enhances the circulation. Conversely, in small beds this flow pattern is depressed or even absent because of friction with the wall surfaces. This gulf stream circulation should be more vigorous with a large excess gas velocity, or $u_o \gg u_{mf}$. On the other hand, bubbles in large particle beds have small wakes, and solid circulation is rather weak; hence, the bubbles rise velocity in these beds should be close to that predicted by the simple two-phase theory.

In order to come up with an equation for bubble rise velocity that covers the whole range of particle sizes from Geldart **A** to **D** and that accounts for the vessel size, Werther [18] proposed the equation

$$u_b = \psi(u_o - u_{mf}) + \alpha u_{br}$$

(9)

where ψ is the fraction of visible bubbles, given by Eq. (3) with Fig. 6, and α is a factor that accounts for the deviation of bed bubbles from single rising bubbles. From his experimental data, reported in Fig. 11, he recommends the following for α:

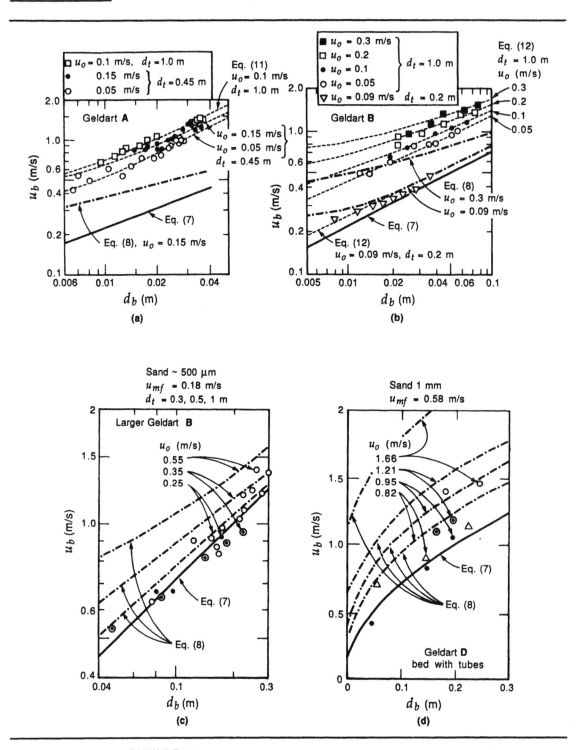

FIGURE 11

Rise velocity of bubbles and a comparison with Eqs. (7) and (8) of the simple two-phase model: (a) Geldart **A**, FCC catalyst, $u_{mf} = 0.002$ m/s; (b) Geldart **B**, sand, $u_{mf} = 0.025$ m/s; (c) Geldart **B**, coarse sand, $u_{mf} = 0.18$ m/s; (d) Geldart **D**, silica sand, $u_{mf} = 0.58$ m/s, in a bed with horizontal tubes; (e) Geldart **A** and Geldart **D**, silica sand, $u_{mf} = 0.58$ m/s. Data for (a), (b), and (c) are from Hilligardt and Werther [16]; data for (d) and (e) are from Glicksman et al. [11].

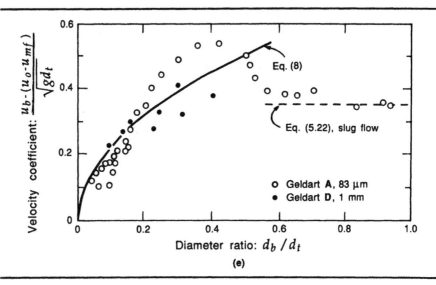

FIGURE 11 (Contd.)

Geldart-type solids	**A**	**B**	**D**
α	$3.2d_t^{1/3}$	$2.0d_t^{1/2}$	0.87
d_t (m)	0.05–1.0	0.1–1.0	0.1–1.0

Morooka et al. [23] tried to account for the contribution of the upflowing emulsion with the expression

$$u_b = u_{e,up} + u_{br} \tag{10}$$

Equation (10) accounts for the emulsion upflow of velocity $u_{e,up}$, caused by vigorous gulf stream circulation of the emulsion.

Analyzing the experimental data in Figs. 11(a) and (b), reported by Werther [13], we propose the following correlations:

for Geldart **A** solids with $d_t \leq 1$ m:

$$u_b = 1.55\{(u_o - u_{mf}) + 14.1(d_b + 0.005)\}d_t^{0.32} + u_{br}, \quad [\text{m/s}] \tag{11}$$

for Geldart **B** solids with $d_t \leq 1$ m:

$$u_b = 1.6\{(u_o - u_{mf}) + 1.13d_b^{0.5}\}d_t^{1.35} + u_{br}, \quad [\text{m/s}] \tag{12}$$

These expressions fit the experimental data well, as seen in Figs. 11(a) and (b). Note that the effect of $u_o - u_{mf}$ is nearly the same for Geldart **A** and **B** solids. On the other hand, d_b and d_t affect $u_{e,up}$ differently. This may be explained in terms of the different fluidity of the emulsion of these two classes of solids. The lower limit to the applicability of Eqs. (9)–(12) should be when

$$[u_b \text{ by Eqs. (9)–(12)}] \cong [u_b \text{ by Eq. (8)}]$$

So, to determine the bubble rise velocity in bubbling beds, calculate u_b, by Eq. (8) and by Eqs. (9)–(12), and take the larger value.

Beds with Internals

In fine particle beds (Geldart **A**) u_b can be very large, even with small bubbles. So, to ensure adequate gas-solid contacting, one may need to use excessively deep beds. This is a serious problem with large-diameter fine particle reactors.

Vertical internals such as heat exchanger tubes can effectively retard the emulsion circulation by increasing the wall effect, and thereby improve reactor performance. For design of vertical tube-filled beds of Geldart **A** solids, we may estimate u_b by using the hydraulic diameter of the bed d_{te} in place of d_t in the calculations, where

$$d_{te} = \frac{4(\text{cross-sectional area available to the fluidized bed})}{(\text{total wetted perimeter of bed and tubes})} \tag{13}$$

For Geldart **B** beds, bubbles usually grow to the size of d_{te}. In this situation u_b should be estimated by Eq. (5.22) for axial slugs. In larger particle (Geldart **D**) beds containing horizontal tube banks, Glicksman et al. [11] found that bubbles quickly grow to roughly 1–1.5 times the tube pitch, l_i, and do not change appreciably with changes in gas velocity. In addition, the rise velocity of these bubbles (see Fig. 11(d)) is the same as for isolated bubbles of that size, given by Eq. (7) or Eq. (8).

EXAMPLE 1

Bubble Size and Rise Velocity in Geldart A Beds

Estimate d_b and u_b at height $z = 0.5$ m in a bed ($d_t = 0.5$ m) of fine catalyst powder ($\rho_s = 1.6$ g/cm^3, $\bar{d}_p = 60$ μm, $u_{mf} = 0.2$ cm/s, $u_o = 20$ cm/s) supported by a perforated plate distributor (square arrangement, $d_{or} = 2$ mm, $l_{or} = 20$ mm).

SOLUTION

Figure 3.9 shows that these solids are square in the middle of the Geldart **A** zone. So we use the estimation method for this system.

Method 1. Procedure Using Eqs. (10) and (11). First estimate the bubble size at $z = 0.5$ m from Figs. 7(a) and (b). This gives

$$d_b = \frac{0.035 + 0.04}{2} = 0.038 \text{ m}$$

From Eqs. (10) and (11),

$$u_b = 1.55\{(0.20 - 0.002) + 14.1(0.038 + 0.005)\}(0.5)^{0.32}$$
$$+ 0.711(9.8 \times 0.038)^{1/2}$$
$$= 0.999 + 0.434 = \underline{1.43 \text{ m/s}}$$

Method 2. Werther's Procedure. Taking $\psi = 0.8$ from Fig. 6 for Geldart **A** solids and replacing the extracted bubble size found above, $d_b = 0.038$ m, into Eq. (9) gives

$$u_b = 0.8(0.2 - 0.002) + 3.2(0.5)^{1/3}(0.711)(9.8 \times 0.038)^{1/2}$$
$$= 0.158 + 1.102 = \underline{1.26 \text{ m/s}}$$

Comment. These two calculation methods give closely similar results. Also, we see that the rise velocity of bubbles relative to the emulsion solids is only 0.434 m/s while the emulsion sweeps upward at a velocity, given by Eq. (10), of

$$u_{e,up} = \frac{1.43 + 1.26}{2} - 0.43 = 0.92 \text{ m/s}$$

The relatively large value for $u_{e,up}$ indicates that gulf stream circulation should be severe in this bed.

EXAMPLE 2

Bubble Size and Rise Velocity in Geldart B Beds

Estimate d_b and u_b at height $z = 0.5$ m in a bed ($d_t = 0.5$ m) of sand ($\rho_s = 2.6$ g/cm^3, $\bar{d}_p = 100$ μm, $u_{mf} = 1$ cm/s, $u_0 = 0.45$ m/s) supported by a perforated plate distributor (triangular arrangement, $d_{or} = 2.0$ mm, $l_{or} = 30$ mm).

SOLUTION
First of all, a check of Fig. 3.9 shows that these are Geldart **B** solids, close to the **AB** boundary.

Part (a). Bubble Size
First determine the initial bubble size in the bed. This we get from the procedure of Chap. 5. Thus

$$N_{or} = \frac{2}{\sqrt{3}\, l_{or}^2} = 1.3 \times 10^3 \text{ m}^{-2}$$

and from Fig. 5.14, with $u_0 - u_{mf} = 0.45 - 0.01 = 0.44$ m/s, we find

$$d_{b0} = 5.5 \text{ cm}$$

We now can proceed in several ways.

Method 1. Werther's Procedure for Finding d_b. First, Eq. (6) relates bubble size with height above a porous plate distributor, or

z (cm)	0	5	10	20	30	50	70
d_b (cm)	2.00	2.86	3.77	5.69	7.73	12.1	16.8

Since we do not have a porous plate distributor and our bubble size starts at $d_{b0} = 5.5$ cm at $z = 0$, we shift the curve accordingly to give, at $z = 0.5$ m,

$$d_b = \underline{16.5 \text{ cm}}$$

This procedure was suggested by Werther [18].

Method 2. Mori and Wen's Procedure for Finding d_b. Here we must find the maximum expected bubble size. Thus, from Eq. (5),

$$d_{bm} = 0.65\left[\frac{\pi}{4}(50)^2(45-1)\right]^{0.4} = 61.3 \text{ cm}$$

Inserting into Eq. (4) then gives

$$\frac{61.3 - d_b}{61.3 - 5.5} = e^{-0.3}, \quad \text{or} \quad \underline{\underline{d_b = 20.0 \text{ cm}}}$$

Comment. These two methods give closely similar predictions.

Part (b) Bubble Velocity

Method 1. Procedure using Eq. (12). Replacing values gives

$$u_b = 1.6\{(0.45 - 0.01) + 1.13(0.163)^{0.5}\}(0.5)^{1.35}$$
$$+ 0.711(9.8 \times 0.163)^{1/2}$$
$$= 0.5625 + 0.8986 = \underline{1.46 \text{ m/s}}$$

Method 2. Werther's Procedure. From Fig. 6 for Geldart **B** solids, we find $\psi = 0.65$. Then Eq. (9) and what follows directly below it gives

$$u_b = 0.65(0.45 - 0.01) + 2(0.5)^{1/2}(0.711)(9.8 \times 0.163)^{1/2}$$
$$= 0.286 + 1.271 = \underline{1.56 \text{ m/s}}$$

Comment. We compare these results with the expressions of the simple two-phase theory, Eqs. (7) and (8). The latter gives

for Werther's bubble size:	$u_{br} = 0.90$ m/s,	$u_b = 1.34$ m/s
for Mori and Wen's bubble size:	$u_{br} = 1.00$ m/s,	$u_b = 1.44$ m/s

In Geldart **B** beds of size $d_t = 0.5$ m, u_b values found by the above two methods are about 13% higher than the values calculated from the simple two-phase theory. Thus, in small beds, $d_t < 0.5$ m, or in large beds with internals such that $d_{te} < 0.5$ m we can reasonably use Eqs. (7) and (8) to calculate u_b. However, in large beds use the above two methods to calculate u_b.

Physical Models: Scale-up and Scale-down

Suppose we have just gotten a rough conceptual design for a commercial-sized system, defining its probable range of operating conditions, particle characteristics, main features of the vessel, and its circulation system and internals, if any. It is a good idea to construct a physical model to get an idea of, to observe, and to confirm the hydrodynamic behavior of the commercial unit. But how large should this model be? And if it is designed for ambient conditions instead of at high temperature and high pressure, would this invalidate the results? Design engineers need answers to such questions. Unfortunately, not much has been done in this area.

Horio et al. [24] stated that similarity between a large bed and its model is achieved if one matched the scaling parameters

$$\frac{u - u_{mf}}{(gd_p)^{0.5}}, \qquad \frac{u_{mf}}{(gd_p)^{0.5}} \tag{14}$$

and Fitzgerald and Crane [25,6] proposed using the more restrictive set of scaling parameters

$$\frac{d_p u \rho_g}{\mu}, \qquad \frac{\rho_s}{\rho_g}, \qquad \frac{u}{(gd_p)^{0.5}}, \qquad \frac{L}{d_p} \tag{15}$$

| Reynolds number | density ratio | Froude number | geometric similarity of distributor, bed, and particle |

The suggested calculation procedure when using Fitzgerald's criteria of Eq. (15) is as follows:

1. Calculate $(\rho_s/\rho_g)_1$ for the system to be modeled.

2. Choose a gas for the model. Most conveniently this would be ambient air. For this gas determine $\rho_{g,2}$. This necessarily fixes $\rho_{s,2}$ and tells what solid to use.

3. Combining the Reynolds and Froude numbers gives the scale factor for the two beds, or

$$m = \frac{L_2}{L_1} = \left(\frac{\rho_{g1}\mu_2}{\rho_{g2}\mu_1}\right)^{2/3} \tag{16}$$

Note that one cannot arbitrarily choose the size of the model; this is necessarily fixed by the choice of ρ_{g2}.

4. Finally, from the Froude and Reynolds numbers, we find the time factor and gas velocity ratio that makes the beds behave similarly. Thus

$$\frac{u_2}{u_1} = \frac{t_2}{t_1} = \left(\frac{L_2}{L_1}\right)^{1/2} = m^{1/2} \tag{17}$$

This modeling strategy was tested with a very wide range of materials. The density of solids was varied by a factor of 1000 (from styrofoam to tungsten), as was the fluid (from air to water), in rather small units but with a size ratio of up to 10 to 1. This criterion seemed to work satisfactorily in that motion pictures of the matched beds showed similar motion of solids, bubbling, and onset of slugging.

Pressure fluctuations were then matched between a $2 \, m \times 2 \, m$ fluidized bed combustor development unit and a small-scale test unit in which copper particles were fluidized with helium. Again, similar behavior was observed.

Glicksman et al. [26] made tests on larger beds and found reasonable agreement with Fitzgerald's similarity relationships. Experiments by Roy and Davidson [27] suggest that the less restrictive criteria of Horio et al. are sufficient to give similarity in behavior when $Re_p < 30$, but that the more restrictive criteria of Fitzgerald and Crane are needed when $Re_p > 30$.

Scale-up, scale-down, and hydrodynamic similarity between different beds are very important problems, especially for the designer, and much more work is needed in this area.

EXAMPLE 3

Scale-down of a Commercial Chlorinator

When zircon sand and coke are contacted by chlorine at the right conditions, they react as follows:

$$ZrSiO_4(s) + 4Cl_2 + 4C(s) \xrightarrow[\text{1 atm}]{\text{1000°C}} 4CO + ZrCl_4 + SiCl_4 \tag{18}$$

A commercial chlorinator operates as follows. A mixture of finely ground zircon sand and coke is fed continuously to a graphite-walled reactor whose gas distributor consists of a layer of 1–2 cm of Bermuda rock. The reactant solids are fluidized by a stream of pure chlorine gas. Particles react and shrink, and hot product gas with some purge nitrogen leaves the system.

About 5% of the entering chlorine leaves the reactor unreacted, and the operating group is not sure why. Is the bed properly fluidized or is it spouting or slugging? Unfortunately, because of the extreme corrosiveness of the environment, it is not practical to insert into the vessel the probes needed to answer this question. So they decide to physically model the reactor and see what the model tells.

Suggest a design for this flow model.

Data

Solids: $\bar{d}_p$ (in bed) $\cong 270$ mesh $= 53 \ \mu m$
$\bar{\rho}_s$ (of the coke–zircon mixture) $= 3200 \ kg/m^3$
$\varepsilon_m = 0.5, \ \varepsilon_f = 0.75$

Gas: $\bar{\rho}_g = 0.64 \ kg/m^3$
$u_0 = 14 \ cm/s$, at bed conditions
$\mu = 5 \times 10^{-5} \ kg/m \cdot s$

Bed: $T = 1000°C$, pressure $= 1$ atm
$d_t = 91.5 \ cm$, slumped height $= 150 \ cm$

SOLUTION

Follow the procedure outlined in the text, and let the commercial reactor and the model be designated by 1 and 2, respectively. Then, for step 1 the density ratio in the commercial unit is

$$\left(\frac{\rho_s}{\rho_g} \right)_1 = \frac{3200}{0.64} = 5000$$

For step 2 first try the most convenient of gases, ambient air, for which

$$\rho_{g2} = 1.2 \ kg/m^3 \quad \text{and} \quad \mu_2 = 1.8 \times 10^{-5} \ kg/m \cdot s$$

Thus, from the requirement of a constant density ratio, we have

$$\rho_{s2} = \rho_{g2} \left(\frac{\rho_s}{\rho_g} \right)_1 = 1.2(5000) = 6000 \ kg/m^3$$

Looking through a handbook, we find that zirconia has a density close to that required, and this material happens to be readily available. So, for steps 3 and 4,

$$m = \frac{L_2}{L_1} = \left[\frac{(0.64)(1.8 \times 10^{-5})}{(1.2)(5.0 \times 10^{-5})} \right]^{2/3} = 0.33$$

$$\frac{u_2}{u_1} = \frac{t_2}{t_1} = m^{1/2} = (0.33)^{1/2} = 0.58$$

Thus, for the model use a 30.5-cm ID bed, a 50-cm slumped bed height, and a packed bed distributor consisting of 3–6 mm rock.

Fluidizing gas: ambient air at 1 atm
Solids: zirconia, $\bar{d}_p = (0.33)(53 \ \mu m) = 18 \ \mu m$
Entering gas: $u_0 = (0.58)(14 \ cm/s) = 8.1 \ cm/s$

If we wish to see what the model suggests is going on in the hot unit, we take movies of the model at 1.73 times normal speed and play back the film at normal speed. We may also want to calculate what is happening in the bed. The next section considers this approach.

Flow Models for Bubbling Beds

The reason for developing a conceptual model for the bubbling bed is to be able to reasonably estimate its main features, such as volume fraction of phases, velocities of gases and solids, contacting regimes, from partial information, such as a few measurements or correlations. The primary use of such models is to predict the performance of bubbling beds for physical and chemical applications. We consider the simple two-phase model and the more realistic K-L model.

General Interrelationship among Bed Properties

First of all, a mass balance for the bed solids gives

$$L_m(1 - \varepsilon_m) = L_{mf}(1 - \varepsilon_{mf}) = L_{mb}(1 - \varepsilon_{mb}) = L_f(1 - \varepsilon_f) \qquad (19)$$

for fixed   at u_{mf} at u_{mb} for bubbling
bed   bed

Next consider a bubbling bed as a two-phase system. Although recent experiments indicate that rising bubbles contain small amounts of solids, we can ignore this for flow models and take the bubble voidage to be $\varepsilon_b = 1$. The volume fraction of the bed in bubbles, δ, and the average bed voidage, ε_f, are then related to the voidage of the emulsion, ε_e, by

$$\varepsilon_f = \delta + (1 - \delta)\varepsilon_e \qquad \text{or} \qquad 1 - \varepsilon_f = (1 - \delta)(1 - \varepsilon_e) \qquad (20)$$

If ε_f and δ are known from experiment, ε_e can be determined by this equation. However, if ε_e cannot be determined, then we must approximate it as follows:

$\varepsilon_e \simeq \varepsilon_{mb}$,   for Geldart **A** solids

$\varepsilon_e \simeq \varepsilon_{mf}$,   for Geldart **B** and **D** solids

The estimation of the flows through bubble and emulsion phases is where the two models differ.

The Simple Two-Phase Model

Toomey and Johnstone [28] first introduced the simple two-phase model, which assumes that all the gas in excess of u_{mf} flows through the bed as bubbles while the emulsion stays stagnant at minimum fluidizing conditions.. With the experimental bubble rise velocity of Eq. (5.2), this model gives

Rise velocity of bubbles: $u_b = 0.711(gd_b)^{1/2}$ (7)

Rise velocity of emulsion gas: $u_e = \dfrac{u_{mf}}{\varepsilon_{mf}}$ (21)

Superficial rise velocity of emulsion gas: u_{mf} (22)

Rise velocity of solids: $u_s = u_{s,up} = u_{s,down} = 0$ (23)

Fraction of bed in bubbles: $\delta = \dfrac{u_o - u_{mf}}{u_b - u_{mf}}$ (24)

Fraction of bed in emulsion: $1 - \delta = \dfrac{u_b - u_o}{u_b - u_{mf}}$ (25)

Our earlier discussion shows that this model does not fit the experimental findings too well, so we do not consider it further.

The K-L Model with Its Davidson Bubbles and Wakes

In the K-L model slow cloudless bubbles and fast clouded bubbles give distinctly different flow patterns of gas about bubbles. In addition, by accounting for bubble wakes, we see that solids are dragged up the bed behind bubbles and drift downward in the emulsion. This downflow of solids, u_s, can be so fast that it overcomes the upflow of gas in the emulsion, resulting in a net downflow of emulsion gas. This condition is frequently met in commercial-scale operations with Geldart **A** and **AB** solids. Figure 12 shows this model and its assumptions.

Bubble Rise Velocities. First we accept that the rise velocity of a single bubble relative to the emulsion solids is given by Eq. (7). Then the presentation given earlier shows that different expressions are needed for different sized beds because gulf streaming must be accounted for in large beds of small particles. So for small laboratory beds of Geldart **A** and **B** solids and any size bed of Geldart **D** solids, use Eq. (8) since gulf streaming is practically absent in these systems. However, for large-diameter beds use Eq. (11) for Geldart **A** solids and Eq. (12) for Geldart **B** solids. Alternatively, u_b may be determined by Werther's procedure of Eq. (9), which has been tested for Geldart **A**, **B**, and **D** solids in beds up to $d_t = 1$ m.

Fraction of the bed in bubbles:

• For slow bubbles, or $u_b < u_e$,

$$\delta = \frac{u_o - u_{mf}}{u_b + 2u_{mf}}, \qquad [m^3 \text{ bubbles/m}^3 \text{ in bed}] \qquad (26)$$

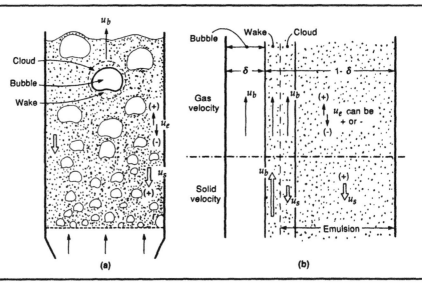

FIGURE 12
The K-L bubbling bed model. This sketch represents the fine particle case, or bubbles with thin clouds. Note that wake solids, cloud gas, and wake gas rise with the bubble; emulsion gas can go up or down, depending on bed conditions; emulsion solids go down.

- For intermediate bubbles with thick clouds, which may well overlap, or $u_{mf}/\varepsilon_{mf} < u_b < 5u_{mf}/\varepsilon_{mf}$, we have a regime that is difficult to represent. Roughly

$$\delta = \begin{cases} \dfrac{u_o - u_{mf}}{u_b + u_{mf}} & \text{when } u_b \cong u_{mf}/\varepsilon_{mf} \\[2ex] \dfrac{u_o - u_{mf}}{u_b} & \text{when } u_b \cong 5u_{mf}/\varepsilon_{mf} \end{cases} \tag{27}$$

- For fast bubbles, or $u_b > 5u_{mf}/\varepsilon_{mf}$, clouds are thin and

$$\delta = \frac{u_o - u_{mf}}{u_b - u_{mf}} \tag{28}$$

- In vigorously bubbling beds where $u_o \gg u_{mf}$, we may take as an approximation

$$\delta = \frac{u_o}{u_b} \tag{29}$$

Cloud volume to bubble volume:

$$f_c = \frac{3}{u_{br}\varepsilon_{mf}/u_{mf} - 1} \tag{30}$$

Wake volume to bubble volume:

$$f_w, \quad \text{found from Fig. 5.8} \tag{31}$$

Fraction of bed in emulsion (not counting bubble wakes)

$$f_e = 1 - \delta - f_w\delta \tag{32}$$

Define the distribution of solids in the various regions by

$$\gamma_b, \gamma_c, \gamma_e = \frac{\text{(volume of solids dispersed in b, c, and e, respectively)}}{\text{volume of bubble}} \tag{33}$$

With δ as the volume fraction of the bed consisting of bubbles, these γ values are related by the expression

$$\delta(\gamma_b + \gamma_c + \gamma_e) = 1 - \varepsilon_f = (1 - \varepsilon_{mf})(1 - \delta) \tag{34}$$

from which

$$\gamma_e = \frac{(1 - \varepsilon_{mf})(1 - \delta)}{\delta} - \gamma_b - \gamma_c \tag{35}$$

With the wake included with the cloud region, we also have

$$\gamma_c = (1 - \varepsilon_{mf})(f_c + f_w) = (1 - \varepsilon_{mf})\left[\frac{3}{u_{br}\varepsilon_{mf}/u_{mf} - 1} + f_w\right] \tag{36}$$

From experiment, γ_b is about 10^{-2} to 10^{-3}, with the actual value uncertain. So for our calculations to come, and until more precise values are known, we take

$$\gamma_b = 0.005 \tag{37}$$

Rise velocity of wake solids:

$$u_{s,wake} = u_b \qquad \text{(upward is "+" for } u_{s,wake}) \tag{38}$$

The above expressions apply to vigorously bubbling beds, even when gulf stream circulation is significant.

In situations where gulf stream circulation can be neglected, namely in small beds of Geldart **A** solids and in any bed of Geldart **B** solids, the following expressions apply.

The downflow velocity of emulsion solids:

$$u_{s,down} = \frac{f_w \delta u_b}{1 - \delta - f_w\delta} \qquad \text{(downward is "+" for } u_{s,down}) \tag{39}$$

Rise velocity of emulsion gas through the bed:

$$u_e = \frac{u_{mf}}{\varepsilon_{mf}} - u_{s,down} \tag{40}$$

The above expressions show that in a vigorously bubbling bed the emulsion gas begins to be dragged down the bed when

$$\frac{u_b}{u_{mf}} > \frac{1 - \delta - f_w\delta}{f_w\varepsilon_{mf}\delta} \tag{41}$$

Depending on bubble size and the nature of solids, this flow reversal occurs at gas flow rates

$$\frac{u_o}{u_{mf}} = 6\text{–}20 \tag{42}$$

In large beds, free of internals and containing Geldart **A** solids, gulf stream circulation becomes significant. This causes the downflow velocity of emulsion solids to be faster than that given by Eq. (39), and flow reversal to occur at a smaller u_o than that given by Eq. (42).

Finally, if the emulsion voidage ε_e is known, use it in place of ε_{mf}. Also for Geldart **A** solids if minimum bubbling conditions are known (but not ε_e) make the following two changes in all of the preceding expressions:

$$u_{mf} \rightarrow u_{mb} \qquad \text{and} \qquad \varepsilon_{mf} \rightarrow \varepsilon_{mb}$$

This model, developed by Kunii and Levenspiel [29] in slightly different form will be used in later chapters for heat transfer, mass transfer, and reaction applications.

EXAMPLE 4

Reactor

Scale-up for

Geldart A

Catalyst

A 1-m ID pilot reactor gives lower conversion of gaseous reactant than does a 20-cm ID bench-scale reactor, both operating at the same static bed height, with the same catalyst, and at the same superficial gas velocity. This worries the reaction engineering group because it suggests that further scale-up to the large commercial unit may lead to still lower conversion. In this situation suggest some reasonable scale-up strategies.

Data

$$\bar{d}_p = 52 \ \mu m, \qquad \varepsilon_m = 0.45, \qquad \varepsilon_{mf} = 0.50, \qquad \varepsilon_{mb} = 0.60,$$
$$u_o = 30 \ cm/s, \qquad L_m = 2 \ m, \qquad u_{mf} = 0.33 \ cm/s, \qquad u_{mb} = 1.0 \ cm/s$$

Equilibrium bubble size in this fine particle system: $d_b = 3$ cm.

Both beds rest on porous plate distributors.

SOLUTION

The rise velocity of 3-cm bubbles relative to stagnant emulsion solids is found from Eq. (7) to be

$$u_{br} = 0.711(9.8 \times 0.03)^{1/2} = 0.386 \ m/s \tag{i}$$

The rise velocity in these beds is then found by either of two procedures: Eqs. (10) with (11) or by Werther's method with Eq. (9).

For the bench unit ($d_t = 0.2$ m), with Eq. (11),

$$u_b = 1.55\{(0.30 - 0.033) + 14.1(0.03 + 0.005)\}0.2^{0.32} + 0.386 = 1.12 \ m/s$$

With Werther's procedure, taking $\psi = 1$ from Fig. 6, Eq. (9) gives

$$u_b = (1)(0.30 - 0.0033) + 3.2(0.2)^{1/3}(0.386) = 1.02 \ m/s$$

Averaging gives

$$u_b = \frac{1.12 + 1.02}{2} = 1.07 \ m/s \qquad \text{for the } d_t = 0.2 \ m \text{ bed} \tag{ii}$$

For the pilot unit ($d_t = 1$ m), with Eq. (11),

$$u_b = 1.55\{(0.30 - 0.0033) + 14.1(0.03 + 0.005)\}(1)^{0.32} + 0.386 = 1.61 \ m/s$$

With Werther's procedure of Eq. (9),

$$u_b = (1)(0.30 - 0.0033) + 3.2(1)^{1/3}(0.386) = 1.53 \ m/s$$

Averaging gives

$$u_b = \frac{1.61 + 1.53}{2} = 1.57 \ m/s \qquad \text{for the } d_t = 1 \ m \text{ bed} \tag{iii}$$

Next, from Eqs. (i)–(iii) we can find the rise velocity of upflowing emulsion. This is the region that contains bubbles. Thus

$$u_{e,up} = u_b - u_{br} = 1.07 - 0.39 = 0.68 \ m/s \qquad \text{for } d_t = 0.2 \ m$$
$$u_{e,up} = 1.57 - 0.39 = 1.18 \ m/s \qquad \text{for } d_t = 1.0 \ m$$

The higher emulsion rise rate and the higher bubble rise velocity in the pilot unit means a smaller contacting time for most of the reactant gas. This is a reasonable explanation for the lowered conversion in the larger pilot unit.

Scale-Up Alternative 1. If the commercial unit is to behave like the 20-cm bench unit, the effective bed diameters of the two units should be close to equal. This can be done by inserting vertical internals into the commercial unit so that its effective bubble diameter d_{te} equals that of the bench unit.

Assuming a rectangular array of tubes of outer diameter d_i and pitch l_i, Eq. (13) gives

$$d_{te} = 20 \text{ cm} = 4 \frac{l_i^2 - (\pi/4)d_i^2}{\pi d_i}$$

The following combinations of d_i and l_i satisfy this equation:

d_i (cm)	5	10	15	20
l_i (cm)	9.91	15.35	20.31	25.07

A suitable arrangement may be

$$d_i = \underline{10 \text{ cm}} \quad \text{with} \quad \frac{l_i}{d_i} = \underline{1.54}$$

With this design the static bed height of the commercial unit can be kept at 2 m. However, the narrow spacing between tubes would make it difficult to inspect or repair any tube within the bundle. This problem is overcome with the next design.

Scale-Up Alternative 2. To have a larger open space between bed internals, we determine the bed height needed to give the conversion of the 20-cm ID bench reactor but with an effective bed diameter of the 1-m ID pilot-plant reactor. Also choose a reasonable geometry for the internals.

With no reaction kinetics given, we argue that roughly the same conversion is obtained in large and small units if the residence time of the bubbles is the same in the two units. This argument assumes that the equilibrium bubble size stays unchanged for the same u_o, which is reasonable for Geldart A solids.

Taking ratios gives

$$\frac{L_m(\text{commercial})}{L_m(d_t = 0.2 \text{ m})} = \frac{u_b(\text{commercial})}{u_b(d_t = 0.2 \text{ m})} = \frac{1.57}{1.07}$$

Thus the static bed height for the commercial unit should be

$$L_m(\text{commercial}) = \frac{1.57}{1.07} (2) = \underline{2.93 \text{ m}}$$

As with alternative 1, possible arrangements of internals to give $d_{te} = 1$ m are found from Eq. (13) as follows:

d_i (cm)	10	15	20	25
l_i (cm)	29.4	36.85	43.4	49.6

Vertical tubes of diameter $d_i = 15$ cm in a rectangular array on 36.8-cm spacing seems to be a reasonable geometry.

Height of Bubbling Beds. Finally, we calculate the average height of the operating beds. For the 20-cm ID bench scale bed, Eq. (28) with $u_{mf} \Rightarrow u_{mb}$ gives

$$\delta = \frac{u_o - u_{mb}}{u_b - u_{mb}} = \frac{30 - 1}{107 - 1} = 0.274$$

From Eq. (20) we find

$$\varepsilon_f = \delta + (1 - \delta)\varepsilon_{mb} = 0.274 + (1 - 0.274)(0.6) = 0.710$$

and Eq. (19) gives

$$L_f = \frac{L_m(1 - \varepsilon_m)}{1 - \varepsilon_f} = \frac{2.0(1 - 0.45)}{1 - 0.710} = 3.79 \text{ m}$$

For the large commercial-sized bed we make a similar calculation. Thus we find

		u_b (m/s)	δ	ε_f	L_m(m)	L_f(m)
Bench unit	$d_t = 0.2$ m	1.07	0.274	0.710	2.00	3.79
Commercial unit	$d_{te} = 1$ m	1.57	0.186	0.674	2.93	4.94

Note: This procedure is good only for rough estimations and requires the same fluidizing conditions through scale-up. When data at different conditions are given (for example, at lower u_o and smaller L_m), then first principles must be used to predict reactor performance. For these more detailed and rigorous procedures, see Chaps. 12 and 17 for catalytic reactors and Chap. 18 for noncatalytic gas-solid reactors.

EXAMPLE 5

**Reactor
Scale-up for
Geldart B
Catalyst**

Take the same problem as in Example 4 except that the catalyst falls into the Geldart **B** group. How does this change affect our scale-up procedure?

Data

$$\bar{d}_p = 200 \ \mu\text{m}, \qquad u_{mb} = u_{mf} = 3 \text{ cm/s}$$

$$u_o = 30 \text{ cm/s}, \qquad \varepsilon_{mb} = \varepsilon_{mf} = 0.50$$

SOLUTION

With these solids, bubbles do not quickly reach an equilibrium size, as in the previous example, but grow continuously, as shown in Fig. 10(a).

In the small bench unit (ID = 20 cm), Fig. 10(a) indicates that bubbles should transform into slugs, for which Eq. (5.22) gives

$$u_b = (1)(0.30 - 0.03) + 0.35(9.8 \times 0.2)^{1/2} = 0.76 \text{ m/s}$$

Equation (5.24) suggests that this transformation from bubble to slug would occur at height

$$z_s = 60(20)^{0.175} = 101 \text{ cm}$$

Thus the bed will slug starting about 1 m above the distributor plate.

In the large pilot unit (ID = 1 m), we again find slugging conditions, with

$$u_b = (0.30 - 0.03) + 0.35(9.8 \times 1)^{1/2} = 1.37 \text{ m}$$

and

$$z_s = 60(100)^{0.175} = 134 \text{ cm} = 1.34 \text{ m}$$

With these results we may expect to have large fast bubbles with thin clouds transforming into thin-clouded slugs about 1.4 m up the bed. This situation represents a bed with very poor gas-solid contacting and much gas bypassing. To ensure the same conversion with the bench unit, employ the same rectangular array of tubes as in Alternative 1 of Example 4.

PROBLEMS

1. Electrical utility companies are looking into the possible use of atmospheric fluidized bed coal combustors (AFBC) as a clean, efficient way to produce high-pressure steam. One design of AFBC consists of a very large (4 m × 25 m) but shallow (1.5 m) fluidized bed traversed with hundreds of 5-cm heat exchange tubes. The bed material is to consist of decomposed limestone, ash, and burning coal, and will be kept at 840°C.

    Many problems, such as nonuniform gas feed, changing solid composition as it slowly moves across the combustor, channeling of bed material between the various tube bundles, and so on, are anticipated with such a unit. All this recommends preliminary studies in a relatively cheap cold model.

    For a cold scale model of this monstrous unit, suggest what solids, fluidizing gas, exchanger tube size, and bed height to use.

    Data

    For the AFBC bed solids: $\bar{\rho}_s = 2500 \text{ kg/m}^3$, $\bar{d}_p = 2 \text{ mm}$
    For the gas passing through the bed at 840°C:
    $\rho_g = 0.37 \text{ kg/m}^3$, $\mu = 4.3 \times 10^{-5} \text{ kg/m} \cdot \text{s}$

2. Ultrapure silicon can be made by the reaction

    $$2Zn(g) + SiCl_4(g) \xrightarrow{\text{1200 K}} 2ZnCl_2(g) + Si \downarrow$$

    In a planned semicommercial process, gaseous reactants fluidize a bed of silicon particles ($\rho_s = 2200 \text{ kg/m}^3$), and the silicon produced by the reaction deposits on the particles, which then grow. The bed is to be 50 cm ID and 1 m high. Small seed particles (200 μm) are fed to the reactor, and larger particles ($\bar{d}_p = 800 \ \mu$m) are discharged. The gases enter in a stoichiometric ratio and are rapidly converted to product. The expected conditions of the gas in the reactor are then

    $$M_{gas} = 0.135 \text{ kg/mol} \qquad T = 1200 \text{ K}$$

    $$\mu = 4.6 \times 10^{-5} \text{ kg/m} \cdot \text{s} \qquad \pi = 1.0 \text{ atm}$$

    We want to model this system at close to room temperature, say 32°C, using common materials in order to be able to study ways of introducing the double gas feed, solid removal methods, and bed hydrodynamics. Suggest a reasonable design for this flow model.

3. A small fluidized bed catalytic reactor ($d_t = 0.15$ m) gives good conversion of gaseous reactant, and at this point we want to scale-up operations and build a commercial unit that will process 400 times as much feed. Determine the dimensions of this commercial unit: the static and fluidized bed heights, the bed size, and amount of catalyst (kg) needed.

Data

Small reactor
  Particles: Geldart **A** type
    $\bar{d}_p = 70~\mu m$, $u_{mf} = 0.25~cm/s$, $u_e = 1~cm/s$
    $\rho_s = 1.5~g/cm^3$, $\varepsilon_m = 0.48$, $\varepsilon = 0.58$
  Gas: $u_o = 35~cm/s$,   same for both units
  Equilibrium bubble size: $d_b = 4~cm$
  Bed: $d_t = 15~cm$, $L_m = 1.5~m$

Commercial reactor
  Vertical internals: $d_i = 10~cm$,   rectangular arrangement
  Equivalent diameter: $d_{te} = 1~m$

4.   A laboratory ore roaster gives good conversion of both gas and solid under conditions where axial slugging is observed throughout most of the unit. Estimate roughly the size of a large roaster free of internals that would be able to treat 100 times as much feed to the same conversion.

Data

Laboratory roaster
  Particles: Geldart **B** type
    $\bar{d}_p = 200~\mu m$, $\varepsilon_m = 0.4$
    $u_{mf} = 4~cm/s$, $\varepsilon_{mf} = 0.45$
  Gas: $u_o = 30~cm/s$
  Bed: $d_t = 15~cm$, $L_m = 1.0~m$

Large roaster
  Same u_o as the small roaster.

5.   You may have found that the design of the previous problem is not very satisfactory, so let us consider an alternative that uses vertical internals $(d_i = 8~cm)$ to give the same equivalent bubble size in the big and small units. For this design find
(a) The bed diameter.
(b) The spacing of internals (use an equilateral triangle arrangement).
(c) The depth of the fluidized bed, L_f.

6.   Coarse Geldart **D** solids are to be processed in a wide shallow fluidized bed. With no bed internals in the bed, estimate
(a) The bubble size in the bed.
(b) The average height of the bed surface.

Data

Particles consist of mineral ore
    $\bar{d}_p = 1~mm$, $u_{mf} = 0.58~m/s$
    $\varepsilon_m = 0.4$, $\varepsilon_{mf} = 0.45$
  Bed: $L_m = 1.5~m$, $u_o = 1~m/s$

7.   Repeat Prob. 6 for a bed that contains horizontal tubes as internals. For the internals, use a triangular arrangement, $d_i = 6$ cm, $l_i = 15$ cm.

REFERENCES

1. J. Werther and O. Molerus, *Int. J. Multiphase Flow*, **1**, 103, 123 (1973); J. Werther, *Trans. Inst. Chem. Eng.*, **52**, 149, 160 (1974); in *Fluidization Technology*, vol. 1, D.L. Keairns, ed., p. 215, Hemisphere, Washington, D.C., 1976; in *Fluidization*, J.F. Davidson and D.L. Keairns, eds., p. 7, Cambridge Univ. Press, New York, 1978; *AIChE Symp. Ser.*, **70(141)**, 53 (1974); *German Chem. Eng.*, **1**, 6 (1977).

2. A.B. Whitehead, in *Fluidization*, J.F. Davidson et al., eds., p. 173, Academic Press, New York, 1985.

3. M. Yamazaki, K. Fukuta, and J. Tokumoto, in *Proc. 3rd World Cong. of Chem. Eng.*, Tokyo, 1986.

4. J.S. Lin, M.M. Chen, and B.T. Chao, *AIChE J.*, **31**, 465 (1985).

5. T. Tsutsui, S. Furusaki, and T. Miyauchi, *Kagaku Kogaku Ronbunshu*, **6**, 501 (1980).

6. T.J. Fitzgerald, in *Fluidization*, J.F. Davidson et al., eds., p. 413, Academic Press, New York, 1985.

7. D. Geldart and R.R. Cranfield, *Chem. Eng. J.*, **3**, 211 (1972); R.R. Cranfield and D. Geldart, *Chem. Eng. Sci.*, **29**, 935 (1974).

8. G.S. Canada, M.H. McLaughlin, and F.W. Staub, *AIChE Symp. Ser.*, **74(176)**, 14 (1978); G.S. Canada and M.H. McLaughlin, *AIChE Symp. Ser.*, **74(176)**, 27 (1978).

9. D. Geldart, J.M. Hurt, and P.H. Wadia, *AIChE Symp. Ser.*, **74(176)**, 60 (1978).

10. G. Miller et al., *AIChE Symp. Ser.*, **77(205)**, 166 (1981).

11. L.R. Glicksman, W.K. Lord, and M. Sakagami, *Chem. Eng. Sci.*, **42**, 479 (1987); L.R. Glicksman and G. McAndrews, *Powder Technol.*, **42**, 159 (1985).

12. A.R. Abrahamson and D. Geldart, *Powder Technol.*, **26**, 35, 47 (1980).

13. J. Werther, in *Fluidization IV*, D. Kunii and R. Toei, eds., p. 93, Engineering Foundation, New York, 1983; K. Hilligardt and J. Werther, in *Proc. 3rd World Cong. of Chem. Eng.*, Tokyo, 1986.

14. A.C. Hoffmann and J.G. Yates, *Chem. Eng. Comm.*, **41**, 133 (1986).

15. A.W. Weimer and G.J. Quarderer, *AIChE J.*, **31**, 1019 (1985).

16. K. Hilligardt and J. Werther, *German Chem. Eng.*, **9**, 215 (1986).

17. R. Yamazaki et al., in *Fluidization '85, Science and Technology*, M. Kwauk et al., eds., p. 63, Science Press, Beijing, 1985; in *Proc. 3rd World Cong. of Chem. Eng.*, Tokyo, 1986.

18. J. Werther, *German Chem. Eng.*, **1**, 166 (1978); W. Bauer, J. Werther, and G. Emig, *German Chem. Eng.*, **4**, 291 (1981).

19. GOLFERS, *Kagaku Kogaku Ronbunshu*, **8**, 464 (1982).

20. S. Mori and C.Y. Wen, *AIChE J.*, **21**, 109 (1975).

21. M. Horio and A. Nonaka, *AIChE J.*, **33**, 1865 (1987).

22. J.F. Davidson and D. Harrison, *Fluidized Particles*, Cambridge Univ. Press, New York, 1963.

23. S. Morooka, K. Tajima, and T. Miyauchi, *Kagaku Kogaku*, **35**, 680 (1971); *Int. Chem. Eng.*, **12**, 168 (1972).

24. M. Horio et al., in *Fluidization V*, K. Østergaard and A. Sørensen, eds., p. 151, Engineering Foundation, New York, 1986; *AIChE J.*, **32**, 1466 (1986).

25. T.J. Fitzgerald and S.D. Crane, in *Proc. 6th Int. Conf. on Fluidized Bed Combustion*, vol. 3, p. 815, Atlanta, GA, 1980.

26. M.T. Nicastro and L.R. Glicksman, *Chem. Eng. Sci.*, **39**, 1381 (1984); T.L. Jones and L.R. Glicksman, paper delivered annual AIChE meeting, November 1984.

27. R. Roy and J.F. Davidson, in *Fluidization VI*, J.R. Grace et al., eds., p. 293, Engineering Foundation, New York, 1989.

28. R.D. Toomey and H.F. Johnstone, *Chem. Eng. Prog.*, **48**, 220 (1952).

29. D. Kunii and O. Levenspiel, *Ind. Eng. Chem. Process Des. Dev.*, **7**, 481 (1968); *Fluidization Engineering*, Wiley, New York, 1969.

7

Entrainment and Elutriation from Fluidized Beds

Freeboard Behavior

In vessels containing fluidized solids, the gas leaving carries some suspended particles. This flux of solids is called *entrainment*, G_s (kg/m$^2 \cdot$ s), or *carryover*, and the bulk density of solids in this leaving gas stream, $\bar{\rho}$ (kg/m^3), is called the *holdup*. For design we need to know the rate of this entrainment and the size distribution of these entrained particles, $\mathbf{p}_e(d_p)$, in relation to the size distribution in the bed, $\mathbf{p}_b(d_p)$, as well as the variation of both these quantities with gas and solid properties, gas flow rate, bed geometry, and location of the leaving gas stream. This chapter considers these questions.

First we define a number of terms used in connection with this aspect of fluidization. A fluidization vessel usually has two zones: a *dense bubbling phase* having a more or less distinct upper surface separating it from an upper *lean* or *dispersed phase* in which the density of solid decreases with height. The section of the vessel between the surface of the dense phase and the exit of the gas stream is called the *freeboard*, and its height is called the *freeboard height* H_f.

Because the density of solids decreases with height in the freeboard, increasing the freeboard decreases the entrainment from the bed. Eventually, a freeboard height is reached above which entrainment does not change appreciably. This is called the *transport disengaging height* (TDH). When the gas stream exits above the transport disengaging height, or $H_f >$ TDH, then both the size distribution and entrainment rate are close to constant.

The *saturation carrying capacity* of the gas stream represents the largest flux of solid that can be entrained out of the vessel whose exit is above the TDH. This depends on the properties of the particles and on the flow conditions of the gas.

The solids thrown up into the freeboard contain the whole spectrum of particle sizes present in the bed. The larger particles fall back to the bed, whereas the smaller may be carried out of the bed. Thus, the size distribution of solids in the lean phase changes with height, and this becomes a zone for separation of particles by size. *Elutriation* refers to the separation or removal of fines from a mixture, and occurs to a lesser or greater extent at all freeboard heights. At small H_f many of the larger particles are entrained by the gas. This is not so at larger H_f. Figure 1 illustrates these terms.

On increasing the gas velocity through the vessel, flow in the bed shifts to the turbulent regime, the demarcation between bed and freeboard, or the *splash zone*, becomes hazy, and entrainment rises sharply. A further increase in gas velocity leads to fast fluidization, as mentioned in Chap. 3. We restrict our discussion here to entrainment and elutriation from bubbling or turbulent beds, and leave consideration of fast fluidization to the next chapter.

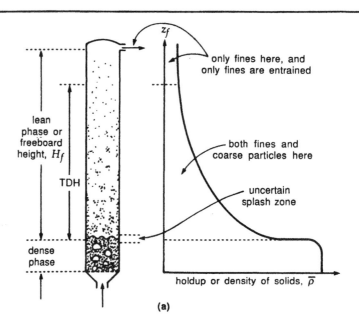

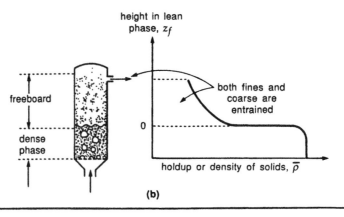

FIGURE 1
Terms used in describing the lean phase above a fluidized bed.

Origin of Solids Ejected into the Freeboard

It is important to know how it is that the freeboard above a bubbling gas-solid fluidized bed contains solids, whereas smoothly fluidized liquid-solid beds do not, except for those fines that are being elutriated from the bed.

Observation and measurements show that it is the bubbles and slugs breaking at the surface of the bed that throw solids into the freeboard. This works in three possible ways, as sketched in Fig. 2.

- Since the bubbles' pressure is higher than bed surface pressure, they "pop" on reaching the surface, spraying solids from the bubble roofs into the freeboard. See Fig. 2(a).
- Since bubbles with their wakes may rise very much faster than the surrounding medium, this wake material may be thrown as a clump into the freeboard. See Fig. 2(b).
- Finally, when two bubbles coalesce just as they break the surface of the bed, one observes an especially energetic ejection of wake solids from the trailing bubble into the freeboard. See Fig. 2(c).

In bubbling beds, it is mainly the wake material that is thrown into the freeboard, whereas in slugging beds it is the roof solids that spray into the freeboard. Also note that the solids thrown into the freeboard are a representative sample of the bed solids, not just the fines. A number of consequences follow from these findings.

- We now see that liquid-solid beds have no freeboard solids because they fluidize smoothly with no bubbling action.

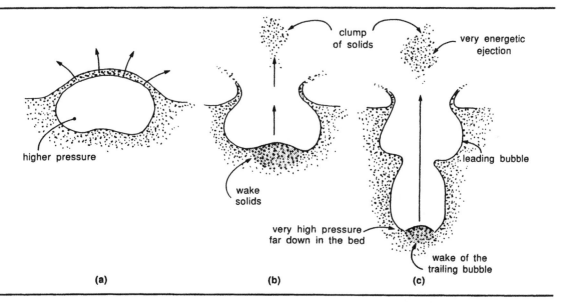

FIGURE 2

Mechanism of ejection of solids from a fluidized bed into the freeboard: (a) from the roof of a bursting bubble; (b) from the bubble wake; (c) from the wake of a trailing bubble just as it coalesces with its leading bubble.

- Entrainment into the freeboard and solid density in the freeboard should be strongly affected by bubble size and bed hydrodynamics. Reducing the size of bursting bubbles by stirrers or bed internals can drastically lower entrainment.
- Elutriation and carryover above TDH should be strongly affected by u_o and the fraction of fines in the bed, but they are unaffected, or at most only slightly affected, by changing the size of the coarse material and by u_{mf}. Bubble size should have some effect, because it determines the quantity of fines thrown into the freeboard to be picked up by the flowing gas.
- Even in beds consisting only of entrainable fines, carryover proceeds rather slowly because it is the saturating carrying capacity of the gas that governs the rate of removal of solids from vessels when $H_f > $ TDH. Thus, in the bed most of the solids move aside and are bypassed as the speedy bubbles rush past carrying most of the gas, whereas in the freeboard below the TDH clumps of ejected solids, acting as larger entities, fall back to the bed.

Experimental Findings

There have been many studies on the lean zone above fluidized beds, and qualitatively they give a coherent picture of the behavior in the freeboard. However, quantitatively we find considerable disagreement between findings, sometimes by an order of magnitude and more. Some reasons for this stem from the different and unaccounted for physical conditions used by researchers, such as large versus small bed diameters, slugging versus bubbling behavior, narrow size cuts versus wide size distributions of solids being used, different, and sometimes unsuitable, measurement techniques, beds that are so shallow that distributor jets punch through to the bed surface, freeboards less than the TDH, which allow coarse solids to be entrained, and so on.

Table 1 lists the steady state entrainment studies for fine particle (Geldart A) systems for which $u_o > u_t$ for most of the particles in the bed. Table 2 lists

TABLE 1 · Experimental Investigations of Steady State Entrainment, Fine Particle

| Investigators | Vessel size (m) | | Distributor Type |
	d_t	L	
Fournol et al. [1] (1973)	0.61 ID	5.8	Perforated plate
Nazemi et al. [2] (1974)	0.61 ID	6.8	Perforated plate
Zhang et al. [3] (1982, 85)	0.5 ID	7, 10	Conical cap
Morooka et al. [4] (1983)	0.066, 0.12	2	Perforated plate
Chen et al. [5] (1986)	0.80	7	—
Kato et al. [6] (1986)	0.15 × 0.15	—	Perforated plate

the steady state studies for large particle (Geldart **B** and **D**) systems, where the gas velocities are much higher and most of the bed solids are not elutriated when $H_f >$ TDH.

Entrainment from Small Particle Beds. Qualitatively, the findings are as follows:

- Vertical internals in the bed do not significantly affect entrainment (Zhang et al. [3], Kato et al. [6]).
- The flux of solids G_s [kg/m$^2 \cdot$s] is practically uniform across the bed, except right at the wall (Fournol et al. [1]). There Morooka et al. [4] found a zone, 4–5 mm thick, of descending particles.

Entrainment from Large Particle Beds. Qualitatively we find

- Horizontal louvers close to the bed surface reduce entrainment about 33% (Martini et al. [7]).
- Horizontal tubes in the bed do not affect the rate of entrainment (George and Grace [10]).
- Vertical internals cause an increase in entrainment (Kato et al. [6]).
- The upward velocity of solids is uniform across the bed except near the wall, where there is a downflow of solids. The thickness of this downflow layer decreases with height above the bed surface from 20 mm to zero (Horio et al. [9]).
- An increase in pressure increases entrainment enormously and changes the size distribution of solids by including more of the larger solids; see Fig. 3 (Chan and Knowlton [12]).

Upflow and Downflow of Solids. Since the clumps of solids thrown into the freeboard by the bursting bubbles are representative of the bed material, and since more of the larger solids in these clumps fall back to the bed, there is an upflow flux G_{su} and downflow flux G_{sd} of solids everywhere in the freeboard

Systems

Bed Particles	Gas	
ρ_s (kg/m^3), d_p (μm)	u_o (m/s)	Remarks
FCC, iron powder $\quad d_p = 58,\ 38$	Air $\quad 0.11–0.23$	—
Catalyst $\quad \rho_s = 840,\ d_p = 59$	Air $\quad 0.09–0.34$	—
FCC $d_p = 58$ silica gel $d_p = 189$	Air $\quad 0.3–0.7$	With and without internals in $\quad$ bed and freeboard
FCC $\quad \rho_s = 1030,\ d_p = 60$	Air $\quad 0.5–2.5$	—
Catalyst $\quad \rho_s = 960,\ d_p = 63$	Air $\quad 0.2–0.55$	—
FCC catalyst $\quad d_p = 58$	Air $\quad 0.2–1.6$	Internals in bed

TABLE 2   Experimental Investigations of Steady State Entrainment, Coarse Particle

Investigators	Vessel size (m)		Distributor Type
	d_t	L	
Martini et al. [7] (1976)	0.61 ID	10	Perforated plate
Gugnoni and Zenz [8] (1980)	0.914 ID	8.6	Sparger
Horio et al. [9] (1980)	0.24 ID	2.6	Perforated plate
George and Grace [10] (1981)	0.254 × 0.432	3	Perforated plate
Bachovchin et al. [11] (1981)	0.15 ID	0.8–3.6	Porous plate
Chan and Knowlton [12] (1983)	0.29 ID	3.92	Bubble cap type
Pemberton and Davidson [13] (1983)	0.1, 0.6 ID	1.5–2.8	—
Geldart and Pope [14] (1983)	0.29 ID	5	—
Ismail and Chen [15] (1984)	0.2 × 0.3	3	Porous plate
Walsh et al. [16] (1984)	0.6 × 0.6	4.4	Tube tuyeres
Horio et al. [17] (1985)	0.26 ID	2.4	—
Hoggen et al. [18] (1986)	10 ID	5	Nozzles

below the TDH. The net flux or carryover out of the vessel G_s is related to these upflows and downflows at any level in the bed by

$$G_s = G_{su} - G_{sd} \tag{1}$$

These fluxes have been measured by various investigators, as follows:

- Hoggen et al. [18] measured G_{su} in a large industrial roaster, as shown in Fig. 4(a).
- Geldart and Pope [14] found that increasing the fraction of elutriable fines in the bed and, hence, in the freeboard, greatly increased the amount of coarse solids present at any level in the freeboard. They attributed this to

Systems

Bed Particles	Gas	
$\rho_s \ (kg/m^3), d_p \ (\mu m)$	$u_o \ (m/s)$	Remarks
Silica $\quad d_p = 40{-}250,$ $\quad \bar{d}_p = 142$	Air, 0.31	Effect of louver close to bed surface
Glass beads $\quad \rho_s = 2400,$ $\quad d_p = 20{-}220$	Air, 0.15–0.64	—
Glass beads $\quad d_p = 57$ from 164–1350	Air, 0.3–0.59	—
Silica $\quad \rho_s = 2630$ $\quad d_p = 30{-}180$	Air 173°C, 0.53–1.13	—
Sand $\quad \rho_s = 2600$ $\quad \bar{d}_p = 120$ from 120–715	Air, 0.61–1.25	—
Sand $\quad \rho_s = 2600$ $\quad d_p = 37{-}350$	N_2 up to 3100 kPa 0.2–0.5	Effect of pressure
Polymer with size distribution $\quad \bar{d}_p = 760$	Air, N_2 up to 20 bar 0.3–0.5	Effect of pressure
Sand with size distribution $\quad \bar{d}_p = 200, 1000, 2500$	Air, 2–5	Effect of recycle of fines
Glass beads $\quad \bar{d}_p = 300{-}850$	Air, 0.17–2.5	—
Silica $\quad \rho_s = 2650$ $\quad \bar{d}_p = 755$	Air, 0.42–0.86	Flux of descending particles
Glass $\quad \bar{d}_p = 41$ from 500	Air, 0.23–0.64	—
Porous hematite $\quad \bar{d}_p = 150{-}600$	Roaster gas 757–827°C, 0.46–0.61	Industrial roaster

the fact that the collision between fast upflowing fines and the larger particles hinders the fall of the larger particles; see Fig. 4(b).

Splashing and Entrainment at the Bed Surface. Lewis et al. [19] proposed that the bursting of bubbles was the means for getting particles into the freeboard (see Fig. 2). We designate the initial upward flux of these solids from the bed surface by G_{su0} [kg/m$^2 \cdot$ s]. Various experiments verify the essential correctness of this view.

• Caram et al. [20] found that the initial velocity of ejected particles

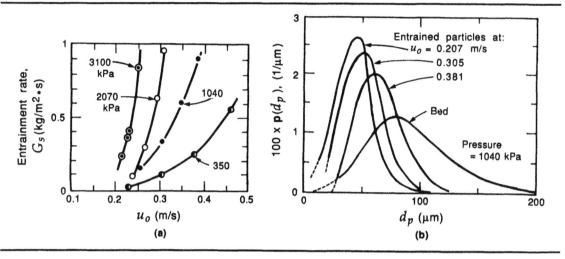

FIGURE 3
Effects of pressure on the entrained solids; values calculated from the reported data of Chan and Knowlton [12]: (a) At high pressure the entrainment rate is more sensitive to gas flow rate. (b) The size distribution of entrained solids approaches that of the bed at higher gas velocities.

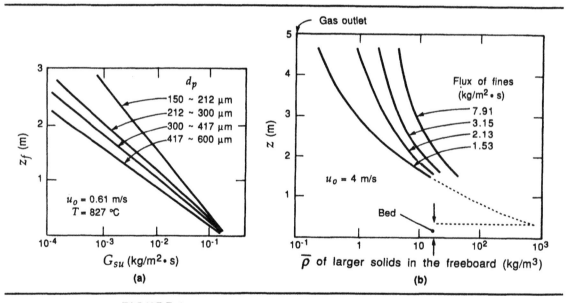

FIGURE 4
Entrainment from beds of coarser particles: (a) upflow flux of solids in the freeboard of a large industrial roaster; adapted from Hoggen et al. [18]; (b) the effect of fines on the holdup of larger solids ($d_p > 500$ μm) in the freeboard; adapted from Geldart and Pope [14].

depended only on the size, hence the rise velocity of the bubble, not on the particle size and density.

- Hoggen et al. [18] found that particle ejection velocities followed a Gaussian distribution with a maximum close to the bubble rise velocity.
- Horio et al. [9] also found that the intensity of turbulence at the bed surface was simply related to bubble size.

Location of the Gas Outlet of a Vessel

Some of the following factors may have to be considered in locating the gas outlet or cyclone inlet for a fluidized bed unit. First of all, there is the TDH. Freeboard heights greater than the TDH will not reduce the carryover of solids but will require taller vessels. However, freeboard heights smaller than TDH will result in more carryover of solids and an increased duty for the solid separation equipment. So if no other factors intrude, the TDH often becomes the economically desirable location for the gas exit port.

When the unit contains a cyclone for continually returning captured fines to the bed, the cyclone dipleg must be long enough or the unit will not work. This could well be the controlling factor in determining the needed freeboard height. Chapter 15 on pressure balances considers this problem.

In reactors, the kinetics may be such that one may need the additional gas residence time, which can be provided by a large freeboard.

Finally, for some chemical reactors, a large entrainment of solids may help to scrape out deposits of minor by-products that would otherwise accumulate and clog the inner surface of the cyclones. In these situations one would need detailed information on entrainment rates at different freeboard heights rather than just the value of the TDH. This information may come from the model to be presented later in this chapter.

Estimation of the TDH

We present two methods for calculating the TDH, both tested only with fine particle (Geldart **A**) systems under conditions where $u_o > u_t$ for most of the bed solids. For beds of larger particles (Geldart **B** and **D**), wherein a large fraction of the solids may be too large to be elutriated ($u_o < u_t$), we must go to the flow model developed at the end of this chapter.

Method 1. Based on their engineering experience, Zenz and Weil [21] proposed the correlation of Fig. 5 for estimating the TDH for 20–150 μm FCC catalyst.

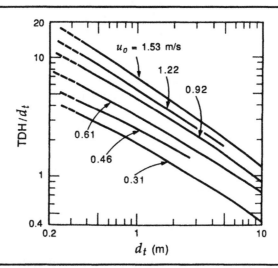

FIGURE 5
Correlation for estimating the TDH for fine particle (Geldart A) beds; adapted from Zenz and Weil [21].

Method 2. Fournol et al. [1] also fluidized fine FCC catalyst ($\bar{d}_p = 58\ \mu$m) in a tall, 0.61-m ID, column and determined the level in the freeboard where the density and size distribution of solids leveled off. From these experiments, they proposed the following expression for the TDH as a function of superficial gas velocity for their fine solids:

$$\text{Froude number:} \qquad \frac{u_o^2}{g(\text{TDH})} = 10^{-3} \tag{2}$$

One may expect this equation to apply to other fine particle systems, with values other than 10^{-3}. So for any fine particle system, if one finds the TDH at any superficial gas velocity, this equation should give the TDH at any other velocity.

Neither of these methods has been tested on larger Geldart **B** and **D** systems, so they should be used there with caution.

Entrainment from Tall Vessels: $H_f > $ **TDH**

We have two somewhat different-looking, but essentially similar, approaches to the determination of elutriation rates from vessels with freeboard heights greater than the TDH. The first method is due to Zenz et al.; the second uses the elutriation constant. Both approaches rest on the assumption that the flux rate of any particular size of solid i is proportional to its weight fraction x_i in the bed, all other factors kept constant, or

$$G_{si} = x_i G_{si}^* \tag{3}$$

where G_{si}^* is the flux rate from an imaginary bed of solids, all of size i. Thus it is the saturation carrying capacity of the gas for that particular size of solid.

Procedure of Zenz et al. [21,8]

1. Divide the size distribution into narrow intervals and find which intervals have $u_t < u_o$. These solids are entrained. For the intervals where $u_t > u_o$, the solids are not entrained.

2. Find G_{si}^* for each interval of elutriable solids from the appropriate curve of Fig. 6. Note here that one curve refers to fine particle (Geldart **A**) systems (more than 90% entrainable), the other to fines removed from larger particle (Geldart **AB**) systems.

3. Under the assumption of Eq. (3), the total entrainment is

$$G_s = \sum_{\substack{\text{all elutriable} \\ \text{size intervals, } i}} x_i G_{si}^* \tag{4}$$

In terms of the continuous size distribution of elutriating particles, $\mathbf{p}_e(d_p)$, Eq. (4) becomes

$$G_s = \int_{\substack{\text{all} \\ \text{particles}}} G_{si}^* \mathbf{p}_e(d_p)\,d(d_p) \tag{5}$$

According to Matsen [22], this approach works well for fine particle systems used in the petroleum industry, but he warns that, to expect reliable scale-up, entrainment data should be taken in reasonably sized columns, at least 0.3–0.6 m ID and 6 m high.

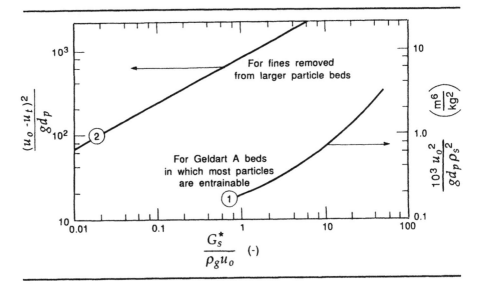

FIGURE 6
Dimensionless plot for estimating the saturation carrying capacity G_s^* of a gas; curve ① from Zenz and Weil [21]; curve ② adapted from Gugnoni and Zenz [8].

The Elutriation Constant Approach

On fluidizing particles of wide size distribution in a vessel with a high enough freeboard ($>$ TDH), the flux of particles of size i out of the bed, from the assumption leading to Eq. (3), may be written as

$$-\frac{1}{A_t}\frac{dW_i}{dt} = \kappa_i^* x_i = \kappa_i^* \left(\frac{W_i}{W}\right) \tag{6}$$

where κ^* [kg/m$^2\cdot$s] is called the *elutriation (rate) constant*.

From the discussion leading to Eq. (3), we see that $\kappa_i^* = G_{si}^*$ refers to the flux rate of solids i if they were alone in the bed. Large κ_i^* means rapid removal of that size of solids from the bed, and $\kappa_i^* = 0$ means that those solids are not removed at all by entrainment.

Another elutriation constant can also be defined as follows:

$$\left(\begin{array}{c}\text{rate of removal} \\ \text{of solids } i\end{array}\right) = \kappa_i \left(\begin{array}{c}\text{weight of that size} \\ \text{of solid in the bed}\end{array}\right)$$

or

$$-\frac{dW_i}{dt} = \kappa_i W_i, \qquad \kappa_i = [\text{s}^{-1}] \tag{7}$$

Comparing definitions, we see that

$$\kappa_i^* = \frac{\kappa_i W}{A_t} = \kappa_i \rho_i (1 - \varepsilon_m) L_m \tag{8}$$

Note that κ_i varies inversely with bed height. Also, in batch or unsteady state experiments, κ_i should change during a run as the bed weight W changes. On the contrary, κ_i^* is unaffected by these changes; it is the true rate constant and is the quantity to use when reporting data and presenting correlations.

The elutriation constant procedure was introduced by Leva [23] and Yagi and Kunii [24].

Relationship between κ and G_s

For a bed containing a single size i of solids that is elutriable, the carryover from the vessel is

$$\kappa_i^* = G_{si}^* = \left(\begin{array}{c} \text{saturation carrying capacity} \\ \text{of gas for solids } i \end{array} \right) \tag{9}$$

If the fluidized bed consists of coarse solids plus only one size of elutriable solids of mass fraction x_1, then the total carryover is

$$G_s = x_1 \kappa_1^* = x_i G_{s1}^* \tag{10}$$

If the bed contains coarse solids plus sizes $1, 2, \ldots, n$ of elutriable solids, then the total entrainment is

$$G_s = \sum_{i=1}^{n} x_i \kappa_i^*, \qquad \sum x_i < 1 \tag{11}$$

If all the bed solids are elutriable, simply put $\Sigma x_i = 1$ in Eq. (11).

Alternatively, if one considers a continuous size distribution in the bed, then the total flux of elutriated solids is

$$G_s = \int_{\text{all sizes}} \kappa^*(d_p) \mathbf{p}_b(d_p) \, d(d_p) \tag{12}$$

Remember that these expressions only refer to fluxes from beds taller than the TDH.

Experimental Methods for Finding κ and κ^*

There are three experimental ways of obtaining elutriation kinetics, as shown in Fig. 7.

Solids Flow Experiment. Figures 7(a) and (b) illustrate two versions of the steady state experiment. Here one measures the carryover rate and weight fraction of size i in the bed. Then from Eq. (6), or in terms of a continuous size distribution of bed solids,

$$\kappa_i^* = \frac{(\text{flux of } i \text{ from the bed, kg/m}^2 \cdot \text{s})}{(\text{weight fraction of } i \text{ in the bed})} = \frac{G_s \mathbf{p}_e(d_p)}{\mathbf{p}_b(d_p)} \tag{13}$$

Batch Experiment. In this approach, shown in Fig. 7(c), we need only know the initial bed composition and the composition at some later time. If the total mass of bed does not change much in this time interval, say $< 20\%$, then integration of Eq. (6) gives

$$\frac{W_i}{W_{i0}} = \exp\left(-\frac{\kappa_i^* A_t t}{W} \right) = \exp(-\kappa_i t) \tag{14}$$

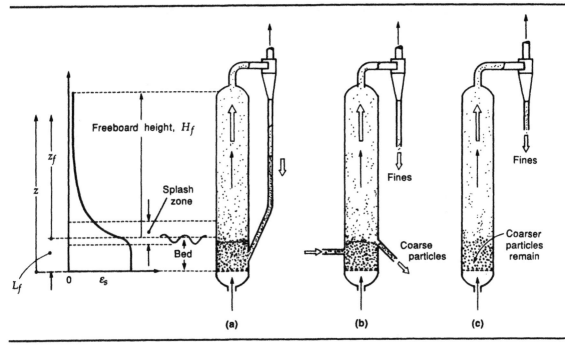

FIGURE 7
Different ways of experimentally studying entrainment from fluidized and turbulent beds: (a) steady state recirculating system; (b) steady state once-through system; (c) batch unsteady state system.

If the bed weight changes significantly during the experimental run—for example, if most of the bed solids are elutriable at the gas velocity being used—then one has to measure all the size fractions before and after the run. Calling the elutriating size fractions $1, 2, \ldots, n$ and the nonelutriating fraction I and integrating Eq. (6) after rearrangement, we get

$$\frac{W_1}{W_{10}} = \left(\frac{W_2}{W_{20}}\right)^{\kappa_1^*/\kappa_2^*} = \left(\frac{W_3}{W_{30}}\right)^{\kappa_1^*/\kappa_3^*} = \cdots \tag{15}$$

and

$$\kappa_1^* A_t t = W_I \ln\left(\frac{W_{10}}{W_1}\right) + \sum_{i=1}^{n} W_{i,0} \frac{\kappa_1^*}{\kappa_i^*}\left[1 - \left(\frac{W_1}{W_{10}}\right)^{\kappa_i^*/\kappa_1^*}\right] \tag{16}$$

Measuring the bed composition before and after a run and inserting into Eq. (15) gives the ratio of κ^* values. Then Eq. (16) gives κ_1^*, from which all other κ^* values can be found.

The batch experiment is much simpler to set up and perform, but the results must be analyzed properly. Some investigators have analyzed their results with Eq. (14) when this was unjustified and when they should have used Eqs. (15) and (16).

Experimental Findings for κ^*

There are practical reasons for having reliable elutriation data. For example, the extent of conversion of solids in gas-solid reactors depends on the residence time

TABLE 3   Correlations for Elutriation Rate Constant

Investigators	Experimental Conditions			
	d_t (m)	$d_{p,coarse}$ (μm)	d_{pi} (μm)	u_o (m/s)
Yagi and Aochi [25] (1955)	0.07–1.0	100–1600	80–300	0.3–1.0
Wen and Hashinger [26] (1960)	0.102	~710	50–150	0.61–0.98
Tanaka et al. [27] (1972)	0.031–0.067	718–1930	106–505	0.9–2.8
Merrick and Highley [28] (1974)	0.91 × 0.91 0.91 × 0.46	63–1000	ash, 8–100	0.61–2.44
Geldart et al. [29] (1979)	0.076 0.30	60–350 ~1500	60–300	0.6–3
Colakyan and Levenspiel [30] (1984)	0.92 × 0.92 0.30 × 0.30	300–1000	36–542	0.901–3.66
Kato et al. [6] (1986)	0.15 × 0.15	58–282	37–150	0.2–1.1

of particles in the vessel, and this in turn depends strongly on the elutriation rate. Thus, a high elutriation rate means a short mean residence time of these solids.

Table 3 lists the experimental findings on elutriation, the experimental conditions used, and the recommended correlations for κ^*. If one needs κ^* values, choose the correlation for the conditions that most closely match those to be used. The size range of the particles as well as gas velocity are the most important factors to check.

In this regard, the correlations of Yagi and Aochi [25], Wen and Hashinger [26], and Geldart et al. [29] are for elutriation of small particles from bubbling or turbulent beds. Merrick and Highley's correlation [28] is for the elutriation of <63-μm fines at high gas velocity, and Colakyan and Levenspiel's correlation [30] deals with elutriation of coarse particles at high gas velocity.

George and Grace [10] took data on the elutriation of four sizes of sand (d_p = 30, 46, 64, 90 μm) from a sand bed (d_p = 30–180 μm) and, in Fig. 8, compared their experimental results with the fine particle correlations of Table 3. The results are similar. In addition, they found that the fine particle curve A of Fig. 6, rather than the coarser particle curve B, fitted their data.

Both Geldart et al. [29], for fine particle systems, and Colakyan and Levenspiel [30], for large particle systems, found no change of κ^* with bed diameter. In addition, the data of Bachovchin et al. [11] show that the

Correlation,
$\text{Re}_t = d_{pi}\,\rho_g u_{\ ti}/\mu$

$$\frac{\kappa_i^* g d_{pi}^2}{\mu(u_o - u_{ti})^2} = 0.0015\,\text{Re}_t^{0.5} + 0.01\,\text{Re}_t^{1.2}$$

$$\frac{\kappa_i^*}{\rho_g(u_o - u_{ti})} = 1.52 \times 10^{-5}\,\frac{u_o - u_{ti}}{(g d_{pi})^{0.5}} \times \text{Re}_t^{0.725}\left(\frac{\rho_s - \rho_g}{\rho_g}\right)^{1.15}$$

$$\frac{\kappa_i^*}{\rho_g(u_o - u_{ti})} = 0.046\,\frac{(u_o - u_{ti})}{(g d_{pi})^{0.5}} \times \text{Re}_t^{0.3}\left(\frac{\rho_s - \rho_g}{\rho_g}\right)^{0.15}$$

$$\frac{\kappa_i^*}{\rho_g u_o} = 0.0001 + 130\exp\left[-10.4\left(\frac{u_{ti}}{u_o}\right)^{0.5}\left(\frac{u_{mf}}{u_o - u_{mf}}\right)^{0.25}\right]$$

$$\frac{\kappa_i^*}{\rho_g u_o} = 23.7\exp\left(-5.4\,\frac{u_{ti}}{u_o}\right)$$

$$\kappa_i^* = 0.011\rho_s\left(1 - \frac{u_{ti}}{u_o}\right)^2, \qquad \rho_s\,(\text{kg/m}^3)$$

$$\frac{\kappa_i^*}{\rho_g(u_o - u_{ti})} = 2.07 \times 10^{-4}\,\text{Fr}^\alpha\,\text{Re}_t^{1.6}\left(\frac{\rho_s - \rho_g}{\rho_g}\right)^{0.61}$$

$\alpha = \text{Re}_t^{-0.6}$, $\text{Fr} = (u_o - u_{ti})^2/g d_{pi}$
for Geldart group **A** particles

assumption leading to Eq. (3) and the whole treatment of this section dealing with the relation of flux rate of a particular size of solid versus its mass fraction in the bed holds well.

Regarding high-temperature data, Yagi and Aochi's correlation [25] includes data on an industrial reactor for the nitrogenation of calcium carbide at 1043°C, whereas Merrick and Highley's [28] is for elutriation from coal combustors between 700 and 800°C.

For high-pressure operations no particular correlations have been reported. However, we may expect the correlation of Table 3 to be as reliable there as at atmospheric conditions.

EXAMPLE 1

Entrainment from Fine Particle Beds with High Freeboard

Calculate the rate of entrainment from a vessel fluidizing fine particles at high pressure. At the gas velocity used all the bed particles are entrainable, and the vessel has a very high freeboard.

Data

$$\rho_g = 5.51\,\text{kg/m}^3, \quad \rho_s = 1200\,\text{kg/m}^3, \quad \bar{d}_p = 130\,\mu\text{m}, \quad u_o = 0.61\,\text{m/s}$$

SOLUTION

Assuming that the freeboard is higher than the TDH, we estimate the entrainment

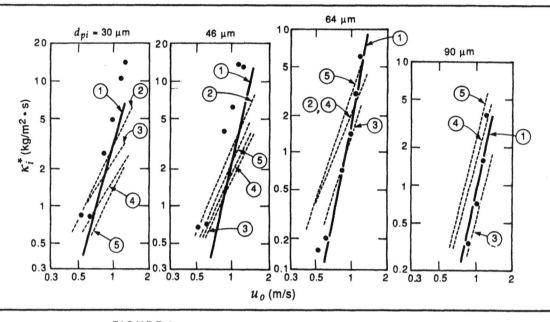

FIGURE 8
Elutriation from beds of silica sand, $d_p = 30$–$180\ \mu m$, $\bar{d}_p = 102\ \mu m$. Comparison of measured elutriation constant (circles) of George and Grace [10] with the literature correlations of ① Zenz and Weil [21]; ② Tanaka et al. [27], ③ Geldart et al. [29], ④ Yagi and Aochi [25], and ⑤ Wen and Hashinger [26].

rate from the saturation carrying capacity by Zenz and Weil's method. Thus,

$$\frac{u_o^2}{g d_p \rho_s^2} = \frac{(0.61)^2}{(9.8)(1.3 \times 10^{-4})(1200)^2} = 0.203 \times 10^{-3}\ m^6/kg^2$$

From Fig. 6, $G_s^*/\rho_g u_o = 1.2$; hence the solid density (or loading) is

$$\frac{G_s^*}{u_o} = (1.2)(5.51) = 6.61\ kg/m^3$$

or

$$G_s^* = (6.61)(0.61) = \underline{\underline{4.03\ kg/m^2 \cdot s}}$$

EXAMPLE 2

Entrainment from Large Particle Beds with High Freeboard

Repeat Example 1 for a bed in which the fines constitute just 20% of the bed solids. The other 80% are larger solids that are not carried out of the bed at the gas velocity used.

SOLUTION

Here only 20% of the solids ejected into the freeboard are entrainable when $H_f >$ TDH. So with a freeboard above the TDH, Eq. (3) with the results of Example 1 gives

$$G_s^* = (0.2)(4.03) = \underline{\underline{0.806\ kg/m^2 \cdot s}}$$

EXAMPLE 3

Entrainment from Beds with a Wide Size Distribution of Solids

Calculate the total entrainment and the solid loading at the exit port of a bed of fine catalyst fluidized under the conditions of Example 1.

Data $d_t = 6\,m$, and the size distribution of bed solids is as follows:

d_p, (μm)	10	30	50	70	90	110	130
$p(d_p)$, (μm^{-1})	0	0.0110	0.0179	0.0130	0.0058	0.0020	0

SOLUTION

Remembering that $H_f > TDH$, first calculate $u_0^2/gd_{pi}\rho_s^2$ for each particle size, and then estimate $G_{si}^*/\rho_g u_0$ from Fig. 6, similar to Example 1. Then, from Eq. (5),

$$\frac{G_s}{\rho_g u_0} = \int_{d_p=10}^{130} \left(\frac{G_s^*}{\rho_g u_0}\right)_{\substack{each \\ size}} p_e(d_p)\,d(d_p) = 5.88$$

Thus the solid loading is .

$$\bar{\rho}_s = \frac{G_s}{u_0} = (5.88)(5.51) = \underline{\underline{32.4\,kg/m^3}}$$

and the total entrainment is

$$A_t G_{s,total}^* = \left(\frac{\pi}{4}\,6^2\right)(32.4)(0.61) = \underline{\underline{559\,kg/s}}$$

EXAMPLE 4

κ^* from Steady State Experiments

Calculate the elutriation constant κ_i^* for 40- to 120-μm particles from the experiments at 1040 kPa and $u_0 = 0.381$ m/s reported by Chan and Knowlton and displayed in Fig. 3.

SOLUTION

From Fig. 3(a), $G_s = 0.9$ kg/m^2 · s. Next, from Fig. 3(b), we find the size distribution functions for bed particles $p_b(d_p)$ and entrained particles $p_e(d_p)$. These are tabulated below. Finally, the κ_i^* values for the various particle sizes are found from Eq. (13) and are added to the table below.

d_{pi} (μm)	40	60	80	100	120
$100 p_b(d_{pi})$ (μm^{-1})	0.45	1.00	1.25	1.00	0.60
$100 p_e(d_{pi})$ (μm^{-1})	1.20	2.00	1.25	0.45	0.10
κ_i^* (kg/m^2 · s)	2.4	1.8	0.90	0.41	0.15

We would expect the elutriation constant to decrease progressively as particle size increases, and this is what the table shows.

EXAMPLE 5

Comparing Predictions for κ^*

Calculate κ^* values from the correlations in the literature (see Table 3) for the fine elutriable fractions present in a bed of coarse and fine particles. In a table compare these predictions.

Data

$$\rho_g = 1.217\,kg/m^3\,, \quad \mu = 1.8 \times 10^{-5}\,kg/m \cdot s\,, \quad u_{mf} = 0.11\,m/s$$

$$\rho_s = 2000\,kg/m^3\,, \quad u_0 = 1.0\,m/s$$

The terminal velocity of particles of various sizes is given as follows:

d_{pi} (μm)	30	40	50	60	80	100	>120
u_{ti} (m/s)	0.066	0.115	0.175	0.240	0.385	0.555	>1.0

SOLUTION

We show the calculation for 60-μm particles.

(a) With *Yagi and Aochi's correlation*, referring to Table 3,

$$\text{Re}_{ti} = \frac{(1.27)(0.240)(60 \times 10^{-6})}{1.8 \times 10^{-5}} = 0.9736$$

and

$$\kappa_i^* = \frac{(1.8 \times 10^{-5})(1 - 0.024)^2}{9.8(60 \times 10^{-6})} [0.0015(0.9736)^{0.6} + 0.01(0.9736)^{1.2}]$$

$$= 3.23 \, \text{kg/m}^2 \cdot \text{s}$$

(b) With *Wen and Hashinger's correlation*

$$\kappa_i^* = \frac{(1.52 \times 10^{-5})(1.217)(1 - 0.024)^2}{[(9.8)(60 \times 10^{-6})]^{0.5}} (0.9736)^{0.725} \left[\frac{2000 - 1.217}{1.217} \right]^{1.15}$$

$$= 2.16 \, \text{kg/m}^2 \cdot \text{s}$$

(c) With *Merrick and Highley's correlation*

$$\kappa_i^* = (1.217)(1) \left[0.001 + 130 \exp \left\{ -10.4 \left(\frac{0.240}{1} \right)^{0.5} \left(\frac{0.11}{1 - 0.11} \right)^{0.25} \right\} \right]$$

$$= 0.834 \, \text{kg/m}^2 \cdot \text{s}$$

(d) With *Geldart's correlation*

$$\kappa_i^* = (23.7)(1.217) \exp(-5.4 \times 0.240) = 7.89 \, \text{kg/m}^2 \cdot \text{s}$$

(e) *Colakyan and Levenspiel's correlation* is for larger sizes, out of the range of these particles being examined, and hence cannot be used.

(f) With *Zenz and Weil's procedure* and noting that $G_{si}^* = \kappa_i^*$, we have

$$\frac{u_o^2}{g d_{pi} \rho_s^2} = \frac{1^2}{(9.8)(60 \times 10^{-6})(2000)^2} = 4.3 \times 10^{-4}$$

Then from Fig. 6, $G_{si}^*/\rho_g u_o = 5.0$, from which

$$\kappa_i^* = G_{si}^* = (1.217)(1)(5) = 6.09 \, \text{kg/m}^2 \cdot \text{s}$$

(g) From *Gugnoni and Zenz's procedure*,

$$\frac{u_o - u_{ti}}{(g d_{pi})^{0.5}} = \frac{1 - 0.240}{[(9.8)(60 \times 10^{-6})]^{0.5}} = 31.3$$

Then from Fig. 6, $G_{si}^*/\rho_g u_o = 1.6$, from which

$$\kappa_i^* = G_{si}^* = (1.217)(1)(1.6) = 1.95 \, \text{kg/m}^2 \cdot \text{s}$$

Similar calculations for the other particle sizes give the following table of values for κ_i^*:

d_p (μm)	30	40	50	60	80	100
Yagi and Aochi	—	—	3.21	3.23	2.87	1.87
Wen and Hashinger	—	—	1.94	2.16	2.12	1.52
Merrick and Highley	10.1	4.19	1.80	0.83	0.21	0.05
Geldart et al.	—	—	—	7.89	3.65	1.44
Zenz and Weil	14.6	10.3	7.3	6.1	3.7	2.2
Gugnoni and Zenz	—	7.3	3.7	1.9	0.49	0.1

The dashes indicate conditions outside the range measured in preparing these correlations.

Note: These tabulated values differ by as much as a factor of 40. This shows that there still is need for good elutriation data and for the development of a reliable correlation for κ_i^*.

Entrainment from Short Vessels: $H_f <$ TDH

Focus attention on the region below the TDH. According to Lewis et al. [19], the density of solids present at various levels z_f as the freeboard height H_f is changed can be represented by the sketch of Fig. 9. Here, curve AGB represents the solids holdup when the freeboard height is at $H_{f,A} >$ TDH. This may be called the condition of *complete reflux*. Curves CD and EF are for freeboard heights $H_{f,C}$ and $H_{f,E}$, respectively, both below the TDH. From this sketch we see that

- At complete reflux (curve AGB), the density of solids in the freeboard falls off exponentially from the value at the bed surface, or

$$\bar{\rho}_R = \bar{\rho}_{R0} e^{-a z_f} \tag{17}$$

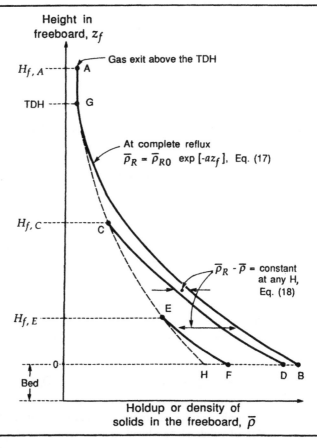

FIGURE 9
Density distribution of solids in the freeboard for three different freeboard heights, according to Lewis et al. [19].

- At smaller H_f, below TDH (curves CD and EF), the solid density is some constant value less than at complete reflux, or

$$\bar{\rho}_R - \bar{\rho} = \text{constant throughout the freeboard} \qquad (18)$$

- The entrainment G_s from the bed is proportional to the density of solids at the gas exit (see dashed line AGCEH). This exit density is also assumed to fall off exponentially, or

$$G_s = G_{s0}e^{-aH_f}, \qquad [\text{kg/m}^2 \cdot \text{s}] \qquad (19)$$

Note that although the density of solids at any level in the freeboard goes down as H_f is lowered, the density at the exit does increase. Hence, entrainment increases as the freeboard height is reduced.

Kunii and Levenspiel [31] presented a simple flow model to represent the complex phenomena occurring in the freeboard. Their model lumped velocities and other quantities, but still incorporated the main features of freeboard behavior, such as the ejection of clumps of solids into the freeboard, followed by the upflow, downflow, and breakup of these clumps. This model explains, and is consistent with, the findings of Lewis et al. [19], given by Eqs. (17)–(19).

We now present a somewhat more generalized version of this model.

Freeboard-Entrainment Model

Consider the freeboard above a bubbling or turbulent fluidized bed, and let x be the fraction of bed solids for which $u_t < u_o$. This is the entrainable fraction, and here we call them the fines.

Postulate 1. Three distinct phases are present in the freeboard.
> *Phase 1:* Gas stream with completely dispersed solids. The fines are carried upward and out of the bed at velocity u_1 while the coarse material rains back into the bed.
> *Phase 2:* Agglomerates, coming from the bed, and moving upward at velocity u_2.
> *Phase 3:* Agglomerates and thin wall layers of particles moving downward at velocity u_3.

Postulate 2. At any level in the freeboard the rate of removal of fines from the agglomerates to form dispersed solid of phase 1 is proportional to the volume fraction (or solid density) of agglomerates at that level.

Postulate 3. Upward-moving agglomerates will eventually reverse direction and move downward, the frequency of change from phase 2 to phase 3 being proportional to the volume fraction of phase 2 at that level.

Figure 10(a) shows the freeboard as viewed by this model.

Now at any level z_f in the freeboard let G_{s1}, G_{s2}, G_{s3} [kg/m$^2 \cdot$ s] be the mass flux of each phase, and let ρ_1, ρ_2, ρ_3 [kg/m^3] be the mass of each phase per unit volume of freeboard. Then at steady state conditions the net upward flux of solids at any level in the freeboard is given by

$$G_s = G_{s1} + G_{s2} - G_{s3} = \rho_1 u_1 + \rho_2 u_2 - \rho_3 u_3, \quad \text{independent of } z_f \qquad (20)$$

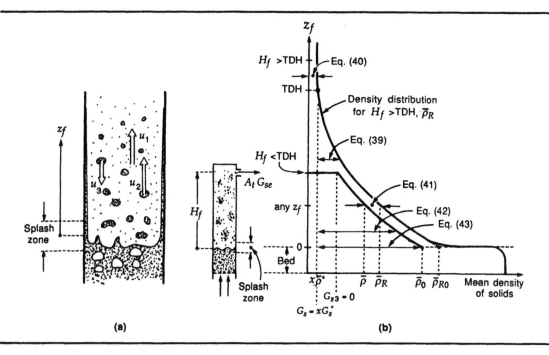

FIGURE 10
Sketch and various terms and expressions of the freeboard model developed in this chapter.

Also, the average (or bulk) density of solids at any level in the bed is

$$\bar{\rho} = \rho_1 + \rho_2 + \rho_3 \tag{21}$$

Relating these quantities to the terms previously used, we have

$$G_{su} = G_{s1} + G_{s2}, \qquad G_{sd} = G_{s3} \tag{22}$$

Now a mass balance for phase 1, and this only concerns the fines, gives

$$\left(\begin{array}{c} \text{increase of solids} \\ \text{in phase 1} \end{array} \right) = \left(\begin{array}{c} \text{transfer of solids from} \\ \text{phases 2 and 3 to 1} \end{array} \right)$$

for phase 2,

$$\left(\begin{array}{c} \text{decrease of solids} \\ \text{in phase 2} \end{array} \right) = \left(\begin{array}{c} \text{transfer of solids from} \\ \text{phases 2 to 1 and 3} \end{array} \right)$$

for phase 3,

$$\left(\begin{array}{c} \text{increase of solids} \\ \text{in phase 3} \end{array} \right) = \left(\begin{array}{c} \text{transfer of solids} \\ \text{from phase 2 to 3} \end{array} \right) - \left(\begin{array}{c} \text{transfer of solids} \\ \text{from phase 3 to 1} \end{array} \right)$$

We now introduce the rate coefficients xK_1 for the transfer of fines from phases 2 and 3 to 1, and K_2 for the transfer from phase 2 to phase 3. In these expressions x is the fraction of fines in the dense fluidized bed. Then the above mass balances at level z_f become

$$u_1 \frac{d\rho_1}{dz_f} = xK_1(\rho_2 + \rho_3) \tag{23}$$

$$-u_2 \frac{d\rho_2}{dz_f} = (xK_1 + K_2)\rho_2 \tag{24}$$

$$-u_3 \frac{d\rho_3}{dz_f} = K_2\rho_2 - xK_1\rho_3 \tag{25}$$

where u_1 and u_2 are the upward velocities of phases 1 and 2, and u_3 is the downward velocity of phase 3.

Since all solids that reach the freeboard height H_f leave the vessel, there is no downflow there, so

$$\rho_3 = 0 \quad \text{at } z_f = H_f \tag{26}$$

Noting that at the bed surface, subscript 0, the only upflow is by clumps of solids projected into the freeboard, we have

$$\rho_1 = 0 \quad \text{and} \quad \rho_2 = \frac{G_{su0}}{u_2} \quad \text{at } z_f = 0 \tag{27}$$

Net Upflow and Carryover of Solids. Solving Eqs. (23)–(25) with the boundary conditions of Eqs. (26) and (27) gives ρ_1, ρ_2, and ρ_3 in terms of exponential functions of z_f. With Eq. (20), the net flux of solids at any freeboard level and at the vessel outlet is then

$$\frac{G_s - xG_s^*}{G_{su0} - xG_s^*} = e^{-aH_f} \tag{28}$$

where G_s^* is the flux of carryover from a very tall vessel fluidizing only entrainable solids—in practical terms, when $H_f > \text{TDH}$.

The ratio of upward flux out of a tall vessel to the upward flux at its surface is

$$\frac{xG_s^*}{G_{su0}} = \frac{(xK_1/K_2)(1 + u_2/u_3)}{1 + (xK_1/K_2)(1 + u_2/u_3)} \tag{29}$$

and

$$a = \frac{K_2}{u_2}\left[1 + \frac{xK_1}{K_2}\left(1 + \frac{u_2}{u_3}\right)\right], \quad [\text{m}^{-1}] \tag{30}$$

In the special case of a vigorously bubbling bed or a turbulent bed with few fines, the amount of solids thrown into the freeboard is very much larger than those eventually removed from the bed. Thus $xG_s^* \ll G_{su0}^*$. In this situation, Eq. (29) indicates that

$$\frac{xK_1}{K_2}\left(1 + \frac{u_2}{u_3}\right) \ll 1, \quad \text{or} \quad \frac{xK_1}{K_2} \ll 1 \tag{31}$$

and, from Eq. (30),

$$a \cong \frac{K_2}{u_2} \tag{32}$$

Thus, Eq. (28) reduces to

$$\frac{G_s - xG_s^*}{G_{su0}} \cong e^{-(K_2/u_2)H_f} \tag{33}$$

These expressions provide a physical interpretation of the parameters reported by Lewis et al. [19].

Upflow Term Alone. The upward mass flux of solids in the freeboard has been measured by Hoggen et al. [18], Geldart and Pope [14], and Bachovchin et al. [11]. In the present model this upward flux is given by

$$G_{su} = \rho_1 u_1 + \rho_2 u_2 \tag{34}$$

Substituting the expressions found for ρ_1 and ρ_2 into Eq. (34) gives

$$\frac{G_{su} - xG_s^*}{G_{su0} - xG_s^*} = e^{-aH_f} + e^{-bz_f} - e^{-[aH_f - (a-b)z_f]} \tag{35}$$

where

$$b = \frac{xK_1 + K_2}{u_2}, \qquad \text{and} \qquad a - b = \frac{xK_1}{u_3} \tag{36}$$

Again, in the special case where $xG_s^* \ll G_{su0}$ we find $xK_1 \ll K_2$; thus, $a - b \ll a$. Since we always have $z_f \leq H_f$, all these inequalities lead us to conclude that

$$(a - b)z_f \ll aH_f \tag{37}$$

Thus, in this special case Eq. (35) reduces to

$$\frac{G_{su} - xG_s^*}{G_{su0}} \cong e^{-az_f} \tag{38}$$

Solids Hold-up in the Freeboard. This model also gives the distribution of bulk density of solids in the freeboard. Substituting the expressions found for ρ_1, ρ_2, and ρ_3 into Eq. (21) gives a bulky expression; however, in the special case where $xG_s^* \ll G_{su0}$, this expression simplifies to two special cases. *For very high freeboard* or, in practical terms, for $H_f >$ TDH, the density of solids $\bar{\rho}_R$ at any height z_f is

$$\bar{\rho}_R - x\bar{\rho}^* \cong G_{su0}\left(\frac{1}{u_2} + \frac{1}{u_3}\right)e^{-az_f} \qquad \text{for } H_f > \text{TDH} \tag{39}$$

where the lowest density of solids in the freeboard, that above TDH and at pneumatic transport conditions, is

$$x\bar{\rho}^* = \frac{xG_s^*}{u_1} \tag{40}$$

For not very high freeboard, or $H_f <$ TDH, the density of solids $\bar{\rho}$ is some constant value less than at reflux conditions. This constant value depends on the freeboard height H_f and is

$$\bar{\rho}_R - \bar{\rho} \cong \frac{G_{su0}}{u_3} e^{-aH_f} \tag{41}$$

In freeboards it is usually found that $G_s/G_{su0} \ll 1$. Then from Eq. (33) one finds

$$e^{-aH_f} \ll 1$$

For this situation, the distribution of solid density in the freeboard is

$$\frac{\bar{\rho} - x\bar{\rho}^*}{\bar{\rho}_0 - x\bar{\rho}^*} \cong e^{-az_f} \qquad \text{for } H_f < \text{TDH} \tag{42}$$

where

$$\bar{\rho}_0 - x\bar{\rho}^* \cong G_{su0}\left[\left(\frac{1}{u_2} + \frac{1}{u_3}\right) - \frac{1}{u_3}e^{-aH_f}\right] \tag{43}$$

Figure 10(b) shows the various quantities and relationships developed by this model. For more details of the derivation see Kunii and Levenspiel [32].

Discussion. Several groups of investigators have presented experimental data of G_{su0} and G_{sd0}, as shown in Fig. 11. We see appreciable variation in these values. This may be because G_{su0} is found by extrapolation back to $z_f = 0$, and different assumptions were made regarding the location of $z_f = 0$ in the vessel. Wen and Chen [33] took the $z_f = 0$ plane at the mean value of the bed surface; others took it somewhere within the splash zone. We suggest that $z_f = 0$ be located at the inflection point of the solid density curve; see Fig. 10(a).

In this model, since $G_{s1} = 0$ at $z_f = 0$, we have

$$G_s = G_{s2} - G_{s3} = G_{su0} - G_{sd0}, \qquad \text{at } z_f = 0 \tag{44}$$

For a vigorously bubbling fluidized bed we also have $xG_s^* \ll G_{su0}$, in

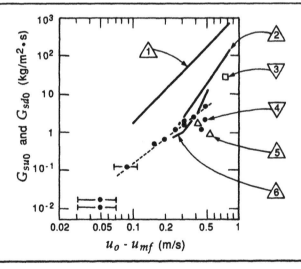

FIGURE 11

Mass flux of particles just above the bed surface; adapted from Walsh et al. [16]. Sources of information: △ G_{su0} from [32], ⃤ G_{su0} from [3], ▽ G_{sd0} from [33], ▿ G_{sd0} from [16], ⃤ G_{su0} from [19], ⃤ G_{su0} from [13].

which case, just above the bed surface

$$G_{su0} \simeq G_{sd0} , \qquad \text{at } z_f = 0 \tag{45}$$

In this special case Fig. 11 can be used to estimate G_{su0} and G_{sd0}.

In almost all cases of elutriation from a bubbling fluidized bed, it is reasonable to assume that $x G_s^* \ll G_{su0}$. Thus, one is justified in using the special case simplifications, or Eqs. (39) and (40) for the distribution of solid density in the freeboard, Eq. (38) for the flux of only upward-moving solids in the freeboard, and Eq. (28) for the carryover as a function of various freeboard heights.

Finally, for fine particles, Lewis et al. [19] found that

$$a u_o \simeq \text{constant} \tag{46}$$

Walsh et al. [16] found that Eq. (46) also holds for coarse particles.

Now if we assume that u_2, the upward velocity of agglomerates, is proportional to u_o, then Eq. (32) gives an immediate meaning to the empirical expression of Eq. (46). Figure 12, prepared from reported literature data, correlates the constant a with particle size and superficial gas velocity u_o. It should only be used in the range of conditions examined, in effect for $u_o \lesssim 1.25$ m/s.

For flow in the fast fluidization regime, which applies to fine particle systems where $u_o = 2\text{--}10$ m/s, we may have to use a different correlation for a; however, the freeboard-entrainment model with $x = 1$ can be used directly to give the vertical distribution of $\bar{\rho}$ or ε_s in such systems. The next chapter considers this flow regime.

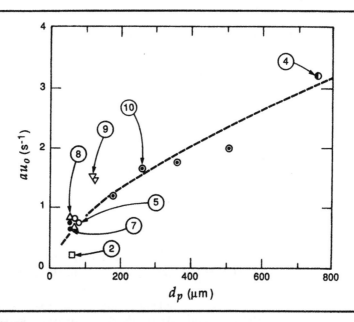

FIGURE 12
Decay constant for the freeboard agglomerates for $H_f < $ TDH and $u_o < 1.25$ m/s; from Kunii and Levenspiel [34]. Reported or calculated value ② from [3], ④ from [16], ⑤ from [19], ⑦ from [2], ⑧ from [5], ⑨ from [11], ⑩ from [18].

EXAMPLE 6

Entrainment

from a Short

Vessel:

$H_f < TDH$

To discourage side reactions in the freeboard of a large-diameter turbulent bed reactor, we plan to use a short freeboard. Thus, the cyclone inlet is to be located 2 m above the middle of the splash zone of the bed. Estimate the entrainment flux from the bed, assuming that at the velocity being used, about $170u_{mf}$, all particles are entrainable.

Data

$$\bar{d}_p = 60\ \mu m, \qquad \rho_g = 1.3\ kg/m^3, \qquad u_{mf} = 0.003\ m/s$$

$$\rho_s = 1500\ kg/m^3, \qquad u_o = 0.503\ m/s$$

SOLUTION

Here we will use Eq. (28) from the freeboard model. But first we must evaluate G_s^*. We use Zenz and Weil's procedure. Thus,

$$\frac{u_o^2}{g d_p \rho_s^2} = \frac{(0.503)^2}{(9.8)(60 \times 10^{-3})(1500)^2} = 1.9 \times 10^{-4}$$

Then, from the fine particle curve of Fig. 6 we get $G_s^*/\rho_g u_o = 1$, from which

$$G_s^* = (1)(1.3)(0.503) = 0.654\ kg/m^2 \cdot s$$

Another term in Eq. (28) that needs to be evaluated is G_{su0}. From Fig. 11 we find the ejection rate of particles into the freeboard to be

$$G_{su0} \approx 5.0\ kg/m^2 \cdot s$$

Still another term to evaluate is a. From Fig. 12, $u_o a = 0.72\ s^{-1}$; thus

$$a = \frac{0.72}{0.503} = 1.4\ m^{-1}$$

Now insert all these terms into Eq. (28) to get

$$\frac{G_s - 0.654}{5.0 - 0.654} = e^{-1.4(2.0)}$$

from which the entrainment rate from the short bed is

$$G_s = \underline{\underline{0.92\ kg/m^2 \cdot s}}$$

Comment. This entrainment is $(0.92-0.65)(100)/(0.65) = 41\%$ higher than it would be if the gas exit were at the TDH.

PROBLEMS

1. A closed-loop solid recirculation system, as shown in Fig. 7, contains a bed of solids at the bottom of a 20.3-cm ID tube 8 m high, which is thought to be higher than the TDH. In a typical experiment the bed is charged with 8 kg of coarse nonentraining solids, 1 kg of 50-μm solids, and 1 kg of 100-μm solids. At steady state, carryover is 0.03 kg/s, and 1 kg of solids are present in the recycle loop at any time. A sample of the entrained solids shows that it consists of two parts 50-μm particles to one part of 100-μm particles. From this information find κ and κ^* for these two sizes of elutriable particles.

2. A batch of solids consisting of sizes A, B, and C is fluidized at high velocity in a tall $(H_f > \text{TDH})$ 45.7-cm ID bed, and after a 10-min run the composition of the bed is measured by screening. The following data are found:

$$\begin{array}{llll}
\textit{Before:} & W_{A0} = 60\,\text{kg}, & W_{B0} = 20\,\text{kg}, & W_{C0} = 20\,\text{kg} \\
\textit{After:} & W_A = 15\,\text{kg}, & W_B = 10\,\text{kg}, & W_C = 20\,\text{kg}
\end{array}$$

Calculate κ and κ^* for these three sizes of solids.

3. A batch of solids (900 kg A, 600 kg B) is to be fluidized in a bed (cross-sectional area 1 m^2) containing a bundle of horizontal tubes. At the planned air velocity, $\kappa_A^* = 0.5\,\text{kg/m}^2 \cdot \text{s}$ and $\kappa_B^* = 0.25\,\text{kg/m}^2 \cdot \text{s}$. Estimate the amount of solids A and B entrained after a 30-min run.

4. In a fluidized bed of wide size distribution, estimate the saturation carrying capacity of the flowing gas.

Data

$$\rho_g = 0.55\,\text{kg/m}^3, \qquad \rho_s = 1800\,\text{kg/m}^3, \qquad u_0 = 0.6\,\text{m/s}$$

Size, d_p (μm)	20	30	40	50	60	70	80	100	120	150	170
$100p_b(d_b)$ (μm)	0.2	1	1.64	1.80	1.54	1.12	0.72	0.38	0.18	0.06	0

5. Estimate the transport disengaging height for beds of single size 50-, 100-, 200-, 300-μm particles fluidized at $u_0 = 1$ m/s. Assume that the TDH corresponds to

$$\frac{G_s - G_s^*}{G_{su0} - G_s^*} = \frac{\bar{\rho} - \bar{\rho}_{sat}}{\bar{\rho}_0 - \bar{\rho}_{sat}} = 0.01$$

REFERENCES

1. A.B. Fournol, M.A. Bergougnou, and C.G.J. Baker, *Can. J. Chem. Eng.*, **51**, 401 (1973).

2. A. Nazemi, M.A. Bergougnou, and C.G.J. Baker, *AIChE Symp. Ser.*, **70**(141), 98 (1974).

3. Zhang Qi et al., in *Proc. CIESC/AIChE Joint Meeting*, p. 374, Chemical Industry Press, Beijing, 1982; in *Fluidization '85, Science and Technology*, M. Kwauk et al., eds., p. 95, Science Press, Beijing, 1985.

4. S. Morooka, T. Kago, and Y. Kato, in *Fluidization IV*, D. Kunii and R. Toei, eds., p. 291, Engineering Foundation, New York, 1983; S. Morooka, K. Kawazuishi, and Y. Kato, *Powder Technol.*, **26**, 75 (1986).

5. G. Chen, G. Sun, and G.T. Chen, in *Fluidization V*, K. Østergaard and A. Sørensen, eds., p. 305, Engineering Foundation, New York, 1986.

6. K. Kato et al., in *Fluidization '85, Science and Technology*, M. Kwauk et al., eds., p. 136, Science Press, Beijing, 1985; in *Proc. 3rd World Cong. Chem. Eng.*, Tokyo, 1986.

7. Y. Martini, M.A. Bergougnou, and C.G.J. Baker, in *Fluidization Technology*, vol. 2, D.L. Keairns, ed., p. 29, Hemisphere, Washington, D.C., 1976.

8. R.J. Gugnoni and F.A. Zenz, in *Fluidization III*, J.R. Grace and J.M. Matsen, eds., p. 501, Plenum, New York, 1980.

9. M. Horio et al. in *Fluidization III*, J.R. Grace and J.M. Matsen, eds., p. 509, Plenum, New York, 1980; in *Fluidization IV*, D. Kunii and R. Toei, eds., p. 307, Engineering Foundation, New York, 1983; in *Fluidization '85, Science and Technology*, M. Kwauk et al., eds., p. 124, Science Press, Beijing, 1985.

10. S.E. George and J.R. Grace, *Can. J. Chem. Eng.*, **59**, 279 (1981).

11. D.V. Bachovchin, J.M. Beer, and A.F. Sarofim, paper, AIChE Annual Meeting, November 1979; *AIChE Symp. Ser.*, **77**(**205**), 76 (1981).

12. I.H. Chan and T.M. Knowlton, in *Fluidization IV*, D. Kunii and R. Toei, eds., p. 283, Engineering Foundation, New York, 1983.

13. S.T. Pemberton and J.F. Davidson, in *Fluidization IV*, D. Kunii and R. Toei, eds., p. 275, Engineering Foundation, New York, 1983.

14. D. Geldart and D.J. Pope, *Powder Technol.*, **34**, 95 (1983).

15. S. Ismail and J.C. Chen, *AIChE Symp. Ser.*, **80**(**234**), 114 (1984).

16. P.M. Walsh, J.E. Mayo, and J.M. Beer, *AIChE Symp. Ser.*, **80**(**234**), 119 (1984).

17. M. Horio et al., in *Fluidization '85, Science and Technology*, M. Kwauk et al., eds., p. 124, Science Press, Beijing, 1985.

18. B. Hoggen, T. Lendstad, and T.A. Engh, in *Fluidization V*, K. Østergaard and A. Sørensen, eds., p. 297, Engineering Foundation, New York, 1986.

19. W.K. Lewis, E.R. Gilliland, and P.M. Lang, *Chem. Eng. Prog. Symp. Ser.*, **58**(**38**), 65 (1962).

20. H.S. Caram, S. Edelstein, and E.K. Levy, in *Fluidization IV*, D. Kunii and R. Toei, eds., p. 265, Engineering Foundation, New York, 1983; H.S. Caram, Z. Efes, and E.K. Levy, *AIChE Symp. Ser.*, **80**(**234**), 106 (1984).

21. F.A. Zenz and N.A. Weil, *AIChE J.*, **4**, 472 (1958).

22. J.M. Matsen, in *Proc. NSF Workshop on Fluidization and Fluid-Particle Systems*, p. 452, Troy, NY, 1979.

23. M. Leva, *Chem. Eng. Prog.*, **47**, 39 (1951).

24. S. Yagi and D. Kunii, *Kagaku Kogaku*, **16**, 283 (1952); in *Proc. 5th Int. Symp. Combustion*, p. 231, Van Nostrand Reinhold, New York, 1955.

25. S. Yagi and T. Aochi, papers presented at meetings of Soc. Chem. Eng. Japan, spring 1955, autumn 1956.

26. C.Y. Wen and R.F. Hashinger, *AIChE J.*, **6**, 220 (1960).

27. I. Tanaka et al., *J. Chem. Eng. Japan*, **5**, 51 (1972).

28. D. Merrick and J. Highley, *AIChE Symp. Ser.*, **70**(**137**), 366 (1974).

29. D. Geldart et al., *Trans. Inst. Chem. Eng.*, **57**, 269 (1979).

30. M. Colakyan et al., *AIChE Symp. Ser.* **77**(**205**), 66 (1981); M. Colakyan and O. Levenspiel, *Powder Technol.*, **38**, 223 (1984).

31. D. Kunii and O. Levenspiel, *J. Chem. Eng. Japan*, **2**, 84 (1969).

32. C.Y. Wen and L.H. Chen, *AIChE J.*, **28**, 117 (1982).

33. F.J.A. Martens, *Proc. Colloquium on Pressurized Fluidized Bed Combustion*, Delft University of Technology, April 1983.

34. D. Kunii and O. Levenspiel, *Powder Technol.*, **61**, 193 (1990).

High-Velocity Fluidization

Chapter 3 discussed the shift in fluidizing regime with increase in gas velocity from bubbling to turbulent to fast fluidization and, finally, to pneumatic transport. Here we deal with design aspects of these high-velocity fluidization regimes. Pneumatic transport is treated in Chap. 15.

Turbulent Fluidized Beds

Let u_k be the superficial gas velocity on entry into the turbulent regime. Yerushalmi and Avidan [1] well summarized the experimental findings on u_k. Despite considerable disagreement between workers, we suggest that this regime is entered under the following conditions:

- For fine (Geldart **A**) particles, $u_k/u_t \cong 2$–11. As an example, for an FCC catalyst, Rhodes and Geldart [2] give $\bar{d}_p = 49 \ \mu$m, $u_t = 0.08$ m/s, and $u_k = 0.3$ m/s; hence, $u_k/u_t \cong 4$.
- For larger particles, u_k/u_t decreases.
- For large (Geldart **B** or **D**) particles, u_k/u_t can be well below 1. Thus, for coarse $\bar{d}_p = 650$-μm particles, Rhodes and Geldart [2] found $u_k/u_t = 0.3$–0.6. *Churning flow* is the name sometimes given to large particle systems in turbulent flow.

The flow map of Fig. 3.16 locates this regime in relation to its neighbors.

Experimental Findings

From typical findings we come up with the following characteristics of the flow in this regime.

- In both small- and large-diameter beds ($d_t = 0.152$ and 3.4 m) of small (Geldart **A**) solids, we find higher-than-average gas flow and bed voidage in the central core of the bed, as shown in Fig. 1. Also, this unevenness in flow becomes more pronounced at higher gas velocities.

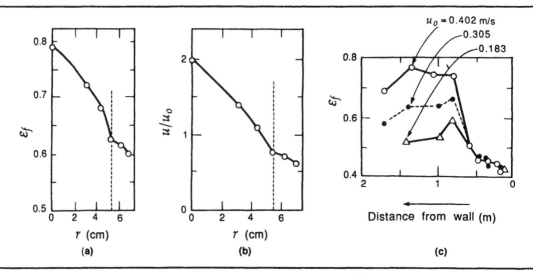

FIGURE 1
Radial distribution of voidage and relative gas velocity; adapted from Abed [3]: (a) and (b) small vessels: $d_t = 0.152$ m, $u_o = 0.55$ m/s; (c) large vessel: $d_t = 3.4$ m, various u_o.

For larger beds, one notices a dip in bed voidage at the centerline of the bed. This may reflect the downflow of emulsion phase due to solid circulation, as illustrated in Fig. 6.1.

- Horizontal baffles increase the bed voidage, but vertical baffles do not; see Fig. 2(a).
- Adding fines under 20 μm to the bed increases the bed voidage; see Fig. 2(b).
- Bed voidages of 0.75–0.80 in the main body of turbulent beds are on the borderline of the fast fluidized regime.

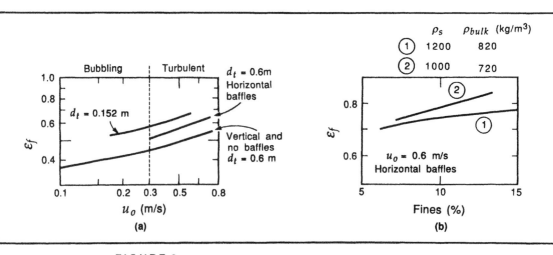

FIGURE 2
Effect of baffles and fines on bed voidage. The line for $d_t = 0.152$ m is taken from Abed [3]; all the other lines are from Avidan et al. [4] for $d_t = 0.6$ m and $d_p = 20$–100 μm.

• The separation between bed and freeboard becomes uncertain, entrainment rises sharply above that of bubbling beds, and the cyclone duty becomes severe enough so that external cyclones may have to be used instead of internal cyclones; see Fig. 1.2.

From these findings we may expect that the entrainment, elutriation, and behavior of the freeboard can reasonably be represented by the freeboard-entrainment model of Chap. 7.

Fast Fluidization

In the fast fluidization regime carryover of solids is very large; hence, fresh solids have to be introduced continuously and at a significant rate to make up for the loss of bed solids and to achieve steady state operations. Figure 3 illustrates a typical fast fluidized bed with its various regions.

• At the bottom one sees a relatively short entry zone having a solid fraction $\varepsilon_s = 0.2–0.4$.

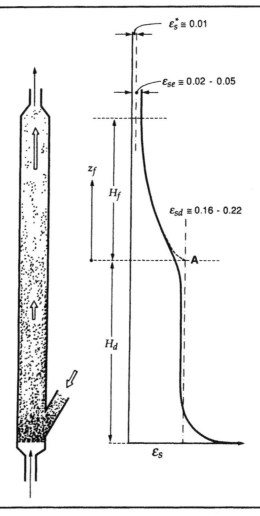

FIGURE 3
The fast fluidized bed and its regions of different fraction of solids.

- Then there is a portion of the vessel of almost constant solid fraction of about $\varepsilon_s = 0.2$. These lower portions may be called the dense region.
- Above this is an upper entrained region where the solid fraction decreases progressively to about $\varepsilon_s = 0.02$–0.05.

These regions correspond somewhat to the dense turbulent bed and its freeboard. Also, the transition between regions is smooth, as shown in Fig. 3.

Because of the large entrainment of solids from fast fluidized beds, inner cyclones and diplegs are too small to handle the solid load. Thus, large cyclones located outside the column are used, and they require careful design for proper operation.

Overall, we may want to run a fast fluidized circulation system in one of four ways:

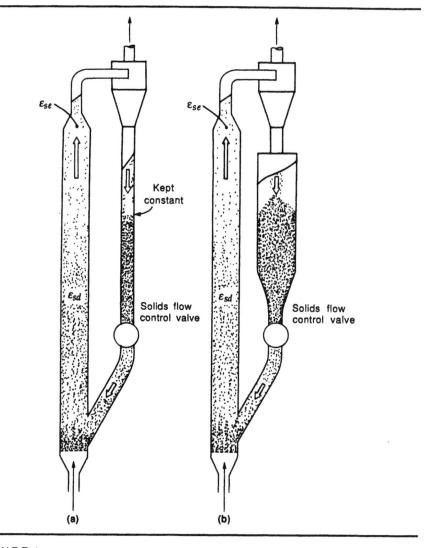

FIGURE 4
Idealized solid circulation systems for fast fluidized operations. Scheme (b) contains a reservoir for solids and is much more flexible than scheme (a).

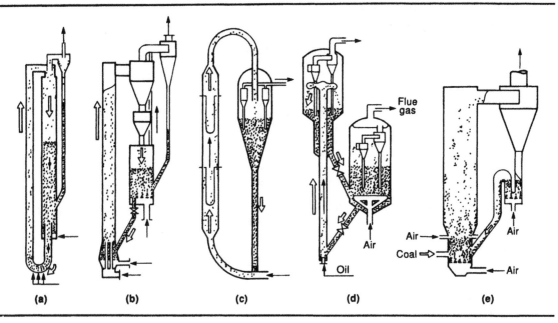

FIGURE 5
Practical designs of fast fluidized bed circulation systems: Sketches (a) and (b) are experimental setups; (c), (d), and (e) are industrial reactors. (a) Yerushalmi and Avidan [1]; (b) Hartge et al. [5]; (c) Sasol process, Fig. 2.9; (d) FCC process, Fig. 2.13; (e) fluid bed combustor, Fig. 2.17.

Mode I. Keep a constant inventory of solids in the bed, even though u_o may change.

Mode II. Keep a constant throughflow of solids G_s (kg/m^2·s), even though u_o may change.

Mode III. Keep a constant gas flow rate u_o while changing the solid throughflow G_s.

Mode IV. G_s and u_o can be changed independently.

Figure 4 illustrates the two basic types of solid circulation systems: those which do not include a reservoir of solids, and those which do. Without the reservoir one can only operate according to Mode I. With the reservoir the system becomes much more flexible in that it can operate in any mode. The sketches in Fig. 4 are only idealized illustrations. In practice, one encounters a variety of designs for getting smooth steady state circulation of solids. Figure 5 illustrates some of these.

Experimental Findings

With the works of Yerushalmi et al. as a catalyst, there has been a sharp increase in interest in investigating the characteristics of the fast fluidization flow regime, and Table 1 lists the experimental conditions of the more recent studies. The findings of these studies can be condensed as follows.

Vertical Distribution of Solids. Li and Kwauk [12], Weinstein et al. [18], and Hartge et al. [5] all found an S-shaped solid fraction curve, as shown

TABLE 1   Experimental Conditions of Fast Fluidized Beds

Investigators	Column d_t (cm)	Particles	ρ_s (kg/m³)	$\bar{d}_p$ (μm)
Kehoe and Davidson [6] (1971)	5–10, 6.2 × 60	cat., glass	1100, 2200	22–55, 22
Massimilla [7] (1973)	15.6	cat.	1000	50
Yerushalmi et al. [8] (1976)	15.2, 5 × 51	cat. alumina	1070–1450 2460	49 103
Canada et al. [9] (1976)	30 × 30, 61 × 61	glass	2480	650
Thiel and Potter [10] (1977)	5 × 22	cat.	930	60
Carotenuto, Crescitelli et al. [11] (1974, 78)	15.2	cat., etc.	940–1550	60–95
Li and Kwuak [12] (1980, 86, 88)	9	cat., iron, pyrite cinder	1780–4510	54–105
Abed [3] (1983, 85)	15.2	cat.	850	˙55
Yang et al. [13] (1983, 85)	11.5	cat., silica gel	2130	68 220
Arena et al. [14] (1986, 88)	4.1 12.0	glass cat.	2600 1700	88 70
Brereton and Stromberg [15] (1986)	20, 30 × 40	sand	2500	170–550
Toda et al. [16] (1983, 85, 86)	10.2	glass, etc.	2300–2500	65–155
Hartge and Werther [5] (1986, 86, 88)	5, 40 40	quartz cat., quartz, ash		56 90, 120, 160
Monceaux et al. [17] (1985, 86)	14.4	cat.	bulk 900	59
Rhodes and Geldart [2] (1986, 87)	15.2	alumina	1020–1800	38–64
Weinstein et al. [18] (1983, 84, 86)	15.2	cat.	1450	59
Lu and Wang [19] (1985)	10	sand, glass	2320–2640	230–365
Fusey et al. [20] (1986)	9 ˙	cat.		60, 119
Horio et al. [21] (1986, 88)	5	glass, iron cat.	2520, 7860 1000	130, 710; 119 60
Takeuchi et al. [22] (1986)	10	cat.	1080	57
Schnitzlein [23] (1987)	15.2	cat.	1070, 1450	49, 59

by the solid line of Fig. 3. This S-shaped curve moved up or down the column depending on the solid and gas flow rates. This behavior is found in both large-diameter (d_t = 0.4 m) and small-diameter (d_t = 0.09 m) columns, is discussed in Chap. 3, and is sketched in Fig. 3.14.

Li and Kwauk [12] fitted this distribution curve by the equation

$$\frac{\varepsilon_s - \varepsilon_s^*}{\varepsilon_{sd} - \varepsilon_s} = e^{-(z-z_t)/z_o} \tag{1}$$

This curve is sketched in Fig. 6 and has been found to reasonably fit the data of other investigators.

Lateral Distribution of Solids. Figure 7 shows typical findings of various gas velocities (Fig. 7(a)), solid flows (Figs. 7(b) and (c)), and at various heights in the column (Fig. 7(d)). These sketches clearly show that the gas and

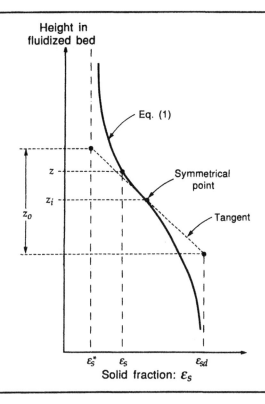

FIGURE 6
Sketch of solid fraction with height according to Eq. (1).

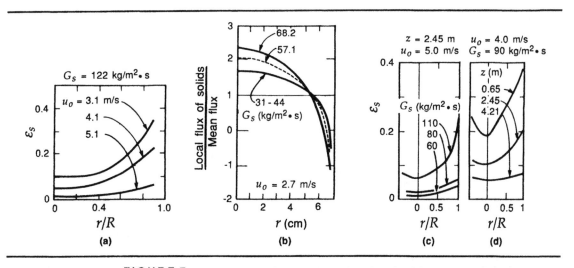

FIGURE 7
Radial distribution of solids and solid flows in fast fluidized beds; $\bar{d}_p \cong 56$–$59\ \mu$m: (a) $d_t = 0.152$ m; from Weinstein et al. [18]; (b) $d_t = 0.144$ m; from Monceaux et al. [17]; (c) and (d) $d_t = 0.4$ m; from Hartge et al. [5].

solid favor flowing in the central core of the column, and that at the wall there is a dense slow-moving layer of solid. In addition, Matsen [24] found an appreciable maldistribution of solids in large-diameter beds—nonsymmetrical and nonreproducible.

Fraction of Solids in the Lower Dense Region, ε_{sd}. If we ignore the short entry zone of Fig. 3, then Fig. 8, prepared from the published data, shows that the flow rates of solids and of gas in the range $u_0 = 1.5$–5.0 m/s do not seem to affect appreciably the fraction of solids in the lower dense region of the vessel.

It is interesting to compare ε_{sd} in the various fluidizing regimes.

Bubbling bed: $\varepsilon_{sd} = 0.55$–0.40
Turbulent bed: $\varepsilon_{sd} = 0.40$–0.22
Fast fluidization: $\varepsilon_{sd} = 0.22$–0.16

Fraction of Solids at the Column Exit, ε_{se}. First note that ε_{se} is somewhat larger than the saturation carrying capacity of the gas ε_s^*. Next, the data of Fig. 9 show that at given solid flux ε_{se} is higher at low u_0 than at high u_0.

This seemingly puzzling finding can be explained in terms of the slip velocity between gas and solid, or $u_p = u_g - u_s$. Thus, letting u_s be the mean velocity of solids at the exit level of the column, a mass balance gives

$$G_s = G_{se} = \rho_s \varepsilon_{se} u_s \tag{2}$$

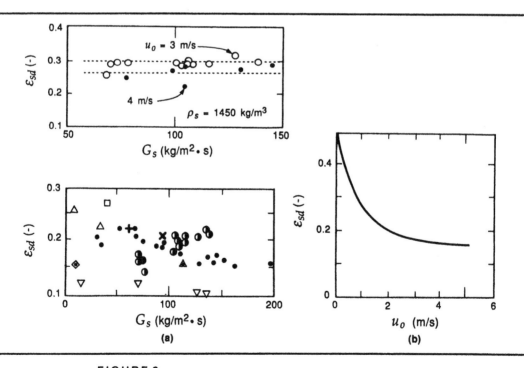

FIGURE 8
Volume fraction of solids in the lower dense region of the fast fluidized bed: (a) various solid flows, see Table 2 for references; (b) various gas flows; from Schnitzlein [23].

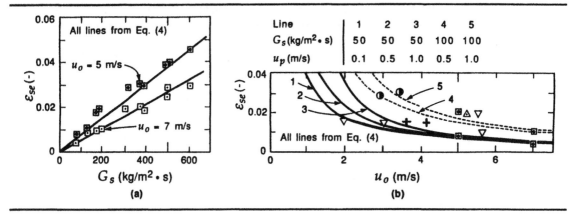

FIGURE 9
Volume fraction of solids at the exit of fast fluidized beds; experiment compared with curves of Eq. (4): (a) ε_{se} at high solid flows, $G_s = 80$–$600\ kg/m^2 \cdot s$; from Arena et al. [14]; (b) ε_{se} at low solid flows, $G_s = 50$–$100\ kg/m^2 \cdot s$; from various sources, see Table 2 for references.

Introducing the slip velocity gives the solid velocity as

$$u_s = \frac{u_o}{1 - \varepsilon_{se}} - u_p \tag{3}$$

For fine particles at high u_o, $\varepsilon_{se} \ll 1$; thus, the Eqs. (2) and (3) become

$$G_s = G_{se} \cong \rho_s \varepsilon_{se}(u_o - u_p) \quad \text{or} \quad \varepsilon_{se} \cong \frac{G_s}{\rho_s(u_o - u_p)} \tag{4}$$

If the fine particles are completely dispersed in the gas stream, we can reasonably assume that $u_p \cong u_t$ and $u_t \ll u_o$. The two lines in Fig. 9(a) are based on Eq. (4) with this assumption and are consistent with the data of this figure.

To find the saturation carrying capacity of gas ε_s^*, values of ε_{se} were taken from reported data at low G_s or high enough columns such that ε_s seemed to level off in the upper portion of the column. Figure 9(b) summarizes this data from various sources and compares it with the curves calculated from Eq. (4), using appropriate values of G_s and u_p. Although there is considerable scatter in the experimental values, Eq. (4) seems to pass through the main body of this data.

The Freeboard-Entrainment Model Applied to Fast Fluidization

Consider a fast fluidization column as having a lower region of constant solid fraction ε_{sd} and an upper leaner region wherein the solid density decreases progressively to its exit value ε_{se}. This leaner region starts at the level given by the intersection of the two extrapolated dashed lines, shown as point A of Fig. 3.

With this as the picture of a fast fluidization column, the decrease or decay of ε_s in the lean entrainment region can be treated with the model of Chap. 7, which was used to represent the freeboard above bubbling and turbulent beds, but with the simplification here that all the solids are entrainable, or $x = 1$.

On the basis of this model, Eqs. (7.42) and (7.43) predict an exponential decay in solid density between ε_{sd} and a limiting value of ε_s^*, with a decay

T A B L E 2   References and Experimental Conditions for Data of Figs. 8(a), 9, and 10

Key	Reporters	d_t (cm)	Type of Solid
+	Hartge et al. [5]	5	Quartz
×		40	Quartz, FCC
▽	Li and Kwauk [12a]	9.0	FCC
			Alumina
			Pyrite cinder
◑	Weinstein et al. [18]	15.2	HFZ-20 cat.
⊕	Rhodes and Geldart [2]	15.2	Alumina, etc.
△	Takeuchi et al. [22]	10.0	FCC
○●	Schnitzlein [23]	15.2	HRZ-33 cat.
□	Kato et al. [25]	6.6, 9.7	FCC, cat.
◇◆	Horio et al. [21]	5.0, 20	FCC
◨	Arena et al. [14]	4.1, 12	FCC, glass
▽	Kwauk et al. [12b]	9.0	Iron
△	Yang et al. [13]	11.5	Silica gel
◙⊡	Arena et al. [14]	4.1	Glass
◉	Brereton and Stromberg [15]	20	Sand
◇	Furchi et al. [26]	7.2	Glass
			Glass
⊕	Lu and Wang [19]	10	Sand, glass

constant a related to gas velocity u_o by

$$au_o = \text{constant} \tag{5}$$

Thus, a higher gas velocity u_o means a slower decrease of solid concentration with height in the entrainment section of the column. This is to be expected.

Figures 10(a) and (b) present values of the decay constant calculated from the literature for fine particle systems ($d_p < 70\ \mu$m) and for coarser particle systems ($d_p > 88\ \mu$m). Lines representing Eq. (5) for different values of the

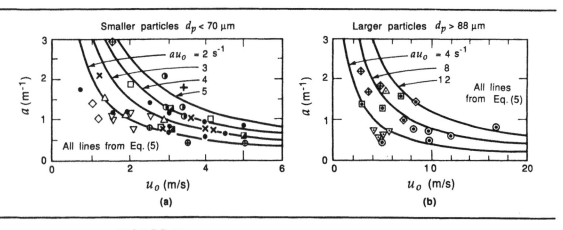

FIGURE 10
Decay coefficient *a*, which represents the changing solid fraction in the entrainment region; experimental data compared with the prediction of Eq. (5). See Table 2 for references.

$\bar{d}_p$ (μm)	u_o (m/s)	G_s (kg/m²·s)	Remarks
56	3.4	72	
56	1.2–4	7–40	
54, 58	0.8–2.1	14–16	
54	2.2–4	73	
56	1.5–2.5	129	
49	2.9, 3.4	71–118	Fine particles,
38–64	2.5–4.5	86–115	$\bar{d}_p < 70\ \mu m$
61	1.7–2.9	8.3–79	
59	1.5–5	89–133	
61	2–4.4	48–50	
60	1.1–1.6	12–19	
70	2.5–5	49, 120	
105	4–5	135	
220	5.3	44–146	
88	3–7	80–600	Larger particles,
170–650	5–16.8	64–146	$\bar{d}_p > 88\ \mu m$
196	7.2	88	
269	8.3	127	
230, 369	2.9–4.9	—	

decay constant are also shown in these figures. Despite considerable scatter in the reported data, this figure suggests the following:

- The decay constant a seems to increase with decreasing d_t. This may be explained by noting that in narrow columns rising agglomerates are more likely to hit the wall surface and then be removed from the rising gas stream.
- The decay constant a seems to increase with increasing d_p. This may be explained by noting that with coarser and denser particles the agglomerates are more likely to change direction and return to the dense region of the vessel. In the model this is represented by a larger K_2 value.
- For the fine particle systems of Fig. 10(a), we see that higher gas velocities give lower values for the decay constant, as predicted by Eq. (5). On the contrary, the data on the coarse particle system reported by Brereton and Stromberg [15] do not seem to fit this relationship.
- The decay constant a for bubbling and turbulent beds and for fast fluidized beds (compare Fig. 7.12 with Fig. 10) all fall in the same range of values, suggesting that the mechanism of decay of solid fraction in the lean region of fast fluidization columns is basically similar to the decay above bubbling and turbulent beds.

The decay constant is important for design. Since reported values of this parameter are scattered and sketchy, more precise data on its proper value and how it changes with the imposed system conditions would be most welcome.

Design Considerations

In design, one needs to know the solid holdup in the vessel as a function of the flow conditions, u_o and G_s. This, in turn, requires knowing the location of the top of the dense region and solid densities throughout. These values can be found from the information given so far in this chapter, as we will now show.

First consider the upper entrainment zone of the column. According to Fig. 3, Eq. (7.42) becomes

$$\frac{\varepsilon_s - \varepsilon_s^*}{\varepsilon_{sd} - \varepsilon_s^*} = e^{-az_f} \tag{6}$$

The fraction of solids at the vessel exit is

$$\varepsilon_{se} = \varepsilon_s^* + (\varepsilon_{sd} - \varepsilon_s^*)e^{-aH_f} \tag{7}$$

and the mean value of ε_s in the upper entrainment region of height H_f is found from the relation

$$\bar{\varepsilon}_s = \frac{1}{H_f} \int_0^{H_f} \varepsilon_s \, dz_f \tag{8}$$

Inserting Eq. (7) into (8) and integrating gives

$$\bar{\varepsilon}_s = \varepsilon_s^* + \frac{\varepsilon_{sd} - \varepsilon_s^*}{aH_f}(1 - e^{-aH_f}) = \varepsilon_s^* + \frac{\varepsilon_{sd} - \varepsilon_{se}}{aH_f} \tag{9}$$

The total inventory of solids in the column of height $H_t = H_f + H_d$ is then

$$\frac{W}{A_t \rho_s} = L_m(1 - \varepsilon_m) = L_{mf}(1 - \varepsilon_{mf}) = H_d \varepsilon_{sd} + H_f \bar{\varepsilon}_s$$

$$= \frac{\varepsilon_{sd} - \varepsilon_{se}}{a} + H_t \varepsilon_{sd} - H_f(\varepsilon_{sd} - \varepsilon_s^*) \tag{10}$$

To use this freeboard-entrainment model, one needs values of a, ε_s^*, and ε_{sd}. We briefly point out where these values can be obtained.

• The value of a is estimated from Fig. 10. Alternatively, if one has solids distribution data fitted to the Li and Kwauk expression (Eq. (1)), one can use this information to find a. To do this, match the slopes of Eqs. (1) and (6) at the midpoint between ε_s^* and ε_{sd}. This relates the parameter z_0 of Eq. (1) with the decay constant a by the equation

$$a = \frac{2}{z_0} \tag{11}$$

• Now, according to Monceaux et al. [17], $\varepsilon_s^* \cong 0.01$ for the pneumatic transport of particles. Phenomenologically, this value may correspond to the transition between dilute and concentrated transportation of solids, as discussed in Chap. 15.

Alternatively, from the definition of mass velocity and with $u_0 \gg u_t$, we get

$$\varepsilon_s^* = \frac{G_s^*}{\rho_s u_0} \tag{12}$$

where G_s^* is found by the method of Zenz and Weil, following Eq. (7.3), or as given by Fig. 9(b).

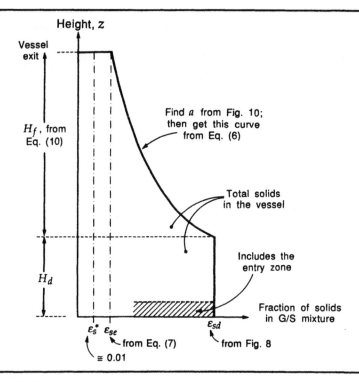

FIGURE 11
Solids distribution in fast fluidization, from the freeboard entrainment model.

• Finally, ε_{sd} is found from Fig. 8.

The various quantities in this model are sketched in Fig. 11.

We now calculate the performance of the fast fluidized bed in the four modes of operation.

Mode I. Constant inventory of solids (with no reservoir of solids)

  a. For given u_o, estimate ε_{sd} from Fig. 8 and estimate a from Fig. 10.
  b. Calculate ε_{se} as a function of H_f, with Eq. (7).
  c. For the desired inventory of solids $L_m(1 - \varepsilon_m)$ and given height of vessel $H_t = H_f + H_d$, determine H_f by substituting all known values into Eq. (10).
  d. Calculate G_{se} from Eq. (4).

The mass flux of circulating particles from bubbling or turbulent fluidized beds through an inner cyclone collector is calculated in the same way.

Modes II, III, and IV. Constant G_s, or u_o, or both changing

  a. Estimate ε_{sd} and a as before.
  b. Calculate ε_{se} for given G_s with Eq. (4).
  c. Determine H_f and $H_d = H_t - H_f$ from Eq. (7).
  d. Determine $L_m(1 - \varepsilon_{mf})$ with Eq. (10).

Example 1 illustrates this calculation procedure, and Fig. 12 displays the results of these calculations. Note that as u_o is changed, Modes I and II give opposite progressions of solid fraction versus height in the bed.

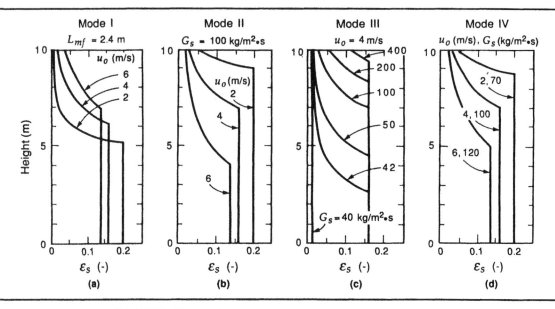

FIGURE 12
Solutions of Example 1; $d_t = 0.4$ m, $H_t = 10$ m: (a) Mode I, (b) Mode II, (c) Mode III, (d) Mode IV.

Pressure Drop in Turbulent and Fast Fluidization

Equation (3.16) and Fig. 3.5 show that the pressure drop through bubbling beds is just about all due to the static head of suspended particles. As the gas velocity is increased to give turbulent fluidization, fast fluidization, and then saturated pneumatic transport, we would expect that the frictional contribution to the Δp_{fr} would become increasingly important, and it is. However, as we show in Chap. 15, even in the extreme of saturated pneumatic transport, well over 50% is still due to the static head. Thus, for turbulent and fast fluidization, we can expect that roughly 80–90% of the pressure drop through the column, exclusive of distributor plate, cyclones, etc., is due to the static head. Thus

$$\Delta p_{fr} = (1.1-1.2)\left(\frac{\text{weight of all the solids in the column}}{\text{cross sectional area of column}}\right) \qquad (13)$$

A system pressure balance combined with entrainment and bed expansion information was used by Rhodes and Geldart [2] in their method of modeling circulating fluidized beds.

EXAMPLE 1

Performance of a Fast Fluidized Vessel

Determine the performance characteristics of a fast fluidized column when operated in the following four modes.

> *Mode I.* Constant solid inventory corresponding to $L_{mf} = 2.4$ m, with variable gas flow of $u_o = 2, 4, 6$ m/s
>
> *Mode II.* Constant solid flow at $G_s = 100$ kg/m²·s, with variable gas flow of $u_o = 2, 4, 6$ m/s
>
> *Mode III.* Constant gas velocity $u_o = 4$ m/s, with changing solid flow $G_s = 42, 50, 100, 200, 400$ kg/m²·s
>
> *Mode IV.* G_s and u_o both vary as follows:

u_o (m/s)	2	4	6
G_s (kg/m²·s)	70	100	120

For Mode I determine the vertical distribution of solids, ε_s. For the other modes determine the solid inventory in the bed as represented by L_{mf}.

Data

Column: $d_t = 0.4$ m, $H_t = 10$ m
Particles: catalyst, $\rho_s = 1000$ kg/m^3, $\bar{d}_p = 55$ μm, $\varepsilon_{mf} = 0.5$
Gas: ambient conditions

SOLUTION

From the discussion of this chapter and in the absence of any specific information on its value, we take the saturation carrying capacity of the gas, ε_s^*, to be 0.01.

Mode I. Follow the four step procedure presented above.

(a) From Fig. 8(b) take $\varepsilon_{sd} = 0.2$, 0.16, and 0.14 for $u_o = 2$, 4, and 6 m/s, respectively. Also, from Fig. 10, estimate that

$$au_o = 3 \, s^{-1}$$

Thus, $a = 1.5$, 0.75, and 0.5 m^{-1} for $u_o = 2$, 4, and 6 m/s, respectively.

(b) At $u_o = 2$ m/s, with Eq. (7),

$$\varepsilon_{se} = 0.01 + (0.2 - 0.01)e^{-1.5H_t} \tag{i}$$

(c) From Eq. (10)

$$(2.4)(1-0.50) = \frac{0.20 - \varepsilon_{se}}{1.5} + (10)(0.20) - H_t(0.20-0.01) \tag{ii}$$

Substituting Eq. (i) into Eq. (ii) and solving graphically gives

$$H_t = 4.8 \, \text{m} ; \quad \text{thus } H_d = 10 - 4.8 = 5.2 \, \text{m}$$

Substituting this value in Eq. (i) gives the fraction of solids in the exit gas stream:

$$\varepsilon_{se} = 0.01 + (0.20 - 0.01)e^{-(1.5)(4.8)} = 0.0101$$

This value is very close to that of ε_s^*.

(d) From Eq. (4), under the assumption $u_o \gg u_p \cong u_t$, we find

$$G_s = \rho_s u_o \varepsilon_{se} = (1000)(2)(0.0101) = 20.2 \, \text{kg/m}^2 \cdot \text{s}$$

Similar calculations with other gas velocities give the following values:

u_o (m/s)	ε_{se} (—)	H_t (m)	H_d (m)	G_s (kg/m$^2 \cdot$s)
2	0.0101	4.8	5.2	20.2
4	0.0187	3.8	6.2	74.7
6	0.0390	3.0	7.0	234.1

Note the rapid rise in the solid circulation rate with increase in gas velocity.

Mode II

(a) Same values for ε_{sd} and a, as in previous case.

(b) From Eq. (4) with the assumption that $u_o \gg u_p \cong u_t$

$$\varepsilon_{se} = \frac{100}{1000 u_o} = 0.050 \quad \text{for } u_o = 2\,\text{m/s}$$

(c) From Eq. (7) for $u_o = 2\,\text{m/s}$

$$\varepsilon_{se} = 0.50 = 0.01 + (0.20-0.01)e^{-1.5H_f}$$

Thus

$$H_f = 1.04\,\text{m}$$

(d) With Eq. (10) for $u_o = 2\,\text{m/s}$

$$L_{mf}(1-0.5) = \frac{0.20-0.050}{1.5} + (10)(0.20) - (1.04)(0.20-0.01)$$

Thus

$$L_{mf} = 3.8\,\text{m}$$

Similar calculations for other gas velocities give the following values:

u_o (m/s)	ε_{se} (—)	H_f (m)	H_d (m)	L_{mf} (m)
2	0.050	1.0	9.0	3.8
4	0.025	3.1	6.9	2.6
6	0.0167	5.9	4.1	1.9

Mode III

(a) At $u_o = 4\,\text{m/s}$ we have $a = 0.75\,\text{m}^{-1}$ and $\varepsilon_{sd} = 0.16$.

(b) From Eq. (4), for $G_s = 100\,\text{kg/m}^2\cdot\text{s}$ and with $u_o \gg u_p \cong u_t$,

$$\varepsilon_{se} = \frac{1000}{(1000)(4)} = 0.025$$

(c) With Eq. (7), for $G_s = 100\,\text{kg/m}^2\cdot\text{s}$,

$$0.025 = 0.01 + (0.16-0.01)e^{-0.75H_f}$$

Thus

$$H_f = 3.07\,\text{m}$$

(d) With Eq. (10), for $G_s = 100\,\text{kg/m}^2\cdot\text{s}$,

$$L_{mf}(1-0.5) = \frac{0.16-0.025}{0.75} + (10)(0.16) - (3.07)(0.16-0.01)$$

Thus

$$L_{mf} = 2.64\,\text{m}$$

Similar calculations with other solid circulation rates give the following values:

G_s (kg/m^2·s)	ε_{se} (—)	H_f (m)	L_{mf} (m)
42	0.0105	7.6	1.3
50	0.0125	5.5	2.0
100	0.025	3.1	2.6
200	0.050	1.8	3.0
400	0.100	0.68	3.2

For $G_s = 40\,\mathrm{kg/m^2 \cdot s}$, we find $H_f = \infty$. Whenever $H_f > H_t$, this means that the flow of solids into the column cannot keep up with the entrainment out of the column. Thus, the column becomes a pneumatic transport tube, as sketched in Fig. 3.13(a).

Mode IV

Calculations similar to those of Mode III give the following values:

$u_o\ (m/s)$	$G_s\ (kg/m^2 \cdot s)$	$\varepsilon_{se}\ (-)$	$H_f\ (m)$	$L_{mf}\ (m)$
2	70	0.035	1.35	3.7
4	100	0.025	3.1	2.6
6	120	0.020	5.1	1.95

Comment: The results of the above calculations are sketched in Fig. 12. In spite of the many simplifications made, the performance characteristics of this model do seem to be reasonable.

PROBLEMS

1. In the fast fluidized bed system of Example 1 the circulation rate of solids is kept constant at $G_s = 100\,\mathrm{kg/m^2 \cdot s}$ while the gas velocity is increased further. Determine the distribution of solids in the vessel; in other words, find ε_s versus z, and L_{mf}. Assume that $\varepsilon_{sd} = 0.13$, 0.12, and 0.10 for $u_o = 7$, 8, and 10 m/s, respectively. Then sketch the lines for these gas velocities in Fig. 12(b).

2. For a particular fast chemical reaction we intend to use very fine catalyst ($\bar{d}_p = 60\,\mu\mathrm{m}$) in a fast fluidized circulation system, somewhat as shown in Fig. 4(b). At the planned operating conditions the feed rate of gas will be $u_o = 6$ m/s to the 0.452-m ID and 10-m high reactor column. For the desired chemical conversion we also find that we need to keep 240 kg of solids in the column at all times. From this information determine the necessary circulation rate of solids (kg/s) in the system. Use $\rho_s = 1200\,\mathrm{kg/m^3}$, $\varepsilon_{mf} = 0.5$, and take ε_{sd} and a from Example 1. Also, since $\bar{d}_p = 60\,\mu\mathrm{m}$, you can reasonably assume that $u_o \gg u_t$.

3. In going through the calculations for Prob. 2, we find that the circulation rate of solids is more than the cyclone can handle. How much higher must the column be if the circulation rate is not to exceed 16 kg/s?

REFERENCES

1. J. Yerushalmi and A. Avidan, in *Fluidization*, 2nd ed., J.F. Davidson et al., eds., p. 226, Academic Press, New York, 1985; J. Yerushalmi, in *Gas Fluidization Technology*, D. Geldart, ed., p. 155, Wiley, New York, 1986; *Powder Technol.*, **24**, 187 (1986).

2. M.J. Rhodes and D. Geldart, in *Fluidization V*, K. Østergaard and A. Sørensen, eds., p. 281, Engineering Foundation, New York, 1986; in *Circulating Fluidized Bed Technology, II*, P. Basu, ed.,

p. 193, Compiegne, 1988; *Powder Technol.*, **53**, 155 (1987).

3. R. Abed, in *Fluidization IV*, D. Kunii and R. Toei, eds., p. 137, Engineering Foundation, New York, 1983; *Ind. Eng. Chem. Fundamentals*, **24**, 78 (1985).

4. A. Avidan, R.M. Gould, and A.Y. Kim, in *Circulating Fluidized Bed Technology*, P. Basu, ed., p. 287, Pergamon, New York, 1986.

5. E.U. Hartge, Y. Li, and J. Werther, in *Circulating*

Fluidized Bed Technology, P. Basu, ed., p. 153, Pergamon, New York, 1986; in *Fluidization V*, K. Østergaard and A. Sørensen, eds., p. 345, Engineering Foundation, New York, 1986; E.U. Hartge et al., in *Circulating Fluidized Bed Technology II*, P. Basu, ed., poster p. 35, Compiegne, 1988.

6. P.W.K. Kehoe and J.F. Davidson, *Inst. Chem. Eng. Symp. Ser.*, **33**, 97 (1971).

7. L. Massimilla, *AIChE Symp. Ser.*, **69(128)**, 11 (1973).

8. J. Yerushalmi et al., paper delivered at AIChE annual meeting, November 1976.

9. G.S. Canada, M.H. McLaughlin, and F.W. Staub, *AIChE Symp. Ser.*, **74(176)**, 27 (1976).

10. W.J. Thiel and O.E. Potter, *Ind. Eng. Chem. Fundamentals*, **16**, 242 (1977).

11. L. Carotenuto, S. Crescitelli, and G. Donsi, *Quad. Ing. Chim. Ital.*, **10**, 185 (1974); S. Crescitelli et al., Chisa Conf., p. 1, Prague, 1978.

12. Y. Li and M. Kwauk, in *Fluidization III*, J.R. Grace and J. Matsen, eds., p. 537, Plenum, New York, 1980; M. Kwauk et al., in *Circulating Fluidized Bed Technology*, P. Basu, ed., p. 33, Pergamon, New York, 1986; in *Circulating Fluidized Bed Technology II*, P. Basu, ed., poster p. 9, Compiegne, 1988.

13. G. Yang, Z. Huang, and L. Zhao, in *Fluidization IV*, D. Kunii and R. Toei, eds., p. 145, Engineering Foundation, New York, 1983; R. Zhang, D. Chen, and G. Yang, in *Fluidization '85, Science and Technology*, M. Kwauk et al., eds., p. 148, Science Press, Beijing, 1985.

14. U. Arena, A. Cammarota, and L. Pistone, in *Circulating Fluidized Bed Technology*, P. Basu, ed., p. 119, Pergamon, New York, 1986; U. Arena et al., in *Circulating Fluidized Bed Technology II*, P. Basu, ed., p. 16, Compiegne, 1988.

15. C. Brereton and L. Stromberg, in *Circulating Fluidized Bed Technology*, P. Basu, ed., p. 133, Pergamon, New York, 1986.

16. M. Toda, S. Satija, and L.S. Fan, in *Fluidization IV*, D. Kunii and R. Toei, eds., p. 153, Engineering Foundation, New York, 1983; S. Satija and L. S. Fan, *Chem. Eng. Sci.*, **40**, 259 (1985); K.D. Wisecarver, K. Kitano, and L.S. Fan, in *Circulating Fluidized Bed Technology*, P. Basu, ed., p. 145, Pergamon, New York, 1986.

17. L. Monceaux et al., in *Circulating Fluidized Bed Technology*, P. Basu, ed., p. 185, Pergamon, New York, 1986; in *Fluidization V*, K. Østergaard and A. Sørensen, eds., p. 337, Engineering Foundation, New York, 1986.

18. H. Weinstein et al., in *Fluidization IV*, D. Kunii and R. Toei, eds., p. 299, Engineering Foundation, New York, 1983; *AIChE Symp. Ser.*, **80(234)**, 52 (1984); **80(241)**, 117 (1984); in *Fluidization V*, K. Østergaard and A. Sørensen, eds., p. 329, Engineering Foundation, New York, 1986.

19. Q. Lu and Y. Wang, private communication, 1985.

20. I. Fusey, C.J. Lim, and J.R. Grace, in *Circulating Fluidized Bed Technology*, P. Basu, ed., p. 409, Pergamon, New York, 1986.

21. M. Horio et al., in *Circulating Fluidized Bed Technology II*, P. Basu, ed., p. 13, Compiegne, 1988; in *Proc. 3rd World Cong. Chem. Eng.*, Tokyo, 1986.

22. H. Takeuchi et al., in *Proc. 3rd World Cong. Chem. Eng.*, p. 477, Tokyo, 1986.

23. M.G. Schnitzlein, Ph.D. dissertation, City University of New York, 1987.

24. J.M. Matsen, in *Fluidization Technology*, D.L. Keairns, ed., p. 135, Hemisphere, Washington, D.C., 1976.

25. K. Kato et al., paper delivered at Soc. Chem. Eng. Japan annual meeting, April 1987.

26. J.C.L. Furchi et al., in *Circulating Fluidized Bed Technology II*, P. Basu, ed., p. 69, Compiegne, 1988.

CHAPTER

9

Solid Movement: Mixing, Segregation, and Staging

As shown in Chap. 6 the channeling of the rising gas bubbles causes the gross circulation of solids in a fluidized bed, while the small-scale intermixing of particles occurs mainly within the wakes that accompany the bubbles up the bed. When solids of wide size distribution and/or of different densities are fluidized, the larger or heavier particles tend to settle to the bottom of bed, but this is countered by the solid circulation, mentioned above. At several multiples of u_{mf} of the largest or heaviest particles, the mixing process dominates. However, as the gas velocity is reduced to and then below u_{mf} of the largest or heaviest particles, these solids progressively concentrate at the bottom of the bed. Thus mixing and segregation of different solids is apparently an equilibrium process that depends on bed conditions. Since vertical segregation of different solids is absent in high-velocity fluidization typical of fast fluidization or pneumatic transport, this chapter only concerns bubbling and turbulent beds wherein u_o is close to u_{mf} of at least some of the bed solids.

The rate of horizontal mixing of solids is also of concern. This is especially so in long shallow beds wherein solids are fed at one end of the bed, react in the bed, and then leave at the other end of the bed.

Overall, there are numerous aspects to the mixing and movement of solids in fluidized beds. In this chapter we consider

- Vertical mixing and segregation of solids
- Horizontal mixing and dispersion of solids
- Mixing-segregation equilibrium
- Large solids in beds of small particles
- Transfer of solids across horizontal baffle plates and leaking through distributor plates

For the design of a number of physical and chemical processes it is important to understand the mechanism and rates of these opposing phenomena of mixing and segregation and related phenomena. In some situations one may even take advantage of and deliberately encourage segregation of solids in the development of improved processes.

Vertical Movement of Solids

In catalytic reactors the large-scale vertical movement of porous particles can carry large amounts of adsorbed reaction components up and down the bed. This type of gas back mixing usually lowers conversion and selectivity. This is one reason why we need to know how much mixing of solids does occur, how to model this mathematically for predictive purposes, and what means are available to depress this movement.

As mentioned in Chap. 2, one often finds vertical or horizontal tubes fitted in catalytic reactors. These are placed there for various reasons: for temperature control, to reduce gulf circulation of solids, to reduce bubble size, to increase the emulsion voidage and thereby increase the overall residence time of reactant gas in the bed. All this raises the conversion of reactant gas and improves the selectivity of the desired product. For this reason various groups have also studied the movement of solids in beds with internals.

Experimental Findings

A variety of techniques have been used to study the vertical moment of solids, for example:

- Following the paths of individual tagged particles for long periods of time as they move about the bed.
- Measuring the extent of intermixing of two kinds of solids, originally located one above the other in the bed.
- Measuring the vertical spread of a thin horizontal slice of tracer solid.
- Finding the residence time distribution of the flowing stream in a bed with a throughflow of solids, using a variety of tracer techniques, such as step or pulse injection.
- Measuring the axial heat flow in a bed with a heated top section and cooled bottom section. This technique assumes that heat transport is caused solely by the movement of solids.

Surveys of the results of these many experimental studies are given by Kunii and Levenspiel [1], Potter [2], and van Deemter [3] in 1969, 1971, and 1985, respectively. These findings are most often reported in terms of the vertical dispersion coefficient D_{sv}. We briefly summarize them.

Beds without Internals. Figure 1(a) shows that the vertical mixing rate in rather small beds is directly related to the gas velocity by

$$D_{sv} = 0.06 + 0.1u_o, \qquad [\text{m}^2/\text{s}] \tag{1}$$

and Fig. 1(b) shows that the vertical mixing of solids is more rapid in large-diameter beds than in smaller beds, the relationship being given by

$$D_{sv} = 0.30d_t^{0.65}, \qquad [\text{m}^2/\text{s}] \tag{2}$$

In fact, in large vigorously bubbling fluidized beds of fine solids, the vertical movement of solids is very rapid. As an example, May [4] found that a slug of

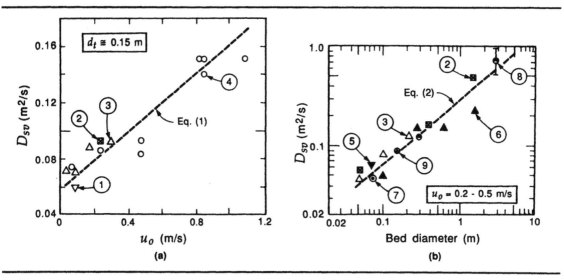

FIGURE 1

Vertical dispersion of solids in fine particle fluidized beds; from Avidan and Yerushalmi [5]; data from de Vries et al. [6] are added, and correlation line is modified somewhat from the original. ① Stemerding (Reman, 1955) [7], ② May (1959) [4], ③ Thiel and Potter (1978) [8], ④ Avidan and Yerushalmi (1985) [5], ⑤ Lewis et al. (1962) [9], ⑥ de Groot (1967) [10], ⑦ Miyauchi et al. (1968) [11], ⑧ de Vries et al. (1972) [6], ⑨ Avidan (1985) [5].

tracer introduced at one location in a large bed ($d_t = 1.52$ m, $L_f = 9.75$ m) of FCC catalyst became uniformly distributed throughout the bed in about 1 min.

Miyauchi et al. [12] studied the vertical mixing of solids in vigorously fluidized ($u_o > 10$ cm/s) beds of fine Geldart **A** solids. Recall from Chap. 3 that in such beds bubbles quickly reach a small limiting size. They found that the mixing data could reasonably be represented by the dispersion model with

$$D_{sv} = 12u_o^{1/2}d_t^{0.9}, \qquad [\text{cm}^2/\text{s}] \tag{3}$$

Example 1 compares values from this equation with the experimental values reported by de Groot [10].

Unfortunately, the dispersion model does not always well represent the vertical movement of solids. For example, May [4] found that for Geldart **A** solids the model well fitted the solid movement in his bed of aspect ratio $L_f/d_t = 9.1$ m$/0.38$ m $= 24$, but was inadequate for this bed of aspect ratio $L_f/d_t = 9.75$ m$/1.52$ m $= 6.4$.

Similarly, Avidan and Yerushalmi [5] found that the dispersion model well represented the mixing during turbulent fluidization where the bed looked close to homogeneous, but fitted the data poorly when the bed was in the bubbling regime.

To summarize, one may expect the dispersion model to reasonably represent the vertical mixing in tall beds in which rather small-scale mixing is taking place. This is characteristic of fine particle (Geldart **A**) systems with only mild gulf streaming. We would not expect it to satisfactorily represent shallow beds, beds with strong solid circulation, or beds with nonuniformly distributed internals.

Where the dispersion model does not fit well (gently bubbling and in not very deep beds), the countercurrent mixing model often is used. This model views the solids moving in two streams, one rising and the other descending, with a crossflow or interchange between streams. This model closely matches the K-L model of Chap. 6 and is taken up in the next section.

Although the dispersion model does not reasonably represent the movement of solids in certain conditions, the results of experiments are invariably forced into this form and are reported in terms of a dispersion coefficient.

Beds with Internals. So far we have presented the findings on solid movement in beds free of internals. However, the presence of internals in a bed strongly hinders this movement. For example, Chen et al. [13] reported the following velocities of solids in the core of the upper vortex (refer to Fig. 6.3(d)) of their experimental bed containing horizontal tubes:

		No tube bundle in bed	Sparse bundle	Dense bundle
Average upward solid velocity (m/s)	at $u_o/u_{mf} = 4$	0.19	0.09	0.01
	at $u_o/u_{mf} = 6$	0.26	0.15	0.02

Admittedly the bed was rather small ($d_t = 0.19$ m, $L_f \simeq d_t$); however, these findings should indicate the general effect of bed internals.

In experiments with larger Geldart **B** solids in a 1.2×1.2 m bed that contained internals, Sitnai [14] noted that if a tube bank was located away from the wall of the bed, then solids would slide down in a thick stream along the wall surface, thereby generating a severe solid circulation pattern. So, to achieve a uniform solid movement, one should take this into account.

As an extreme of bed internals, Claus et al. [15] reported on the behavior of a fluidized bed ($d_t = 9.2$ cm, $L_m = 9.2$ m) packed with 2-cm wire screen Raschig rings. With Geldart **B** solids they found remarkably uniform fluidization with small bubbles throughout the bed. However, with fine Geldart **A** solids, fluidization was unsatisfactory, with considerable agglomeration of fine solids on the packing.

We next discuss the various models that have been used to interpret the experimental findings on the vertical moment of solids.

Dispersion Model

The *dispersion model* is a diffusion-type model represented by the differential equation

$$\frac{\partial C_s}{\partial t} = D_{sv} \frac{\partial^2 C_s}{\partial z^2} \tag{4}$$

where C_s is the concentration of tagged particles at position z at time t, and D_{sv} is the vertical dispersion coefficient averaged over the entire cross section of the bed.

The solution of Eq. (4) may take several forms. For a step input of tracer introduced into the stream of solids entering the bottom of a fluidized bed and leaving at the top, or vice versa, the use of the appropriate boundary and initial conditions gives

$$\frac{C_s \text{ (at exit)}}{C_s \text{ (at } t = \infty)} = f\left(\frac{D_{sv}}{\bar{u}_{s,up}z_f}, \frac{t}{\bar{t}_s}\right) \tag{5}$$

For a pulse of tracer introduced into a bed with no throughflow of solids,

$$\frac{C_s(t)}{C_s \text{ if well mixed in the bed}} = f\left(\frac{D_{sv}t}{L_f^2}, \frac{z}{L_f}\right) \tag{6}$$

Verloop et al. [16] gave several solutions to Eq. (5) for solids throughflow, and also told where additional solutions may be found. For the batch situation, May [4] gave the solution for one initial condition. Others can be extracted from Carslaw and Jaeger [17].

As understanding of the hydrodynamics of fluidized beds grew, attempts were made to relate the dispersion model to more mechanistic models so that more fundamental measurements could be used for the design of large-scale units. We now look at some of these developments.

Counterflow Solid Circulation Models

In the bubbling bed models sketched in Fig. 6.12, we see some solids flowing up the bed and others flowing down the bed. This upflow and downflow with an interchange between streams is the basis for various counterflow models that have been proposed to account for the vertical mixing of solids.

The simplest version, introduced by van Deemter [18], divides the solids into two streams: one flowing up at a velocity u_{su}, the other flowing down at u_{sd}, with f_u and f_d ($= m^3$ solids/m^3 bed) being the bed fractions consisting of these streams (see Fig. 2). Consider the movement of some labeled or tagged

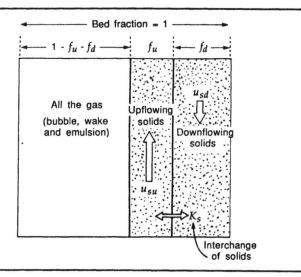

FIGURE 2
Counterflow solid circulation model for the vertical movement of solid in a bubbling fluidized bed.

solids that constitute a fraction X_{su} and X_{sd} ($= m^3$ tagged solids/m^3 total solid stream) of the up- and downflowing streams. The differential equation describing the vertical movement of these tagged solids and their interchange is then

$$f_d \frac{\partial C_{sd}}{\partial t} + f_d u_{sd} \frac{\partial C_{sd}}{\partial z} + K_s(C_{sd} - C_{su}) = 0 \tag{7}$$

and

$$f_u \frac{\partial C_{su}}{\partial t} + f_u u_{su} \frac{\partial C_{su}}{\partial z} + K_s(C_{su} - C_{sd}) = 0 \tag{8}$$

where the solids interchange coefficient K_s (m^3 tracer/m^3 bed·s) represents the transfer of tagged solid from one stream to the other.

For a tall enough bed of fine particles and sufficiently large values of elapsed time, van Deemter showed that the changes in concentration of labeled solids could be represented by an effective dispersion coefficient given by

$$D_{sv} = \frac{f_d^2 u_{sd}^2}{K_s(f_d + f_u)} = \frac{f_d^2 u_{sd}^2}{K_s(1 - \delta)(1 - \varepsilon_f)}, \qquad [m^2/s] \tag{9}$$

He then applied this relationship to the data on vertical mixing of silica sand ($u_{mf} = 0.15$ cm/s) reported by de Groot [10], and found values of D_{sv} ranging from 0.03 to 0.23 m^2/s.

Relating the Counterflow to the Dispersion Model

Kunii et al. [19] proposed using the Davidson bubble plus wake as the basis for developing an expression for the interchange coefficient between up- and downflowing solids in beds of fast-rising, hence clouded, bubbles.

Consider the movement of solids around a clouded bubble as shown in Fig. 3. Since the circulation of cloud gas is rapid, Kunii et al. assumed that all the solids from the lower part of the cloud are swept into the wake, mix with the

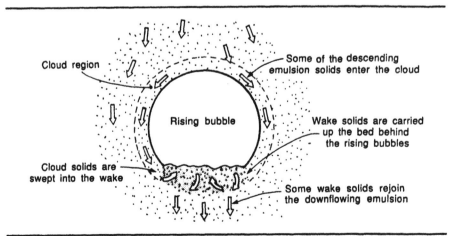

FIGURE 3
Model for the mechanism of interchange of solids between downflowing emulsion solids and upflowing wake solids; from Kunii et al. [19].

solids already there, and eventually leak back into the emulsion. By this process, slowly downflowing emulsion solids are swept into the rising bubble wake and then return to the downflowing emulsion. From this mechanism the interchange coefficient for solids in beds with clouded bubbles is

$$K_s = \frac{\text{(volume of solids transferred from the emulsion to the wake)}}{\text{(volume of bubble)(time)}}$$

$$\simeq \frac{3(1 - \varepsilon_{mf})}{(1 - \delta)\varepsilon_{mf}} \frac{u_{mf}}{d_b}, \qquad [s^{-1}] \tag{10}$$

With a somewhat similar model, Chiba and Kobayashi [20] derived the following expression for the interchange coefficient:

$$K_s = \frac{3}{2}\left(\frac{f_w}{1 + f_w}\right)\frac{u_{mf}}{\varepsilon_{mf}d_b} \tag{11}$$

Introducing K_s from Eq. (10) into Eq. (9) and simplifying leads to the following expression for the vertical dispersion coefficient in terms of measurable bubble and bed properties:

$$\boxed{D_{sv} \cong \frac{f_w^2 \varepsilon_{mf} \delta d_b u_b^2}{3 u_{mf}}} \tag{12}$$

Potter [2] developed a similar expression with the term $1 - \delta$ multiplied on the right-hand side, and with u_o in place of u_b.

As mentioned, van Deemter [18] extracted values of D_{sv} from the experimental results reported by de Groot [10]. Example 1 compares the predictions of Eq. (12) with these reported D_{sv} values. By estimating appropriate values of the bubble rise velocity, one obtains a good fit. In particular, Eq. (12) predicts that D_{sv} should be larger for beds of smaller particles. This is consistent with experimental findings.

Coarse Particle Beds

So far we have only considered the movement of fine solids in tall beds, thus beds with small equilibrium bubbles. When coarse particles are fluidized in beds of aspect ratio of unity, or shallower still, the application of the dispersion model to vertical mixing is not justified unless internals are positioned exactly uniformly across the bed, in which case the spacing of the internals governs the scale of solid mixing.

Such positioning is not really practical; hence one may find a thick stream of solids descending at vessel walls or in the spaces between neighboring tube banks. Sitnai [14] proposed another counterflow model, which accounts for this third phase of solids that descends along these walls. This model is sketched in Fig. 4. On fitting his model to tracer data from a 1.2 m × 1.2 m tube-filled bed, he gave the mean velocity of descending solids as 1.7–1.8 cm/s in the main part of the bed and 4.8–5.8 cm/s along the wall for $u_o = 0.6$–0.9 m/s and $u_{mf} = 0.3$ m/s. Van Deemter [3] surveyed these counterflow solid circulation models.

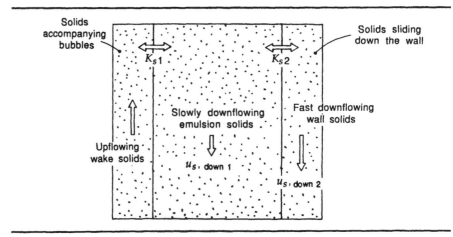

FIGURE 4
Three-region counterflow model for vertical solid movement in large particle shallow beds with internals; from Sitnai [14].

EXAMPLE 1

Vertical

Movement

of Solids

Calculate the vertical dispersion coefficient from Eqs. (3) and (12) and compare these values with the values extracted by van Deemter [18] from the experiments of de Groot [10] in various sized vessels.

Data $u_{mf} = 0.015 \, \text{m/s}$, $\varepsilon_{mf} = 0.5$, $u_o = 0.1 \, \text{m/s}$, $\delta = 0.2$, $d_b = 0.06 \, \text{m}$

d_t (m)	0.1	0.3	0.6	1.5
u_b (m/s)	0.40	0.75	0.85	1.1
Reported D_{sv} (m²/s)	0.030	0.11	0.14	0.23

SOLUTION

Using Eq. (3) we find

$$D_{sv} = 12(10)^{0.5}d_t^{0.9} = 38d_t^{0.9} , \qquad [\text{cm}^2/\text{s}]$$

The main problem in using Eq. (12) is choosing a reliable value for the wake fraction f_w. Hamilton et al. [21] report $f_w = 1-2.9$ for this range of particle size (mean value of 2), whereas Fig. 5.8 gives $f_w = 0.32$. Although not very satisfactory, we average these f_w values. Thus

$$f_w = \frac{0.32 + 2}{2} = 1.16 \cong 1.2$$

Then Eq. (12) becomes

$$D_{sv} = \frac{(1.2)^2(0.5)(0.2)(0.06)u_b^2}{3(0.015)} = 0.192u_b^2$$

Thus we find

d_t	0.1	0.3	0.6	1.5 m
D_{sv}, from experiment	0.030	0.11	0.14	0.23 m²/s
D_{sv}, from Eq. (3)	0.030	0.08	0.15	0.35 m²/s
D_{sv}, from Eq. (12)	0.031	0.11	0.14	0.23 m²/s

Horizontal Movement of Solids

Experimental Findings

The horizontal movement of solids was first studied by Brötz [22] in a shallow rectangular bed, as shown in Fig. 5. Measuring the rate of approach to uniformity after removal of the dividing plate then gave the information needed to evaluate the horizontal dispersion coefficient D_{sh}. A similar approach was used by other investigators [23–25].

Heertjes et al. [26] suggested that the wake material scattered into the freeboard by the bursting bubbles could contribute significantly to the horizontal movement of solids. Hirama et al. [24] and Shi and Gu [27] used partition plates in the freeboard just above the bed to study this effect.

All of these investigators used rather shallow beds of height between 5 and 35 cm. In contrast, Bellgardt and Werther [28] made measurements in a much larger bed, namely a $2 \text{ m} \times 0.3 \text{ m}$ bed about 1 m deep. Quartz sand ($d_p = 450 \ \mu\text{m}$) was fluidized, and careful measurements confirmed that vertical mixing was much faster than horizontal mixing, thus justifying the use of a one-dimensional dispersion model in the horizontal direction. They found $D_{sh} = 6$–$25 \text{ cm}^2/\text{s}$ for $u_0 = 0.23$–0.73 m/s. Applying their model to coal combustion and gasification, Bellgardt et al. [29] presented a performance model for solid movement that was tested in TVA's 20-MW FBC pilot plant.

Table 1 gives information on the reported studies of horizontal movement of solids. We summarize these findings as follows:

- Comparing the dispersion coefficient for the horizontal movement with the vertical movement of solids (compare Table 1 with Fig. 1), we see that D_{sh} is roughly an order of magnitude smaller than D_{sv}.
- D_{sh} increases with bed width. For example, Hirama et al. [24] in their very small units found a 60% increase for a doubling in bed width.
- The scattering of solids into the freeboard contributes significantly to D_{sh} in shallow beds.

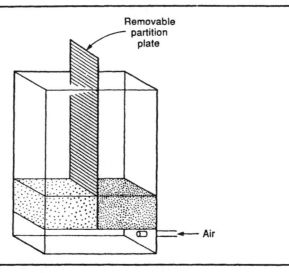

FIGURE 5
Experimental setup used by Brötz [22] to study the horizontal movement of solids.

TABLE 1 Range of Experimental Data for the Horizontal Movement of Solid

Investigators	Bed (m)	Particles
Mori and Nakamura [23] (1965)	0.9 × 0.3	Polyvinyl chloride
Hirama et al. [24] (1975)	0.4 × (0.042, 0.08, 0.2)	Glass beads, cracking cat.
Bellgardt and Werther [28] (1984)	2.0 × 0.3 $L_f \simeq 1$	Quartz sand
Kato et al. [30] (1985)	0.5 × 0.2 Vertical tubes 0.032 Horizontal tubes 0.018	Activated carbon
Shi and Gu [27] (1986)	0.57 × 0.05	Resin, silica gel

- Vertical and horizontal internals reduce D_{sh} significantly. From Kato et al. [30] we find this effect to be as follows:

	No tubes present	Effective bed diameter with tubes present, d_{te} (cm)				
		15.9	10.6	7	5	~2.6
D_{sh} (cm²/s) with vertical tubes	20	—	10	—	6	3
D_{sh}, (cm²/s) with horizontal tubes	20	8	—	6	—	3

This table shows no appreciable difference between the effect of horizontal and vertical tubes. Kato et al. [30] also found that for their operating conditions D_{sh} was unaffected by bed height.

Mechanistic Model Based on the Davidson Bubble

Consider the following mechanism for the horizontal movement of solids in a fine particle bed of fast rising bubbles, as sketched in Fig. 6.

Postulate. As a bubble rises, it pushed emulsion aside. However, the solids passing close to the bubble enter its cloud and are then drawn into the wake, whose diameter is roughly α times the bubble diameter. Solids mix uniformly in the wake and leave the wake from random positions, thereby giving rise to horizontal mixing. Solids further from the bubble move aside as the bubble passes, but return close to their original position.

For this mechanism it is simplest to evaluate the horizontal dispersion coefficient D_{sh} in terms of the Einstein random walk equation:

$$D_{sh} = \frac{\text{(fraction of solids that mix)(mean square distance moved)}}{4\text{(time interval considered)}}$$

$$= \frac{1}{4}\left(\begin{array}{c}\text{fraction of bed solids that enter bubble wakes} \\ \text{to mix there per unit time}\end{array}\right)\overline{\Delta r^2} \quad (13)$$

Particles

d_p (μm)	u_{mf} (m/s)	D_{sh} (m^2/s) at u_0 (m/s) and d_b (m)
595	0.295	$1-7 \times 10^{-3}$ at $u_o = 0.4-0.7$
150, 460	0.024, 0.15	$0.7-2 \times 10^{-3}$ at $u_b d_b = 1-2 \times 10^{-2}$
75	0.0055	$0.6-3 \times 10^{-3}$ at $u_b d_b = 1-6 \times 10^{-2}$
		$d_b = 0.02-0.06$ at $u_o - u_{mf} = 0.05-0.4$
450	0.17	$0.6-2.5 \times 10^{-3}$ at $u_o - u_{mf} = 0-0.5$
394–1073	0.035–0.27	$0.1-1 \times 10^{-3}$ at $u_o - u_{mf} = 0.1-0.6$
		$0.2-1 \times 10^{-3}$ at $u_o - u_{mf} = 0.15-0.6$
		$d_e = 0.026-0.159$
450, 750	0.076, 0.20	$0.3-0.8 \times 10^{-3}$ at $L_{mf} = 0.02-0.07$
620	0.0905	

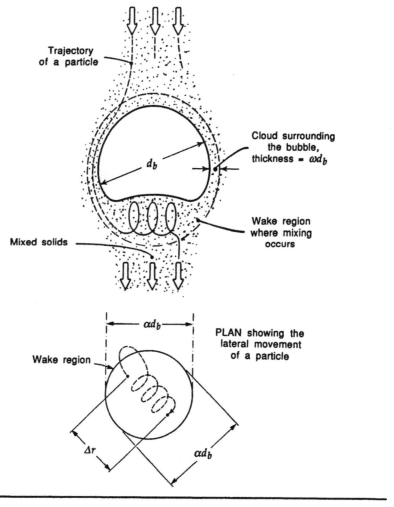

FIGURE 6
The horizontal movement of solids according to the model of Kunii and Levenspiel [31].

Kunii and Levenspiel [31] evaluated the terms of this expression. Thus for clouded Davidson bubbles the term in parentheses depends on bubble size, cloud thickness, and bubble density in the bed. Next, from probability theory the mean square horizontal shift of a particle on passing through the bubble wake is given by

$$\overline{\Delta r^2} = \frac{(\alpha d_b)^2}{4}$$

where αd_b is the effective diameter of the wake.

Replacing all quantities into Eq. (13) gives, in general, for both fast and intermediate bubbles,

$$D_{sh} = \frac{3}{16} \frac{\delta}{1-\delta} \alpha^2 d_b u_{br} \left[\left(\frac{u_{br} + 2u_f}{u_{br} - u_f} \right)^{1/3} - 1 \right] \tag{14}$$

For fast bubbles with thin clouds typical of fine particle systems, or $u_{br} \gg u_f$, Eq. (14) simplifies to

$$\boxed{D_{sh} = \frac{3}{16} \frac{\delta}{1-\delta} \frac{\alpha^2 u_{mf} d_b}{\varepsilon_{mf}}} \tag{15}$$

For fine Geldart **A** and **AB** solids ($d_p = 60$ and 150 μm), Kunii and Levenspiel found that Eq. (15) with $\alpha = 1$ fitted their data, while for larger Geldart **BD** solids (quartz, $d_p = 450$ μm), Bellgardt and Werther [28] found that $\alpha = 0.77$ well represented their data.

Equations (14) and (15) do not account for the scattering of solids at the bed surface, and Shi and Gu [27] presented a model that does account for it. However, since this freeboard scattering is restricted to the top layer of bed solids, its effect on D_{sh} for the bed as a whole becomes smaller for deeper beds.

Equation (15) can be used for approximate prediction of D_{sh} for deep beds, but more importantly it suggests how changes in system variables such as u_{mf}, u_o, and d_b will affect D_{sh}. Example 2 concerns D_{sh}.

EXAMPLE 2

Horizontal

Drift of

Solids

Bellgardt and Werther [28] presented the following data on the horizontal dispersion coefficient for quartz solids in fluidized beds. Compare the predicted D_{sh} from Eq. (14) or (15) with their findings.

Data

Bed size: 0.3×2 m, $L_{mf} = 0.83$ m
Quartz sand: $d_p = 450$ μm, $\varepsilon_{mf} = 0.42$, $u_{mf} = 0.17$ m/s

u_o (m/s)	0.37	0.47	0.57	0.67
Reported D_{sh} (m²/s)	0.0012	0.0018	0.0021	0.0025

From Fig. 6.10(a) we estimate $d_b = 0.10–0.14$ m for this range of gas velocities.

SOLUTION

Let us show the solution for the first data point, $u_o = 0.37$ m/s, and for the smallest estimated bubble size, $d_b = 0.10$ m. Then the first problem is to decide whether to use Eq. (14) or (15). For this determine whether $u_{br} \gg u_f$. Now

$$u_{br} = 0.711(d_b g)^{1/2} = 0.711(0.1 \times 9.8)^{1/2} = 0.70 \, \text{m/s}$$

$$u_b = u_o - u_{mf} + u_{br} = 0.37 - 0.17 + 0.70 = 0.90 \, \text{m/s}$$

$$u_f = \frac{u_{mf}}{\varepsilon_{mf}} = \frac{0.17}{0.42} = 0.40 \, \text{m/s}$$

Since we do not have $u_b \gg u_f$, we use Eq. (14) instead of Eq. (15) to calculate D_{sh}. In addition, since we are not dealing with fine Geldart **A** or **AB** solids, we take $\alpha = 0.77$ in this equation (see just after Eq. (15)).

The last quantity needed before using Eq. (14) is δ. Since u_{br} is close to u_f, we use Eq. (6.27) to give

$$\delta \cong \frac{u_o - u_{mf}}{u_b + u_{mf}} = \frac{0.37 - 0.17}{0.90 + 0.17} = 0.187$$

Substituting all into Eq. (14) gives

$$D_{sh} = \frac{3}{16} \frac{0.187}{1 - 0.187} (0.77)^2 (0.1)(0.70) \left[\left(\frac{0.70 + 2 \times 0.40}{0.70 - 0.40} \right)^{1/3} - 1 \right]$$

$$= 0.0013 \, \text{m}^2/\text{s}$$

Similar calculations are made for other u_o values. The final results and comparison with the experimental values gives

u_o		0.37	0.47	0.57	0.67 m/s
D_{sv} calculated $\begin{cases} \text{at } d_b = 0.10 \text{ m} \\ \text{at } d_b = 0.14 \text{ m} \end{cases}$		1.3	1.9	2.5	$3.2 \times 10^{-3} \, \text{m}^2/\text{s}$
		1.4	2.2	2.9	$3.6 \times 10^{-3} \, \text{m}^2/\text{s}$
D_{sv} from experiment		1.2	1.8	2.1	$2.5 \times 10^{-3} \, \text{m}^2/\text{s}$

The calculated values are somewhat high.

Segregation of Particles

Numerous processes require fluidizing a mixture of solids of very different density. As an example, in one step in the production of titanium or zirconium a mixture of the metal oxide (high-density solid) and coke (very low density) is fluidized by chlorine gas at high temperature. To achieve close to 100% conversion of chlorine, bed bubbling should not be too vigorous; hence the gas velocity should not be too high. On the other hand, the gas velocity should not be too low or solids will separate out. What size ratio of solids and what gas velocity should be used in such situations? The whole question of the mixing-segregation equilibrium is of vital concern in these situations.

Mixing-Segregation Equilibrium

Much has been reported in recent years on the mixing-segregation phenomenon in gas fluidized beds, especially on binary systems of different size and/or density; see [32]. Here, particle segregation occurs at close to u_{mf} of the biggest or heaviest particles in the bed. Also, this whole question mainly concerns large particle systems.

Cooperative investigations on particle segregation were carried out by Rowe et al. [33,34], Nienow et al. [35,36], Chiba [37–39] and others, in which the following special terminology was used for the fluidized bed components:

jetsam: component that ultimately sinks
flotsam: component that floats to the top of the bed

We summarize those findings as follows:

With solids of the same size but different density, the bed segregates readily. When this occurs, the dense material forms a relatively pure bottom layer. The upper layer always contains some of the denser solids, more or less uniformly dispersed.

Particles of different size but the same density will also segregate, but not easily. Even particles an order of magnitude different in diameter will mix fairly uniformly at moderate bubbling conditions. With a wide size distribution of particles rather than a sharp cut of two distinct sizes, we may expect much less segregation.

When the gas velocity is close to u_{mf}, the segregation of the jetsam can be severe. At higher gas velocities it is less severe. Figure 7 illustrates the segregation pattern in binary mixtures; Fig. 8 illustrates this behavior for mixtures of commercial powders. Thus, Fig. 8(a) shows a sharp segregation of denser material (density ratio $\simeq 1.8 : 1$), and Fig. 8(b) shows that even with very different particle sizes (size ratio $= 3.5 : 1$) segregation is minor. Both systems display less segregation when the gas velocity is raised.

Rowe et al. [33] proposed using a solids mixing index defined as

$$M = \left(\frac{\text{fraction of jetsam in the top portion of the bed}}{\text{fraction in a well-mixed bed}} \right) = \frac{X_{sJ,\,\text{top}}}{\bar{X}_{sJ}} \qquad (16)$$

$M = 0$ and 1 correspond to complete segregation and complete mixing, respectively.

Noting that the jetsam fraction in Fig. 8 is practically constant in a large portion of the bed, we can use this value to get an approximation for M; thus

$$M \cong \frac{X_{sJ,\,\text{straight-line portion}}}{\bar{X}_{sJ}} \qquad (17)$$

Factors affecting particle segregation were studied by Rowe and Nienow [34,35] in terms of this index.

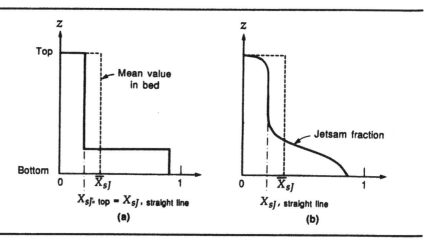

FIGURE 7
Distribution of solids in strongly segregating binary mixtures: (a) idealized segregation at low fluidizing velocity; (b) pattern of segregation at vigorous bubbling conditions.

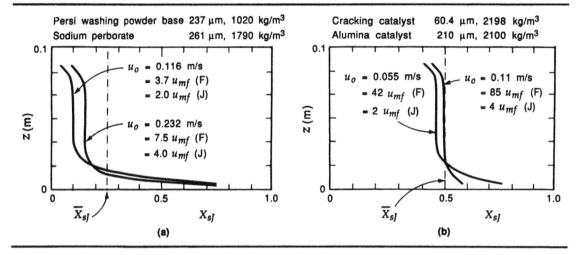

FIGURE 8
Vertical segregation of commercial solids; $d_t = 0.141$ m, $L_m = 0.1{-}0.15$ m; adapted from Rowe et al [33]: (a) different density materials; (b) different sized materials.

Rowe et al. [33] were the first to suggest that rising bubbles are the vehicle for particle segregation. Thus, all solids, both flotsam and jetsam, are carried up the bed in the bubble wakes. However, only the larger denser particles preferentially move down the bed as a bubble passes by. They emphasized that this upflow in the bubble wake is the only way that the smaller less dense particles can reach the top of the bed.

Mathematical models to account for the axial distribution of solids at a stable mixing-segregation equilibrium have been proposed by several research groups. Chiba et al. [39] reviewed and summarized these works. They also explained the preferential downflow of denser solids as follows: the denser larger particles tend to fall preferentially through the temporarily disturbed region (the wake) behind the bubbles. This downward movement of jetsam is clearly analogous to the jiggling separation technique.

Chiba et al. [38] found that an increase in pressure promoted solid mixing, and explained this in terms of increased wake fractions. This observation suggests that for their system the mixing of bed solids by the upflow of wakes was more important than the preferential downflow of jetsam from the wakes.

Steady State Separation of Particles

A number of operations are being explored that require fluidization with separation of two different kinds of solids: for example, driers in which fine wet solids are mixed with a hot solid heat carrier, and some advanced combustor-gasifiers that fluidize dolomite and char. Some researchers [35,40–43] have studied these systems.

Note that with one inflow and one outflow stream for a mixture of solids that readily segregate, the bed composition will adjust itself so that at steady state the outflow composition will equal the inflow. Of course, the location of the outflow will influence the bed composition. Figure 9 illustrates this. With two outflow streams, one should be able to get a good separation of components, and Chiba et al. [36] showed how to predict the composition of the two outflow

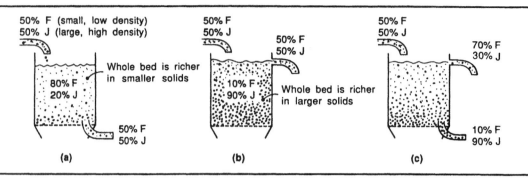

FIGURE 9
Sketches show that the location of the exit pipe can be used to control the bed composition. The percentages are for illustration only. In (c) the exiting compositions are related to their flow rates.

streams of Fig. 9 for a binary system, knowing the equilibrium distribution of solids in a batch operation such as shown in Fig. 8.

Large Solids in Beds of Smaller Particles

The movement of large objects in freely bubbling beds has been studied by several groups of investigators whose experimental conditions are summarized in Table 2.

Aiming to develop bubble breakers for reactors, Keillor and Bergougnou [45] introduced large thin cylinders into Geldart **B** beds. Their effect on bubble size and rise velocity was measured. When the density of the cylinders exceeded the mean bed density by more than 10%, the cylinders began to settle on the distributor.

Masson et al. [47] systematically studied the effect of the density of large spheres and cylinders on their fluidization behavior. Their observations are

TABLE 2 Experimental Conditions for the Movement of Large Solids in Fine

Investigators	Bed (m) L_m (m)	Bed Particles	ρ_s (kg/m^3)
Pruden et al. [44] (1975)	0.247 $L_m = 0.32$	Sand Glass beads	2649 2435
Nienow et al. [36] (1978)	0.14 $L_m = 0.25$	Sand Glass beads Malachite	2720 2940 1800 (bulk)
Bergougnou et al. [45] (1975, 81)	0.89 × 0.0254 5 m tall	Sand	2610
Nienow and Cheesman [46] (1980)	0.14	Alumina	1450
Masson et al. [47] (1983, 86)	0.18 × 0.18 $L_m = 0.6$	Glass beads	2570
Bemrose et al. [48] (1986)	0.4 × 0.15 $L_m = 0.11–0.13$	Sand	2600

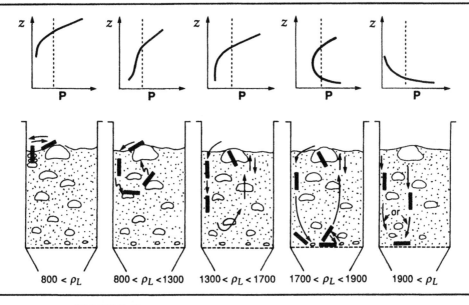

FIGURE 10
Movement of large cylinders (11 mm OD, 45 mm long) of various densities in fine particle beds ($u_{mf} = 0.096$ m/s, $u_o = 2.8u_{mf}$); adapted from Masson et al [47]: ρ_L = density of cylinder; P = probability of having the cylinder present at level z in bed; $\rho_{bulk} = 1490$ kg/m^3.

illustrated in Fig. 10, and their results are summarized below. Note that the bulk density of the bed solids (glass beads) is 1490 kg/m^3.

- For a large body of very low density, $\rho_L < 800$ kg/m^3, the large body is swept up by the roof of the bubbles and remains near the top of the bed.
- At a density $800 < \rho_L < 1300$, the body occasionally descends into the bed where rising bubble roofs push it back up. Sometimes several bubbles are needed to return it to the top of the bed.

Particle Beds

$\bar{d}_p$ (μm)	u_o (m/s)	Large Solids	ρ_L (kg/m^3)	d_L (mm)
210–707	0.12–0.23	Cylinder	609–2707	12.7, 50.8,
177–250				25.4
160–3030	$u_o - u_{mf} = 0.02-0.24$	Char,	1270–1420	6.4–13.9
243, 461		anthracite,		
338		etc.		
120–595	0.64	Disk	930–1190	51, 16 thick
265	$u_o - u_{mf} = 0.1-0.3$	Foam rubber,	370–809	9.8–25.4
		cardboard,		$(0.85, 2.8) \times 25.4$
		etc.		
300, 550	$u_o = 0.269 = 2.8u_{mf}$	Cylinder,	890–1900	11 OD, 45
	$1-4.1u_{mf}$	sphere		8, 20
590	$u_o - u_{mf} = 0.61-0.69$	Marble,	2580	16.5
	$u_{mf} = 0.35$	pebble,	2560	0.55–10.5
		shale	1600	

- For a density comparable to the bed density, or $1300 < \rho_L < 1700$, the large body penetrates the bubble roofs; however, the bubble wakes are still able to push the body up.
- For density between 1700 and 1900, the large body falls to the distributor from time to time. It stays there until a swarm of bubbles exert a lifting effort great enough to push it toward the top of the bed. Once caught by this swarm of bubbles, the large body is often conveyed rapidly all the way to the top of the bed.
- For very dense objects, $\rho_L > 1900$, or with a density ratio greater than $1.3:1$, the large body falls to the bottom of the bed and stays there.

Masson et al. reported that the downward velocities for both large spheres and large cylinders were the same as for the downflowing emulsion solids.

Large Solids Resting on Distributors

Bemrose et al. [48] observed the movement of larger solids resting on perforated plate distributors. On inclined distributors, they found that these particles did not slide downward along the plate but upward. They attributed this to the gulf circulation of bed particles. These findings suggest that it is important to consider gulf circulation of solids in the region of the distributor when bed solids are to be continually removed there.

To protect metal distributors from very corrosive atmospheres in some high-temperature fluidized bed reactors, a layer of larger pebbles or rocks is sometimes laid on the distributor plate. With gulf circulation of bed solids or uneven fluidization, these pebbles may shift across the distributor to leave a bare spot or relatively thin layer of pebbles at the center of the bed. Fine particles are then apt to penetrate into this thin layer of large pebbles, reinforce the percolating effect, and even damage the distributor. For practical design this sort of effect should be tested for beforehand.

Staging of Fluidized Beds

Here we consider the control of solid movement about the fluidized bed by use of leaky perforated plates.

Batch of Solids

When reactant gas contacts a batch of solids, say catalyst, in a fluidized bed, vertical staging of solids may have definite advantages because the gas may more closely approach plug flow and a desirable temperature profile can be achieved in the reactor. Horizontal perforated plates are probably the simplest way to get such staging. However, since vigorous fluidization of fine solids always leads to freeboard carryover, the holes in the perforated plates must be large enough to allow the passage of solids; otherwise solids will plug the holes from below. But with large enough holes, solids are liable to leak from above. Thus, one must be careful to match these two solid flows. We now consider this question of solid interchange rate between stages.

The literature on multistage fluidization has been reviewed by Grace and Harrison [49], followed by some more recent studies [50–52]. Unfortunately, there is little interest by the research community in this subject and no general quantitative findings reported in the literature. However, qualitatively we can come to some useful conclusions.

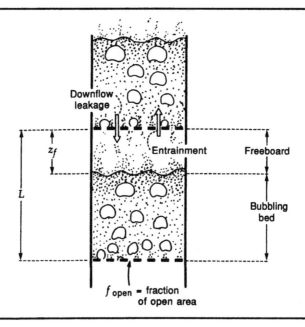

FIGURE 11
Two stages of a multistage fluidized bed showing the flow of solids across a baffle plate.

Consider two stages of a multistage fluidized bed, as shown in Fig. 11. When equilibrium is established in terms of solid flow, the downflow leakage rate $G_{s,down}$ must just match the upflow freeboard entrainment rate $G_{s,up}$. Thus, the solid interchange rate per unit cross-sectional area of the baffle plate is

$$G_{s,up} = G_{s,down} = I_s f_{open} , \qquad [kg/m^2 \text{ column·s}] \qquad (18)$$

where I_s is the flux based on the open area of holes (kg/m² holes·s). As a first approximation, one may expect the solids upflow through the plate, $G_{s,\,up}$, will be related to the entrainment rate of solids from an ordinary fluidized bed having the same freeboard height z_f. Thus the most important parameters to influence $G_{s,\,up}$ would be the freeboard height z_f and u_o.

The downflow leakage of solids should increase with both the hole diameter and the fraction of open area of the plate. Another important factor is the thickness of the baffle plate, thicker plates giving less leakage. Kono and Huang [52] even suggested using long downcomer pipes at the baffle plates to reduce the downflow.

Figure 12 illustrates how changes in the different variables will influence the interchange rate of solids between stages.

Finally, the study by Guigon et al. [51] is indicative of the findings on particle interchange at perforated plates. Figure 13(a) shows their experimental setup. Here 135- and 210-μm sand was fluidized by hot air in the bottom stage of a two-stage unit. The top stage was cooled, and the particle interchange was determined by a heat balance. Baffle plates at various spacing, with hole diameters $d_{or} = 12.7$ and 19.1 mm, and with 12–26% open area were tested. Representative results are shown in Fig. 13(b) and (c).

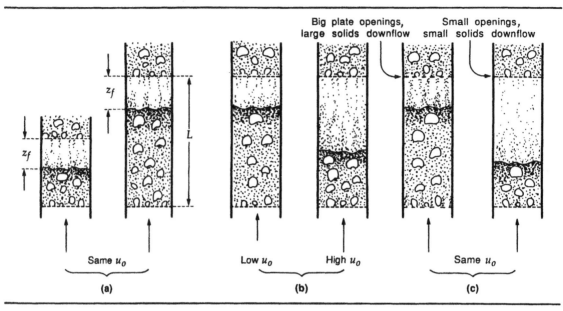

FIGURE 12
Effect of system variables on the equilibrium distribution of solids in multistage beds. (a) For different baffle spacing L, the same freeboard height gives the same solid interchange. (b) High gas velocity must be matched with an increased freeboard. (c) Bigger baffle openings must be matched by smaller freeboard.

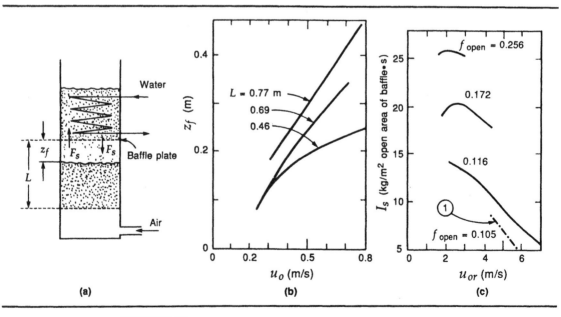

FIGURE 13
Experiments with horizontal baffle plates; from Guigon [51]: (a) Experimental setup: $d_t = 0.28$ m, $L = 0.46$–0.77 m, $d_{or} = 19.1$ mm, $d_p = 210$ μm; (b) L_m in upper bed = 0.44–0.96 m; (c) $L = 0.46$ m, line 1 from Martyushin and Golovin [53].

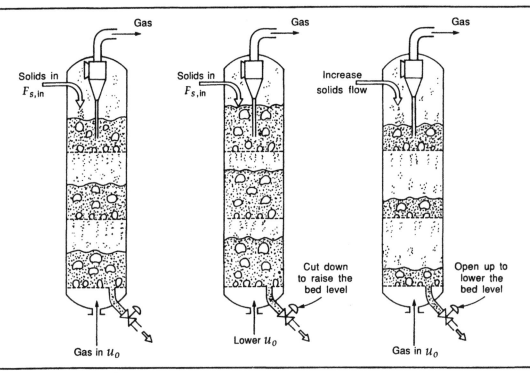

FIGURE 14
Bed levels must be adjusted to accommodate changes in gas flow rate and solid flow rate.

Throughflow of Solids

In many solid treatment processes, such as the fluidized reduction of metal ores, or catalytic processes with rapidly deactivating catalysts, it is highly desirable to have countercurrent contacting of gas and solid as well as temperature gradients along the flow path of reactants. Again, staged fluidized beds are the way to go. But here the downflow of solids will be greater than the upflow. Thus, with a throughflow flux of solids $G_{s,through}$, we can write

$$G_{s,through} = G_{s,down} - G_{s,up} \qquad (19)$$

In a solids throughflow system with reasonably designed baffle plates, Fig. 14 shows how to adjust the solid inventory to meet changes in gas and solid flows. The reasoning behind these changes is that the leakage of solids is close to constant for a given baffle plate, but the entrainment is strongly dependent on the freeboard height above the dense bed.

Leakage of Solids through Distributor Plates

Perforated plate distributors for fluidized beds are simple, cheap, and convenient to use, and Chap. 4 considers their design. However, substantial backflow of particles into the plenum below the bed is undesirable because it can lead to grid erosion or plugging and particle attrition. Briens et al. [54] carried out experiments to discover the important factors influencing grid leakage for the following system.

Bed: 0.61 m ID, 10 m high.
Solid: Geldart **A** FCC powder 65 μm.
Distributor: Sixteen 12.7-mm holes in a square pitch. Tubes of various lengths were fitted to the bottom of the grid holes to simulate various thicknesses of grid plate.

Bubble breakers were placed at the bed surface to reduce the effect of bursting bubbles.

Solids leakage was shown to be caused by pressure fluctuations due to bubble formation at the distributor and by the sloshing of the bed solids. Bubble breakers were found to reduce leakage drastically, and selecting the proper distributor thickness reduced leakage up to four orders of magnitude.

EXAMPLE 3

Design of

Baffle Plates

Determine the hole spacing needed in a baffle plate with $d_{or} = 19.1$-mm holes to give a solid interchange rate of 1.5 kg/m^2 plate·s across the plate for the solids of Fig. 13.

Data

$$d_p = 210 \ \mu\text{m}, \ u_o = 0.4 \ \text{m/s}$$

SOLUTION
From Eq. (18),

$$G_{s,up} = G_{s,down} = I_s f_{open} = 1.5 \ \text{kg/m}^2 \ \text{plate·s} \qquad \text{(i)}$$

Also

$$u_o = u_{or} f_{open} = 0.4 \ \text{m/s} \qquad \text{(ii)}$$

For $f_{open} = 0.12, 0.17, 0.26$ find I_s from Eq. (i) and u_{or} from Eq. (ii). Then, from Fig. 13(c) determine I_s, the flux of solids through the holes, for each of these u_{or} values.

f_{open}	0.12	0.17	0.26
u_{or} (m/s)	3.33	2.35	1.54
I_s from Eq. (i)	12.9	8.7	5.9
I_s from Fig. 13(c)	12	20	25

An open area fraction $f_{open} = 0.12$ gives a reasonable fit. For a square pitch with spacing I_{or} (mm),

$$\frac{(\pi/4)(19.1 \ \text{mm})^2}{I_{or}^2} = f_{open} = 0.12$$

Thus, the orifice spacing should be $I_{or} = \underline{\underline{49 \ \text{mm}}}$.

PROBLEMS

1. Figure 1 shows that the vertical dispersion of fine (Geldart **A**) solids is strongly affected by bed size. Using Eq. (12), estimate the bubble rise velocity in beds of different size, given D_{sv} in these beds.

Data

$$\varepsilon_{mf} = 0.5, \ u_{mf} = 0.01 \text{ m/s}, \ f_w = 1.2, \ u_o = 0.25 \text{ m/s}, \ \bar{d}_b = 0.05 \text{ m}$$

Bed diameter, d_t (m)	0.2	0.5	1	2	5
D_{sv} (m²/s)	0.1	0.2	0.3	0.5	0.8

2. Mori and Nakamura [23] measured the rate of horizontal movement of coarse Geldart **B** solids in broad shallow beds, as reported below. Compare these experimental results with the predictions of this chapter.

Data

Bed section: 0.9×0.3 m, $L_{mf} = 0.3$ m
PVC particles: $d_p = 595 \ \mu$m, $\varepsilon_{mf} = 0.5$, $u_{mf} = 0.295$ m/s

u_o	0.4	0.5	0.6	0.7
D_{sh} (m²/s)	0.0012	0.0021	0.0032	0.0041

From Fig. 6.11 we estimate the bubble size to be $d_b = 0.08$–0.12 m.

3. Show that Eq. (14) simplifies to Eq. (15) for situations where $u_{br} \gg u_f$. (*Hint:* Rearrange the term in brackets and then expand.)

4. In a multistage batch fluidized bed with perforated baffle plates, determine the hole spacing (square arrangement) needed to get a solid interchange rate of 5 kg/m²·s across each baffle plate.

Data

$d_p = 210 \ \mu$m, $d_{or} = 19.1$ mm, $u_o = 0.6$ m/s
Baffle plate spacing $L = 0.46$ m

Note: This is an exothermic reaction, and a heat balance calculation tells us that we need this high solid interchange rate to properly control the temperature of the different stages.

REFERENCES

1. D. Kunii and O. Levenspiel, *Fluidization Engineering*, Chap. 5, Wiley, New York, 1969.
2. O.E. Potter, in *Fluidization*, J.F. Davidson and D. Harrison, eds., p. 293, Academic Press, New York, 1971.
3. J.J. van Deemter, in *Fluidization*, 2nd ed., J.F. Davidson et al., eds., p. 331, Academic Press, New York, 1985.
4. W.G. May, *Chem. Eng. Prog.*, **55**, 49 (1959).
5. A. Avidan, Ph.D. Dissertation, City College of New York, 1980; A. Avidan and J. Yerushalmi, *AIChE J.*, **31**, 835 (1985).
6. R.J. de Vries, et al., in *Proc. 5th European Symp. Chem. React. Eng.*, B9, p. 56, 1972.
7. G.H. Reman, *Chem. and Ind.*, No. 3, 46 (15 Jan. 1955).
8. W.J. Thiel and O.E. Potter, *AIChE J.*, **24**, 561 (1978).
9. W.K. Lewis, E.R. Gilliland, and H. Girouand, *Chem. Eng. Prog. Symp. Ser.*, **58(38)**, 87 (1962).
10. J.H. de Groot, in *Proc. Int. Symp. on Fluidization*, A.A. Drinkenburg, ed., p. 348, Netherlands Univ. Press, Amsterdam, 1967.
11. T. Miyauchi, H. Kaji, and K. Saito, *J. Chem. Eng. Japan*, **1**, 72 (1968); S. Morooka, Y. Kato, and T. Miyauchi, *J. Chem. Eng. Japan*, **5**, 161 (1972).
12. T. Miyauchi et al., *Adv. Chem. Eng.*, **11**, 275 (1981).

13. M.M. Chen, B.T. Chao, and J. Liljegren, in *Fluidization IV*, D. Kunii and R. Toei, eds., p. 203, Engineering Foundation, New York, 1983.

14. O. Sitnai, *Ind. Eng. Chem. Process Des. Dev.*, **20**, 533 (1981).

15. G. Claus, F. Vergnes, and P. Le Goff, in *Fluidization Technology*, vol. 2, D.L. Keairns, ed., p. 87, Hemisphere, Washington, D.C., 1975.

16. J. Verloop, L.H. de Nie, and P.M. Heertjes, *Powder Technol.*, **2**, 32 (1968/69).

17. H.S. Carslaw and J.C. Jaeger, *The Conduction of Heat in Solids*, 2nd ed., Oxford, Univ. Press, New York, 1959.

18. J.J. van Deemter, in *Proc. Int. Symp. on Fluidization*, A.A.H. Drinkenburg, ed., p. 334, Netherlands Univ. Press, Amsterdam, 1967.

19. D. Kunii, K. Yoshida, and O. Levenspiel, *Inst. Chem. Eng. Symp. Ser.*, No. 30, 79 (1968).

20. T. Chiba and H. Kobayashi, *J. Chem. Eng. Japan*, **10**, 206 (1977).

21. C. Hamilton, C. Fryer, and O.E. Potter, *Chemeca '70*, Melbourne and Sydney, 1970.

22. W. Brötz, *Chem.-Ing.-Tech.*, **28**, 165 (1956).

23. Y. Mori and K. Nakamura, *Kagaku Kogaku*, **29**, 868 (1965).

24. T. Hirama, M. Ishida, and T. Shirai, *Kagaku Kogaku Ronbunshu*, **1**, 272 (1975).

25. V.A. Borodulya, Y.G. Epanov, and Y.S. Teplitskii, *J. Eng. Phys.*, **42**, 528 (1982).

26. P.M. Heertjes, L.H. de Nie, and J. Verloop, in *Proc. Int. Symp. on Fluidization*, A.A.H. Drinkenburg, ed., p. 476, Netherlands Univ. Press, Amsterdam, 1967.

27. Y. Shi and M. Gu, in *Proc. 3rd World Cong. Chem. Eng.*, Tokyo, 1986.

28. D. Bellgardt and J. Werther, in *Proc. 16th Int. Symp. on Heat and Mass Transfer*, Dubrovnik, 1984.

29. D. Bellgardt, M. Schoessler, and J. Werther, paper delivered at AIChE Annual Meeting, November 1986.

30. K. Kato et al., *J. Chem. Eng. Japan*, **18**, 254 (1985).

31. D. Kunii and O. Levenspiel, *J. Chem. Eng. Japan*, **2**, 122 (1969).

32. S. Chiba et al., *Powder Technol.*, **26**, 1 (1980); A.W. Nienow and T. Chiba, in *Fluidization*, 2nd ed., J.F. Davidson et al., eds., p. 357, Academic Press, New York, 1985.

33. P.N. Rowe, A.W. Nienow, and A.J. Agbim, *Trans. Inst. Chem. Eng.*, **50**, 310, 324 (1972); L.G. Gibilaro and P.N. Rowe, *Chem. Eng. Sci.*, **29**, 1403 (1974).

34. P.N. Rowe and A.W. Nienow, *Powder Technol.*, **15**, 141 (1976); A.W. Nienow, P.N. Rowe, and L.Y.L. Cheung, in *Fluidization*, J.F. Davidson and

D.L. Keairns, eds., p. 146, Cambridge Univ. Press, New York, 1978.

35. A.W. Nienow and N.S. Naimer, *Trans. Inst. Chem. Eng.*, **58**, 181 (1980); A.W. Nienow, N.S. Naimer, and T. Chiba, in *Proc. 3rd World Cong. Chem. Eng.*, Tokyo, 1986.

36. A.W. Nienow, P.N. Rowe, and T. Chiba, *AIChE Symp. Ser.*, **74**(176), 45 (1978); T. Chiba, S. Chiba, and A.W. Nienow, in *Fluidization V*, K. Østergaard and A. Sørensen, eds., p. 185, Engineering Foundation, New York, 1986.

37. T. Chiba and H. Kobayashi, in *Fluidization and Its Applications*, p. 468, Cepadues, Toulouse, 1973; *J. Chem. Eng. Japan*, **10**, 206 (1977).

38. S. Chiba et al., *Powder Technol.*, **22**, 255 (1979); H. Tanimoto et al., in *Fluidization III*, J.R. Grace and J.M. Matsen, eds., p. 381, Plenum, New York, 1986.

39. T. Chiba et al., in *Fluidization Science and Technology*, M. Kwauk and D. Kunii, eds., pp. 69, 79, Science Press, Beijing, 1982; in *Fluidization V*, K. Østergaard and A. Sørensen, eds., p. 185, Engineering Foundation, New York, 1986.

40. A.P. Baskakov, G.A. Malykh, and I.I. Shishiko, *Int. Chem. Eng.*, **15**(2), 286 (1975).

41. J.L.P. Chen and D.L. Keairns, *Can. J. Chem. Eng.*, **53**, 395 (1975); *Ind. Eng. Chem. Process Des. Dev.*, **17**, 135 (1978).

42. J.M. Beekmans and T. Smith, *Can. J. Chem. Eng.*, **55**, 993 (1977); J.M. Beekmans and A. Jeffs, *Can. J. Chem. Eng.*, **56**, 286 (1978); J.M. Beekmans, in *Fluidization IV*, D. Kunii and R. Toei, eds., p. 177, Engineering Foundation, New York, 1983.

43. A.B. Whitehead, D.C. Dent, and R. Close, in *Fluidization IV*, D. Kunii and R. Toei, eds., p. 515, Engineering Foundation, New York, 1983.

44. B.R. Pruden, D. Crosbie, and B.J.P. Whalky, in *Fluidization Technology*, vol. 2, D.L. Keairns, ed., p. 65, Hemisphere, Washington, D.C., 1975.

45. S.A. Keillor and M.A. Bergougnou, in *Fluidization Technology*, vol. 2, D.L. Keairns, ed., p. 95, Hemisphere, Washington, D.C., 1975; E.A.M. Grodzoe, W. Bulani, and M.A. Bergougnou, *AIChE Symp. Ser.*, **77**(205), 1 (1981).

46. A.W. Nienow and D.J. Cheesman, in *Fluidization III*, J.R. Grace and J.M. Matsen, eds., p. 373, Plenum, New York, 1980.

47. H.A. Masson et al., in *Fluidization IV*, D. Kunii and R. Toei, eds., p. 185, Engineering Foundation, New York, 1983; G.M. Rios, K.D. Tran, and H.A. Masson, *Chem. Eng. Comm.*, **47**, 247 (1986).

48. C.R. Bemrose et al., in *Fluidization V*, K. Østergaard and A. Sørensen, eds., p. 201, Engineering Foundation, New York, 1986.

49. J.R. Grace and D. Harrison, *Chem. Proc. Eng.*, 232 (June 1970).

50. H. Bauer, J. Muhle, and M. Schmidt, *Chem.-Ing.-Tech.*, **42**, 494 (1970).

51. P. Guigon, M.A. Bergougnou, and C.G.J. Baker, *AIChE Symp. Ser.*, **70**(**141**), 63 (1974); P. Guigon et al., in *Fluidization*, J.F. Davidson and D.L. Keairns, eds., p. 134, Cambridge Univ. Press, New York, 1978.

52. H. Kono and J.J. Huang, *AIChE Symp. Ser.*, **79**(**222**), 37 (1983); **80**(**241**), 169 (1984).

53. I.G. Martyushin and V.N. Golovin, *Tr. Mosk. Inst. Khim. Mashiostr.*, **26**, 23 (1964).

54. C.L. Briens, M.A. Bergougnou, and C.G.J. Baker, in *Fluidization*, J.F. Davidson and D.L. Keairns, eds., p. 38, Cambridge Univ. Press, New York, 1978; in *Fluidization III*, J.R. Grace and J.M. Matsen, eds., p. 413, Plenum, New York, 1980.

10 Gas Dispersion and Gas Interchange in Bubbling Beds

As the result of the movement of solids and the bubbling action, the fluidizing gas passes in a complex manner through the bed. Prediction of bed behavior for various operations, particularly catalytic reactions, requires knowing how the gas passes through the bed, its dispersion, and its interchange between bubble and emulsion phases. This chapter deals with these matters. We deal first with the dispersion phenomena, both vertical and horizontal, and conclude with measures of gas interchange between regions.

Dispersion of Gas in Beds

Vertical and horizontal dispersion of gas in bubbling beds has been examined using a variety of steady and unsteady state tracer techniques, and the results of these studies have been interpreted in several ways:

- By running steady state gas tracer experiments and adopting a diffusion-type model with vertical and horizontal dispersion coefficients D_{gv} and D_{gh} to represent the deviation of flow from the ideal of plug flow in the bed.
- By running stimulus-response experiments and fitting a diffusion-type model with dispersion coefficient D_{gv} to the response curve.
- By injecting tracer bubbles into incipiently fluidized beds, following the loss of tracer from these bubbles and thereby finding the interchange coefficient between phases.
- By running stimulus-response experiments and fitting these results with a two-region model that includes gas interchange between regions.

Steady State Tracer Studies

The steady state tracer experiment, first used by Gilliland and Mason [1] and sketched in Fig. 1, introduces a steady flow of tracer gas at a horizontal plane in a tall, narrow fluidized bed and measures the upstream diffusion of the tracer.

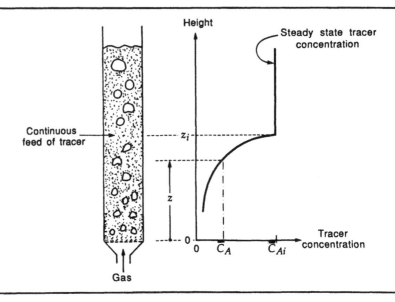

FIGURE 1
Steady state experiment for finding the vertical dispersion coefficient of gas, D_{gv}, in a fluidized bed.

Figure 2 shows typical experimental results on the backmixing of gas. Here $\bar{C}_A$ is the mean concentration of tracer at level z in the bed (upward is $+$) and $\bar{C}_{Ai}$ is the concentration at the injection plane z_i. Note the greatly increased back mixing of gases that are adsorbed by the porous bed solids and are thereby carried down the bed by these solids.

When represented by the dispersion model, the differential equation for this vertical dispersion process is

$$D_{gv} \frac{d^2 C_A}{dz^2} - u_o \frac{dC_A}{dz} = 0 \qquad (1)$$

and with reasonable boundary conditions its solution is

$$\frac{\bar{C}_A}{\bar{C}_{Ai}} = \exp\left[-\frac{u_o(z_i - z)}{D_{gv}}\right] \qquad (2)$$

Values of D_{gv} thus determined by Yoshida et al. [4] are shown in Fig. 3(a). They indicate that D_{gv} is approximately proportional to u_o. Miyauchi et al. [7] summarize the reported data on the effect of bed size on D_{gv} in Fig. 4. Figures 3(b) and 4 also present D_{gv} values determined by the stimulus-response method, considered in the next section.

A second steady state technique introduces a continuous stream of tracer at a point in the bed, usually at the axis, while measuring this tracer at various neighboring positions in the bed. Solving the diffusion equation for cylindrical coordinates with the appropriate boundary conditions gives the horizontal dispersion coefficient for gas, D_{gh}. Table 1 and Fig. 5(a) give values of D_{gh} found this way for fine particle systems. Note that at gas flows near to and below u_{mf}, the measured D_{gh} in these fine particle beds is close to the molecular diffusivity of the tracer gas.

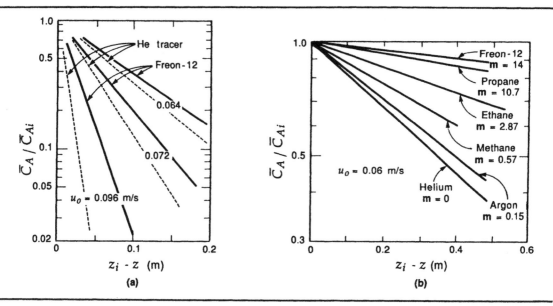

FIGURE 2
Vertical back-mixing profiles of tracer gas introduced as in Fig. 1. Note the difference between the profiles of adsorbed and nonadsorbed gases: (a) $d_t = 0.3$ m, $\bar{d}_p = 145$ μm, from Nguyen and Potter [2]; (b) $d_t = 0.1$ m, $\bar{d}_p = 70$ μm, from Bohle and van Swaaij [3].

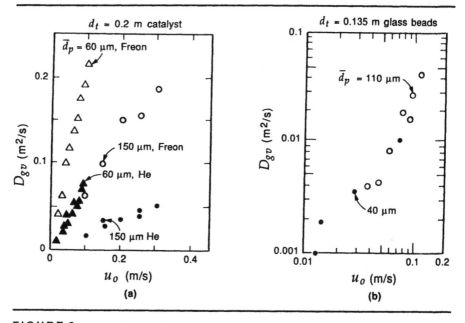

FIGURE 3
Vertical dispersion of gas in fluidized beds: (a) steady backmixing experiments, circular points for microspherical catalyst $\bar{d}_p = 150$ μm; triangular points for FCC catalyst 60 μm; adapted from Yoshida et al. [4]; (b) stimulus-response experiments by Schügerl [5], adapted from Zuiderweg [6].

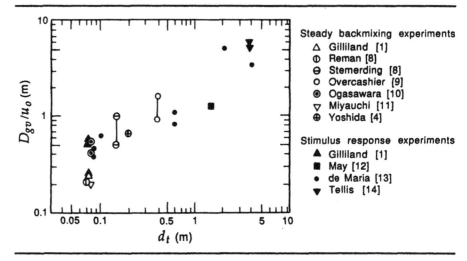

FIGURE 4
Effect of bed diameter on the vertical dispersion of gas in beds of Geldart **A** solids (FCC catalyst); data taken mainly from Miyauchi et al. [7].

TABLE 1 Horizontal Dispersion Coefficient of Gas D_{gh} (m^2/s) in Fluidized Beds

Observer	d_t (m)	Particles	d_p (μm)	Gas velocity (m/s)	Tracer	D_{gh} (m^2/s)
Baerns et al. [15] (1963)	0.075	Sand	75	$u_o/\varepsilon_f = 0.1{-}0.3$	H_2	$0.4{-}2.2 \times 10^{-3}$ (75 μm)
			175			$0.1{-}0.4 \times 10^{-3}$ (175 μm)
Hiraki et al. [16] (1968/69)	0.20	Cat.	150	$u_o = 0.1{-}0.3$	H_2	$1{-}5 \times 10^{-3}$
					CCl_2F_2	$8{-}7 \times 10^{-3}$

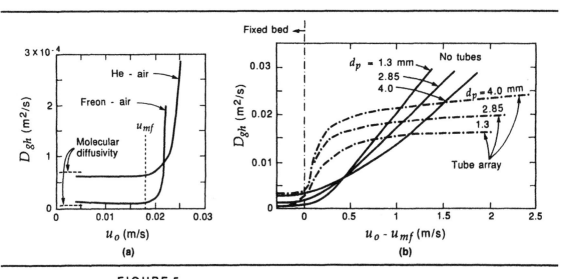

FIGURE 5
Horizontal dispersion of gas in beds: (a) near u_{mf} in beds of fine solids; $d_t = 0.2$ m, microspherical catalyst, $d_p = 150$ μm; from Hiraki et al. [16]; (b) effect of tube array in beds of coarse particles; 0.48×0.13 m, $u_{mf} = 0.73{-}1.83$ m/s; from Jovanović et al. [17].

Jovanović et al. [17] studied the horizontal spread of gas in large particle Geldart **BD** and **D** beds by introducing tracer gas continuously at one location on the distributor plate of a two-dimensional bed and noting where it appeared at the bed surface. The time-averaged tracer concentration curve was found to be bell-shaped with its maximum located directly above the tracer feed. This suggests a diffusional spread of tracer, from which D_{gh} can be found. Experiments, summarized in Fig. 5(b), show that horizontal tube arrays enhance dispersion at low gas velocity. However, at higher velocity D_{gh} levels off for beds with tube bundles but increases steadily in beds without internals.

Examining these results more carefully, they noted instantaneous tracer readings as shown in Fig. 6(a), in which the tracer missed the probe completely much of the time. This behavior cannot be explained by simple diffusion theory, so they attributed this phenomenon to the meandering of a plume of tracer, somewhat as a flickering candle flame. This is sketched in Fig. 6(b). Developing a model of this sort, they found the following expression for the horizontal dispersion coefficients:

$$D_{gh} = D_{gm} + D_{gt} \tag{3}$$

where D_{gm} represents the meandering of the plume and D_{gt} represents the turbulent or actual intermixing of gases about the axis of the plume. By measuring the time-average spread of tracer and the spread of the root mean square concentration of tracer, Jovanović et al. showed how to evaluate the individual dispersion coefficients. For 0.46-m high beds of 4-mm solids at gas velocities of 0.5–2 m/s, they found the following values:

	D_{gh} (m^2/s)	D_{gt} (m^2/s)
Without tubes	0.008–0.030	0.004
With tubes	0.020–0.025	0.015–0.018

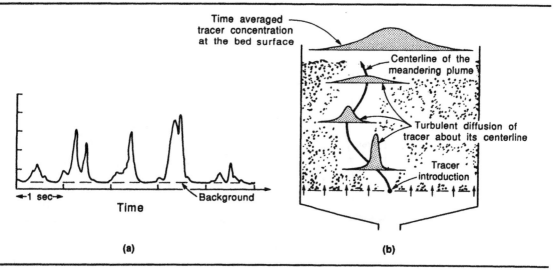

(a)   (b)

FIGURE 6
Horizontal spread of tracer in large particle beds. (a) Fluctuating tracer concentration at a point on the bed surface. This suggests a meandering plume of tracer. (b) Main aspects of the meandering plume model.

These results indicate that in the absence of bed internals the meandering plume is the main mechanism for horizontal spread of gas, whereas in beds with tube bundles actual turbulent mixing dominates.

Note that the turbulent dispersion coefficient D_{gt} is the pertinent measure of contacting or mixing, and the meandering coefficient D_{gm} does not contribute to the actual mixing of gas from the standpoint of chemical reaction. Thus, for coarse particles, tube-filled beds have a greater "useful" horizontal dispersion.

Stimulus-Response Studies

The stimulus-response technique has been used extensively to explore the flow behavior of gas in fluidized beds. A brief outline of the method can be found in Chap. 6 of [18], with a more detailed explanation in [19].

Figure 7 shows typical output curves obtained from pulse and step inputs. Thus, Figs. 7(a) and (b) show the response for the ideal of plug flow and for the ideal of mixed flow (or backmix flow). Here the mean residence time of fluid in

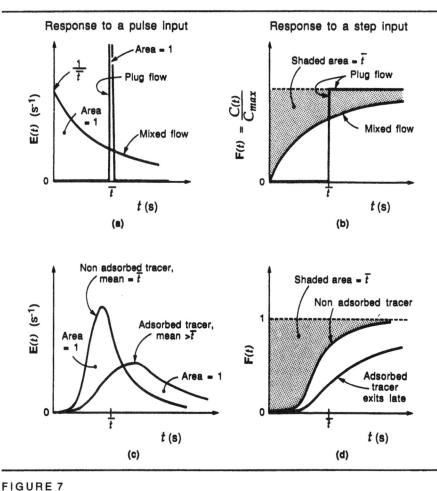

FIGURE 7
(a) and (b) Pulse and step response for plug flow and mixed flow of gas. (c) and (d) Comparison of response curves for adsorbed and nonadsorbed tracer gases.

the bed is

$$\bar{t} = \frac{\text{volume of void space in the bed}}{\text{volumetric feed rate of gas}} = \frac{\varepsilon_f L_f}{u_o} \tag{4}$$

and the concentration measure is normalized such that the area under the pulse-response curve is unity. These normalized curves are called the $\mathbf{E}(t)$ curve for the pulse response, and the $\mathbf{F}(t)$ curve for the step response. In addition, the $\mathbf{E}(t)$ and $\mathbf{F}(t)$ curves are related. Thus, at any time t after introduction of the tracer,

$$\frac{d\mathbf{F}(t)}{dt} = \mathbf{E}(t) \qquad \text{or} \qquad \mathbf{F}(t) = \int_0^t \mathbf{E}(t)\,dt \tag{5}$$

Figures 7(c) and (d) compare the response curves of ordinary (nonadsorbed) tracer with tracer that is adsorbed on the bed solids. Note that the mean of the ordinary tracer curve is at $\bar{t}$. However, the absorbing tracer is held back by the solids and leaves the bed later than expected.

If the extent of adsorption is represented by an equilibrium constant defined by

$$\mathbf{m} = \frac{\text{concentration of tracer in the solid (mol/m}^3\text{ solid)}}{\text{concentration of tracer in the gas (mol/m}^3\text{ gas)}} \tag{6}$$

then for a bed of adsorbing solids

$$\bar{t} < \bar{t}_{\text{measured}} < (1 + \mathbf{m})\bar{t} \tag{7}$$

and if equilibrium is rapidly established, then

$$\bar{t}_{\text{measured}} \rightarrow (1 + \mathbf{m})\bar{t} \tag{8}$$

Figure 8 shows that the measured response curves for nonadsorbed tracer

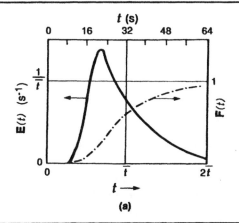
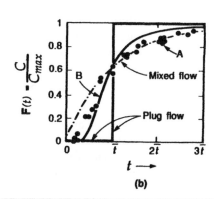

FIGURE 8

Residence time distribution curves for nonadsorbed tracer gas: (a) $\mathbf{E}(t)$ and $\mathbf{F}(t)$ curves for the regenerator of a commercial FCC unit; from Danckwerts et al. [20]; (b) points A: $\mathbf{F}(t)$ curve, $d_t = 0.076$ m, $L_f = 0.114$ m, $u_o = 0.134$–0.305 m/s, from Gilliland and Mason [1]; solid curve: $\mathbf{F}(t)$ curve, $d_t = 1.53$ m, $u_o = 0.244$ m/s, from May [12].

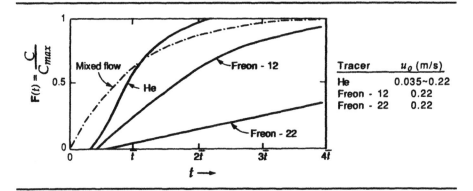

FIGURE 9
F(t) curves for adsorbed and nonadsorbed tracer gases, $d_t = 0.1$ m, microspherical catalyst, $\bar{d}_p = 150$ μm; from Yoshida and Kunii [21].

gas in both small and large fluidized beds lie somewhere between plug flow and mixed flow. Figure 9 shows that adsorption of a component of the gas stream can result in serious holdback of that material in the bed. Reviewing the reported data such as shown in Figs. 8 and 9 for a variety of fluidizing conditions, we find that gas is closer to plug flow in larger particle beds and in larger-diameter beds.

The one-dimensional diffusion-type model often reasonably represents flows that do not deviate much from plug flow, and its differential equation relating the response curve with the dispersion coefficient is

$$\frac{\partial C_A}{\partial t} = D_{gv} \frac{\partial^2 C_A}{\partial z^2} - u_o \frac{\partial C_A}{\partial z} \tag{9}$$

Solving for a pulse input gives a simple expression for D_{gv} in terms of σ^2, the variance of the $E(t)$ curve, as follows:

$$\sigma^2 = 2\bar{t}^{-2}\left(\frac{D_{gv}\varepsilon_f}{u_o L_f}\right), \qquad [s^2] \tag{10}$$

More generally, for any "one-shot" upstream tracer input into the bed, the increase in variance of the response curves between outputs 1 and 2, σ_1^2 and σ_2^2, respectively, is related to D_{gv} by

$$\sigma_2^2 - \sigma_1^2 = 2(\bar{t}_2 - \bar{t}_1)^2\left(\frac{D_{gv}\varepsilon_f}{u_o L_f}\right), \qquad [s^2] \tag{11}$$

where $\bar{t}_1$ and $\bar{t}_2$ are the means of these output curves, measured from any starting time.

The above expressions and similar expressions can be used to obtain D_{gv} from experiment. Examples of such data reported by Zuiderweg [6], using Schügerl's experimental results [5], are shown in Fig. 3(b). Miyauchi et al. [7] compared values of D_{gv}/u_o obtained by several workers, and these are shown as the solid data points in Fig. 4. The open points in this figure are from steady state experimentation mentioned earlier.

For gaseous components that can be adsorbed by the bed solids, the vertical dispersion of tracer gas should also include the material carried about

T A B L E 2 Effect of Adsorptive Tracer Gas on the Vertical
Dispersion Coefficient of Fluid Cracking Catalyst

Tracer	Adsorption constant, **m**	$D_{gv}(m^2/s)$ at u_o (m/s)			
		0.2	0.3	0.4	0.5
He	0.6	0.03	0.06	0.075	—
CO_2	4.5	0.10	0.13	0.15	0.18
CCl_2F_2	10	0.15	0.18	0.20	0.24

D_{gv} (m^2/s), taken from Miyauchi and Kaji [11];
$d_t = 0.079$ m; $\bar{d}_p = 58$ μm.

the bed by these solids. Miyauchi and Kaji [11] assumed that equilibrium existed
between gas and solid at all points in the bed and came up with the following
expression to account for adsorbed gases:

$$D_{gv} = (D_{gv})_{nonads} + mD_{sv} \tag{12}$$

Thus, in addition to the diffusive flux of gas, they included the term **m** to
account for the diffusive flux of the gaseous component while it is adsorbed by
the solid. Table 2 shows the values of **m** found.

Gas Interchange between Bubble and Emulsion

Preceding chapters showed that bubbles in fluidized beds have sharply defined
boundaries and that gas moves across these boundaries. In slow cloudless
bubbles (see Fig. 5.3) gas flows directly from the main body of the emulsion into
the bubble and then back again into the emulsion, whereas in fast clouded
bubbles (see same figure) one may view the transfer as occurring in two steps:
namely, between bubble and cloud, and between cloud and emulsion. We now
consider this gas interchange.

Definitions of Gas Interchange

First we introduce the various measures of gas interchange.

Interchange Coefficients K_{bc}, K_{ce}, and K_{be}. Consider the removal of
material A from a bubble of volume V_b. Based on unit volume of bubble, the
interchange coefficient between bubble and cloud (K_{bc}), between cloud and
emulsion (K_{ce}), and the overall coefficient between bubble and emulsion (K_{be})
can be defined by the rate equations

$$-\frac{1}{V_b}\frac{dN_{Ab}}{dt} = -u_b\frac{dC_{Ab}}{dz} = K_{be}(C_{Ab} - C_{Ae})$$

$$= K_{bc}(C_{Ab} - C_{Ac})$$

$$= K_{ce}(C_{Ac} - C_{Ae}) \tag{13}$$

where the K values have dimensions of s^{-1} and C_{Ab}, C_{Ac}, and C_{Ae} are the
mean concentrations of gaseous component A in the bubble, in the gas cloud
and wake, and in the emulsion phase, respectively. The relationship between

interchange coefficients is then

$$\frac{1}{K_{be}} = \frac{1}{K_{bc}} + \frac{1}{K_{ce}} \tag{14}$$

From a physical standpoint, the interchange coefficient K_{be} can be looked upon as a flow of gas from bubble to emulsion with an equal flow in the opposite direction:

$$K_{be} = \left[\frac{\text{volume of gas going from bubbles to emulsion or from emulsion to bubbles}}{(\text{volume of bubbles in the bed})(\text{time})} \right], \quad [s^{-1}] \tag{15}$$

The two other interchange coefficients, K_{bc} and K_{ce}, have similar meanings. These coefficients are sometimes called the *crossflow rates*.

Crossflow Ratio X_b. The gas interchange between bubble and the rest of the bed may also be expressed as a dimensionless crossflow ratio, defined as

$$X_b = \left(\begin{array}{c} \text{number of times the bubble gas is replaced} \\ \text{as the bubble passes through the bed} \end{array} \right) = \frac{K_{be}}{u_b/L_f}, \quad [-] \tag{16}$$

Note that for uniform bed conditions K is independent of bed height and X_b varies linearly with bed height.

Mass Transfer Coefficient from Bubble to the Dense Region, k_{be}. The net flux of tracer A, k_{be} (m/s), from a bubble of volume V_b and surface S_{be} is given by

$$-\frac{dN_{Ab}}{dt} = -u_b V_b \frac{dC_{Ab}}{dz} = S_{be} k_{be}(C_{Ab} - C_{Ae}) \tag{17}$$

If a_b is the bubble-emulsion interfacial area per unit volume of bed, then the volumetric mass transfer coefficient is

$$k_{be} a_b = k_{be} \frac{6\delta}{d_b}, \quad [s^{-1}] \tag{18}$$

Interrelationship between Transfer Coefficients. By comparing the above equations, we find

$$K_{be} = \frac{S_{be}}{V_b} k_{be} = \frac{6}{d_b} k_{be} = \frac{k_{be} a_b}{\delta}, \quad [s^{-1}] \tag{19}$$

For vigorously bubbling beds where $\delta = u_o/u_b$ (see Eq. (6.29)), we also have

$$K_{be} \cong k_{be} a_b \frac{u_b}{u_o} \tag{20}$$

Experimental Methods

There are two experimental approaches: first, to analyze the loss of tracer gas from single bubbles injected into a fluidized bed otherwise at minimum fluidizing conditions and, second, to analyze tracer concentrations in ordinary bubbling beds. Also, this whole section applies to fine particle beds containing bubbles with their thin clouds of recirculating gas, because it is here where bubble-emulsion transfer is slow enough to cause difficulties.

Single-Bubble Method. Consider a single clouded bubble containing tracer A at concentration C_{Ai} injected at level z_i into a fluidized bed that contains A at C_{Ae}. Normally $C_{Ae} = 0$. With the following boundary condition for the bubble gas,

$$C_{Ab} = C_{Ai} \qquad \text{at } z = z_i$$

Equation (13) integrates simply to give

$$\frac{C_{Ab} - C_{Ae}}{C_{Ai} - C_{Ae}} = \exp\left[-\frac{K_{be}(z - z_i)}{u_b} \right] \tag{21}$$

K_{be} is then found by measuring the changing concentration of A in the rising bubble.

Bubbling Bed Method. Pulse- or step-response tracer measurements put in the framework of the two-region model of Chap. 6 can be used to yield values for the interchange coefficient. A simple version of this model for fast clouded bubble beds with either upflow or downflow of emulsion gas is shown in Fig. 10. The differential equations representing the movement of tracer introduced

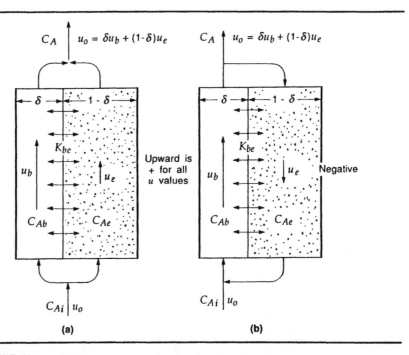

FIGURE 10
Two-region model to obtain the gas interchange coefficient between bubble and the rest of the bed: (a) upflow of emulsion; (b) downflow of emulsion.

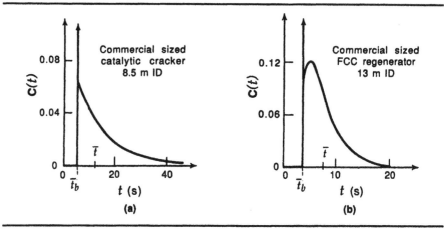

FIGURE 11
Calculated pulse-response curves for the two-region flow model of Fig. 10; adapted from Dayan and Levenspiel [22].

uniformly across the bottom of the bed are

$$\frac{\partial C_{Ab}}{\partial t} + u_b \frac{\partial C_{Ab}}{\partial z} = K_{be}(C_{Ab} - C_{Ae})$$

$$\frac{\partial C_{Ae}}{\partial t} + \frac{u_e}{\varepsilon_e} \frac{\partial C_{Ae}}{\partial z} = \frac{\delta}{1 - \delta} K_{be}(C_{Ab} - C_{Ae}) \qquad (22)$$

where

$$u_o = \delta u_b + (1 - \delta) u_e \qquad (23)$$

Knowing u_b, u_e/ε_e, and δ and matching the measured tracer response to the curves derived from this model will then yield K_{be}.

Dayan and Levenspiel [22] evaluated the $E(t)$ curves for both up and down emulsion flow by various methods—characteristics, Riemann function, Monte Carlo and Markov chain—and Fig. 11 displays some of their results. Compare these with the $E(t)$ curve of Fig. 8(a) for a very large commercial FCC regenerator.

De Groot [23] and van Deemter [24] also developed this type of model. In addition, van Deemter made a parametric comparison between the above model and a more complicated version that included a diffusion term. He concluded that the simpler model was more realistic for beds where the scale of the circulation currents was comparable to the dimensions of the bed.

Yamazaki and Miyauchi [25] and Morooka et al. [26] extended this method to account for tracer gases that are adsorbed on the bed solids.

Experimental Findings on Interchange Coefficients

Figure 12 shows values of K_{be} obtained from the single injected bubble technique. These values are only for nonadsorbed or negligibly adsorbed gases.

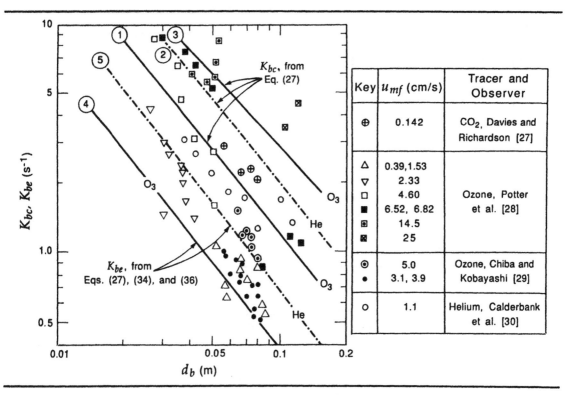

FIGURE 12
Experimental values for the gas interchange coefficient determined by the single injected bubble method. For the calculated lines see Example 1.

For tracer gas adsorbed on the bed solids, the measured K_{be} values are much higher than the values given in Fig. 12. For example, for the adsorption by microspherical catalyst particles of water vapor from bubbles of humid air, Wakabayashi and Kunii [31] found that

$$K_{be} = 15\text{--}10 \, s^{-1} \qquad \text{for } d_b = 0.04\text{--}0.11 \, m$$

These values are one order of magnitude higher than for nonadsorbed tracer gas, as may be seen from Fig. 12. Similar enhancement was found by Toei et al. [32] with Freon-12 on activated alumina and by Rietema et al. [33,34] with hydrocarbon tracer on spent cracking catalyst.

Experimental gas interchange findings obtained by the bubbling bed method are presented in Figs. 13 and 14. Figure 13 shows that the volumetric flux increases with gas flow rate and with increased adsorption characteristics of the system, and Fig. 14 shows a gradual decrease of flux in larger beds. This is probably a consequence of having faster and larger bubbles in these beds. As can be seen in these two figures, the volumetric mass transfer coefficient $k_{be}a_b$ was the measure used by experimenters to evaluate the interchange rate, not K_{be}. However, Eqs. (19) and (20) relate these quantities. (Note that the lines in Fig. 13(b) come from an analysis developed in the next chapter (see Example 11.2), so ignore them for now.)

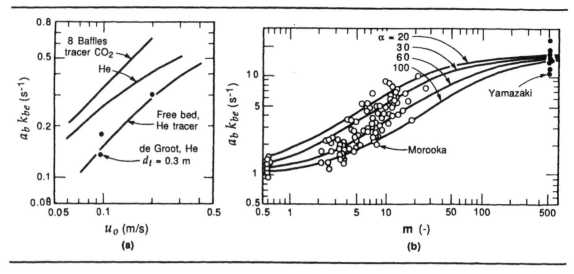

FIGURE 13
Experimental values for the bubble-emulsion transfer coefficient; mostly from Morooka et al.
[26]. (a) Influence of baffles and of adsorption, the CO_2 curve. Free bed data points are from
de Groot [23]. (b) Effect of adsorption equilibrium; from Yamazaki and Miyauchi [25]. For the
calculated lines see Example 11.2.

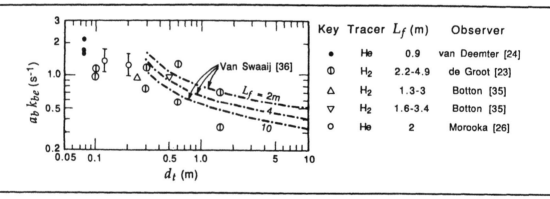

FIGURE 14
Effect of bed size on the volumetric mass transfer coefficient in beds of fine particles, mainly
FCC; adapted from Miyauchi et al. [7].

Estimation of Gas Interchange Coefficients

From the detailed behavior of gas about bubbles we can estimate these
interchange coefficients. First consider the interchange between bubble and
cloud for fast clouded bubbles, $u_{br} > 5u_{mf}/\varepsilon_{mf}$. This involves both bulk flow
and diffusion across the boundary. So, referring to Fig. 15, we have for the
removal of tracer A in a single rising bubble

$$-\frac{dN_{Ab}}{dt} = (q + k_{bc}S_{bc})(C_{Ab} - C_{Ac}) \tag{24}$$

where q is the volumetric gas flow into or out of a single bubble and k_{bc} is the
mass transfer coefficient between bubble and cloud (see Eq. (17)). From the

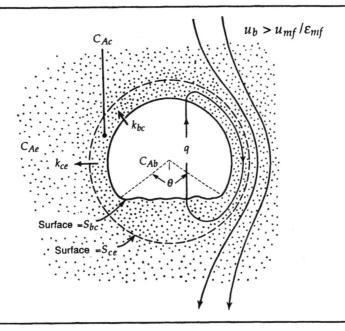

FIGURE 15
Sketch of a clouded bubble and the symbols used in deriving the individual interchange coefficients of Eqs. (27) and (34).

Davidson bubble the value of q is

$$q = \frac{3\pi}{4} u_{mf} d_b^2, \qquad [m^3/s] \tag{25}$$

Assuming a spherical cap bubble of $\theta = 100°$ and the Higbie penetration model with diffusion limited to a thin layer at the interface, Davidson and Harrison [37] derived the following expression for the mass transfer coefficient between bubble and cloud:

$$k_{bc} = 0.975 \mathscr{D}^{1/2} \left(\frac{g}{d_b}\right)^{1/4}, \qquad [m/s] \tag{26}$$

Substituting these two expressions in Eq. (24) and matching with Eq. (13) gives the interchange coefficient between bubble and cloud:

$$K_{bc} = 4.5\left(\frac{u_{mf}}{d_b}\right) + 5.85\left(\frac{\mathscr{D}^{1/2} g^{1/4}}{d_b^{5/4}}\right), \qquad [s^{-1}] \tag{27}$$

Next, estimate the coefficient between cloud and emulsion. Because there is no flow of gas between these regions, diffusion will be the only acting mechanism, so we have

$$-\frac{dN_{Ac}}{dt} = S_{ce} k_{ce}(C_{Ac} - C_{Ae}) \tag{28}$$

where k_{ce} is the mass transfer coefficient between cloud and emulsion and S_{ce} is the cloud-emulsion interfacial area of a bubble. Since the exposure time is the same for all elements of interface moving from the top to the bottom of the bubble, this process is best represented by the Higbie penetration model. Analogous to the contacting of a bubble by liquid, the characteristics of this system are equivalent to the contacting of a vertical cylinder with the same diameter and height as the spherical cloud (see Higbie [38] for details). Thus

$$k_{ce} \cong \left(\frac{4\mathscr{D}_e \varepsilon_{mf}}{\pi t}\right)^{1/2}, \quad [m/s] \tag{29}$$

For these bubbles with thin clouds (see Fig. 5.3) we can take

$$d_c \cong d_b \quad \text{and} \quad \frac{S_c}{V_b} \cong \frac{6}{d_b} \tag{30}$$

Thus the exposure time of an element of bubble surface with the emulsion is

$$t = \frac{d_c}{u_{br}} \cong \frac{d_b}{u_{br}} \tag{31}$$

Inserting Eqs. (30) and (31) into Eq. (29) and matching with Eq. (19) gives

$$K_{ce} \cong \left(\frac{4\mathscr{D}_e \varepsilon_{mf}}{\pi} \frac{u_{br}}{d_b}\right)^{1/2}\left(\frac{S_c}{V_b}\right) \cong 6.77\left(\frac{\mathscr{D}_e \varepsilon_{mf} u_{br}}{d_b^3}\right)^{1/2} \tag{32}$$

Utilizing the stream function for gas around a spherical cap bubble, Chiba and Kobayashi [29] solved the fundamental equation governing diffusion through the cloud-emulsion interface. For the special case of a spherical bubble, without assuming that $d_c \cong d_b$, but requiring that

$$\mathscr{D}_e = \varepsilon_{mf}\mathscr{D}$$

their analytical result reduces precisely to Eq. (32).

Results of experiments in fine particle systems, such as shown in Fig. 5(a), suggest that the effective diffusion coefficient of gas in the emulsion phase of a fluidized bed is better approximated by

$$\mathscr{D}_e \cong \mathscr{D} \tag{33}$$

where $\mathscr{D}$ is the diffusion coefficient of the gas alone. With this approximation and with $u_{br} = 0.711(gd_b)^{1/2}$, Eq. (32) becomes

$$K_{ce} = 6.77\left(\frac{\mathscr{D}\varepsilon_{mf}(0.711)(gd_b)^{1/2}}{d_b^3}\right)^{1/2} = 6.77\left(\frac{\mathscr{D}\varepsilon_{mf}u_{br}}{d_b^3}\right)^{1/2}, \quad [s^{-1}] \tag{34}$$

Comments about the Measured K_{be} by the Bubble Injection Technique. Consider the period just after injection of a tracer-laden bubble into a tracer-free bed. The cloud has little or no tracer; hence, measurements of K_{be} under these unsteady state conditions really reflect the bubble-cloud interchange. As a result, close to the time of bubble injection

$$K_{be, \text{ measured}} \cong K_{bc} \tag{35}$$

Thus, although Eq. (14) properly gives

$$\frac{1}{K_{be}} = \frac{1}{K_{bc}} + \frac{1}{K_{ce}} \tag{14) or (36}$$

the short-time measured K_{be} will lie between the true K_{be} and the true K_{bc}, the closeness to one or other of these values depending on the experimental conditions.

Figure 12 shows that the measured K_{be} values do lie between the predicted K_{bc} from Eq. (27) and K_{be} obtained by combining Eqs. (27) and (34) according to Eq. (36).

We emphasize that these interchange expressions only properly represent nonadsorbed gases. For kinetic processes, such as sublimation or solid-catalyzed gas-phase reactions in beds of porous particles, Eq. (14) or (36) should not be used directly. The next chapter shows how these interchange coefficients for nonadsorbed gases can be incorporated into a "bubbling bed" model that accounts for the kinetic processes involving gaseous species that are adsorbed or somehow processed by the bed solids.

EXAMPLE 1   Calculate K_{bc} and K_{be} for the operating conditions of Fig. 12.

Estimate Interchange Coefficients In Bubbling Beds

Data

Fine particles	$u_{mf} = 0.01$ m/s, $\varepsilon_{mf} = 0.5$
Tracer gas, ozone	$\mathscr{D} = 2 \times 10^{-5}$ m^2/s
Tracer gas, helium	$\mathscr{D} = 7 \times 10^{-5}$ m^2/s
Coarser particles	$u_{mf} = 0.045$ m/s, $\varepsilon_{mf} = 0.5$
Tracer gas, ozone	

SOLUTION

Calculate K_{bc}. For the fine particles with ozone tracer, Eq. (27) gives

$$\begin{aligned} K_{bc} &= \frac{4.5 \times 0.01}{d_b} + 5.85 \frac{(2 \times 10^{-5})^{1/2}(9.8)^{1/4}}{d_b^{5/4}} \\ &= \frac{0.045}{d_b} + \frac{0.046}{d_b^{5/4}} \end{aligned} \tag{i}$$

This equation is shown as line 1 in Fig. 12. Similarly,

K_{bc} for fine particles and helium: line 2 in Fig. 12
K_{bc} for coarser particles and ozone: line 3 in Fig. 12

Calculate K_{be}. Equation (14) gives K_{be} as

$$\frac{1}{K_{be}} = \frac{1}{K_{bc}} + \frac{1}{K_{ce}} \tag{ii}$$

K_{bc} has been evaluated above. To evaluate K_{ce}, apply Eq. (34), which gives

$$K_{ce} = 6.77 \left[\frac{(2 \times 10^{-5})(0.5)0.711(9.8 \times d_b)^{1/2}}{d_b^3} \right]^{1/2} = \frac{0.0319}{d_b^{5/4}} \tag{iii}$$

Replacing Eqs. (i) and (iii) in (ii) gives K_{be}. Thus, for long residence time we obtain

K_{be} *for fine particles and ozone:*   line 4 in Fig. 12
K_{be} *for fine particles and helium:*   line 5 in Fig. 12

EXAMPLE 2

Compare the Relative Importance of K_{bc} and K_{ce}

In fluidized bed processes where the rate-controlling step is the transfer of gaseous component from bubble gas to the bed solids, some workers propose models in which the bubble-cloud interchange is rate controlling (see Fig. 16(a)), others take the cloud-emulsion interchange to be rate controlling (see Fig. 16(b)), and still others say that both should be considered (see Fig. 16(c)).

From the values of K_{bc} and K_{ce} for the ammonia-air system ($\mathscr{D} = 0.69$ cm^2/s) in beds of nonabsorptive particles ($u_{mf} = 1$ cm/s, $\varepsilon_{mf} = 0.5$), determine the relative importance of these two transfer resistances and determine what error is introduced when the minor resistance is ignored.

(a) Consider beds with 5-cm bubbles.
(b) Consider high-velocity operations where beds have 15-cm bubbles.

SOLUTION

From Eq. (27) we have

$$K_{bc} = 4.5\,\frac{1.0}{d_b} + 5.85\,\frac{(0.69)^{1/2}(980)^{1/4}}{d_b^{5/4}} = \frac{4.5}{d_b} + \frac{27.19}{d_b^{5/4}} \tag{i}$$

From Eq. (34)

$$K_{ce} = 6.77\left[\frac{(0.69)(0.5)0.711(980d_b)^{1/2}}{d_b^3}\right]^{1/2} = \frac{18.76}{d_b^{5/4}} \tag{ii}$$

Comparing these coefficients for 5-cm and 15-cm bubbles, we obtain the following information:

d_b (cm)	*Calculated* K_{bc}	K_{ce}	K_{be}, *from Eq.* (14)	*Error in K_{be} when minor resistance is ignored*
5	4.54	2.51	1.62	55% high
15	1.22	0.64	0.422	51% high

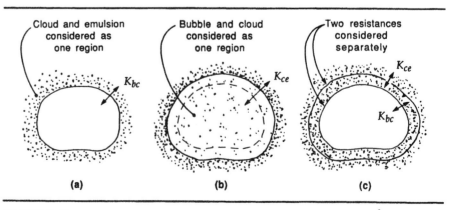

(a) Cloud and emulsion considered as one region K_{bc}

(b) Bubble and cloud considered as one region K_{ce}

(c) Two resistances considered separately K_{ce} K_{bc}

FIGURE 16
Sketches for Example 2.

A comparison of Eqs. (i) and (ii), as well as the results of the above table, shows that the cloud-emulsion interchange provides the major resistance. However, both resistances are of the same order of magnitude. Therefore, in modeling these transfer processes, one should consider both resistances if this will not unduly complicate matters.

EXAMPLE 3

Compare Interchange Rates for Adsorbed and Nonadsorbed Gases

Drinkenburg and Rietema [33] reported finding $k_{be} = 0.028$–0.05 m/s for adsorbed hydrocarbon tracer in a 0.9-m ID fluidized bed of cracking catalyst. Compare this with the interchange rate for nonadsorbed tracer gas.

Data

$$u_o = 0.30 \text{ m/s}, \qquad \bar{d}_b = 0.13 \text{ m}, \qquad m = 7, \qquad \varepsilon_{mf} = 0.5$$
$$u_{mf} = 0.0018 \text{ m/s}, \qquad \mathscr{D} = 9\text{–}22 \times 10^{-6} \text{ m}^2/\text{s}$$

SOLUTION

We compare K_{be} values for these two cases. From Eq. (19), from experiment,

$$(K_{be})_{m=7} = \frac{6}{\bar{d}_b} k_{be} = \frac{6}{0.13} (0.028\text{–}0.05) = 1.29\text{–}2.31 \text{ s}^{-1}$$

Now for nonadsorbed gases Eq. (27) gives

$$K_{bc} = 4.5\left(\frac{0.0018}{0.13}\right) + 5.85 \frac{(9\text{–}22 \times 10^{-6})^{1/2}(9.8)^{1/4}}{(0.13)^{5/4}}$$
$$= 0.46\text{–}0.68 \text{ s}^{-1}$$

Also from Eq. (34)

$$K_{ce} = 6.77\left[\frac{(9\text{–}22 \times 10^{-6})(0.5)0.711(9.8 \times 0.13)^{1/2}}{(0.13)^3}\right]^{1/2} = 0.27\text{–}0.43 \text{ s}^{-1}$$

Combining according to Eq. (14) gives

$$\frac{1}{(K_{be})_{m=0}} = \frac{1}{K_{bc}} + \frac{1}{K_{ce}} = \frac{1}{0.46\text{–}0.68} + \frac{1}{0.27\text{–}0.43}$$

or

$$(K_{be})_{m=0} = 0.17\text{–}0.26$$

Comparing gives

$$\frac{(K_{be})_{m=7}}{(K_{be})_{m=0}} = \frac{1.29}{0.17} - \frac{2.31}{0.26} = \underline{7.6\text{–}8.8}$$

Thus, the interchange with adsorbed gas is roughly eight times that of nonadsorbed gas.

PROBLEM

1. Calculate the interchange coefficients based on bubble volume K_{bc}, K_{ce}, and K_{be} for a helium tracer in a bubbling fluidized bed of nonadsorptive particles. Also determine the overall interchange coefficient based on bed volume $a_b k_{be}$.

Data

$$d_p = 105 \ \mu m, \qquad u_{mf} = 1.8 \ cm/s, \qquad \mathcal{D} = 0.7 \ cm^2/s$$
$$d_b = 9 \ cm, \qquad u_o = 40 \ cm/s, \qquad \varepsilon_{mf} = 0.5$$

REFERENCES

1. E.R. Gilliland and E.A. Mason, *Ind. Eng. Chem.*, **41**, 1191 (1949); **44**, 218 (1952).
2. H.V. Nguyen and O.E. Potter, *Adv. Chem. Ser.*, **133**, 290 (1974).
3. W. Bohle and W.P.M. van Swaaij, in *Fluidization*, J.F. Davidson and D.L. Keairns, eds., p. 167, Cambridge Univ. Press, New York, 1978.
4. K. Yoshida, D. Kunii, and O. Levenspiel, *Ind. Eng. Chem. Fund.*, **8**, 402 (1969).
5. K. Schügerl, in *Proc. Intern. Symp. on Fluidization*, A.A.H. Drinkenburg, ed., p. 782, Netherlands Univ. Press, Amsterdam, 1967.
6. F.J. Zuiderweg, in *Proc. Intern. Symp. on Fluidization*, A.A.H. Drinkenburg, ed., p. 739, Netherlands Univ. Press, Amsterdam, 1967.
7. T. Miyauchi et al., *Adv. Chem. Eng.*, **11**, 275 (1981).
8. G.H. Reman, *Chem. and Ind.*, 46 (Jan. 1955).
9. R.H. Overcashier, D.B. Todd, and R.B. Olney, *AIChE J.*, **5**, 54 (1959).
10. S. Ogasawara, T. Shirai, and K. Morikawa, *Kagaku Kogaku*, **28**, 59 (1964).
11. T. Miyauchi and H. Kaji, *J. Chem. Eng. Japan*, **1**, 72 (1968).
12. W.G. May, *Chem. Eng. Prog.*, **55**(12), 49 (1959).
13. F. De Maria and J.E. Longfield, *Chem. Eng. Prog. Symp. Ser.*, **58**(38), 16 (1962).
14. C.B. Tellis and H.M. Hulburt, in *Proc. PACHEC I*, p. 178, Kyoto, 1972.
15. M. Baerns, F. Fetting, and K. Schügerl, *Chem.-Ing.-Techn.*, **35**, 609 (1963).
16. I. Hiraki, D. Kunii, and O. Levenspiel, *Powder Technol.*, **2**, 247 (1968/69).
17. G.N. Jovanović et al., in *Fluidization III*, J.R. Grace and J.M. Matsen, eds., p. 325, Plenum, New York, 1980.
18. D. Kunii and O. Levenspiel, *Fluidization Engineering*, Chap. 6, Wiley, New York, 1969.
19. O. Levenspiel, *Chemical Reaction Engineering*, 2nd ed., Wiley, New York, 1972; O. Levenspiel and K.B. Bischoff, *Adv. Chem. Eng.*, **4**, 95 (1963).
20. P.V. Danckwerts, J.W. Jenkins, and G. Place, *Chem. Eng. Sci.*, **3**, 26 (1954).
21. K. Yoshida and D. Kunii, *J. Chem. Eng. Japan*, **1**, 11 (1968).
22. J. Dayan and O. Levenspiel, *Chem. Eng. Prog. Symp. Ser.*, **66**(101), 28 (1970).
23. J.H. de Groot, in *Proc. Intern. Symp. on Fluidization*, A.A.H. Drinkenburg, ed., p. 348, Netherlands Univ. Press, Amsterdam, 1967.
24. J.J. van Deemter, *Chem. Eng. Sci.*, **13**, 143 (1961); in *Proc. Int. Symp. on Fluidization*, A.A.H. Drinkenburg, ed., p. 334, Netherlands Univ. Press, Amsterdam, 1967.
25. M. Yamazaki and T. Miyauchi, *J. Chem. Eng. Japan*, **4**, 324 (1971).
26. S. Morooka et al., *Kagaku Kogaku Ronbunshu*, **2**, 71 (1976); *Int. Chem. Eng.*, **17**, 254 (1977).
27. L. Davies and J.F. Richardson, *Trans. Inst. Chem. Eng.*, **44**, T293 (1966).
28. G.K. Stephens, R.J. Sinclair, and O.E. Potter, *Powder Technol.*, **1**, 157 (1967); O.E. Potter, in *Fluidization*, J.F. Davidson and D. Harrison, eds., p. 293, Academic Press, New York, 1971.
29. T. Chiba and H. Kobayashi, *Chem. Eng. Sci.*, **25**, 1375 (1970).
30. P.H. Calderbank, J. Pereira, and J.M. Burgess, in *Fluidization Technology*, vol. 1, D.L. Keairns, ed., p. 115, Hemisphere, Washington, D.C., 1975.
31. T. Wakabayashi and D. Kunii, *J. Chem. Eng. Japan*, **4**, 226 (1971).
32. R. Toei et al., *Kagaku Kogaku*, **32**, 565 (1968); *J. Chem. Eng. Japan*, **5**, 273 (1972).
33. A.A.H. Drinkenburg and K. Rietema, *Chem. Eng. Sci.*, **28**, 259 (1973).
34. K. Rietema and J. Hoebink, in *Fluidization Technology*, vol. 1, D.L. Keairns, ed., p. 279, Hemisphere, Washington, D.C., 1975.
35. R.J. Botton, *Chem. Eng. Prog. Symp. Ser.*, **66**(101), 8 (1970).
36. W.P.M. van Swaaij and F.J. Zuiderweg, in *Fluidization and Its Applications*, p. 454, Cepadues, Toulouse, 1973.
37. J.F. Davidson and D. Harrison, *Fluidized Particles*, App. C, Cambridge Univ. Press, 1963.
38. R. Higbie, *Trans. Am. Inst. Chem. Eng.*, **31**, 365 (1935).

11

Particle-to-Gas Mass and Heat Transfer

The previous chapter dealt with the movement of gas in fluidized beds and the interchange of gas between bubble and emulsion phases. Here we proceed to mass and heat transfer phenomena in fluidized beds, in particular the transfer rates between bed particles and the throughflowing fluidizing gas.

Mass Transfer: Experimental

Single Spheres and Fixed Beds

The mass transfer coefficient k_d^* (m/s) for a single sphere of diameter d_{sph} moving through a fluid at relative velocity u_o is given by Froessling [1] as

$$Sh^* = \frac{k_d^* d_{sph} y}{\mathscr{D}} = 2 + 0.6\, Re_{sph}^{1/2}\, Sc^{1/3}$$

$$Re_{sph} = \frac{d_{sph} u_o \rho}{\mu}, \qquad Sc = \frac{\mu}{\rho \mathscr{D}}$$

(1)

where y is the logarithmic mean fraction of the inert or nondiffusing component, $\mathscr{D}$ is the gas-phase diffusion coefficient, and Sh, Re, and Sc are the Sherwood, Reynolds, and Schmidt numbers, respectively. For fine particles u_o becomes small, so $Sh^* \to 2$. In addition, for nonspherical but isometric particles, replace d_{sph} with the screen size d_p and use Eq. (1) as a reasonable approximation. Note that Eq. (1) only applies to single or widely dispersed particles falling through fluids.

For fixed beds of particles of size d_p and sphericity ϕ_s, based on the studies of Ranz [2], we may write

$$Sh^* = 2 + 1.8\, Re_p^{1/2}\, Sc^{1/3}, \qquad \text{for } Re_p > 80$$

(2)

Gas Fluidized Beds

The mass transfer coefficient between particles and fluidizing gas is difficult to evaluate because the large specific area of solids leads to rapid attainment of equilibrium and because the bubbling behavior of these beds makes it difficult to determine the proper driving force for mass transfer. Nonetheless, several groups of workers using a variety of experimental techniques have measured these coefficients, and these studies are shown in Fig. 1 and Table 1.

Now, if smaller and smaller solids are completely dispersed in flowing gas, the Sherwood number should approach 2, the theoretical minimum for diffusion into a stagnant medium. However, Fig. 1 clearly shows that the experimental results fall further and further below the theoretical minimum of 2 as the Reynolds number is lowered. We explain this in the next section.

Some of these studies were made in very shallow beds, just a few particles deep. Various other types of mass transfer measurements have also been made, for example from single bubbles injected into beds and from standing jets. The literature is diffuse in this area, and the reader is referred to the various sections in Davidson et al. [8] for references.

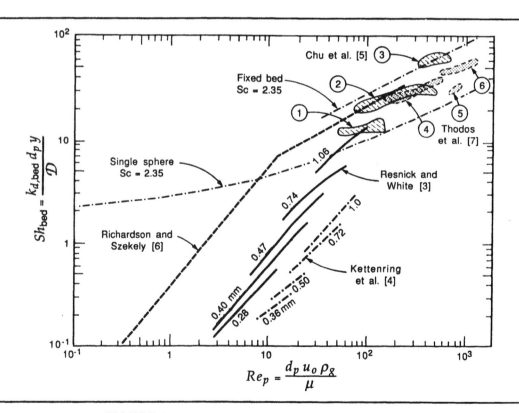

FIGURE 1
Experimental findings on mass transfer in fluidized beds.
① glass, lead, ~0.72 mm; ② rape seed, 1.96 mm; ③ lead shot, 1.98 mm; ④ alumina, 0.93–1.95 mm; ⑤ alumina, 2.95 mm (with water); ⑥ alumina, 2.95 mm (with nitrobenzene).

T A B L E 1 Mass Transfer Data Shown in Fig. 1

Investigators	Gas	Process*	Particles	$\bar{d}_p$ (μm)	d_t (cm)	L_m (cm)
Resnick & White [3] (1949)	Air, H_2, CO_2	s. naphthalene	Naphthalene	210–1700	2.2, 4.4	1.3–2.5
Kettenring et al. [4] (1950)	Air	v. water	Silica gel Alumina	360–1000	5.9	10–15
Chu et al. [5] (1953)	Air	s. naphthalene	Glass, Lead, Rape seed	710–1980	10.2	0.3–9.2
Richardson & Szekely [6] (1961)	Air, H_2	a. CCl_4 a. water	Active carbon, Silica gel	88–2580	3.0	$5d_p$
Thodos et al. [7] (1961, 69, 72)	Air	s. dichlorobenzene v. water v. nitrobenzene v. n-decane	Alumina, Celite	1800–3100	3.8 9.5 11.3	0.6–7.0

*s. = sublimation, v. = vaporization, a. = adsorption.

Interpretation of Mass Transfer Coefficients

In fluidized beds one can define two distinctly different mass transfer coefficients:

$k_{d, bed}$ the overall or whole bed coefficient
$k_{d,p}$ the single particle or local coefficient

Knowing the differences between these coefficients will make sense of the findings in Fig. 1.

Meaning of $k_{d,p}$. Introduce a single particle containing removable material A into a fluidized bed that is free of A—for example, a particle of naphthalene into a naphthalene-free bed. This particle is surrounded by gas and other particles that are free of A, and $k_{d,p}$ represents the mass transfer coefficient of this particle in its environment. More generally, if the concentration of A encountered by this particle as it wanders about the bed is $C_{A,bed}$, then

$$-\frac{1}{S_{particle}} \frac{dN_{A,particle}}{dt} = k_{d,p}(C_{A,p} - C_{A,bed}) \tag{3}$$

Meaning of $k_{d,bed}$. Consider a bed of particles all containing material A, which is being removed by A-free fluidizing gas—for example, a bed of naphthalene particles fluidized by air.

In *fine particle beds* (bubbles with thin clouds) most of the gas passes through the bed as bubbles, and flow through the emulsion is very minor. In this situation material A must go from particle to emulsion gas to cloud gas to bubble gas before it can get out of the bed. Thus, if one considers the bed as a whole, the mass transfer coefficient for A must account for the overall resistance of all these transfer steps.

In fine particle systems, investigators found it impractical to follow the

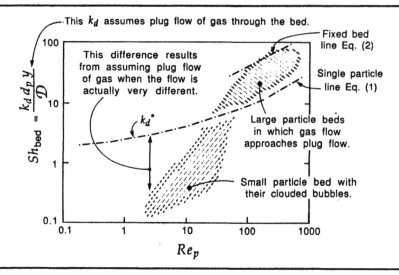

FIGURE 2
Flow regimes in fluidized bed mass transfer.

history of single particles, so they reported whole bed coefficients in Fig. 1. Typically, they noted the amount of A picked up by the gas stream passing through the bed, and they assumed plug flow of gas through the bed. This meant that they assumed much better contacting than what was really taking place. As a result, their reported coefficients for fine particle systems are *lower* than the true coefficients for single particles. Thus

$$k_{\mathrm{d,bed}} \text{ (assuming plug flow)} < k_{\mathrm{d,p}} \tag{4}$$

By introducing a flow model that accounts for the bubbling bed behavior, we can properly relate these two coefficients. We do this in the next section.

In *large particle beds* (unclouded bubbles) the fluidizing gas passes close to plug flow through the bed solids, all without bubble-cloud or cloud-emulsion resistance. Thus, by measuring the inlet and outlet concentrations of A in the gas and assuming plug flow of gas, the investigators chose a flow model that does reflect the actual flow in the bed. As a result

$$k_{\mathrm{d,bed}} \text{ (assuming plug flow)} \cong k_{\mathrm{d,p}} \tag{5}$$

As a reasonable approximation, if we estimate $k_{\mathrm{d,p}}$ by k_{d}^* of Eq. (1), then Fig. 2 sketches the various regimes and summarizes the above discussion. We now account for the difference between $k_{\mathrm{d,p}}$ and $k_{\mathrm{d,bed}}$ for fine particle systems.

Mass Transfer Rate from the Bubbling Bed Model

We now relate $k_{\mathrm{d,bed}}$ with $k_{\mathrm{d,p}}$ in vigorously bubbling fine particle clouded bubble systems. Consider the vaporization or sublimation of A from all particles in the bed. Since the flow of gas in the emulsion is small, and maybe even downward, we ignore its contribution to the total flow of gas. Thus, we assume that fresh gas enters the bed only as bubbles, and that at steady state the measure of sublimation of A is given by the increase in C_A with height in the bubble phase. Then, with the nomenclature of Fig. 3, for a slice of bed of height dz, we may write

$$\frac{1}{S_{\text{particles}}} \frac{dN_A}{dz} = k_{\text{d,bed}}(C_{A,p} - C_{A,b}) \tag{6}$$

or, in terms of the increasing concentration of A in the rising bubbles,

$$\begin{aligned}
\frac{dC_{A,b}}{dt} &= u_b \frac{dC_{A,b}}{dz} \\
&= k_{\text{d,bed}} \left(\frac{S_{\text{particles}}}{V_{\text{bubble}}}\right)_{\text{in slice}} (C_{A,p} - C_{A,b}) \\
&= \frac{k_{\text{d,bed}}(1 - \varepsilon_f)a'}{\delta} (C_{A,p} - C_{A,b}) \tag{7}
\end{aligned}$$

where a' is the specific surface area, given by Eq. (3.4a).

Assume that equilibrium is rapidly established between C_A at the particle surface and its surroundings. Then, since particles rapidly enter and leave the rising clouds and since C_{Ap} only changes slowly with time, we conclude that $C_{Ap} \cong C_{Ae} \cong C_{Ac}$. Thus, in terms of an interchange coefficient for mass transfer K_d (s^{-1}), we may write

$$\frac{1}{(\text{bubble phase volume})} \frac{dN_A}{dt} = u_b \frac{dC_{Ab}}{dz} = K_d(C_{A,p} - C_{A,b}) \tag{8}$$

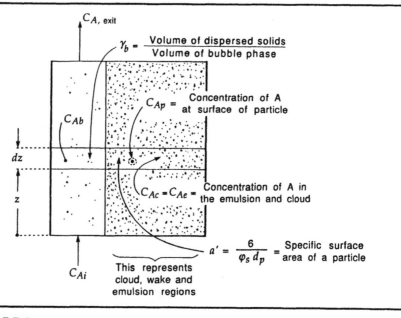

FIGURE 3
Nomenclature used in developing a model to explain the experimental findings in mass transfer in fluidized beds.

Comparing Eqs. (7) and (8) gives

$$K_d = \frac{k_{d,bed}(1 - \varepsilon_f)a'}{\delta} , \quad [s^{-1}]$$

So, in terms of K_d we have, for spherical and nonspherical particles,

$$Sh_{bed} = \frac{k_{d,bed}d_p y}{\mathscr{D}} = \frac{y\phi_s d_p^2 \delta}{6\mathscr{D}(1 - \varepsilon_f)} K_d , \qquad d_p \cong d_{sph} \qquad (9)$$

Now K_d represents the addition of A to the bubble gas from two sources. thus,

$$K_d = \begin{pmatrix} \text{added from particles} \\ \text{dispersed in the bubbles} \end{pmatrix} + \begin{pmatrix} \text{transferred across the} \\ \text{bubble-cloud boundary} \end{pmatrix}$$

$$= \gamma_b a' k_d^* + K_{bc} = \gamma_b \frac{6}{\phi_s d_p} k_d^* + K_{bc}$$

$$= \gamma_b \frac{6(Sh^*)\mathscr{D}}{\phi_s d_p^2 y} + K_{bc} \qquad (10)$$

where k_d^* (m/s) represents the mass transfer coefficient for single particles that

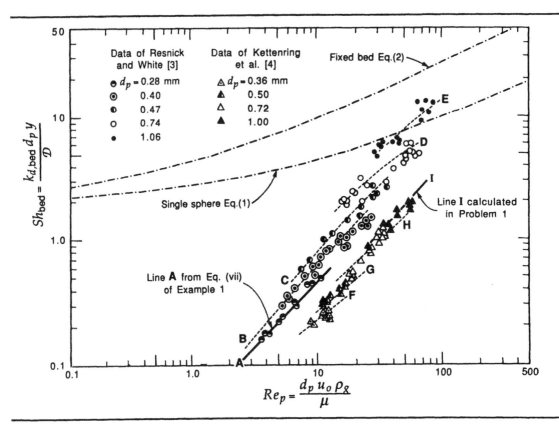

FIGURE 4
Details of the central portion of Fig. 1. Lines A to I come from the K-L bubbling bed model

are widely dispersed in bubble gas. Combining Eqs. (9) and (10) then gives

$$\text{Sh}_{\text{bed}} = \frac{\delta}{1 - \varepsilon_f}\left[\gamma_b(\text{Sh}^*) + \frac{\phi_s d_p^2 y}{6\mathscr{D}}\, K_{\text{bc}}\right] \tag{11}$$

where K_{bc} is given by Eq. (10.27). For a given bed of solids and constant bubble size, Eq. (11) reduces to an expression of the form

$$\text{Sh}_{\text{bed}} = A\, \text{Re}_p - B \tag{12}$$

Lines A and I in Fig. 4 were calculated and drawn according to this procedure, with d_b chosen to fit the data. Example 1 calculates line A in Fig. 4; Prob. 1 asks the reader to verify line I. In a similar fashion lines B to H, shown dashed in Fig. 4, can be constructed.

What is needed today is good mass transfer data in larger beds of fine particles in which bubble size is measured. This will represent a better test of the above two equations and of this analysis.

Effect of Adsorption on the Interchange Coefficient K_d

Figure 10.13 shows that changes in the adsorption equilibrium constant **m** greatly affect the interchange coefficient K_d between the bubble and dense phases. We now show that this can be explained in terms of the extra mass transfer between bubble gas and the particles dispersed in the bubbles, even though gas bubbles contain a very small volume fraction of particles, $\gamma_b = 0.01 - 0.001$.

Consider a bubble containing tracer of concentration C_{Ab} passing up a bed of adsorbing particles, as shown in Fig. 5. In general, we may write

$$C_{As} = \text{m}C_{Ap} \tag{13}$$

where C_{As} is the concentration of tracer A within the particle in equilibrium with the concentration C_{Ap} of tracer gas at the surface of the particle. In the emulsion and cloud regions of the bed we assume equilibrium, as before, in

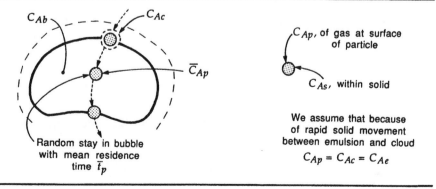

FIGURE 5
Notation used in deriving the expression that accounts for the takeup of tracer A from bubble gas by solids passing through the bubbles.

which case

$$C_{Ap} = C_{Ac} = C_{Ae} \tag{14}$$

We now focus on a particle of sphericity ϕ_s passing through the bubble. The transfer of tracer A from bubble gas to particle is then given by

$$\begin{pmatrix} \text{accumulation of} \\ \text{A in the particle} \end{pmatrix} = \begin{pmatrix} \text{transfer of A from} \\ \text{bubble gas} \end{pmatrix}$$

or, with $d_p \cong d_{sph}$,

$$\left(\frac{\pi}{6} d_p^3\right) dC_{As} = \left(\frac{\pi d_p^2}{\phi_s}\right) k_d^*(C_{Ab} - C_{Ap}) dt \tag{15}$$

At time $t = 0$, when the particle just enters the bubble

$$C_{Ap} = C_{Ac} \tag{16}$$

Integrating Eq. (15) with this condition gives

$$\frac{C_{Ab} - C_{Ap}}{C_{Ab} - C_{Ac}} = \exp\left(-\frac{6k_d^*}{m\phi_s d_p} t\right) \tag{17}$$

For a random stay with mean residence time $\bar{t}_p$, the residence time distribution of particles in the bubble is

$$\mathbf{E}(t) = \frac{1}{\bar{t}_p} \exp\left(-\frac{t}{\bar{t}_p}\right) \tag{18}$$

The mean concentration of tracer A at the surface of a particle just leaving the bubble and reentering the emulsion is then $\bar{C}_{Ap}$, given by

$$C_{Ab} - \bar{C}_{Ap} = \int_{t=0}^{\infty} (C_{Ab} - C_{Ap})\mathbf{E}(t) \, dt \tag{19}$$

Combining the above three equations and integrating then gives

$$C_{Ab} - \bar{C}_{Ap} = \frac{C_{Ab} - C_{Ac}}{1 + 6k_d^*\bar{t}_p/m\phi_s d_p} \tag{20}$$

Tracer A goes from bubble gas to the bed solids via two paths: capture by solids passing through the bubble and by transfer across the bubble-cloud interface, or

$$K_d(C_{Ab} - C_{Ac}) = \frac{\gamma_b}{\bar{t}_p} (m\bar{C}_{Ap} - mC_{Ac}) + K_{bc}(C_{Ab} - C_{Ac}) \tag{21}$$

Substituting Eqs. (1) and (20) into (21) and simplifying gives

$$K_d = \gamma_b \frac{6(Sh^*)\mathscr{D}}{\phi_s d_p^2 y} \eta_d + K_{bc} \tag{22}$$

where the fitted adsorption efficiency factor, because $\bar{t}_p$ is unknown, is

$$\eta_d = \frac{1}{1 + \alpha/m}, \qquad \text{where} \qquad \alpha = \frac{6k_d^*\bar{t}_p}{\phi_s d_p} \tag{23}$$

Thus, inserting Eq. (22) in Eq. (9) gives

$$\mathrm{Sh_{bed}} = \frac{\delta}{1 - \varepsilon_f} \left[\gamma_b (\mathrm{Sh^*}) \eta_d + \frac{\phi_s d_p^2 y}{6\mathscr{D}} K_{bc} \right] \qquad (24)$$

Important Note: For nonporous nonadsorbing particles, $m = 0$ and $\eta_d = 0$; hence, no tracer gas is picked up by particles passing through the bubbles, and $K_d = K_{bc}$. For porous but nonadsorbing particles of voidage ε_p, we have $m = \varepsilon_p$. For highly adsorbing or sublimable particles, m is on the order of thousands, in which case $\eta_d \to 1$. In this extreme, Eq. (22) reduces to Eq. (10), and Eq. (24) reduces to Eq. (11).

This expression, Eq. (24) with α as parameter, can explain the findings of Fig. 10.13. For detailed calculations, see Example 2. Also, a look at Eq. (21) shows that gas interchange between bubble and dense phases can be greatly enhanced by adsorption (or desorption) of tracer gas in particles that pass through the rising bubbles. For example, Wakabayashi and Kunii [9] found these values:

$\bar{d}_b$ (m)	0.04	0.06	0.08	0.11
$K_{d,\text{measured}}$ (s^{-1})	11–16	9–18	8–14	7–11
K_d { calculated, with $\gamma_b \eta_d = 0$	5	3	2	1.6
{ calculated, with $\gamma_b \eta_d = 4 \times 10^{-4}$	13	12	12	11

These values indicate the importance for mass transfer of particles dispersed in bubbles, no matter how small is $\gamma_b \eta_d$.

Rietema and Hoebink [10] found that their measured gas interchange coefficients increased with increase in bubble size and commented that this was just the reverse of the predictions of Eq. (10.27) for K_{bc}. Since the first term on the right-hand side of Eq. (21) dominates for large bubbles, this may explain their findings in terms of larger γ_b for large bubbles, caused by more vigorous splitting and coalescing of bubbles.

In conclusion, we make three points: first, particles dispersed in bubbles should be taken into account when kinetic processes, such as mass transfer, are carried out in fluidized beds; second, when dealing with a gaseous component that is adsorbed or somehow captured by the bed solids (or else desorbed), then K_{be} should be used carefully to represent the movement of these adsorbed gaseous components. Finally, the mass transfer coefficient measured for the bed as a whole, $k_{d,bed}$, is model-dependent. Where the model closely matches the flow conditions in the bed (for large particle cloudless bubble beds), the bed coefficient should match the single particle coefficient $k_{d,p}$. Where it does not, these coefficients will differ. This is the case for fine particle clouded bubble beds.

EXAMPLE 1

Fitting Reported Mass Transfer Data with the Bubbling Bed Model

Given the estimated effective bubble size in the bed, $d_b = 0.37$ cm, determine the Sherwood versus Reynolds number relationship for $d_p = 0.028$-cm naphthalene particles for the experimental conditions of Resnick and White [3]. Show that you obtain line A in Fig. 4 and that this line reasonably represents the reported data points.

Data

$\rho_s = 1.06$ g/cm^3, $\varepsilon_{mf} = 0.5$, $\phi_s = 0.4$, $\gamma_b = 0.005$ (estimated)
$\rho_g = 1.18 \times 10^{-3}$ g/cm^3, $\mu = 1.8 \times 10^{-4}$ g/cm·s
$\mathscr{D}_e \cong \mathscr{D} = 0.065$ cm^2/s, Sc = 2.35

Because of sublimation take $\eta_d = 1$. Because of the very low vapor pressure of naphthalene, about 0.1 mm Hg, take $y = 1$. For these 0.028-cm particles, calculations give $u_{mf} = 1.21$ cm/s and $u_t = 69$ cm/s.

SOLUTION

These experimental results represent the $\mathbf{m} \to \infty$ extreme (see after Eq. (24)); hence, the analysis leading to Eqs. (11) and (12) applies. Equation (11) is the required relationship, so let us evaluate terms. First focus on the Sh* term.

$$Re_{p,t} = \frac{d_p u_t \rho_g}{\mu} = \frac{(0.028)(69)(0.00118)}{0.00018} = 12.7$$

Then Eq. (1) gives

$$Sh^* = 2 + 0.6(12.7)^{1/2}(2.35)^{1/3} = 4.8 \tag{i}$$

For the last term in Eq. (11), we find, from Eq. (10.27),

$$K_{bc} = 4.5\left(\frac{1.21}{0.37}\right) + 5.85\left(\frac{(0.065)^{1/2}(980)^{1/4}}{0.37^{5/4}}\right) = 43.6 \text{ s}^{-1} \tag{ii}$$

From Eqs. (6.27) and (6.20) we have

$$\delta = \frac{u_o - u_{mf}}{u_b} \quad \text{and} \quad 1 - \varepsilon_f = (1 - \varepsilon_{mf})(1 - \delta)$$

Combining, and noting for these small beds that $u_b = u_o - u_{mf} + u_{br}$, gives

$$\frac{\delta}{1 - \varepsilon_f} = \frac{u_o - u_{mf}}{u_{br}(1 - \varepsilon_{mf})} = \frac{u_o - 1.21}{0.711(980 \times 0.37)^{1/2}(1 - 0.5)} = \frac{u_o - 1.21}{6.77} \tag{iii}$$

Also

$$\frac{y\phi_s d_p^2}{6\mathscr{D}} = \frac{1(0.4)(0.028)^2}{6(0.065)} = 8.04 \times 10^{-4} \tag{iv}$$

Putting Eqs. (i)–(iv) into Eq. (11) gives

$$Sh_{bed} = \frac{u_o - 121}{6.77}[(0.005)(4.8) + (0.000804)(43.6)]$$

$$= 0.0088u_o - 0.011 \tag{v}$$

Also

$$Re_p = \frac{d_p u_o \rho_g}{\mu} = \frac{(0.028)u_o(0.00118)}{0.00018} = 0.184u_o \tag{vi}$$

Combining Eqs. (v) and (vi) gives the desired result

$$\underline{\underline{Sh_{bed} = 0.048 \, Re_p - 0.011}} \tag{vii}$$

Comment. Equation (vii) is line A in Fig. 4. It has a slope close to unity, and it correctly fits the mass transfer data for this size of solid. The small bubble size was chosen because the experimental bed was very shallow (1.3–2.5 cm) and because mass transfer is a rapid process; hence, the major portion of mass transfer takes place in the region just above the distributor, where bubbles have not had time to grow large.

EXAMPLE 2

The Effect of m on Bubble-Emulsion Interchange

Show how changes in the adsorption equilibrium constant **m** affect the whole bed bubble-emulsion interchange coefficient

$$a_b k_{be} = \left[\frac{m^3 \text{ transferred from bubble to emulsion}}{m^3 \text{ of bed} \cdot s} \right] \cong \frac{u_o}{u_b} K_d \qquad \text{(i)}$$

Compare your calculated results with the experimental findings reported in Fig. 10.13(b).

Data

$$u_{mf} = 0.12 \text{ cm/s}, \quad u_o = 40 \text{ cm/s}, \quad u_b = 120 \text{ cm/s}, \quad \mathscr{D} = 0.7 \text{ cm}^2/s$$

From the extremes in Fig. 10.13(b), experiments show the following

$$a_b k_{be} = \begin{cases} 1 \text{ s}^{-1} & \text{for nonadsorbing particles } (m = 0) \\ 18 \text{ s}^{-1} & \text{for highly adsorbing particles } (m = \infty) \end{cases} \qquad \text{(ii)}$$

SOLUTION

In general, for any value of the adsorption equilibrium constant, putting Eq. (22) in Eq. (i) gives

$$a_b k_{be} = \frac{u_o}{u_b} \gamma_b \frac{6(Sh^*)\mathscr{D}}{\phi_s d_p^2 y} \eta_d + \frac{u_o}{u_b} K_{bc} = M\eta_d + \frac{u_o}{u_b} K_{bc} \qquad \text{(iii)}$$

First we find the bubble size that fits the nonadsorbing extreme, **m** = 0. This means that solids passing through bubbles do not adsorb any gas. Hence, from Eq. (23), $\eta_d = 0$, and Eq. (iii) reduces to

$$a_b k_{be} = \frac{u_o}{u_b} K_{bc} \qquad \text{(iv)}$$

Using Eq. (10.27) then gives

$$a_b k_{be} = \frac{u_o}{u_b} \left[4.5 \frac{u_{mf}}{d_b} + 5.85 \frac{\mathscr{D}^{1/2} g^{1/4}}{d_b^{5/4}} \right]$$

and with Eq. (ii),

$$1 = \frac{40}{120} \left[4.5\left(\frac{0.12}{d_b}\right) + 5.85 \frac{0.7^{1/2} 980^{1/4}}{d_b^{5/4}} \right]$$

from which $d_b = 6$ cm; and with Eq. (iv) $K_{bc} = 3 \text{ s}^{-1}$.

Next, for strongly adsorbing particles, **m** $\to \infty$, and from Eq. (23), $\eta_d = 1$; hence, Eq. (iii) gives

$$a_b k_{be} = M + \frac{u_o}{u_b} K_{bc}$$

Substituting obtained values gives

$$18 = M + \frac{40}{120} (3) \quad \text{or} \quad M = 17$$

Finally, consider some intermediate condition, say $\alpha = 100$ and **m** = 10. Then the adsorption efficiency is

$$\eta_d = \frac{1}{1 + \alpha/\mathbf{m}} = \frac{1}{1 + 100/10} = \frac{1}{11}$$

Hence, substituting in Eq. (iii) gives the adsorption rate constant

$$a_b k_{be} = 17\left(\frac{1}{11}\right) + 1 = 2.55 \text{ s}^{-1}$$

By choosing different values for α and m, we draw the lines shown in Fig. 10.13(b). These lines are consistent with the reported data.

Heat Transfer: Experimental

Single Spheres and Fixed Beds

The heat transfer coefficient h^* at the surface of a sphere of diameter d_{sph} passing at velocity u_0 through a gas is correlated by Ranz [2] as follows:

$$Nu^* = \frac{h^* d_{sph}}{k_g} = 2 + 0.6 \, Re_{sph}^{1/2} \, Pr^{1/3} \tag{25}$$

with

$$Re_{sph} = \frac{d_{sph} u_0 \rho}{\mu}, \qquad Pr = \frac{C_p \mu}{k_g}$$

For nonspherical particles we approximate the heat transfer with Eq. (25), with d_p replacing d_{sph}.

For a gas passing at a superficial velocity u_0 through a fixed bed of large isometric particles of sphericity ϕ_s, the data of Ranz [2] suggest that

$$Nu^* = 2 + 1.8 \, Re_p^{1/2} \, Pr^{1/3} \tag{26}$$

Gas Fluidized Beds

Many investigators have studied heat transfer between fluidizing gas and the bed solids. Whole bed coefficients were evaluated by using a variety of steady state and unsteady state techniques.

Typical of the steady state method of finding h, hot gas enters a bed that is kept cool by internal heat exchange, by heat removal at the walls, or by replacement of hot solid by fresh cool solids. By measuring bed temperatures close to the gas inlet or by measuring temperature gradients in beds containing heaters and coolers, one can find values of h.

Typical of the unsteady state method, the temperature of the hot entering gas is changed in a known manner, the temperature of the exit gas is followed with time, and a heat balance around the bed gives the temperature of the solids at any time, from which h is found. At this point most researchers using the unsteady state approach assumed complete mixing of gas throughout the bed. Given the heterogeneous structure of fluidized beds, this assumption does not seem realistic; and these studies are not cited here. The few studies that did not make this assumption, but assumed plug flow of gas up the bed instead, are referred to.

Figure 6, prepared from the charts given by Gelperin and Einstein [11] and by Kunii and Levenspiel [12], summarizes the results of 22 investigations on heat transfer in gas fluidized beds. Comparing this figure with Fig. 1, we see a clear similarity between mass and heat transfer behavior. Thus,

For $Re_p > 100$, Nu falls between the values for single particles and for fixed beds.

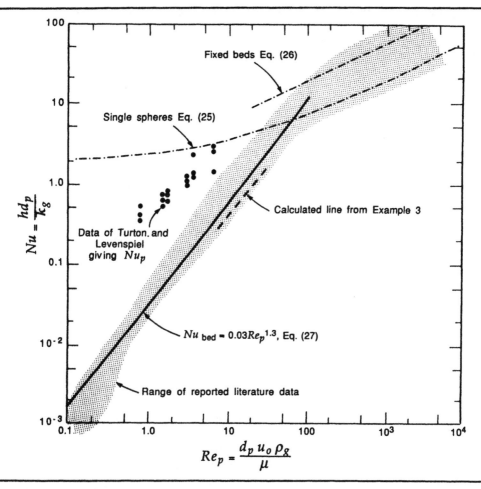

FIGURE 6
Shaded area represents the experimental findings of 22 studies, prepared from information collected by Kunii and Levenspiel [12] and by Gelperin and Einstein [11].

For $Re_p < 10$, Nu decreases drastically with Re_p to values far smaller than 2, the theoretical minimum given by Eq. (25).

In the region of rapidly falling Nusselt number, the empirical expression of Kothari [13] fits all the reported data; thus

$$\mathrm{Nu_{bed}} = \frac{hd_p}{k_g} = 0.03\,\mathrm{Re_p^{1.3}}, \qquad Re_p = 0.1\text{--}100 \tag{27}$$

In the next section we explain this behavior.

One set of experimental results, the recent measurements of Turton and Levenspiel [14], does not fit in with all the rest of the findings of Fig. 6. The reason is that their reported h values were not whole bed coefficients, but were coefficients for individual particles. We consider these findings later in this chapter.

Interpretation of Heat Transfer Coefficients

As with mass transfer, with heat transfer we also define two different heat transfer coefficients:

h_{bed} = overall or whole bed coefficient

h_p = single particle or local coefficient

Meaning of h_p. Introduce a single hot particle at temperature T_p into a cold fluidized bed at T_{bed}. Then the rate of cooling of this hot particle represented by

$$\frac{1}{S_{particle}}\frac{dQ}{dt} = -\frac{\rho_s C_{p,s} V_{particle}}{S_{particle}}\frac{dT_p}{dt} = h_p(T_p - T_{bed}) \qquad (28$$

gives the single-particle heat transfer coefficient h_p. We may expect this coefficient to fall somewhere between the values of Eqs. (25) and (26) for single falling particles and particles in fixed beds.

Meaning of h_{bed}. Consider a cold gas entering and fluidizing a bed of hot particles. The heat transfer coefficient measured in this situation represents the whole bed coefficient, h_{bed}. However, this coefficient is model-dependent in that to evaluate it one must decide on the flow pattern of gas and of particles in the bed. All researchers chose the solids to be well mixed in the bed, and this seems quite reasonable. However, for the gas some investigators chose mixed flow, whereas others chose plug flow. Naturally, the calculated coefficients will differ, often greatly, depending on the flow pattern chosen. We feel that the assumption of mixed flow of gas is not a useful choice; hence we do not include here the reported coefficients based on this assumption.

For *coarse particle beds*, bubbles are cloudless and gas passes straight through the bed, bubbles and all. In this extreme the plug flow assumption seems reasonable, and we may expect

$$h_{bed} \text{ (assuming plug flow)} \to h_p \qquad (29$$

Figure 6 shows that for high Re_p the fluidized bed heat transfer data fall between the single particle and the fixed bed lines.

For *fine particle beds* with their clouded bubbles, the plug flow assumption gives much better contacting of gas with solid than what actually occurs in these beds. Hence, the calculated whole bed coefficient will be *lower* than the single particle coefficient, or

$$h_{bed} \text{ (assuming plug flow)} < h_p \qquad (30)$$

At very low Re_p one sees that h_{bed} is as much as three orders of magnitude below the single-particle coefficient, meaning that gas-solid contacting is that much poorer than for plug flow of gas through all the bed solids.

Consequences. If we use the wrong coefficient, we will get meaningless results. For example, suppose we have a process where very small (say 100-μm) copper particles are fed into a hot fluidized bed. If we use h_{bed} to estimate how long it would take to heat up these particles, we will find times on the order of a minute when it actually takes about a second; see Example 4.

Attempt at Measuring h_p Directly in Fine Particle Beds

By extrapolating back to smaller particles, we may expect h_p to lie somewhere in the neighborhood of the values given by Eqs. (25) and (26), or

$$\boxed{\mathrm{Nu}_p \cong 2 + (0.6\text{--}1.8)\,\mathrm{Re}_p^{1/2}\,\mathrm{Pr}^{1/3}} \tag{31}$$

Turton and Levenspiel [14] recently attempted to measure h_p directly for very fine ($\sim$100-μm) particles. They did this by injecting small cold particles into a hot fluidized bed ($\sim$95°C). The characteristic of these particles is that when cold they are magnetic, but they lose their magnetic properties when heated above 70°C; hence by noting when the bed becomes nonmagnetic, one can tell when the temperature of the particles rises past 70°C.

Since the injection of a single 100-μm particle did not give a strong enough signal, a clump of particles had to be injected into the bed. Unfortunately, the time needed to disperse the clump into the cool bed was roughly the same as the heating time for the particles ($\sim$1 s); hence, the two factors were confounded, and the measured coefficients were lower than would be found if each particle of the clump was well surrounded by only cold bed solids.

Nevertheless, their data lay distinctly above all the other reported measurements, which were whole bed measurements. These results suggest that the heat transfer coefficient for particles in a fluidized bed is best represented by Eq. (31).

Heat Transfer from the Bubbling Bed Model

The treatment here will be brief since it closely parallels the discussions and derivations for mass transfer. We are concerned only with bubbling beds of fine particles in which bubbles are surrounded by thin clouds and where just about all the gas passes through the bed as bubbles. We consider only this situation because here h_{bed} drops drastically from the expected of Eq. (31).

We will develop the expression for the heat transfer between bubble gas and dense bed. The approach is analogous to the derivation of the expression for K_{bc} in Eq. (10.27). Thus, based on unit volume of bubble phase, the total heat interchange across the bubble-cloud boundary is

$$H_{bc} = \left(\begin{array}{c}\text{transfer by bulk}\\\text{flow of gas}\end{array}\right) + \left(\begin{array}{c}\text{transfer}\\\text{by convection}\end{array}\right) = \frac{vC_{pg} + h_{bc}S_{bc}}{V_b}$$

$$= 4.5\left(\frac{u_{mf}\rho_g C_{pg}}{d_b}\right) + 5.85\,\frac{(k_g \rho_g C_{pg})^{1/2} g^{1/4}}{d_b^{5/4}}, \quad [\text{W/m}^3\text{ bubble·K}] \tag{32}$$

where v represents the volumetric flow rate of gas from bubble to cloud, and where the heat transfer coefficient at the bubble-cloud interface, analogous to the derivation of k_{bc} in Eq. (10.26), is given by

$$h_{bc} = 0.975\rho_g C_{pg}\left(\frac{k_g}{\rho_g C_{pg}}\right)^{1/2}\left(\frac{g}{d_b}\right)^{1/4}, \quad [\text{W/m}^2\text{·K}] \tag{33}$$

The quantity h_{bc} is analogous to k_{bc}. For a detailed derivation of these equations, see [15].

Now H_{bc} does not consider the possible pickup of heat by particles in the bubble phase. In fact, the particles dispersed in and passing through the bubble phase play an important role in transferring heat from bubble gas to the bed solids. Similar to the calculations of Eqs. (15)–(24) for mass transfer, we can show for heat transfer that the overall transfer rate, including the heat picked up by particles dispersed in the gas bubbles, is given by

$$H_{total} = \gamma_b \frac{6(Nu^*)k_g}{\phi_s d_p^2} \eta_h + H_{bc}, \qquad [W/m^3 \text{ bubble} \cdot K] \tag{34}$$

From the definition of the heat transfer coefficient, we have

$$\frac{6}{\phi_s d_p}(1 - \varepsilon_f)h_{bed} = \delta H_{total} \tag{35}$$

Combining Eqs. (34) and (35) gives

$$\boxed{Nu_{bed} = \frac{h_{bed}d_p}{k_g} = \frac{\delta}{1-\varepsilon_f}\left[\gamma_b(Nu^*)\eta_h + \frac{\phi_s d_p^2}{6k_g}H_{bc}\right]} \tag{36}$$

Finally, we estimate the value of the adsorption efficiency for heat transfer η_h. For mass transfer, Eqs. (1) and (23) with $m = 1$ give

$$\eta_d = \left(1 + \frac{6k_d^*\bar{t}_p}{\phi_s d_p}\right)^{-1} = \left(1 + \frac{6(Sh^*)}{Sc}\frac{\mu}{\rho_g}\frac{\bar{t}_p}{\phi_s d_p^2}\right)^{-1} \tag{37}$$

and for heat transfer, Eqs. (25) and (26) give

$$\eta_h = \left(1 + \frac{6h^*\bar{t}_p}{\phi_s d_p \rho_s C_s}\right)^{-1} = \left[1 + \frac{\rho_g C_{pg}}{\rho_s C_{ps}}\left(\frac{6(Nu^*)}{Pr}\frac{\mu}{\rho_g}\frac{\bar{t}_p}{\phi_s d_p^2}\right)\right]^{-1} \tag{38}$$

But from the analogy between heat and mass transfer, Eqs. (1) and (25), we have

$$\frac{Sh^*}{Sc} \cong \frac{Nu^*}{Pr} \tag{39}$$

Combining Eqs. (37)–(39) with Eq. (23) then gives

$$\eta_h = \frac{1}{1 + \alpha(\rho_g C_{pg}/\rho_s C_{ps})} \tag{40}$$

Now $\rho_g C_{pg}/\rho_s C_{ps} = 10^{-3}$, and for fine particles $\alpha \cong 20\text{--}1000$. Thus

$$\eta_h = \frac{1}{1 + (20\text{--}100)10^{-3}} = 0.98\text{--}0.91 \tag{41}$$

This rough estimate shows that we can reasonably assume that $\eta_h \cong 1$.

Equation (36), with H_{bc} given by Eq. (32) and $\eta_h = 1$, relates the whole bed coefficient with the single-particle coefficient in the left-hand region of Fig. 6, where these coefficients differ significantly. As shown in Example 3, for a given bed of solids Eq. (36) reduces to the form

$$\text{Nu}_{\text{bed}} = A' \text{Re}_p - B' \tag{42}$$

EXAMPLE 3

Fitting Reported Heat Transfer Data with the Bubbling Bed Model

Given the effective bubble size, determine the Nusselt versus Reynolds number relationship for a fluidized bed of $d_p = 0.036$-cm particles for the experimental conditions of Heertjes and McKibbins [16]. Show that you obtain the dashed line in Fig. 6 and that this line well represents the reported data.

Data

$$\rho_s = 1.3 \text{ g/cm}^3, \qquad \phi_s = 0.806, \qquad \gamma_b = 0.001 \text{ (estimated)}$$

$$\rho_g = 1.18 \times 10^{-3} \text{ g/cm}^3, \quad \text{Pr} = 0.69, \qquad \mu = 1.8 \times 10^{-4} \text{ g/cm·s}$$

$$C_{pg} = 1.00 \text{ J/g·K}, \qquad \varepsilon_{mf} = 0.45, \qquad k_g = 2.61 \times 10^{-4} \text{ W/cm·K}$$

$$d_p = 0.036 \text{ cm}, \qquad u_{mf} = 6.5 \text{ cm/s}, \qquad u_t = 150 \text{ cm/s}$$

Estimate $d_b = 0.4$ cm and take $\eta_h = 1$.

SOLUTION

Equation (36) is the required relationship, so we evaluate terms. First, from Eq. (25),

$$\text{Nu}^* = 2 + 0.6 \left[\frac{(0.036)(150)(0.00118)}{0.00018} \right]^{1/2} (0.69)^{1/3} = 5.15$$

From Eq. (32),

$$H_{bc} = \frac{4.5(6.5)(0.00118)(1.00)}{0.4} + \frac{5.85[(0.000261)(0.00118)(1.00)]^{1/2}(980)^{1/4}}{(0.4)^{5/4}}$$

$$= 0.1434 \text{ W/cm}^3 \cdot \text{K}$$

Next

$$\frac{\phi_s d_p^2}{6k_g} = \frac{(0.806)(0.036)^2}{6(0.000261)} = 0.67 \text{ cm}^3 \cdot \text{K/W}$$

and, as with Example 1,

$$\frac{\delta}{1 - \varepsilon_f} = \frac{u_o - u_{mf}}{u_{br}(1 - \varepsilon_{mf})} = \frac{u_o - 6.5}{0.711(0.4 \times 980)^{1/2}(1 - 0.45)} = \frac{u_o - 6.5}{7.75}$$

Putting all these terms into Eq. (36) gives

$$\text{Nu}_{\text{bed}} = \frac{u_o - 6.5}{7.75} [0.001(5.15)(1) + 0.67(0.1434)] = 0.0131 u_o - 0.085 \tag{i}$$

$$\text{Re}_p = \frac{d_p u_o \rho_g}{\mu} = \frac{(0.036) u_o (0.00118)}{0.00018} = 0.236 u_o \tag{ii}$$

Combining Eqs. (i) and (ii) gives

$$\underline{\underline{\text{Nu}_{\text{bed}} = 0.0555 \text{ Re}_p - 0.085}} \tag{iii}$$

This equation is shown as the dashed line in Fig. 6.

EXAMPLE 4

Heating a

Particle in a

Fluidized Bed

A small (100-μm) cold copper sphere is dropped into a hot fluidized bed of similar solids. Estimate the time needed for the particle to approach within 1% of the bed temperature.

(a) Use a whole bed coefficient from Fig. 6.
(b) Use the single-particle coefficient of Eq. (25), also shown in Fig. 6.

Data

Gas: $\rho_g = 1.2 \text{ kg/m}^3$, $\mu = 1.8 \times 10^{-5} \text{ kg/m·s}$, $k_g = 2.6 \times 10^{-2} \text{ W/m·K}$
Solid: $d_p = 10^{-4}$ m, $\rho_s = 8920 \text{ kg/m}^3$, $C_{ps} = 390 \text{ J/kg·K}$
Bed: $\varepsilon_f = 0.5$, $u_o \cong u_{mf} = 0.1 \text{ m/s}$

SOLUTION

Consider the heat gained Q by the cold particle of mass m_p, originally at t_o, as it heats up to the bed temperature T. It can be shown (see note below) that resistance at the particle surface is controlling; hence, we can write

$$\frac{dQ}{dt} = hA(T - t), \quad \text{with } Q = m_p C_{ps}(t - t_o)$$

Combining and integrating gives the heating time as

$$t = \frac{m_p C_{ps}}{hA} \ln \frac{T - t_o}{T - t} \tag{i}$$

Evaluating terms gives

$$m_p = \frac{\pi}{6} d_p^3 \rho_s = \frac{\pi}{6} (10^{-4})^3 (8920) = 4.67 \times 10^{-9} \text{ kg}$$

$$A = \pi d_p^2 = \pi (10^{-4})^2 = 3.14 \times 10^{-8} \text{ m}^2$$

To evaluate h we need to know the Reynolds number. Thus

$$\text{Re}_p = \frac{d_p u_o \rho_g}{\mu} = \frac{(10^{-4})(0.1)(1.2)}{1.8 \times 10^{-5}} = 0.67$$

(a) Basing the calculation on the whole bed coefficient, Fig. 6 gives

$$\text{Nu}_{bed} = \frac{h_{bed} d_p}{k_g} = 0.0178$$

or

$$h_{bed} = \frac{\text{Nu}_{bed} k_g}{d_p} = \frac{(0.0178)(0.026)}{10^{-4}} = 4.63 \text{ W/m}^2\text{·K}$$

Putting all these values into Eq. (i), we have

$$t = \frac{(4.66 \times 10^{-9})(390)}{(4.63)(3.14 \times 10^{-8})} \ln \frac{100}{1} = \underline{\underline{58 \text{ s}}}$$

(b) From Fig. 6, Nu_p is about 140 times as large as Nu_{bed}. Thus, the h value is 140 times as large as h_{bed}, and the time needed is that much smaller, or

$$t = \underline{0.41 \text{ s}}$$

Comment. Experience suggests that a sphere of copper whose diameter is that of a human hair will not take about a minute to heat up close to its surroundings. The 1-s value seems more reasonable.

Note. The Biot number tells us where the major resistance to heat transfer lies; thus

$$Bi = \frac{hd_p}{k_{solid}} = \frac{\text{interior resistance}}{\text{surface resistance}}$$

with $k_{copper} = 3.84$ W/m·K, $d_p = 10^{-4}$ m, and for either of the above h values we find $Bi \ll 1$; hence, the major resistance to heat transfer lies at the surface of the particle.

PROBLEMS

1. Calculate the overall bed Sherwood number for the desorption of water from air-fluidized porous alumina spheres for the experimental conditions of Kettenring et al. [4] for the largest particle size used and for $u_o = 60$ cm/s. See if your result falls on line I of Fig. 4.

    Data

$\rho_g = 0.00118$ g/cm^3,	$y = 0.9$,	$\gamma_b = 0.001$
$\mu = 1.8 \times 10^{-4}$	$\phi_s = 0.806$,	$Sc = 0.60$
$\mathcal{D} = 0.256$ cm^2/s,	$\varepsilon_{mf} = 0.45$,	$\bar{\rho}_s = 1.37$ g/cm^3
$d_p = 0.10$ cm,	$u_{mf} = 37.9$ cm/s,	$u_t = 334$ cm/s
$d_b = 1.0$ cm (estimated)		

2. Bohle and van Swaaij [17] measured the bubble-emulsion interchange for a fine particle bed using a number of tracer gases. From their results, estimate the interchange for a very highly adsorbed tracer gas.

    Data

Tracer gas	m	K_d (s^{-1})
Helium	0	0.30
Methane	0.57	0.38
Propane	7.07	0.60

3. If gas flows upward at v cm^3/s from a small injector tube into a just-fluidized bed, bubbles will form and grow at the mouth of the tube, then detach and rise without further growth through the bed. Let t_1 be the time of bubble formation and t_2 the time of bubble rise.

    a. Noting that the relationship between size and time for a forming bubble is

    $$\frac{\pi}{6} d_b^3 = vt \quad \text{or} \quad \frac{\pi}{6} d_{b1}^3 = vt_1$$

    show that the interchange coefficient K_{bc} of a growing bubble based on the volume of detaching bubble is

    $$K_{bc} = 2.7\left(\frac{u_{mf}}{d_{b1}}\right) + 3.7\left(\frac{\mathcal{D}^{1/2}g^{1/4}}{d_{b1}^{5/4}}\right)$$

    which is about 60% of an ordinary rising bubble.

b. Assuming that the velocity of rise of a growing bubble can be given by the translation velocity of the center of the bubble, or

$$u_b = \frac{1}{2}\frac{d(d_b)}{dt} = \frac{v}{\pi d_b^2} = \frac{d_{b1}^3}{6t_1 d_b^2}$$

show that the interchange coefficient K_{ce} for the period of bubble formation, again basing this on the volume of detaching bubble, is

$$K_{ce} = 2.37\left(\frac{\varepsilon_{mf}\mathcal{D}_e}{t_1 d_{b1}^2}\right)^{1/2}$$

4. For the conditions of the mass transfer experiment of Prob. 1, using the same estimate for bubble size, develop an expression for Nu_{bed} versus Re_p. Check whether your results are consistent with the reported data in Fig. 6. Additional data: $k_g = 2.6\,W/cm\cdot K$, $C_{pg} = 1.0\,J/gm\cdot K$, $\gamma_b\,\eta_h = \gamma_b\,\eta_d$.

REFERENCES

1. N. Froessling, *Gerland Beitr. Geophys.*, **52**, 170 (1938).

2. W.E. Ranz, *Chem. Eng. Prog.*, **48**, 247 (1952).

3. W.E. Resnick and R.R. White, *Chem. Eng. Prog.*, **45**, 377 (1949).

4. K.N. Kettenring, E.L. Manderfield, and J.M. Smith, *Chem. Eng. Prog.*, **46**, 139 (1950).

5. J.C. Chu, J. Kalil, and W.A. Wetteroth, *Chem. Eng. Prog.*, **49**, 141 (1953).

6. J.F. Richardson and J. Szekely, *Trans. Inst. Chem. Eng.*, **39**, 212 (1961).

7. G. Thodos et al., *AIChE J.*, **7**, 442 (1961); **15**, 47 (1969); in *Proc. Int. Symp. on Fluidization*, A.A.H. Drinkenburg, ed., p. 586, Netherlands Univ. Press, Amsterdam, 1967; *Chem. Eng. Sci.*, **27**, 1549 (1972).

8. J.F. Davidson, R. Clift, and D. Harrison, *Fluidization*, 2nd ed., Academic Press, New York, 1984.

9. T. Wakabayashi and D. Kunii, *J. Chem. Eng. Japan*, **4**, 226 (1971).

10. K. Rietema and J. Hoebink, in *Fluidization Technology*, vol. 1, D.L. Keairns, ed., p. 279, Hemisphere, Washington, D.C., 1975.

11. N.I. Gelperin and V.G. Einstein, in *Fluidization*, J.F. Davidson and D. Harrison, eds., p. 471, Academic Press, New York, 1971.

12. D. Kunii and O. Levenspiel, *Fluidization Engineering*, Chap. 7, Wiley, New York, 1969.

13. A.K. Kothari, M.S. thesis, Illinois Institute of Technology, Chicago, 1967.

14. R. Turton and O. Levenspiel, *Int. J. Heat Mass Transfer*, **32**, 289 (1989).

15. D. Kunii and O. Levenspiel, *Ind. Eng. Chem. Process Des. Dev.*, **7**, 481 (1968).

16. P.M. Heertjes and S.W. McKibbins, *Chem. Eng. Sci.*, **5**, 161 (1956).

17. W. Bohle and W.P.M. van Swaaij, in *Fluidization*, J.F. Davidson and D.L. Keairns, eds., p. 167, Cambridge Univ. Press, New York, 1978.

12

Conversion of Gas in Catalytic Reactions

The problems of predicting performance and scale-up of fluidized bed reactors are very important for rational research and development of new chemical processes, and this chapter presents reactor models for dealing with these problems. These models are of four types, depending on the system at hand, namely those that account for conversion of gas:

- In bubbling beds of very fine (often Geldart **A**) particles. Here bubbles are surrounded by very thin clouds, and carry most if not all the gas through the bed.
- In bubbling beds of intermediate (often Geldart **AB** and **B**) particles. Here bubbles rise somewhat faster than the emulsion (up to five times as fast) and are surrounded by thick clouds that make up most of the emulsion.
- In bubbling beds of large (often Geldart **B** and **BD**) particles. Here the bubbles rise more slowly than the emulsion gas and are cloudless.
- In the lean freeboard zone above dense bubbling beds. This particular model can be extended to fast fluidized operations.

These models are based on the findings of the earlier chapters of this book, and the examples at the end of this chapter show how to use them.

Measures of Reaction Rate and Reactor Performance

For a first-order, solid-catalyzed, gas-phase reaction, the rate is usefully expressed in various ways. Thus, per unit volume of catalyst V_s, we have

$$-\frac{1}{V_s}\frac{dN_A}{dt} = K_r C_A, \qquad K_r = [\text{m}^3 \text{ gas/m}^3 \text{ solid·s}] \tag{1}$$

where K_r is the reaction rate constant and V_s considers the solid as nonporous. This measure of rate constant is independent of bed voidage and particle size if pore diffusion effects do not intrude; hence, it is useful for linking reactor performance of fixed beds to fluidized beds.

For a feed rate v (m³/s) of reactant gas C_{Ai} (mol/m³) to a catalyst bed containing solids of volume V_s (m³), integration of the performance expression gives the outlet concentration C_{Ao}, or the outlet fractional conversion X_A, as

$$\text{For plug flow:} \quad 1 - X_A = \frac{C_{Ao}}{C_{Ai}} = \exp(-K_r \tau) \tag{2}$$

$$\text{For mixed flow:} \quad 1 - X_A = \frac{C_{Ao}}{C_{Ai}} = \frac{1}{1 + K_r \tau} \tag{3}$$

where the reactor ability measure is

$$\tau = \left(\frac{\text{volume of catalyst}}{\text{volumetric flow rate of gas}} \right) = \frac{V_s}{v}, \qquad [\text{m}^3 \text{ cat}/\text{m}^3 \text{ feed/s}] \tag{4}$$

and where the dimensionless reaction rate group is

$$K_r \tau = K_r \frac{L_i(1 - \varepsilon_i)}{u_o}, \qquad i = \text{m, f, or mf} \tag{5}$$

For reversible reactions replace C_A by $C_A - C_{A,\text{equilib}}$ and X_A by $X_A/X_{A,\text{equilib}}$ in Eqs. (2) and (3) and elsewhere in this chapter, and the development throughout follows without further change. Also, the performance expression developed here assumes isothermal flow with negligible density change on reaction. When these assumptions are not reasonable, Levenspiel [1] shows how to account for this.

The findings of the previous chapters suggest that plug flow and mixed flow are both poor representations of gas flow in fluidized beds. Bubble bypassing, bubble-emulsion interchange, and other mass transfer resistances all enter the picture to slow down the overall rate; hence, the problem addressed in this chapter is to develop performance equations that reasonably account for the chemical rate and all these physical resistances in fluidized beds. In addition, since the contacting is so different in fine particle or large particle or fast fluidized systems, the models and performance equations for these regimes will be distinctly different.

Previous Findings

Experimental Investigations

Experience has shown that a fluidized reactor requires more catalyst than does a fixed bed to achieve a given conversion, and it has long been recognized that this is due to the poor gas-solid contacting in the reactor. This behavior has received much attention in the open literature, and Table 1 lists the various systems that have been used to investigate both conversion and selectivity in these reactors. In addition, industry has also studied reactions of special interest to them without reporting the results in the open literature.

For rational design of commercial-scale fluidized reactors, we need to be able to predict the behavior of large units from experimental data obtained from smaller units. Since the 1950s, a variety of models have been proposed for this purpose. In order to test the reasonableness of these models, well-known reactions such as the decomposition of ozone [24–37] and the oxidation of carbon monoxide [23] were run in relatively large experimental units.

As mentioned in Chap. 6, bubbles tend to grow into slugs in small-diameter beds to give somewhat different behavior from freely bubbling

T A B L E 1   Experimental Studies on Catalytic Reactions in Fluidized Beds

Reaction, Temp. (°C)	Investigators (d_t, m) [reference]
Dealkylation of cumene, 510	Mathis (0.05–0.10) [2]
Hydrogenation of ethylene, 113–140	Lewis (0.05) [3]
	Gilliland (0.05) [4]
	Miyauchi (0.051) [5]
Acrylonitrile from acetylene, 500–600	Ogasawara (0.038–0.203) [6]
Methylation of phenol, 325	Katsumata (0.1–0.212) [7]
Oxidation of ammonia, 220–250	Johnstone (0.11) [8]
	Massimilla (0.11) [9]
Oxidation of hydrogen chloride, 310–370	Furusaki (0.05) [10]
Oxidation of ethylene, 200–290	Dŏgu (0.036) [11]
Oxidation of propylene, 390–470	Miyauchi (0.08) [5]
Oxidation of o-xylene, 350	Yates (0.10) [12]
Oxidation of benzene, 430–475	Kizer (0.18) [13]
	Jaffrès (0.10) [14]
Oxidation of naphthalene, 325–375	Hirooka (0.10) [15]
Ammoxidation of propylene, 490–580	Stergiou (0.17) [16]
Oxidehydration of butenes, 420	Ellis (0.10) [17]
Dehydration of isopropanol, 200–500	Mohr (0.03) [18]
Isomerization of butene-1, 100–500	Yates (0.15) [12]
	Verkooijen (0.1) [19]
Isomerization of cyclopentane, 150–200	Ishii (0.05, 0.15) [20]
Decomposition of cumene, 400–500	Gomezplata (0.08) [21]
	Iwasaki (0.08) [22]
Oxidation of carbon monoxide, 200–500	van den Aarsen (0.03, 0.30) [23]
Decomposition of ozone, 27–90	Frye (0.05–0.08) [24]
	Orcutt (0.10–0.15) [25]
	Grekel (0.20) [26]
	Kobayashi (0.08, 0.20) [27]
	Calderbank, (0.15–0.46) [28]
	Botton (0.50) [29]
	Hovmand (0.46) [30]
	Potter (0.23, 0.61) [31, 32]
	Walker (2D, 0.016 × 0.38) [33]
	Chavarie (2D, 0.01 × 0.56) [34]
	van Swaaij (0.05–0.60) [35]
	Werther (0.20, 1.0) [36, 37]

commercial-scale reactors. Thus experimental studies with large-diameter reactors [23, 28–30, 32, 34–37] should be especially useful. Figures 1 and 2 present experimental conversion found by various investigators for fine particle (Geldart **A** and **AB**) bubbling beds without baffles or internals. Figure 3 reports the findings on larger (Geldart **B**) particle systems. These findings are compared with the predictions of the fine particle model (curves 1), the intermediate particle model (curve 2), and the large particle model (curve 3), which are presented later in this chapter.

Various other related studies have been reported as follows. Regarding bed geometry, Werther et al. [36] studied the effect of bed size and distributor type, and Botton [29] studied the effect of two different types of vertical tube banks placed in the bed.

Regarding bed composition, Chavarie and Grace [34] measured vertical concentration profiles in both bubble and emulsion phases in a small particle

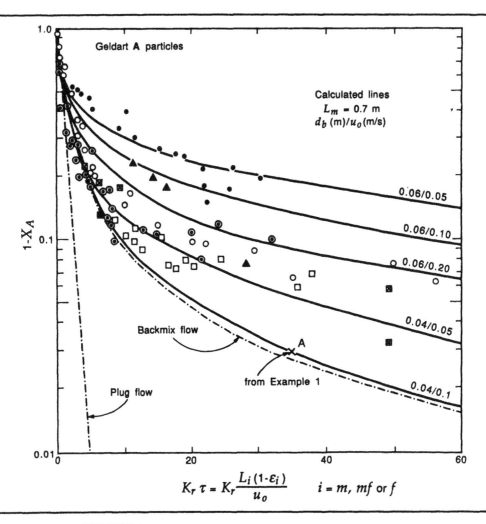

FIGURE 1

Conversion in beds of very fine Geldart **A** catalyst; from [77]. Lines are calculated from Eqs. (14) and (16) of the fast bubble model, with $u_{mf} = 0.006$ m/s; see Example 1 for a sample calculation.

	d_p (μm)	u_{mf} (m/s)	d_t (m)	L_m (m)	u_o (m/s)
Ozone decomposition					
○ Orcutt [25]	20–60	0.0043	0.10–0.15	0.71	0.09–0.15
⊙ Calderbank [28] (porous plate)	68	0.006	0.458	0.69	0.043
● Calderbank [28] (perforated plate)	68	0.006	0.458	0.69	0.043
□ Van Swaaij [35]	~70	0.005	0.23	—	0.12
⊠ Van Swaaij [35]	~70	0.005	0.23	—	0.20
Acetonitrile synthesis					
▲ Ogasawara [6]	74–150	0.008	0.204	0.35	0.04–0.18

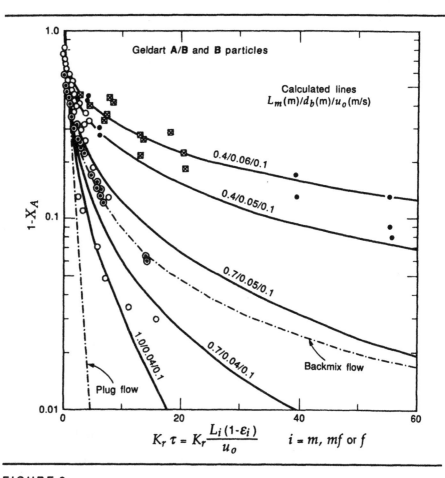

FIGURE 2
Conversion in beds of **AB** and **B** catalyst; from [77]. Lines are calculated from Eqs. (14) and (16) of the fast bubble model with $u_{mf} = 0.02$ m/s.

		d_p (μm)	u_{mf} (m/s)	d_t (m)	L_m (m)	u_o (m/s)
Ozone decomposition						
O	Kobayashi [27]	194	0.021	0.20	0.1–1.0	0.05–0.18
●	Potter [31]	117	0.017	0.23	0.1–0.4	0.06–0.08
⊙	Calderbank [28]	192	0.037	0.46	0.69	0.086
⊠	Van Swaaij [35]	~200	0.010	0.30	—	0.15

bed. Figure 4(a) shows how different are the concentrations in these phases. On the other hand, Potter, with Fryer [31, 39], measured the average concentration of reactant at various positions within the bed. As their measurements in Fig. 4(b) show, when a considerable percentage of reactant is still present in the leaving bubble gas, then the average reactant concentration in the bed is lowest some distance below the bed surface instead of at the bed surface. Potter attributes this somewhat surprising finding to the downward movement of the emulsion phase, which then draws in reactant-rich gas from above the bed into the emulsion. These investigators also sampled gas from three radial positions in a $d_t = 0.22$-m ID bed and found that radial variations in gas concentration were not serious.

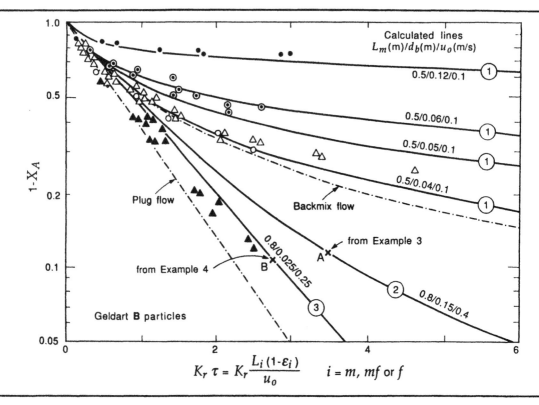

$$K_r \tau = K_r \frac{L_i (1-\varepsilon_i)}{u_0} \qquad i = m, \, mf \text{ or } f$$

FIGURE 3

Conversion in beds of Geldart **B** catalyst; from [77]. Lines ① are calculated from the fast bubble model, Eqs. (14) and (16). Line ② is calculated from the intermediate bubble model, Eq. (49); see Example 3. Line ③ is calculated from the slow bubble model, Eq. (57); see Example 4.

	d_p (μm)	u_{mf} (m/s)	d_t (m)	L_m (m)	u_0 (m/s)	Distributor
Decomposition of ozone, sand, Werther [36]						
●	108	0.016	1.0	0.5	0.15	perforated, $d_{or} = 6.7$ mm
⊙	108	0.016	1.0	0.5	0.1–0.25	porous
○	108	0.016	0.2	0.5	0.1	porous
Oxidation of CO, dense alumina, van den Aarsen [23]						
△	80	0.018	0.30	0.21–0.49	0.1	
▲	325	0.21	0.30	0.79	0.3, 0.5	

To investigate the effect of pressure on the mass transfer rate between bubble and emulsion, Verkooijen et al. [19] carried out the isomerization of 1-butene to 2-butene in a 0.1-m ID bed of silica-alumina catalyst at various pressures up to 10 bar, with the following results:

$$\text{1-butene} \underset{K_{r2}}{\overset{K_{r1}}{\rightleftarrows}} \text{2-butene}$$

Pressure (bar)	1	2	4	6	8	9	10
$k_{be} a_b \dfrac{L_f}{u_0}$ (−)	1.00	0.66	0.50	0.50	0.80	1.20	1.65

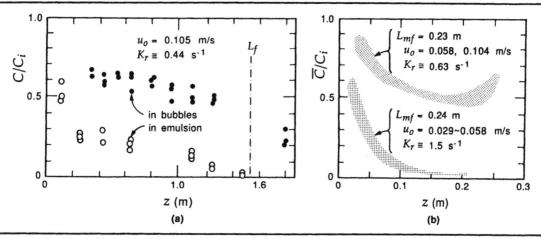

FIGURE 4
Vertical concentration profiles of reactant gas in fine particle fluidized beds: (a) adapted from Chavarie and Grace [34]; 2-D, $0.56 \times 0.01 \, m^2$, mixture of glass beads and sand, $u_{mf} = 0.053 \, m/s$; (b) adapted from Potter, with Fryer [31, 39], $d_t = 0.229 \, m$, sand $d_p = 117 \, \mu m$, $u_{mf} = 0.017 \, m/s$, $\varepsilon_{mf} = 0.48$.

This shows an appreciable effect of pressure on the volumetric mass transfer coefficient $k_{be}a_b$ (s^{-1}), as given by Eq. (10.18).

Miyauchi et al. [5] injected reactant ethylene gas through small nozzles into a fluidized bed at various levels to study the contacting efficiency at different levels in the reactor. They found good contacting just above the distributor (up to 0.05 m), in the splash zone at the bed surface, and also in the freeboard, but very poor contacting in the bubbling bed itself. Figure 5 illustrates their results.

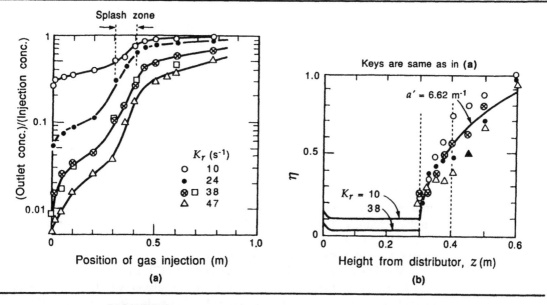

FIGURE 5
Hydrogenation of ethylene, which is injected at different levels into a vigorously bubbling bed of Geldart **A** catalyst ($d_p = 62 \, \mu m$, $u_{mf} = 0.002 \, m/s$, $u_o = 0.305 \, m/s$); from Miyauchi et al. [5].

For reaction in lean-phase pneumatic transport, de Lasa and Gau [40] decomposed ozone in a $d_t = 0.02$-m tube in which silica gel ($\bar{d}_p = 220$ μm) was pneumatically transported upward.

In a number of industrial operations, good prediction of product distribution is crucial. Nonetheless, few experimental results on product distribution have been reported in the literature, except for data in very small reactors. These include the findings by Jaffrès et al. [14] on the oxidation of benzene, by Hirooka et al. [15] on the oxidation of naphthalene, and by Katsumata and Dozono [7] on the methylation of phenol to give o-cresol and 2,6-xylenol. From the last-mentioned study a commercial plant was directly designed and constructed.

Reactor Models

Early models of dense bubbling fluidized bed catalytic reactors [2, 3, 8, 41, 42] were generally based on the two-phase concept of fluidization, originally proposed by Toomey and Johnstone [43], which assumed that all gas in excess of that required for incipient fluidization passed through the bed as bubbles. These models differ only in respect to the assumed flow patterns of gas in the emulsion phase and the form of the phase interchange coefficient.

In a different approach, attempts were made to explain the lower contacting efficiency in terms of the gas residence time distribution. These models [20, 25, 41, 44–46] considered the fluidized bed either as a single-phase homogeneous reactor, ignoring the presence of bubbles, or as consisting of two phases with an interchange coefficient between phases and a mixing parameter for the dense phase.

In the early 1960s, Davidson et al. determined the gas flow in the vicinity of rising bubbles while Rowe and others experimented to demonstrate the general properties of single rising bubbles. This is all presented in Chap. 5. These breakthroughs stimulated the development of a new class of model, the *hydrodynamic* model. Since these models try to account for the behavior of the fluidized bed in physical terms, such as bubble action, they form a sound basis for the rational scale-up and design of fluidized bed reactors.

We now consider the many variations of these models. Some assume two regions, bubble and emulsion, with an interchange K_{be} between them. Others use the height of transfer unit (HTU) to represent the interchange. Noting from Eqs. (10.16) and (10.20) that

$$\text{HTU} = \frac{u_o}{a_b k_{be}} = \frac{u_b}{K_{be}}, \quad [\text{m}] \tag{6}$$

we see that these two measures K_{be} and HTU are essentially equivalent.

Some of these two-region models consider the cloud as part of the emulsion; others combine it with the bubble. Still others assume three regions—bubble, cloud, and emulsion—with the wake as part of the cloud or part of the emulsion. There are even four-region models. All these permutations, plus others, lead to many variations of these models.

Van Deemter [45], Kunii and Levenspiel [47], and Fryer and Potter [31] take into account the downflow of gas in the emulsion, which can then explain quantitatively the seemingly strange data shown in Fig. 4(b) (see [39]). In addition, some of these models neglect the contribution of emulsion flow to total conversion [47–49], arguing that for fine particle catalytic reactors $u_o \gg u_{mf}$, thus very little gas, if any, passes upward through the emulsion. The role of particles dispersed within bubbles is taken into account in some models [47, 48, 50], and others ignore it. Table 2 summarizes the assumptions of these hydrodynamic models, as applied to a narrow horizontal section of bed.

Now consider the bed as a whole. Kato and Wen [49] proposed the "bubble assemblage model," which divided the bed into horizontal slices of the same height as the size of bubble at that level in the bed. Each slice contained a bubble, a cloud and an emulsion region, with gas interchange between regions. Too et al. [56] extended this approach to stochastic compartment sizes.

Van Swaaij and Zuiderweg [35] found an increase in the phase interchange rate with increase in the first-order reaction rate constant. Miyauchi and Morooka [48] and Werther and Schoessler [37] viewed this as a rate enhancement in the phase boundary of the emulsion and introduced the Hatta number, commonly used in gas-liquid systems, to account for this effect.

Analyzing previously reported experimental data, Miyauchi et al. [5] pointed out that particles in the freeboard could contribute significantly to the conversion of reactants and, for these situations, proposed a successive contact model.

Behie and Kehoe [57] proposed a two-region model that considered a jet region devoid of particles in the distributor zone below a bubbling bed. This model was extended by Grace and de Lasa [58] to account for particles entrained in the distributor jets. Thermal effects in this jet region were considered by Errazu et al. [59].

Tigrel and Pyle [60] applied Davidson's model to interpret the performance of an FCC reactor under steady state operations, and Fan and Fan [61] studied the transient behavior of reactors with nonlinear kinetics.

When we can reasonably assume average properties of bubbles throughout the bed, then the differential equations of these hydrodynamic models can easily be integrated to give the overall conversion of reactant leaving the bed surface. However, in all cases, we must be careful to choose reasonable boundary conditions.

Van Swaaij [38] classified all reactor models into two groups: those that assumed a constant bubble size, and those that accounted for bubble growth. Note, however, that all constant-size models can be extended to account for bubble growth by a usual stepwise numerical integration. In this sense, all the hydrodynamic models of Table 2 are on the same level of complexity.

Models for Complex Reactions

Reactor models developed for simple, one-step, first-order kinetics can be extended to deal with the more complex reactions normally encountered in industry. Table 3 gives some examples of these, including a comparison of findings of experimental, pilot-plant, and commercial-scale operations. Based on extensive development works, Botton [67] concludes that these bubble models correctly predict the mass transfer between phases in large pilot-plant and commercial-scale reactors.

Ikeda [62] successfully applied the bubbling bed model of Kunii and Levenspiel [47] for direct scale-up from laboratory to a large commercial-sized unit for the complex reaction scheme of producing acrylonitrile from the ammoxidation of propylene.

In connection with the development of the fluidized reactor for the

T A B L E 2 Comparison of the Postulates of Various Hydrodynamic-Type Reactor

Worker	Particles Dispersed in Bubbles	Gas-Solid Contact in Cloud
Davidson and Harrison [51] (1963)	—	—
Van Deemter [45] (1961, 67)	—	—
Orcutt et al. [25] (1962)	—	—
Partridge and Rowe [52] (1966)	—	With bubble gas
Kunii and Levenspiel [47] (1968)	Small fraction	Cloud gas
Fryer and Potter [31] (1972)	—	Cloud gas
Kato and Wen [49] (1969)	—	Bubble gas
Kobayashi et al, [53] (1973)	—	Bubble gas
Miyauchi and Morooka [48] (1969)	Small fraction	Cloud gas
Werther [36, 37] (1977, 84)	—	—
Krishna [54] (1981)	—	—
Chen et al. [55] (1982)	Small fraction	Upward and downward emulsion phases
Grace [50] (1986)	Small fraction	—

methanol-to-gasoline process, Krambeck et al. [65] applied van Deemter's reactor model to predict reactor efficiency in their 0.6-m ID demonstration unit, which contained horizontal tube internals.

Johnsson et al. [66] tested three reactor models against performance data of an industrial phthalic anhydride reactor of 2.13 m ID. For the conditions of operation, they concluded that all three models gave good overall predictions of conversion and selectivity, as shown in Table 4, and that the grid and freeboard effects appear to play relatively minor roles for this type of reactor system. Hirooka [15] and Iwasaki and Tashiro [68] also used the bubbling bed model to analyze and design reactors for the oxidation of naphthalene.

Multiple Stability

When highly exothermic reactions are carried out in fluidized beds, then, in principle, more than one steady state can result. To maintain stable operations at the desired steady state, one should know the safe limits on "permissible"

Models

Gas-Solid Contact In Wake	Interchange Coefficient	Simultaneous Reaction and Diffusion in Emulsion	Gas Flow in Emulsion
—	b/e	—	Well mixed, and plug flow up
—	b/e	—	Down
—	b/e	—	Up
—	c/e	—	Up
Cloud gas	b/c and c/e	—	None
Cloud gas	b/c and c/e	—	Down at $u_o \gg u_{mf}$
—	b/e	—	None
—	c/e	—	Up
—	b/e	Hatta number	None, conversion in freeboard
—	b/e	Hatta number	Up
—	b/e	—	Up
	b/upward e, and upward e/downward e	—	Up
Cloud gas	b/e	—	None

disturbances, since the reactor may become unstable if the disturbances are sufficiently large. Elnashaie and Cresswell [69] investigated the stability and dynamic behavior of fluidized reactors in terms of a two-region model, allowing rapid exploration of the effects of system parameters and methods of start-up. Bukur et al. [70] indicated from theory that multiplicity was possible.

In a fluidized reactor severe temperature differences, well over 100°C, can exist between the particles in the bubble and in emulsion phases for highly exothermic reactions, as demonstrated visually by Kunii et al. [71]. Kulkarni et al. [72] used a two-region model to derive a criterion to aid in predicting the region of multiplicity for a first-order reaction. Furusaki et al. [73] pointed out the important role of particles in the freeboard in maintaining reactor stability. On the premise that the temperature rise caused by catalytic reactions in the freeboard may trigger violent noncatalytic reactions, they calculated when temperature runaway was likely to take place. Finally, de Lasa and Errazu [74] developed a non-steady state model of a FCC regenerator and showed that the freeboard region could trigger ignition of the regenerator.

TABLE 3   Application of Reactor Models for Analysis of Experimental Data

Worker	Reaction System	Basic Reactor Model	Reactor, d_t (m)
Ellis et al. [17] (1968)	Oxidehydration of butenes	Partridge-Rowe	0.102
Ikeda [62] (1970)	Ammoxidation of propylene	Kunii-Levenspiel	0.2, 3.60, 7.20
Yates and Constans [12] (1973)	Isomerization of butene-1	Partridge-Rowe	0.15
Van Swaaij and Zuiderweg [35] (1973)	Decomposition of ozone	van Deemter	0.05–0.60
Bauer and Werther [63] (1982)	Synthesis of maleic anhydride	Werther	0.04, 0.15, 0.45
Stergiou and Laguerie [16] (1983)	Ammoxidation of propylene	Davidson, Partridge-Rowe, Kunii-Levenspiel, Kato-Wen	0.165
Jaffrès et al. [14] (1983)	Oxidation of benzene	Orcutt, Kato-Wen	0.10
Perrier et al. [64] (1984)	Oxidation of benzene	Werther	0.05
Dŏgu and Sözen [11] (1985)	Oxidation of ethylene	Equivalent to Orcutt	0.036
Krambeck et al. [65] (1987)	Methanol to gasoline	van Deemter	0.60, horizontal tube baffles
Hirooka et al. [15] (1987)	Oxidation of naphthalene	Kunii-Levenspiel	0.083
Johnsson et al. [66] (1987)	Oxidation of naphthalene	Orcutt, Grace, Kunii-Levenspiel, Kato-Wen	2.13

TABLE 4   Comparison of Reactor Models with Performance Data of an Industrial 2.13-m ID Phthalic Anhydride Reactor

Model	$\dfrac{Outlet\ Concentration}{Inlet\ Concentration} \times 100$			
	NA	NQ	PA	OP
Kunii and Levenspiel [47] (1968)	2.13	0.95	86.2	10.7
Kato and Wen [49] (1969)	0.14	0.11	88.8	10.9
Grace [50] (1984)	1.28	0.55	87.5	10.7
Measured	~0 (<2)	1.31	88.9	~9.8

NA = naphthalene, NQ = naphthaquinone, PA = phthalic anhydride, OP = oxidation products.
From Johnsson et al. [66].

Design and Scale-up Procedures

Along with the testing of realistic reactor models, several workers have presented scale-up procedures for the development of commercial-sized reactors [36–38, 50, 67, 75, 76].

Until about two decades ago, it was commonly believed that scale-up of fluidized reactors was extremely risky. Nonetheless, stories of successful scale-

T A B L E 5   Comparison of Performance Data with Model Predictions: Ammoxidation of Propylene to Acrylonitrile

Catalyst	Reactor diam. (m) u_o (m/s)	0.205 0.25	3.6 0.46	7.2 0.50
Low activity	d_b estimated	0.037	0.068	
	γ_b assumed	0	0	
	(X_A) calcd.	0.971	0.964	
	(X_A) obs.	0.951	0.956	
	(S) calcd.	0.619	0.592	
	(S) obs.	0.611	0.569	
High activity	d_b estimated	0.037	0.068	0.080
	γ_b assumed	0.03	0.03	0.03
	(X_A) calcd.	0.981	0.977	0.976
	(X_A) obs.	0.981	0.981	0.982
	(S) calcd.	0.719	0.736	0.685
	(S) obs.	0.752	0.724	0.722

X_A = conversion of propylene
$$S = \left(\begin{array}{c} \text{selectivity of} \\ \text{acrylonitrile} \end{array} \right) = \frac{\text{mols acrylonitrile formed}}{\text{mols propylene reacted}}$$
Taken from Ikeda [62].

up, sometimes from bench scale directly to commercial-sized units, have been reported, despite the curtain of secrecy that often surrounds these industrial developments. Some of these developments used the above reactor models with little modification (see [7]).

Consider Ikeda's experience [62] in designing the commercial plant for producing acrylonitrile from oxygen, ammonia, and propylene. He extended the bubbling bed model of Kunii and Levenspiel to his parallel and successive reaction system and confirmed that it was able to explain the performance of an existing 3.6-m ID commercial reactor. He and his collaborators then developed a more active catalyst, made tests in two small reactors, 0.081 m and 0.205 m ID, and then used this catalyst in their 3.6-m ID reactor, confirming the applicability of their model. From these results they then designed a larger commercial reactor, 7.2 m ID, for use with this new catalyst. Start-up was smooth, and conversion and selectivity were in the range of their predictions, as shown in Table 5. Note that Ikeda took safe values for bubble size, thus giving conservative estimates for reactor performance.

Reactor Model for Fine Particle Bubbling Beds

Catalytic reactions in dense bubbling fluidized beds usually use fine Geldart **A** solids that have a very small minimum fluidizing velocity. Consequently, industrial operations are usually run at many multiples of u_{mf}, or with $u_o/u_{mf} \gg 1$ and $u_b/u_{mf} \gg 1$. For this situation, Kunii and Levenspiel [47, 77] proposed a simple "bubbling bed model." Its features are shown in Fig. 6, and it is based on the following assumptions:

1. Fresh feed gas containing reactant A at C_{Ai} enters the bed and, on contact with the fine catalyst powder, reacts there according to a first-order reaction.

2. The bed consists of three regions: bubble, cloud, and emulsion, with the wake region considered to be part of the cloud. We designate these regions

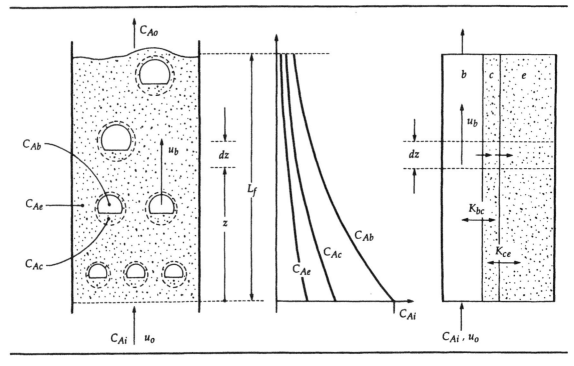

FIGURE 6
Features of the bubbling bed model for a vigorously bubbling fast bubble, thin cloud bed of fine particles, u_o and $u_b \gg u_{mf}$.

by the letters b, c, and e; we designate the reactant concentration at any level in these regions as C_{Ab}, C_{Ac}, and C_{Ae}, respectively.

3. Since $u_o \gg u_{mf}$, all the feed gas passes through the bed as bubbles, and flow through the emulsion is negligible.

4. The gas interchange rate between bubble and cloud and between cloud and emulsion are given by K_{bc} and K_{ce}, respectively.

In developing this model, we must first know the distribution of solid in the various regions of the bed. This is measured by γ_b, γ_c, and γ_e with values given by Eqs. (6.33)–(6.37).

First-Order Reaction

Consider this case first. An accounting for reactant A in the three regions at any level z in the bed gives

$$\begin{pmatrix} \text{overall disappearance} \\ \text{in bubble} \end{pmatrix} = \begin{pmatrix} \text{reaction} \\ \text{in bubble} \end{pmatrix} + \begin{pmatrix} \text{transfer to} \\ \text{cloud-wake} \end{pmatrix} \tag{7}$$

$$\begin{pmatrix} \text{transfer to} \\ \text{cloud-wake} \end{pmatrix} \cong \begin{pmatrix} \text{reaction in} \\ \text{cloud-wake} \end{pmatrix} + \begin{pmatrix} \text{transfer to} \\ \text{emulsion} \end{pmatrix} \tag{8}$$

$$\begin{pmatrix} \text{transfer to} \\ \text{emulsion} \end{pmatrix} \cong \begin{pmatrix} \text{reaction in} \\ \text{emulsion} \end{pmatrix} \tag{9}$$

In symbols these expressions become

$$-\frac{dC_{Ab}}{dt} = -u_b \frac{dC_{Ab}}{dz} = \gamma_b K_r C_{Ab} + K_{bc}(C_{Ab} - C_{Ac}) \tag{10}$$

$$K_{bc}(C_{Ab} - C_{Ac}) \cong \gamma_c K_r C_{Ac} + K_{ce}(C_{Ac} - C_{Ae}) \tag{11}$$

$$K_{ce}(C_{Ac} - C_{Ae}) \cong \gamma_e K_r C_{Ae} \tag{12}$$

where the interchange coefficients K_{bc} and K_{ce} are calculated with Eqs. (10.27) and (10.34) for the estimated bubble size d_b at that position in the bed.

Combining Eqs. (10)–(12) to eliminate C_{Ac} and C_{Ae} gives,

$$-u_b \frac{dC_{Ab}}{dz} = K_f C_{Ab} \tag{13}$$

where the overall rate constant K_f for the fluidized bed with all its mass transfer resistances is

$$K_f = \left[\gamma_b K_r + \cfrac{1}{\cfrac{1}{K_{bc}} + \cfrac{1}{\gamma_c K_r + \cfrac{1}{\cfrac{1}{K_{ce}} + \cfrac{1}{\gamma_e K_r}}}} \right], \qquad [\text{s}^{-1}] \tag{14}$$

For these fine particle beds, bubbles quickly reach an equilibrium size not far above the distributor. Thus, when reaction is not extremely fast and occurs significantly throughout the bed, one can reasonably use an average bubble size with constant interchange coefficients K_{bc} and K_{ce} to represent the reactor. In this situation, integrating Eq. (13) gives the bubble concentration at height z as

$$\frac{C_{Ab}}{C_{A,inlet}} = \frac{C_{Ab}}{C_{Ai}} = \exp\left[-K_f \frac{z}{u_b} \right] \tag{15}$$

and for the reactor as a whole,

$$1 - X_A = \frac{C_{A,outlet}}{C_{A,inlet}} = \frac{C_{Abo}}{C_{Ai}} = \exp\left[-K_f \frac{L_f}{u_b} \right] \tag{16a}$$

For vigorously bubbling conditions, or $u_o \gg u_{mf}$, we may use the approximation of Eq. (6.29). Thus, with Eq. (5), we get, in slightly different form,

$$1 - X_A = \exp\left(-K_f \frac{\delta L_f}{u_o} \right) = \exp\left[-K_f \frac{\delta \tau}{1 - \varepsilon_f} \right] \tag{16b}$$

Where reaction is very fast, most of the conversion occurs near the bottom of the bed with its rapid bubble growth. In this situation, Eq. (16) may have to be integrated numerically using bubble size profiles obtained from Chap. 6, with changing K_{bc} and K_{ce} values corresponding to these changing bubble sizes.

Equation (16a) or (16b) gives the conversion in fluidized beds as a function of bed conditions and reaction rate constant as evaluated from fixed bed experiments. It applies whenever $u_o \gg u_{mf}$. Bubble size in the bed is estimated from Chap. 6, K_{bc} and K_{ce} from Eqs. (10.27) and (10.34), and Example 1 illustrates the calculation procedure.

Special Cases of the Conversion Equation

The five terms in Eq. (14) represent the first-order rate constants of the various resistance steps to reaction. We examine these by referring to Fig. 6.

Reactant A in the feed enters and passes through the bed in bubbles, and for it to react on the surface of the catalyst,

1. It either contacts and reacts on the solids dispersed in the bubbles, or
2. It transfers to the cloud-wake region surrounding the bubbles.

In the cloud-wake region,

3. It either contacts and reacts on the solids, or
4. It further transfers to the emulsion region.

In the emulsion

5. It contacts and reacts with these solids.

For fast reaction, K_r is large, and one can see from Eq. (14) that the early steps (1 and 2) dominate. For slow reaction, K_r is small, so the late steps (4 and 5) dominate. In any situation, the dominant terms are found by comparing the numerical values of the five rate constants in Eq. (14).

Fast Reaction Extreme. In heat and mass transfer where the gaseous component has only to be transported to the outer surface of the particles, only steps 1 and 2 have to be considered, as shown in Chap. 11. We now look at the extreme of a very fast catalytic reaction where the value of $\gamma_b K_r$ compares with that of K_{bc}, even though $\gamma_b \cong 10^{-3}$. In this situation, Eq. (14) reduces to

$$K_f = [\gamma_b K_r + K_{bc}] \tag{17}$$

in which case Eq. (16) becomes

$$1 - X_A = \exp\left[-(\gamma_b K_r + K_{bc})\frac{\delta\tau}{1-\varepsilon_f}\right] = \exp\left[-(\gamma_b K_r + K_{bc})\frac{\delta L_f}{u_o}\right] \tag{18}$$

$$\underset{\substack{\text{reaction} \\ \text{in bubble}}}{} \quad \underset{\substack{\text{transfer} \\ \text{to cloud-wake}}}{}$$

A comparison of this simplified expression with Eqs. (11.11) and (11.36) shows the equivalence in equation forms for chemical and physical rate phenomena.

Slow Reaction Extreme. If reaction is slow and the bubbles are not too large, $K_r \ll K_{bc}$ and K_{ce}; hence, Eq. (14) with Eq. (6.34) reduces to

$$K_f = (\gamma_b + \gamma_c + \gamma_e)K_r = K_r \frac{1-\varepsilon_f}{\delta} \tag{19}$$

in which case, Eq. (16) becomes

$$1 - X_A = \exp\left[-K_r \frac{(1 - \varepsilon_f)L_f}{u_o}\right] = \exp[-K_r \tau] \tag{20}$$

which is the expression for plug flow, Eq. (2).

Conversion Efficiency Compared to Plug Flow

Define the reactor efficiency η for given feed and flow rate of gaseous reactant as follows:

$$\eta_{\text{bed}} = \left(\frac{\begin{array}{c}\text{amount of catalyst}\\\text{needed in a}\\\text{plug flow reactor}\end{array}}{\begin{array}{c}\text{amount of catalyst}\\\text{needed in a}\\\text{fluidized bed}\end{array}}\right)_{\substack{\text{for same}\\\text{conversion}}} = \left(\frac{\begin{array}{c}\text{effective overall first-order}\\\text{rate constant in}\\\text{fluidized bed}\end{array}}{\begin{array}{c}\text{true first-order}\\\text{rate constant}\end{array}}\right) \tag{21}$$

Then for a vigorously bubbling bed of fine particles, a comparison of Eqs. (2) and (16) gives

$$\eta_{\text{bed}} = \left[\gamma_b + \cfrac{1}{\cfrac{K_r}{K_{bc}} + \cfrac{1}{\gamma_c + \cfrac{1}{\cfrac{K_r}{K_{ce}} + \cfrac{1}{\gamma_e}}}}\right] \frac{\delta}{1 - \varepsilon_f} = \frac{K_f \delta}{K_r(1 - \varepsilon_f)} \tag{22}$$

Note that the efficiency is a function of reaction rate constant. For a slow reaction, $\eta_{\text{bed}} \to 1$; for a very fast reaction, $\eta_{\text{bed}} \to \gamma_b \delta/(1 - \varepsilon_f)$.

EXAMPLE 1

Fine Particle (Geldart A) Bubbling Bed Reactor

Estimate the conversion for a first-order irreversible reaction with rate constant $K_r = 10 \text{ m}^3 \text{ gas}/\text{m}^3 \text{ cat·s}$ taking place in a fluidized bed, and plot your result in Fig. 1. Ignore the possible presence of solids in the splash zone and in the freeboard and the extra conversion that this may give. Example 5 considers this extra factor.

Data

Gas: $\mathscr{D} = 2 \times 10^{-5} \text{ m}^2/\text{s}$
Particles: $\bar{d}_p = 68 \ \mu\text{m}$
Bed: $\varepsilon_m = 0.50$, assume $\gamma_b = 0.005$
            $\varepsilon_{mf} = 0.55$, $u_{mf} = 0.006 \text{ m/s}$, $d_b = 0.04 \text{ m}$
            $L_m = 0.7 \text{ m}$, $u_o = 0.1 \text{ m/s}$, $d_{\text{bed}} = 0.26 \text{ m}$

SOLUTION

First, a check of Fig. 3.9 shows that these solids lie in the A′ zone of Geldart A particles. Next, since $u_o \cong 16u_{mf}$ and, as we shall see, $u_b \cong 90u_{mf}$, we are dealing here with a fine particle bed of fast bubbles with very thin clouds. Thus, the analysis leading to Eqs. (14) and (16) can safely be used. So we evaluate the physical quantities needed to calculate the conversion.

If this were a large bed free of internals, one would expect to have considerable gulf streaming, in which case one would estimate the bubble rise velocity from

Eqs. (6.11) and (6.12). However, here the vessel is not very large ($d_t = 0.26$ m), so we use Eqs. (6.7) and Eq. (6.8). Thus,

$$u_{br} = 0.711(9.8 \times 0.04)^{1/2} = 0.445 \text{ m/s}$$
$$u_b = 0.1 - 0.006 + 0.445 = 0.539 \text{ m/s}$$

For the interchange coefficients, we have from Eqs. (10.27) and (10.34),

$$K_{bc} = \frac{(4.5)(0.006)}{0.04} + \frac{5.85(2 \times 10^{-5})^{1/2}(9.8)^{1/4}}{(0.04)^{5/4}} = 3.26 \text{ s}^{-1}$$

$$K_{ce} = 6.77 \left[\frac{(2 \times 10^{-5})(0.55)(0.445)}{(0.04)^3} \right]^{1/2} = 1.873 \text{ s}^{-1}$$

Since $u_b/u_{mf} \cong 90$, we certainly have a vigorously bubbling bed. Hence, we use Eq. (6.29) to find δ:

$$\delta \cong \frac{u_o}{u_b} = \frac{0.1}{0.539} = 0.186$$

From Eq. (6.36) with Fig. 5.8, for catalyst particles,

$$\gamma_c = (1 - 0.55)\left[\frac{3}{(0.445)(0.55)/0.006 - 1} + 0.60 \right] = 0.304$$

From Eq. (6.35),

$$\gamma_e = (1 - 0.55)\frac{1 - 0.186}{0.186} - 0.005 - 0.304 = 1.668$$

and from Eqs. (6.20) and (6.19),

$$1 - \varepsilon_f = (1 - 0.55)(1 - 0.186) = 0.366$$
$$L_f = \frac{(1 - \varepsilon_m)L_m}{1 - \varepsilon_f} = \frac{(1 - 0.50)0.7}{0.366} = 0.956 \text{ m}$$

We are now ready to consider the reaction. First, the dimensionless reaction rate group, from Eq. (5), is

$$K_r\tau = \frac{K_r L_m(1 - \varepsilon_m)}{u_o} \cong \frac{10(0.7)(1 - 0.50)}{0.1} = 35 \tag{i}$$

and for the fluidized bed Eqs. (14) and (16) give

$$K_f = \left[0.005(10) + \cfrac{1}{\cfrac{1}{3.26} + \cfrac{1}{0.304(10) + \cfrac{1}{\cfrac{1}{1.873} + \cfrac{1}{1.668(10)}}}} \right] = 1.979 \text{ s}^{-1} \tag{ii}$$

and

$$1 - X_A = \exp\left(-\frac{K_f L_f}{u_b} \right) = \exp\left(-\frac{(1.979)(0.956)}{0.539} \right) = \underline{\underline{0.030}} \tag{iii}$$

This value of $1 - X_A$ is plotted as point A in Fig. 1. Taking other values for K_r gives the lowest line on this figure. The other lines on this figure are similarly constructed with other values of d_b, u_o, and K_r.

Application to Multiple Reactions

As an example of the use of the bubbling bed model for multiple reactions, consider the rather general reaction scheme known as the Denbigh reactions:

$$A \xrightarrow{K_{r1}} R \xrightarrow{K_{r3}} S \quad \text{with } K_{r12} = K_{r1} + K_{r2}, \quad K_{r34} = K_{r3} + K_{r4} \quad (23)$$
$$\searrow K_{r2} \quad \searrow K_{r4}$$
$$\quad T \qquad U$$

Calculating the conversion and product distribution for this system is a direct extension of the method used for the one-step reaction considered above. Thus, referring to Fig. 7, the material balance equations are, for reactant A,

$$-u_b \frac{dC_{Ab}}{dz} = \gamma_b K_{r12} C_{Ab} + K_{bc,A}(C_{Ab} - C_{Ac})$$

$$K_{bc,A}(C_{Ab} - C_{Ac}) = \gamma_c K_{r12} C_{Ac} + K_{ce,A}(C_{Ac} - C_{Ae}) \quad (24)$$

$$K_{ce,A}(C_{Ac} - C_{Ae}) = \gamma_e K_{r12} C_{Ae}$$

and for R,

$$-u_b \frac{dC_{Ab}}{dz} = \gamma_b K_{r34} C_{Rb} - \gamma_b K_{r1} C_{Ab} + K_{bc,R}(C_{Rb} - C_{Rc})$$

$$K_{bc,R}(C_{Rb} - C_{Rc}) = \gamma_c K_{r34} C_{Rc} - \gamma_c K_{r1} C_{Ac} + K_{ce,R}(C_{Rc} - C_{Re}) \quad (25)$$

$$K_{ce,R}(C_{Rc} - C_{Re}) = \gamma_e K_{r34} C_{Re} - \gamma_e K_{r1} C_{Ae}$$

When the feed gas contains no R, the boundary conditions for these equations are

$$C_{Ab} = C_{Ai} \quad \text{and} \quad C_{Rb} = 0 \quad \text{at } z = 0$$

Eliminating the intermediates C_{Ac} and C_{Ae} in Eq. (24) and C_{Rc} and C_{Re} in Eq. (25) and then integrating gives, after much manipulation, the concentration of reaction components leaving the bed, denoted by subscript o, as

$$\frac{C_{Ao}}{C_{Ai}} = \frac{C_{Abo}}{C_{Abi}} = \exp(-K_{f12}\tau) \quad (26)$$

$$\frac{C_{Ro}}{C_{Ai}} = \frac{K_{fAR}}{K_{f34} - K_{f12}} \left[\exp(-K_{f12}\tau) - \exp(-K_{f34}\tau) \right] \quad (27)$$

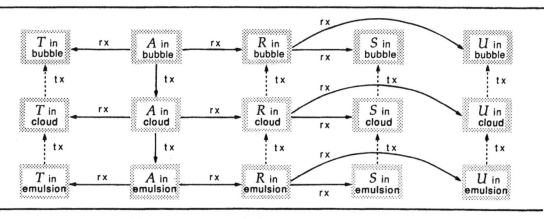

FIGURE 7
Sketch showing the 22 reaction and mass transfer steps representing the Denbigh reaction scheme taking place in a fluidized bed.

and, by material balance,

$$C_{So} = \frac{K_{r1}K_{r3}}{K_{r12}K_{r34}}(C_{Ai} - C_{Ao}) \tag{28}$$

$$C_{To} = \frac{K_{r2}}{K_{r12}}(C_{Ai} - C_{Ao}) \tag{29}$$

$$Cu_o = \frac{K_{r1}K_{r4}}{K_{r12}K_{r34}}(C_{Ai} - C_{Ao}) \tag{30}$$

In these expressions, as with Eq. (4),

$$\tau = \frac{V_s}{v} = \frac{L_f(1 - \varepsilon_f)}{u_o} = \frac{L_f(1 - \varepsilon_f)}{u_b\delta} \tag{31}$$

and

$$K_{f12} = \left[\gamma_b K_{r12} + \cfrac{1}{\cfrac{1}{K_{bc,A}} + \cfrac{1}{\gamma_c K_{r12} + \cfrac{1}{\cfrac{1}{K_{ce,A}} + \cfrac{1}{\gamma_e K_{r12}}}}} \right] \frac{\delta}{1 - \varepsilon_f} \tag{32}$$

$$K_{f34} = \left[\gamma_b K_{r34} + \cfrac{1}{\cfrac{1}{K_{bc,R}} + \cfrac{1}{\gamma_c K_{r34} + \cfrac{1}{\cfrac{1}{K_{ce,R}} + \cfrac{1}{\gamma_e K_{r34}}}}} \right] \frac{\delta}{1 - \varepsilon_f} \tag{33}$$

$$K_{fAR} = \frac{K_{r1}}{K_{r12}}K_{f12} - K_{fA} \tag{34}$$

where

$$K_{fA} = \frac{\left[\frac{K_{bc,R}K_{ce,A}}{\gamma_c^2} + \left(K_{r12} + \frac{K_{ce,A}}{\gamma_c} + \frac{K_{ce,A}}{\gamma_e} \right)\left(K_{r34} + \frac{K_{ce,R}}{\gamma_c} + \frac{K_{ce,R}}{\gamma_e} \right) \right] \frac{\delta K_{bc,A}K_{r1}K_{r34}}{1 - \varepsilon_f}}{\left[\left(K_{r12} + \frac{K_{bc,A}}{\gamma_c} \right)\left(K_{r12} + \frac{K_{ce,A}}{\gamma_e} \right) + \frac{K_{r12}K_{ce,A}}{\gamma_c} \right]\left[\left(K_{r34} + \frac{K_{bc,R}}{\gamma_c} \right)\left(K_{r34} + \frac{K_{ce,R}}{\gamma_e} \right) + \frac{K_{r34}K_{ce,R}}{\gamma_c} \right]} \tag{35}$$

In situations where $K_{r34} \ll K_{r1}$, the term K_{fA} in Eq. (34) is very small and can be dropped with little error to give

$$K_{fAR} \cong \frac{K_{r1}}{K_{r12}}K_{f12}, \qquad \text{for } K_{r34} \ll K_{r1} \tag{36}$$

The maximum amount of intermediate obtainable in a fluidized bed and the amount of catalyst required to achieve this are

$$\frac{C_{Rmax}}{C_{Ai}} = \frac{K_{fAR}}{K_{f12}}\left(\frac{K_{f12}}{K_{f34}} \right)^{K_{f34}/(K_{f34} - K_{f12})} \tag{37}$$

and

$$\tau \text{ (at } C_{Rmax}) = \frac{V_s}{v} = \frac{\ln(K_{f34}/K_{f12})}{K_{f34} - K_{f12}} \tag{38}$$

For plug flow of gas through the reactor, K_{bc} and $K_{ce} \to \infty$, and the above expressions, Eqs. (26)–(38), simplify as follows:

$$\frac{C_{Ao}}{C_{Ai}} = \exp(-K_{r12}\tau) \tag{39}$$

$$\frac{C_{Ro}}{C_{Ai}} = \frac{K_{r1}}{K_{r34} - K_{r12}} [\exp(-K_{r12}\tau) - \exp(-K_{r34}\tau)] \tag{40}$$

For C_{So} and C_{To}, Eqs. (28) and (29) apply.

$$\frac{C_{Rmax}}{C_{Ai}} = \frac{K_{r1}}{K_{r12}} \left(\frac{K_{r12}}{K_{r34}}\right)^{K_{r34}/(K_{r34}-K_{r12})} \tag{41}$$

$$\tau \text{ (at } C_{Rmax}) = \frac{\ln(K_{r34}/K_{r12})}{K_{r34} - K_{r12}} \tag{42}$$

These expressions for the Denbigh reactions were developed by Levenspiel et al. [78]. They also represent the special cases shown in Fig. 8, and can be extended directly to the reaction scheme

$$A \longrightarrow rR \longrightarrow sS$$
$$\searrow \qquad \searrow$$
$$tT \qquad uU$$

simply by replacing C_R, C_S, C_T, and C_U by C_R/r, C_S/s, C_T/t, and C_U/u, respectively.

Ikeda [62], earlier and independently, developed the analysis for the scheme of Fig. 8(c) to represent the ammoxidation of propylene to form acrylonitrile. He and his collaborators then designed and successfully built large commercial plants on the basis of the predictions of this model.

Irani et al. [79] applied this type of analysis to the reaction schemes of Fig. 9. These also reduce to some special cases.

Hirooka et al. [15] used the same approach, but used the scheme of Fig. 9(f) for the phthalic anhydride reaction. They report good prediction of performance of their commerical reactor with this model.

Models other than the bubbling bed have also been used successfully in design. Thus Bauer and Werther [63] applied Werther's two-region model to the synthesis of maleic anhydride from a C_4 feed in the Mitsubishi fluidized bed process. They indicate that their model was able to describe the behavior of the reactors, including scale-up effects.

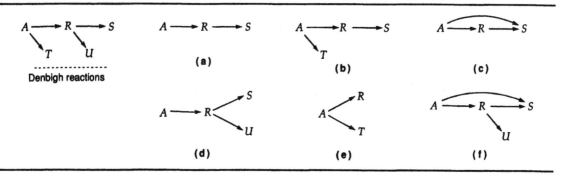

FIGURE 8
The Denbigh reaction scheme and its special cases, analyzed by Kunii and Levenspiel [77].

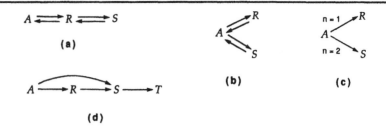

(a)

(b)

(c)

(d)

FIGURE 9
Reaction schemes and their special cases, analyzed by Irani et al. [79].

EXAMPLE 2

Commercial-Sized Phthalic Anhydride Reactor

Phthalic anhydride is produced by the reaction

$$\text{Naphthalene} \xrightarrow{K_{r1}} \text{phthalic anhydride} \xrightarrow{K_{r3}} \text{oxidation product}$$
$$\quad\;\;(A) \qquad\qquad\qquad (R) \qquad\qquad\qquad\;\; (S)$$

Calculate the conversion to desired product and the selectivity, defined as $R_{produced}/A_{reacted}$ if the reaction is run in a large fluidized bed fitted with vertical heat exchange tubes to remove the large exothermic reaction heat.

Data

Catalyst: V_2O_5–K_2SO_4–SiO_2, d_p (50%) = 50–70 μm, 25–35% < 44 μm, $u_{mf} = 0.005$ m/s, $\varepsilon_m = 0.52$, $\varepsilon_{mf} = 0.57$

Gas: 2.27% A in air, $\mathscr{D}_A = 8.1 \times 10^{-6}$ and $\mathscr{D}_B = 8.4 \times 10^{-6}$ m²/s

Bed: $T = 350°C$, $\pi = 252$ kPa, $u_o = 0.45$ m/s
$L_m = 5$ m and, to account for bed internals, $d_{te} = 1$ m

Reaction rates: $K_{r1} = 1.5$ m³ gas/m³ cat·s, $K_{r3} = 0.01$ m³ gas/m³ cat·s, obtained in fixed bed studies by Iwasaki and Tashiro [68]

Assume: $\gamma_b = 0.005$.

In a large bed there sould be considerable circulation of the emulsion, and from the data of Chap. 6 we estimate the bubble size and bubble velocity at $u_o = 0.45$ m/s and 2.52 bar to be

$d_b = 0.05$ m, | from Fig. 6.8
$u_b = 1.5$ m/s, | from Fig. 6.11(a)

SOLUTION
From Eq. (6.7) the relative velocity between bubble and emulsion is

$$u_{br} = 0.711(9.8 \times 0.05)^{1/2} = 0.4977 \text{ m/s}$$

Then for the very high multiples of u_{mf} used, Eq. (6.29) gives

$$\delta = \frac{u_o}{u_b} = \frac{0.45}{1.5} = 0.300$$

and Eq. (6.20) gives

$$1 - \varepsilon_f = (1 - 0.300)(1 - 0.57) = 0.3010$$

From Eq. (6.36) with Fig. 5.8,

$$\gamma_c = (1 - 0.57)\left[\frac{3}{(0.4977)(0.57)/0.005 - 1} + 0.60\right] = 0.2811$$

From Eq. (6.35),

$$\gamma_e = (1 - 0.57)\frac{1 - 0.3}{0.3} - (0.2811 + 0.005) = 0.7172$$

The interchange coefficients for naphthalene are then, from Eqs. (10.27) and (10.34),

$$K_{bcA} = \frac{(4.5)(0.005)}{(0.05)} + \frac{(5.85)(8.1 \times 10^{-6})^{1/2}(9.8)^{1/4}}{(0.05)^{5/4}} = 1.696\ s^{-1}$$

$$K_{ceA} = 6.77\left[\frac{(8.1 \times 10^{-6})(0.57)(0.4977)}{(0.05)^3}\right]^{1/2} = 0.9179\ s^{-1}$$

Similarly, for phthalic anhydride,

$$K_{bcR} = 1.719\ s^{-1}, \qquad K_{ceR} = 0.9347\ s^{-1}$$

From Eq. (6.19),

$$L_f = \frac{(1 - \varepsilon_m)L_m}{(1 - \varepsilon_{mf})(1 - \delta)} = \frac{(1 - 0.52)5}{(1 - 0.57)(1 - 0.3)} = 7.973\ m$$

Now, to the effective rate constants for the fluidized bed: since the reactions of this problem are a special case of the Denbigh scheme in which $K_{r2} = K_{r4} = 0$, we put $K_{r12} = K_{r1}$, $K_{f12} = K_{f1}$, and $K_{f34} = K_{f3}$. Thus, Eqs. (32) and (33) give

$$K_{f1} = \left[(0.005)(1.5) + \cfrac{1}{\cfrac{1}{1.696} + \cfrac{1}{(0.2811)(1.5) + \cfrac{1}{\cfrac{1}{0.9179} + \cfrac{1}{(0.7172)(1.5)}}}}\right]\frac{0.300}{0.301} = 0.6007$$

$$K_{f3} = \left[(0.005)(0.01) + \cfrac{1}{\cfrac{1}{1.719} + \cfrac{1}{(0.2811)(0.01) + \cfrac{1}{\cfrac{1}{0.9347} + \cfrac{1}{(0.7172)(0.01)}}}}\right]\frac{0.300}{0.301} = 0.0099$$

Equation (35) becomes

$$K_{fA} = \frac{\left[\frac{(1.719)(0.9179)}{(0.2811)^2} + \left(1.5 + \frac{0.9179}{0.2811} + \frac{0.9179}{0.7172}\right)\left(0.01 + \frac{0.9347}{0.2811} + \frac{0.9347}{0.7172}\right)\right]\frac{(0.3)(1.696)(1.5)(0.01)}{0.3010}}{\left[\left(1.5 + \frac{1.696}{0.2811}\right)\left(1.5 + \frac{0.9179}{0.7172}\right) + \frac{1.5(0.9179)}{0.2811}\right]\left[\left(0.01 + \frac{1.719}{0.2811}\right)\left(0.01 + \frac{0.9347}{0.7172}\right) + \frac{0.01(0.9347)}{0.2811}\right]} = 0.0058$$

Putting this in Eq. (34) gives

$$K_{fAR} = 0.6007 - 0.0058 = 0.5949$$

Note that, since $K_{r1} \gg K_{r3}$, we could have bypassed calculating K_{fA} and simply put $K_{fAR} = K_{f1}$, with negligible error.

Next, from Eq. (5),

$$\tau = \frac{L_f(1 - \varepsilon_f)}{u_o} = \frac{(7.973)(0.301)}{0.45} = 5.333 \text{ s}$$

The conversion expressions, Eq. (26) and (27), then become

$$\frac{C_{Ao}}{C_{Ai}} = \exp(-0.6007 \times 5.333) = 0.0406$$

and

$$\frac{C_{Ro}}{C_{Ai}} = \frac{0.5949}{0.6007 - 0.0099} [\exp(-0.0099 \times 5.333) - \exp(-0.6007 \times 5.333)]$$

$$= 0.9143$$

Finally, the selectivity is

$$S_{Ro} = \frac{C_{Ro}}{C_{Ai} - C_{Ao}} = \frac{0.9143}{1 - 0.0406} = 0.953$$

To summarize: the conversion of naphthalene = <u>96%</u>
              the selectivity of phthalic anhydride = <u>95%</u>

Reactor Model for Bubbling Beds of Intermediate-Sized Particles or $u_{mf}/\varepsilon_{mf} < u_b < 5u_{mf}/\varepsilon_{mf}$

With fairly large particles, roughly Geldart **B** solids, the bubbling bed may behave somewhere between the extremes of very fast and slow bubbles. Here bubbles surrounded by large overlapping clouds rise faster, but not much faster, than the emulsion gas. These overlapping clouds may well constitute the entire emulsion phase.

A model to represent this situation was developed by Kunii and Levenspiel [77] and is sketched in Fig. 10. It views the bed as consisting of two regions, bubble and emulsion, with just one interchange coefficient K_{be} to represent the transfer of gas between regions. In contrast to the fine particle model, the upflow of gas through the emulsion is not ignored.

For a first-order irreversible catalytic reaction, an accounting of reactant gas A as it rises through these two side-by-side regions gives

$$\begin{pmatrix} \text{disappearance} \\ \text{in bubble} \end{pmatrix} = \begin{pmatrix} \text{reaction} \\ \text{in bubble} \end{pmatrix} + \begin{pmatrix} \text{transfer to} \\ \text{emulsion} \end{pmatrix}$$

$$\begin{pmatrix} \text{disappearance} \\ \text{in emulsion} \end{pmatrix} = \begin{pmatrix} \text{reaction} \\ \text{in emulsion} \end{pmatrix} + \begin{pmatrix} \text{transfer to} \\ \text{bubble} \end{pmatrix}$$

In symbols the equations become

$$-\delta u_b^* \frac{dC_{Ab}}{dz} = \delta \gamma_b K_r C_{Ab} + \delta K_{be}(C_{Ab} - C_{Ae}) \tag{43}$$

and

$$-(1 - \delta)u_{mf} \frac{dC_{Ae}}{dz} = (1 - \delta)(1 - \varepsilon_{mf})K_r C_{Ae} - \delta K_{be}(C_{Ab} - C_{Ae}) \tag{44}$$

In these expressions the rise velocity of bubble gas, not just the bubble, is

$$u_b^* = u_b + 3u_{mf} \tag{45}$$

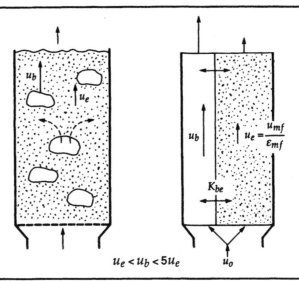

FIGURE 10
Features of the bubbling bed reactor model for intermediate-sized particles, or when $u_f < u_b < 5u_f$.

and, from Eq. (6.27),

$$\delta = \begin{cases} \dfrac{u_o - u_{mf}}{u_b} & \text{at } u_b \approx \dfrac{5u_{mf}}{\varepsilon_{mf}} \\[3mm] \dfrac{u_o - u_{mf}}{u_b + u_{mf}} & \text{at } u_b \approx \dfrac{u_{mf}}{\varepsilon_{mf}} \end{cases} \qquad (6.27) \text{ or } (46)$$

In addition, since the emulsion consists primarily of cloud gas, the interchange coefficient between phases can be approximated by the bulk flow term of K_{bc} of Eq. (10.27). Thus

$$K_{be} \cong 4.5\left(\dfrac{u_{mf}}{d_b}\right) \qquad (47)$$

Solving Eqs. (43) and (44) simultaneously subject to the boundary condition

$$C_{Ab} = C_{Ae} = C_{Ai} \qquad \text{at } z = 0 \qquad (48)$$

gives the overall conversion of reactant gas as

$$\begin{aligned} 1 - X_A &= \frac{\delta u_b^* C_{Ab0} + (1-\delta)u_{mf}C_{Ae0}}{u_o C_{Ai}} \\[2mm] &= \left(\frac{\delta}{1-\delta}\right)\frac{1}{u_o\Phi}\left[(1-\psi_2)\{\psi_1\delta u_b^* + (1-\delta)u_{mf}\}\, e^{-q_1 L_f} \right. \\[2mm] &\quad \left. + (\psi_1 - 1)\{\psi_2\delta u_b^* + (1-\delta)u_{mf}\}e^{-q_2 L_f}\right] \end{aligned} \qquad (49)$$

where

$$q_1, q_2 = \frac{1}{2}\frac{K_r}{u_{mf}}\left\{(1-\varepsilon_{mf}) + \gamma_b\left(\frac{u_{mf}}{u_b^*}\right)\right\} + \frac{1}{2}\frac{K_{be}}{u_{mf}}\left\{\left(\frac{\delta}{1-\delta} + \frac{u_{mf}}{u_b^*}\right) \mp \Phi\right\},$$

$$q_2 > q_1 > 0 \qquad (50)$$

$$\psi_1, \psi_2 = \frac{1}{2} - \frac{1}{2}\frac{1-\delta}{\delta}\left[\frac{u_{mf}}{u_b^*} - \frac{K_r}{K_{be}}\left\{(1-\varepsilon_{mf}) - \gamma_b\left(\frac{u_{mf}}{u_b^*}\right)\right\} \mp \Phi\right] \qquad (51)$$

and

$$\Phi = \left[\left(\frac{K_r}{K_{be}}\right)^2\left\{(1-\varepsilon_{mf}) - \gamma_b\left(\frac{u_{mf}}{u_b^*}\right)\right\}^2 + \left(\frac{\delta}{1-\delta} + \frac{u_{mf}}{u_b^*}\right)^2\right.$$
$$\left. + 2\left(\frac{K_r}{K_{be}}\right)\left\{(1-\varepsilon_{mf}) - \gamma_b\left(\frac{u_{mf}}{u_b^*}\right)\right\}\left(\frac{\delta}{1-\delta} - \frac{u_{mf}}{u_b^*}\right)\right]^{1/2} \qquad (52)$$

Under normal fluidizing conditions, q_2 is much larger than q_1. This justifies dropping the last term of Eq. (49). Comparing with Eqs. (2) and (5) shows that the bubbling bed approaches plug flow behavior when reaction becomes very much slower than mass transfer.

Experimental data on the catalytic oxidation of carbon monoxide in a bed of dense 325-μm alumina are reasonably fitted by Eq. (49). This is shown in Fig. 3 and calculated in Example 3.

EXAMPLE 3

Bubbling Bed Reactor for Intermediate-Sized Particles

Determine the conversion in a bubbling bed of Geldart **B** particles for a rate constant $K_r = 3\,m^3$ gas/m^3 solid·s and for a bubble size of $d_b = 0.12\,m$. Plot your result on Fig. 3.

Data

Gas: $\mathcal{D} = 9 \times 10^{-5}\,m^2/s$
Particles: $\bar{d}_p = 325\,\mu m$
Bed: $\varepsilon_m = 0.42$, $u_0 = 0.4\,m/s$, $L_m = 0.8\,m$
           $\varepsilon_{mf} = 0.45$, $u_{mf} = 0.21\,m/s$, $\gamma_b \cong 0$

SOLUTION

First calculate bubble velocities to determine the existing flow regime. So from Eqs. (6.7) and (6.8) we have

$$u_{br} = 0.711(9.8 \times 0.12)^{1/2} = 0.771\,m/s$$
$$u_b = 0.4 - 0.21 + 0.771 = 0.961\,m/s$$

Now compare the upflow in the emulsion with the bubble rise velocity. Thus,

$$\frac{u_b}{u_f} = \frac{0.961}{0.21/0.45} = 2.1$$

This certainly is not in the fast bubble regime ($u_b \gg u_f$); neither is it in the slow bubble regime ($u_b < u_f$). Thus we apply the intermediate regime model.

Next, evaluate the physical quantities needed from Eqs. (45) to (47).

$$u_b^* = 0.961 + 3(0.21) = 1.591\,m/s$$
$$\delta = \frac{0.40 - 0.21}{0.961 + 0.21} = 0.1623$$

$$\frac{\delta}{1-\delta} = \frac{0.1623}{1-0.1623} = 0.1937$$

$$K_{be} = 4.5\left(\frac{0.21}{0.12}\right) = 7.875 \text{ s}^{-1}$$

Also

$$L_f \cong \frac{L_{mf}}{1-\delta} = \frac{L_m(1-\varepsilon_m)}{(1-\delta)(1-\varepsilon_{mf})} = \frac{0.8(1-0.42)}{(1-0.1623)(1-0.45)} = 1.0071$$

Next, with $\gamma_b = 0$, evaluate the conversion from Eqs. (49) to (52):

$$\Phi = \left[\begin{array}{l} \left(\frac{K_r}{7.875}\right)^2 (1-0.45)^2 + \left(0.1937 + \frac{0.21}{1.591}\right)^2 \\ \\ \quad + 2\left(\frac{K_r}{7.875}\right)(1-0.45)\left(0.1937 - \frac{0.21}{1.591}\right) \end{array} \right]^{1/2}$$

$$= [0.00488 K_r^2 + 0.1061 + 0.00862 K_r]^{1/2} = 0.4194$$

$$\psi_1, \psi_2 = \frac{1}{2} - \frac{1}{2(0.1937)}\left[\frac{0.21}{1.591} - \left(\frac{K_r}{7.845}\right)(1-0.45) \mp \Phi\right]$$

$$= 0.1594 + 0.1803 K_r \pm 2.5813\Phi$$

$$= 1.7828, -0.3823$$

$$q_1, q_2 = \frac{1}{2}\left(\frac{K_r}{0.21}\right)(1-0.45) + \frac{1}{2}\left(\frac{7.875}{0.21}\right)\left[\left(0.1937 + \frac{0.21}{1.591}\right) \mp \Phi\right]$$

$$= 1.3095 K_r + 6.1067 \mp 18.75\Phi$$

$$= 2.1715, 17.899$$

$$1 - X_A = \frac{0.1937}{0.4\Phi}\left[(1-\psi_2)\{\psi_1(0.1623)(1.591) + (1-0.1623)(0.21)\}e^{-1.0071 q_1}\right.$$

$$\left. + (\psi_1 - 1)\{\psi_2(0.1623)(1.591) + (1-0.1623)(0.21)\}e^{-1.0071 q_2}\right]$$

$$= 1.1546[0.0988 + 10^{-18}]$$

$$= \underline{0.114}$$

Finally, the dimensionless reaction rate group, from Eq. (5), is

$$K_r\tau = K_r \frac{L_m(1-\varepsilon_m)}{u_o} = (3)\frac{0.8(1-0.42)}{0.4} = \underline{3.48}$$

This operating condition is shown as point A in Fig. 3. Line 2 gives the locus of conversions for different values of the reaction rate group for this fluidized contacting.

Reactor Model for Large Particle Bubbling Beds

In vigorously bubbling beds of fine particles, fast-rising clouded bubbles with rapidly recirculating gas carry most of the feed gas through the bed, and flow through the emulsion is negligible. In contrast, in bubbling large particle beds the upflow of emulsion gas is both rapid and faster than the rise velocity of bubbles, or

$$\frac{u_{mf}}{\varepsilon_{mf}} > u_b \tag{53}$$

This is called the *slow bubble bed*. Kunii and Levenspiel [77] developed a simple model to represent a reactor in this flow regime.

In overtaking the slower-rising bubbles, the emulsion gas uses these

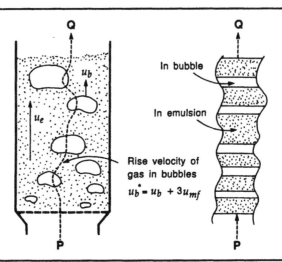

FIGURE 11
Features of the large particle bubbling bed reactor model, or $u_b < u_f$. Path **P–Q** shows the behavior of a typical element of gas.

bubbles as a convenient low-resistance shortcut through the bed. Therefore, a typical element of entering reactant gas rises through emulsion, then bubble, and so on, as shown in Fig. 11. In addition, we can reasonably assume plug flow of gas through the bed.

At any level of the bed the distribution of flows between phases is

$$u_o A = \left(\begin{array}{c} \text{flow through} \\ \text{emulsion} \end{array} \right) + \left(\begin{array}{c} \text{flow through} \\ \text{bubble} \end{array} \right)$$

$$= (1 - \delta)u_{mf}A + \delta(u_b + 3u_{mf})A \tag{54}$$

from which the bubble fraction is

$$\delta = \frac{u_o - u_{mf}}{u_b + 2u_{mf}} \tag{6.26 or 55}$$

Also, the fraction of gas that passes through the emulsion and contacts solids is, with Eqs. (54) and (55),

$$\text{Frac} = \frac{(1 - \delta)u_{mf}}{(1 - \delta)u_{mf} + \delta(u_b + 3u_{mf})} = \frac{u_{mf}}{u_o}(1 - \delta) \tag{56}$$

Ignoring the conversion in the bubble phase, since so little solid is present there, Eq. (56) represents the fraction of bed solids contacted by every element of gas as it passes through the bed. So, comparing Eq. (56) with the plug flow expression of Eq. (2), we obtain the performance expression for a fluidized bed in this regime:

$$1 - X_A = \frac{C_{Ao}}{C_{Ai}} = \exp[-K_r \tau(\text{Frac})]$$

or

$$\boxed{1 - X_A = \exp\left[-K_r \frac{L_f(1 - \varepsilon_f)}{u_o} \frac{u_{mf}}{u_o} (1 - \delta) \right]} \tag{57}$$

Compared to plug flow the reactor efficiency is simply

$$\eta_{\text{bed}} = \frac{u_{\text{mf}}}{u_o}(1 - \delta) \tag{58}$$

Since $\varepsilon_{\text{mf}} \cong 0.5$ and $u_b \leq u_{\text{mf}}/\varepsilon_{\text{mf}}$, and since δ cannot reasonably exceed 0.25–0.35, Eq. (55) shows that this contacting regime only exists when $u_o < (2-2.5)u_{\text{mf}}$. At higher superficial gas velocities, channeling, slugging, and explosive bubbling replace ordinary bubbling behavior.

EXAMPLE 4

Reaction in the Slow Bubble Regime

Suppose the solids of Example 3 are fluidized at $u_o = 0.25$ m/s in a baffled bed wherein bubbles are kept small at $d_b = 0.025$ m. Calculate the conversion for $K_r = 1.5$ m³ gas/m³ cat·s and plot the result in Fig. 3. Additional data are given in Example 3.

SOLUTION

First determine the flow regime. Thus, from Eqs. (6.7) and (6.8),

$$u_{br} = 0.711(9.8 \times 0.025)^{1/2} = 0.3519 \text{ m/s}$$
$$u_b = 0.25 - 0.21 + 0.3519 = 0.3919 \text{ m/s}$$

Comparing bubble and emulsion velocities gives

$$\frac{u_b}{u_f} = \frac{0.3919}{0.21/0.45} = 0.84$$

This ratio is smaller than 1, showing that the reactor is operating in the slow bubble regime; hence, from Eq. (55),

$$\delta = \frac{0.25 - 0.21}{0.3919 + 2(0.21)} = 0.0493$$

and from Eq. (57),

$$1 - X_A = \exp\left[(-1.5)\frac{0.8(1 - 0.42)}{0.25}\frac{0.21}{0.25}(1 - 0.0493)\right] = \underline{0.1083}$$

The dimensionless reaction rate group, from Eq. (5), is

$$K_r\tau = (1.5)\frac{0.8(1 - 0.42)}{0.25} = \underline{2.784}$$

This operating condition is shown as point B in Fig. 3, and line 3 through this point represents the locus of conversions for this contacting regime.

Reactor Model for the Freeboard Region above Fluidized Beds

The data of Miyauchi et al. [5] shown in Fig. 5 clearly indicate that the contact efficiency between catalyst and reactant gas increases progressively as the gas moves up into the freeboard from the bed surface. On the basis of this finding and the development of Chaps. 7 and 8, Kunii and Levenspiel [77] present a reactor model for estimating the added conversion in the freeboard region above turbulent or bubbling fluidized beds, and that can be extended to the relatively dilute contacting of fast fluidized beds. Figure 12 sketches these regimes.

Letting the density of particles at any level z_f in the splash and freeboard regions be $\bar{\rho}$, Eq. (7.42) reduces to

$$\bar{\rho} \cong \bar{\rho}_0 \exp(-az_f), \qquad \text{for } \bar{\rho} \gg \overline{\rho^*} \tag{59}$$

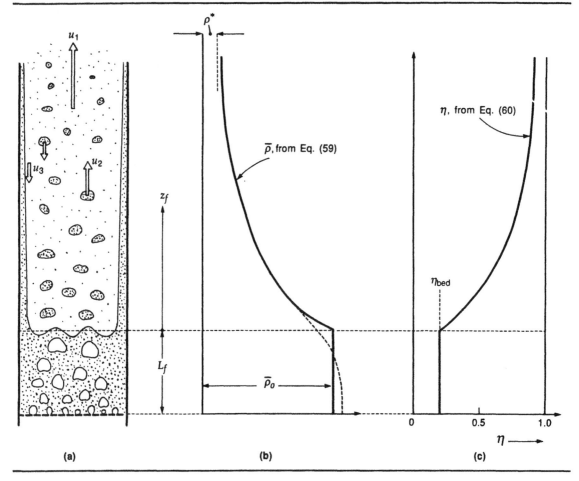

FIGURE 12
Features of the reactor model for the freeboard region above bubbling fluidized beds.

As for contact efficiency, data such as shown in Fig. 5 can be represented by

$$\frac{1-\eta}{1-\eta_{bed}} = \exp(-a'z_f) \tag{60}$$

where η_{bed} is the contact efficiency at the top of bubbling bed. This is given by Eq. (22) for fine particle systems, by Eq. (49) with Eq. (2) for intermediate systems, and by Eq. (58) for large particle systems.

For a first-order reaction taking place in a plug flow reactor of voidage ε, Eq. (1) becomes

$$-\bar{u}\,\frac{dC_A}{dz} = (1-\varepsilon)K_rC_A \tag{61}$$

However, in the freeboard the contacting efficiency is not 100%, and the solid fraction is low. Accounting for these two additional factors, the first-order expression at any height z_f becomes

$$-\bar{u}\,\frac{dC_A}{dz_f} = \frac{\bar{\rho}}{\rho_s}\,\eta K_rC_A \tag{62}$$

Substituting Eqs. (59) and (60) into Eq. (62) and integrating from just above the bed surface, C_{Ao}, $\bar{\rho}_0$, and η_{bed} at $z_f = 0$, to the top of the freeboard, C_{Aex} at H_f, gives

$$\ln \frac{C_{Ao}}{C_{Aex}} = \frac{\bar{\rho}_0 K_r}{\rho_s \bar{u} a} \left[(1 - e^{-aH_f}) - \frac{1 - \eta_{bed}}{1 + a'/a} (1 - e^{-(a+a')H_f}) \right] \qquad (63)$$

In this equation $\bar{u}$ is the average gas velocity in the freeboard, often reasonably approximated by u_o, $\bar{\rho}_0/\rho_s$ can be approximated by $(1 - \varepsilon_f)_{bed}$, a is found from Fig. 7.12, and a' is found from Fig. 5(b). Thus

$$\boxed{\ln \frac{C_{Ao}}{C_{Aex}} \cong \frac{(1 - \varepsilon_f)_{bed} K_r}{u_o a} \left[(1 - e^{-aH_f}) - \frac{1 - \eta_{bed}}{1 + a'/a} (1 - e^{-(a+a')H_f}) \right]} \qquad (64)$$

EXAMPLE 5

Conversion in
the Freeboard
of a Reactor

Suppose the 2.3-m-high reactor of Example 1 is operated at a higher gas velocity, $u_o = 0.3$ m/s, at which condition considerable solids are present in the splash zone and freeboard above the bubbling bed. The bubbling bed is $L_f = 1.1$ m high, the splash zone and freeboard are $H_f = 1.2$ m high, and any entrained solids from the reactor exit are trapped by a cyclone and are returned to the reactor.

(a) In terms of the feed concentration C_{Ai}, find the concentration of reactant C_{Ao} at the top of the bubbling bed.
(b) Determine the concentration C_{Aex} at the reactor exit.

Additional data on this system are given in Example 1.

SOLUTION
In the dense bed. We follow the procedure of Example 1. Thus

Eq. (6.8): $\quad u_b = 0.30 - 0.006 + 0.445 = 0.739$ m/s

Eq. (6.29): $\quad \delta \cong \dfrac{u_o}{u_b} = \dfrac{0.3}{0.739} = 0.4060$

and with Eqs. (6.20) and (6.35),

$$1 - \varepsilon_f = (1 - 0.55)(1 - 0.4060) = 0.2673$$

$$\gamma_e = (1 - 0.55) \frac{1 - 0.4060}{0.4060} - (0.005 + 0.3040) = 0.349$$

All other physical values, u_{br}, K_{bc}, K_{ce}, and γ_c remain unchanged from Example 1. Next, consider the reaction. From Eqs. (14) and (16) we have

$$K_f = \left[(0.005)(10) + \cfrac{1}{\cfrac{1}{3.263} + \cfrac{1}{(0.304)(10) + \cfrac{1}{\cfrac{1}{1.873} + \cfrac{1}{(0.349)(10)}}}} \right] = 0.7056 \ \text{s}^{-1}$$

and

$$\frac{C_{Ao}}{C_{Ai}} = \exp\left[-\frac{K_f L_f}{u_b} \right] = \exp\left[-\frac{(0.7056)(1.1)}{0.739} \right] = 0.350$$

and from Eq. (22),

$$\eta_{bed} = \frac{K_f \delta}{K_r(1 - \varepsilon_f)} = \frac{(0.7056)(0.4060)}{(10)(0.2673)} = 0.1072$$

This bubbling bed efficiency compares well with the experimental values reported in Fig. 5(b).

In the freeboard. Here we plan to use Eq. (64); however, we must first have values for the decay constants and a and a'. For 68-μm particles, Fig. 7.12 gives

$$u_0 a = 0.6\,\text{s}^{-1}, \quad \text{or} \quad a = \frac{0.6}{0.3} = 2\,\text{m}^{-1}$$

For a' the only information available today is that shown in Fig. 5. Thus

$$a' = 6.62\,\text{m}^{-1}$$

Substituting in Eq. (64) gives

$$\ln \frac{C_{Ao}}{C_{Aex}} = \frac{(0.2673)(10)}{(0.3)(2)} \left[\{1 - e^{-(2)(1.2)}\} - \frac{1 - 0.1072}{1 + 6.62/2} \{1 - e^{-(8.62)(1.2)}\} \right]$$

or

$$\frac{C_{Aex}}{C_{Ao}} = 0.0438$$

Thus

$$\frac{C_{Aex}}{C_{Ai}} = \frac{C_{Aex}}{C_{Ao}} \frac{C_{Ao}}{C_{Ai}} = (0.0438)(0.350) = 0.0153$$

In summary, in terms of the conversion

At the top of the dense bed:	$X_{Ao} = 1 - 0.350 = \underline{65.0\%}$
At the reactor exit:	$X_{Aex} = 1 - 0.015 = \underline{98.5\%}$

Comment. This example shows that when the bed height is insufficient for high conversion and when the freeboard contains a considerable amount of solid, then a significant amount of reaction can take place above the dense bed. This situation exists here because the flow of gas through the bed is many multiples of u_{mf}, the bed is probably approaching turbulent flow, much solid is present in the freeboard, and the catalyst is very active ($K_r = 10\,\text{s}^{-1}$).

With highly exothermic reactions the dense bed provides good temperature control while the freeboard region can experience large temperature variations. So, especially when selectivity is an important consideration, the reactor should be designed so that very little reaction occurs in the freeboard.

Turbulent Bed Reactors

To achieve high gas throughput, many commercial reactors are designed for and operated in the tubulent flow regime. Although the bubbling bed model, developed above, is not strictly applicable in this flow regime, examples of its successful use in the design and scale-up to commercial-sized turbulent bed reactors suggest that it can reasonably approximate the behavior of these systems.

The very high gas velocities (relative to u_{mf}) in turbulent beds generate a large dense splash zone of emulsion clusters at the bed surface, plus considerable solids in the freeboard, as explained in Chap. 8. This gives good gas-solid contacting and additional conversion of reactant in the freeboard. This action may be the reason for the steep change in reactant concentration above the bed surface shown in Fig. 5(a). Therefore one may need to use both the dense bubbling bed model and the freeboard model to reasonably represent the turbulent fluidized bed reactor. Example 5 illustrates this situation.

**Fast
Fluidized
Bed
Reactors**

As seen in Fig. 3.14 and in Chap. 8, fast fluidized beds contain a denser zone with voidage $\varepsilon_f = 0.75$–0.85. The contacting here is somewhat like that in turbulent beds in that gas voids accompanied by clumps of emulsion rise rapidly in the core of the vessel while some emulsion descends along the wall of the vessel.

A simple approach to the modeling of fast fluidized beds would first divide the reactor into two zones: a lower dense zone of height L_f, and an upper freeboard zone. *For the lower dense zone*, we define a coefficient K'_{be} to represent the interchange of gas between the fast-rising core region, which is lean in solids, and the slower-moving emulsion clumps and wall region. Then Eq. (14) for the bubbling bed model simplifies to give

$$K_{ff} = \left[\gamma_{core}K_r + \frac{1}{1/K'_{be} + 1/\gamma_{wall}K_r} \right], \qquad [\text{s}^{-1}] \tag{65}$$

and the performance equation for this lower zone becomes, analogous to Eq. (16b),

$$1 - X_A = \exp\left(-K_{ff}\, \frac{\delta L_f}{u_o} \right) \tag{66}$$

For the upper freeboard zone, simply apply the freeboard model that leads to the conversion expression of Eq. (63). When used for fast fluidized beds, the various physical quantities may fall in different ranges of values. Thus

$$\eta_{bed} = \left(\gamma_{core} + \frac{1}{K_r/K'_{be} + 1/\gamma_{wall}} \right) \frac{\delta}{(1 - \varepsilon_f)} \tag{67}$$

$\delta = 0.6$–0.9, representing the core fraction of the reactor.
$\bar{\rho}_0/\rho_s = 0.15$–$0.25$, $\bar{u} \cong u_o$.
$\gamma_{core} = $ order of 0.01 (estimated), $\gamma_{wall} = 0.1$–0.2 (estimated).
$a = $ decay constant in the freeboard; see Chap. 8.

The fast fluidized bed is a complex system to study experimentally; hence, there are few reported measurements of the physical quantities needed to apply this or any other model, particularly δ and K'_{be}. However, preliminary estimates suggest that $K'_{be} > 11 \text{ s}^{-1}$. Research in the next few years should provide firmer information on these factors as researchers focus on this type of reactor.

PROBLEMS

1. For the reaction of Example 1, consider bed heights of $L_m = 0.5$ and 1 m. With all other values unchanged, calculate X_A at these two conditions and plot the corresponding points on Fig. 1. Note the alignment of these points with point A.

2. For the reaction of Example 1, we used Eq. (6.8) to find the bubble rise velocity because this was a small-diameter bed. For a large-diameter bed free of internals, circulation of bed solids (gulf streaming) becomes

important and requires use of a different equation to calculate u_b, and this would certainly modify the overall bed behavior.

Calculate X_A in a large $(d_{bed} = 1 \text{ m})$ fluidized reactor, all other values unchanged, and compare this value with that found in Example 1.

3. Calculate X_A for the decomposition of ozone in a fluidized reactor $(d_{bed} = 0.2 \text{ m})$ and compare your value with the corresponding curve in Fig. 2.

Data

Particles: $\bar{d}_p = 120 \ \mu\text{m}$, $\rho_s = 2200 \text{ kg/m}^3$

Gas: $\mathscr{D} = 2 \times 10^{-5} \text{ m}^2/\text{s}$, $u_o = 0.1 \text{ m/s}$

Bed: $L_m = 0.7 \text{ m}$, $\varepsilon_m = 0.5$, $\varepsilon_{mf} = 0.55$, $u_{mf} = 0.02 \text{ m/s}$

$d_b = 0.05 \text{ m}$, $f_w = 0.40$, $\gamma_b = 0.005$.

4. In the commercial naphthalene reactor of Example 2, we anticipate that the size distribution of the catalyst will narrow with time to give more intense gulf streaming and larger bubbles. Calculate the conversion of naphthalene and selectivity of phthalic anhydride if the bubble size increases to $d_b = 0.07 \text{ m}$, if the bubble rise velocity increases to $u_b = 2.0 \text{ m/s}$, while all other values in Example 2 remain unchanged.

REFERENCES

1. O. Levenspiel, *Chemical Reaction Engineering*, 2nd ed., Wiley, New York, 1972.

2. J.F. Mathis and C.C. Watson, *AIChE J.*, **2**, 518 (1956).

3. W.K. Lewis, E.R. Gilliland, and W. Glass, *AIChE J.*, **5**, 419 (1959).

4. E.R. Gilliland and C.W. Knudsen, *Chem. Eng. Prog. Symp. Ser.*, **67**(116), 168 (1971).

5. T. Miyauchi, Furusaki et al., *J., Chem. Eng. Japan*, **7**, 207 (1974); *AIChE J.*, **20**, 1087 (1974); **22**, 354 (1976); in *Fluidization III*, J.R. Grace and J.M. Matsen, eds., p. 571, Plenum, New York, 1980; *Adv. Chem. Eng.*, **11**, 275 (1981).

6. S. Ogasawara, T. Shirai, and K. Morikawa, *Bull. Tokyo Inst. Tech.*, **B3**, 21 (1959).

7. T. Katsumata and T. Dozono, *AIChE Symp. Ser.*, **83**(255), 86 (1987).

8. H.F. Johnstone et al., *AIChE J.*, **1**, 319 (1955); **1**, 349 (1955).

9. L. Massimilla and H.F. Johnstone, *Chem. Eng. Sci.*, **16**, 105 (1961).

10. S. Furusaki, *AIChE J.*, **19**, 1009 (1973).

11. D. Dŏgu and Z.Z. Sözen, *Chem. Eng. J.*, **31**, 145 (1985).

12. J.G. Yates and J.A.P. Constans, in *Fluidization and Its Applications*, p. 516, Cepadues, Toulouse, 1973; J.G. Yates and J.Y. Gregoire, in *Fluidization III*, J.R. Grace and J.M. Matsen, eds., p. 581, Plenum, New York, 1980.

13. O. Kizer, C. Laguerie, and H. Angelino, *Chem. Eng. J.*, **14**, 205 (1977).

14. J.F. Jaffrès et al., in *Fluidization IV*, D. Kunii and R. Toei, eds., p. 565, Engineering Foundation, New York, 1983.

15. N. Hirooka et al., paper presented at Annual Meeting of SChE, Japan, April 1987.

16. L. Stergiou and C. Laguerie, in *Fluidization IV*, D. Kunii and R. Toei, eds., p. 557, Engineering Foundation, New York, 1983.

17. J.E. Ellis, B.A. Partridge, and D.I. Lloyd, in *Fluidization*, Tripartite Chem. Eng. Conf., Montreal, p. 43, Inst. Chem. Eng., 1968.

18. K.H. Mohr and F. Lunge, *Chemtech*, **16**, 457 (1964).

19. A.H.M. Verkooijen et al., in *Fluidization IV*, D. Kunii and R. Toei, eds., p. 541, Engineering Foundation, New York, 1983.

20. T. Ishii and G.L. Osberg, *AIChE J.*, **11**, 279 (1965).

21. A. Gomezplata and W.W. Shuster, *AIChE J.*, **6**, 454 (1960).

22. M. Iwasaki et al., *Kagaku Kogaku*, **29**, 892 (1965).

23. F.G. van den Aarsen et al., in *Proc. 16th Int. Symp. on Heat and Mass Transfer*, Dubrovnik, 1984.

24. C.G. Frye, W.C. Lake, and H.C. Eckstrom, *AIChE J.*, **4**, 403 (1958).

25. J.C. Orcutt, J.F. Davidson, and R.L. Pigford, *Chem. Eng. Prog. Symp. Ser.*, **58(38)**, 1 (1962).

26. H. Grekel, K.L. Hujsak, and R. Mungen, *Chem. Eng. Prog.*, **60(1)**, 56 (1964).

27. H. Kobayashi et al., *Kagaku Kogaku*, **30**, 656 (1966), **33**, 274 (1969).

28. P.H. Calderbank, F.D. Toor et al., in *Proc. Int. Symp. on Fluidization*, A.A.H. Drinkenburg, ed., pp. 652, 373, Netherlands Univ. Press, Amsterdam, 1967; in *Fluidization*, Tripartite Chem. Eng. Conf., Montreal, p. 12, Inst. Chem. Eng., 1968; in *Fluidization*, J.F. Davidson and D. Harrison, eds., p. 383, Academic Press, New York, 1971.

29. R.J. Botton, *Chem. Eng. Prog. Symp. Ser.*, **66(101)**, 8 (1970).

30. S. Hovmand, W. Freedman, and J.F. Davidson, *Trans. Inst. Chem. Eng.*, **49**, 149 (1971).

31. C. Fryer and O.E. Potter, *Ind. Eng. Chem. Fundamentals*, **11**, 338 (1972); *AIChE J.*, **22**, 38 (1976).

32. Q.M. Mao and O.E. Potter, *AIChE Symp. Ser.*, **80(241)**, 65 (1984); in *Fluidization V*, K. Østergaard and A. Sørensen, eds., p. 449, Engineering Foundation, New York, 1986.

33. D.V. Walker, *Trans. Inst. Chem. Eng.*, **53**, 255 (1975).

34. C. Chavarie and J.R. Grace, *Proc. 5th European Symp. on Chemical Reaction Engineering*, B9, Amsterdam, 1972; *Ind. Eng. Chem. Fundamentals*, **14**, 75, 79, 86 (1975).

35. W.P.M. van Swaaij and F.J. Zuiderweg, *Proc. 5th European Symp. on Chemical Reaction Engineering*, B9 p. 25, Amsterdam, 1972; in *Fluidization and Its Applications*, p. 454, Cepadues, Toulouse, 1973.

36. J. Werther et al., *Chem.-Ing.-Tech.*, **49**, 777 (1977); in *Proc. 2nd World Congr. Chem. Eng.*, Montreal, 1981; *German Chem. Eng.*, **4**, 291 (1981).

37. J. Werther and M. Schoessler, in *Proc. 16th Int. Symp. on Heat and Mass Transfer*, Dubrovnik, 1984.

38. W.P.M. van Swaaij, in *A.C.S. Symp. Ser.*, **72**, 193 (1978); in *Fluidization*, 2nd ed., J.F. Davidson et al., eds., p. 595, Academic Press, New York, 1985.

39. O.E. Potter, *Cat. Rev.-Sci. Eng.*, **17(2)**, 155 (1978).

40. H.I. de Lasa and G. Gau, *Chem. Eng. Sci.*, **28**, 1875 (1973).

41. E.R. Gilliland et al., *Ind. Eng. Chem.*, **44**, 218 (1952); **45**, 1177 (1953).

42. W.F. Pansing, *AIChE J.*, **2**, 71 (1956).

43. R.D. Toomey and H.F. Johnstone, *Chem. Eng. Prog.*, **48**, 220 (1952).

44. R.H. Overcashier, D.B. Todd, and R.B. Olney, *AIChE J.*, **5**, 54 (1959).

45. J.J. van Deemter, *Chem. Eng. Sci.*, **16**, 143 (1961); in *Proc. Int. Symp. on Fluidization*, A.A.H. Drinkenburg, ed., p. 334, Netherlands Univ. Press, Amsterdam, 1967.

46. K. Heidel et al., *Chem. Eng. Sci.*, **20**, 557 (1965).

47. D. Kunii and O. Levenspiel, *Ind. Eng. Chem. Fundamentals*, **7**, 466 (1968); *Ind. Eng. Chem. Process Des. Dev.*, **7**, 481 (1968).

48. T. Miyauchi and S. Morooka, *Kagaku Kogaku*, **33**, 369 (1969); *Int. Chem. Eng.*, **9**, 713 (1969).

49. K. Kato and C.Y. Wen, *Chem. Eng. Sci.*, **24**, 1351 (1969).

50. J.R. Grace, in *Recent Advances in Engineering Analysis of Chemical Reaction Systems*, L.K. Doraiswamy, ed., Wiley Eastern, New Delhi, 1984; in *Gas Fluidization Technology*, D. Geldart, ed., p. 285, Wiley, New York, 1986.

51. J.F. Davidson and D. Harrison, *Fluidized Particles*, Chap. 6, Cambridge Univ. Press, New York, 1963.

52. B.A. Partridge and P.N. Rowe, *Trans. Inst. Chem. Eng.*, **44**, T335, T349 (1966); P.N. Rowe, B.A. Partridge, and J.G. Yates, in *Proc. Int. Symp. on Fluidization*, A.A.H. Drinkenburg, ed., p. 711, Netherlands Univ. Press, Amsterdam, 1967.

53. H. Kobayashi, H. Arai, and T. Sunakawa, *Kagaku Kogaku*, **31**, 239 (1967).

54. R. Krishna, NATO Adv. Study Inst., Ser. E52, 1981.

55. G.T. Chen, J.Y. Shang, and C.Y. Wen, in *Fluidization, Science and Technology*, M. Kwauk and D. Kunii, eds., p. 12, Science Press, Beijing, 1982.

56. J.R. Too et al., *AIChE J.*, **31**, 992 (1985).

57. L.A. Behie and P. Kehoe, *AIChE J.*, **19**, 1070 (1973).

58. J.R. Grace and H.I. de Lasa, *AIChE J.*, **24**, 364 (1978); **25**, 984 (1979).

59. A.F. Errazu, H.I. de Lasa, and F. Sarti, *Can. J. Chem. Eng.*, **57**, 191 (1979).

60. A.Z. Tigrel and D.L. Pyle, *Chem. Eng. Sci.*, **26**, 133 (1971).

61. L.S. Fan and L.T. Fan, *AIChE J.*, **26**, 139 (1980).

62. Y. Ikeda, *Kagaku Kogaku*, **34**, 1013 (1970); Dr. Eng. thesis, Univ. of Tokyo, 1972, private communication.

63. W. Bauer and J. Werther, *A.C.S. Symp. Ser.*, **196**, 121 (1982).

64. M. Perrier et al., *Chem. Eng. J.*, **28**, 79 (1984).

65. F.J. Krambeck et al., *AIChE J.*, **33**, 1727 (1987).

66. J.E. Johnsson, J.R. Grace, and J.J. Graham, *AIChE J.*, **33**, 619 (1987).

67. R. Botton and F. Vergnes, *Chem.-Ing.-Tech.*, **53**(6), 481 (1981); R. Botton, in *Fluidization IV*, D. Kunii and R. Toei, eds., p. 575, Engineering Foundation, New York, 1983.

68. M. Iwasaki and M. Tashiro, *Kagaku Kogaku*, **52**, 147 (1988).

69. S.S. Elnashaie and J.G. Yates, *Chem. Eng. Sci.*, **28**, 515 (1973); S.S. Elnashaie and D.L. Cresswell, in *Fluidization and Its Applications*, p. 501, Cepadues, Toulouse, 1973.

70. D.B. Bukur, N.R. Amundson, and C.V. Wittman, *Chem. Eng. Sci.*, **29**, 1173 (1974); **30**, 847, 1159 (1975).

71. M. Aoyagi and D. Kunii, *Chem. Eng. Comm.*, **1**, 191 (1974); K. Yoshida, T. Mii, and D. Kunii, in *Fluidization and Its Applications*, p. 512, Cepadues, Toulouse, 1973.

72. B.D. Kulkarni, P.A. Ramachandran, and L.K. Doraiswamy, in *Fluidization III*, J.R. Grace and J.M. Matsen, eds., p. 589, Plenum, New York, 1980.

73. S. Furusaki, M. Takahashi, and T. Miyauchi, *J. Chem. Eng. Japan*, **11**, 309 (1978).

74. H.I. de Lasa and A. Errazu, in *Fluidization III*, J.R. Grace and J.M. Matsen, eds., p. 563, Plenum, New York, 1980; H.I. de Lasa, in *Proc. 2nd World Congr. Chem. Eng.*, Montreal, 1981; H.I. de Lasa, *Can. J. Chem. Eng.*, **59**, 549 (1981).

75. D. Geldart, *The Chemical Engineer*, **239**, CE147 (1970).

76. A. Avidan and M. Edwards, in *Fluidization V*, K. Østergaard and A. Sørensen, eds., p. 457, Engineering Foundation, New York, 1986.

77. D. Kunii and O. Levenspiel, *Ind. Eng. Chem. Research*, **29**, 1226 (1990).

78. O. Levenspiel, N. Baden, and B.D. Kulkarni, *Ind. Eng. Chem. Process Des. Dev.*, **17**, 478 (1978).

79. R.K. Irani, B.D. Kulkarni, and L.K. Doraiswamy, *Ind. Eng. Chem. Process Des. Dev.*, **19**, 24 (1980).

CHAPTER

13

Heat Transfer between Fluidized Beds and Surfaces

One of the remarkable features of the fluidized bed is its temperature uniformity. In practice, this uniformity exists in both the radial and axial directions, even in beds as large as 10 m in diameter. To maintain a given temperature level in the bed requires removing (or adding) a definite amount of heat by contact with an appropriate heat exchange surface. Consequently, a quantitative value of the heat transfer coefficient between surface and bed is needed in design for chemical and physical operations where temperature control is required.

Experimental Findings

Heat Transfer Coefficient

The bed-wall* heat transfer coefficient h (W/m^2·K) is defined by

$$q = A_w h \, \Delta T \tag{1}$$

where q is the heat transfer rate (W), A_w is the area of the heat exchanger surface, and ΔT is the mean temperature difference between the bed and surface.

Bed-wall coefficients in gas fluidized beds have been found to be one or two orders of magnitude larger than for gases alone, and since the bed represents a complex interaction of gas and solid, many factors influence the value of h.

Numerous experimental studies and correlations for h have been reported in the literature, most limited to a narrow range of conditions. Because of the complex nature of fluidized contacting, these correlations are far from universal. Furthermore, most early experiments were made in small-diameter units for which the flow behavior differs greatly from beds of commercial size.

In connection with catalytic reactors, early reports of h were mainly for

---

*Let the word *wall* designate all heat exchange surfaces, whether they be the walls of the vessel or the surface of tubes immersed in the bed.

beds of fine particles. More recently, fluidized bed combustion of coal has attracted much attention, and this led to numerous studies on coarse particle high-temperature beds. We are thus in a position to correlate h in a wide range of fluidizing conditions.

Data and correlations until 1970 are well summarized by Gelperin and Einstein [1]. In 1975 Botterill [2] made a comprehensive study of the many investigations until that time, and in 1984 Xavier and Davidson [3] and Baskakov [4] reviewed more recent heat transfer studies.

Vertical Tubes and Bed Walls

Typical of fairly large-sized equipment, Bock and Molerus [5] measured h for a tube bundle in a $d_t = 1$-m bed of fine particles, with results shown in Fig. 1. Their findings show how gas velocity, distributor type, and radial position in the bed affect the measured h values. Note that $h = 300$–$400 \ \text{W/m}^2 \cdot \text{K}$ for $u_0 = 0.2$–0.4 m/s, excluding the wall region.

Similar experiments were performed by Seko et al. [6] with $d_t = 0.55$ m and $d_p = 160 \ \mu$m, and by Piepers et al. [7] with $d_t = 0.7$ m and $d_p = 66 \ \mu$m. For fine particles and a velocity range of $u_0 = 0.2$–0.4 m/s, they reported results similar to those found by Bock and Molerus, or $h = 200$–$400 \ \text{W/m}^2 \cdot \text{K}$.

Horizontal Tubes

Many experimental studies have been done on heat transfer from fluidized beds to single tubes and tube bundles. Typical of these findings, Fig. 2 displays the

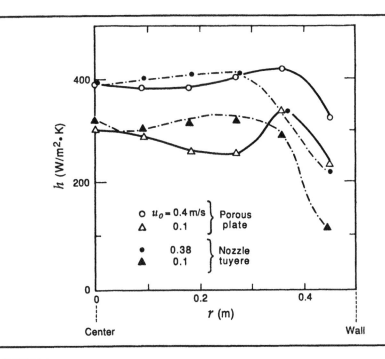

FIGURE 1
h measured at $z = 0.85$ m on a vertical tube bundle in a $d_t = 1$-m, $L_m = 1.36$-m bed of quartz sand, $d_p = 96 \ \mu$m, supported on two types of distributor plates; from Bock and Molerus [5].

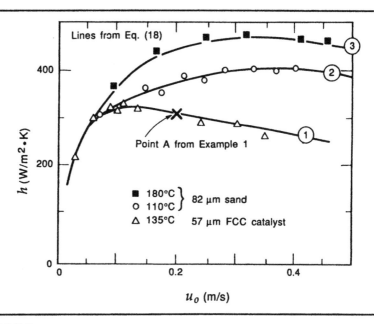

FIGURE 2
h on a horizontal tube bundle in a 0.3 m × 0.3 m bed; data from Beeby and Potter [8]. Line ① calculated from Example 2; line ② from Prob. 1; line ③ from Prob. 2.

data reported by Beeby and Potter [8] in beds of fine solids and shows the effect of particle size, gas velocity, and bed temperature. Note that h goes through a maximum at some intermediate gas velocity. The decreasing h at higher u_o may be attributed to more contact time with bubbles with their very low h values.

Several groups measured the local but time-averaged h around the horizontal tubes. At mild fluidizing conditions, namely for u_o close to u_{mf}, they found low h values at the bottom and at the top of the tube, and much higher values at the sides of the tubes. This finding can be attributed to voids frequently forming below the tube and by stagnant particles resting on the top of the tube. At higher gas velocities, fluidization becomes more agitated, and this gives a more uniform distribution of h around the tube. At the usual operating conditions, u_o is much higher than u_{mf}; hence, it is reasonable to use a mean value for h.

In beds of large and coarse particles and at $T < 200°C$ the effect of particle size on h is reported by Carlomagno et al. [9] in Fig. 3. A narrow band of h values well represents their data and the data they collected from other investigators.

As described in Chap. 6, fluidization becomes smoother at high pressure; consequently, h should increase with pressure. Figure 4, from Bock and Schweinzer [10], shows that this is so. Staub and Canada [11] found similar high-pressure behavior.

In Fig. 5, Glicksman and Decker [12] correlated the results of six groups of investigators, who used $d_p = 0.65–4.0$ mm and pressures up to 10 bar. As may be seen, the h values all fall in a narrow range.

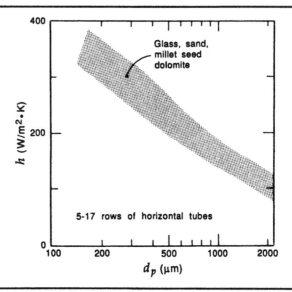

FIGURE 3
Summary of six studies that relate h to d_p in coarse particle beds containing horizontal tube bundles: $L_m = 0.4–0.7$ m, $d_{tl} = 16–32$ mm, $u_o = 0.2–4$ m/s; from Carlomagno et al. [9].

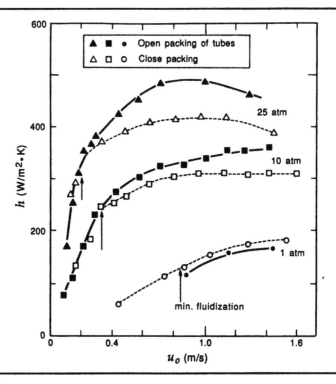

FIGURE 4
Effect of pressure on h on horizontal tube bundles in large particle beds, quartz sand ($d_p = 970$ μm); adapted from Bock and Schweinzer [10].

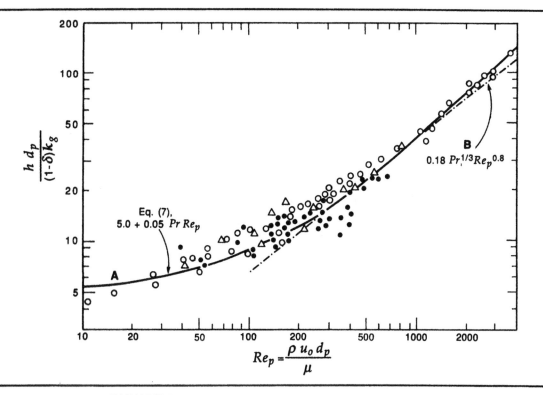

FIGURE 5
Correlation for h in large particle beds at low temperature, d_p up to 4 mm, pressure up to 10 atm; from Glicksman and Decker [12]. Line A from Eq. (7), line B from Baskakov [13].

Splash Zone

In fluidized combustors containing horizontal tube banks, the total heat transfer rate is often adjusted and controlled not by changing the bed temperature but by raising or lowering the bed level. This changes the fraction of tubes immersed in the bed. How well this procedure works depends on the difference in h between tubes immersed in the bed and those in the splash zone above the bed surface; hence, one must know the h values in these zones.

In another application, shallow bed exchangers with horizontal tube banks are being tested as a means for avoiding the fouling of exchanger surfaces with carbonaceous materials, for example, Diesel exhausts.

Interest in these types of applications has led to research on h in shallow beds and in the splash zone above fluidized beds. Figure 6 shows the range of experimental results of the studies as reported by Kortleven et al. [14] and Grewal et al. [15].

Temperature Effect

Comparing the reported data at high temperature with that at ambient conditions, such as reported in Fig. 3, shows that h is 100–200 W/m^2·K higher at 800°C than at ambient conditions. This may be attributed partly to the increase

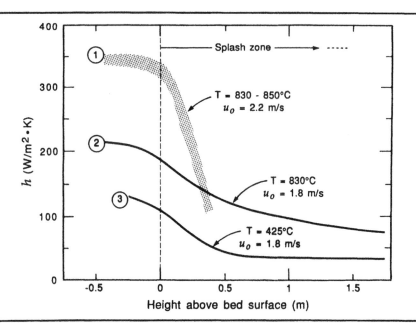

FIGURE 6

h for a horizontal tube bank in and above high-temperature, large particle (~1 mm), fluidized bed coal combustors. Line ① from Kortleven et al. [14], lines ② and ③ from Grewal et al. [15].

in gas thermal conductivity and partly to the increase in radiant heat transfer at the higher temperatures.

Objects Immersed in Bubbling Beds

As a tool for studying the mechanism of heat transfer in fluidized beds, workers have measured h on the surface of single cylinders or spheres immersed in

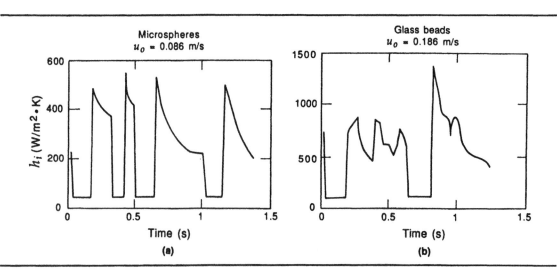

FIGURE 7

Instantaneous h on a vertical $d_{ti} = 6.35$-mm heater in a $d_t = 0.1$-m fluidized bed; adapted from Mickley et al. [16].

fluidized beds of fine or intermediate-sized particles. Mickley et al. [16] measured the instantaneous heat transfer coefficient h_i at a point on a vertical 6.35-mm tube located along the axis of a 0.1-m fluidized bed of 43–320 μm particles and found sharply varying h_i values, as shown in Fig. 7. Similar data were also reported by Baskakov et al. [13]. Such findings suggest that the exchanger surface is being bathed alternately by gas bubbles (very low h_i values) and emulsion packets (high h_i values). In addition, this means that h as defined by Eq. (1) only represents a time-averaged value. Mickley et al. [16] also measured the effect of the gas thermal conductivity on the time-averaged h_{max}. Their results are shown in Fig. 8(a).

Martin [17] correlated h with particle size, as shown in Fig. 8(b), using

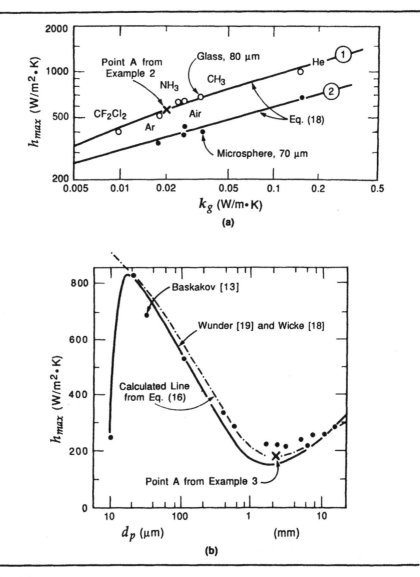

FIGURE 8
Effect of gas thermal conductivity and particle size on h_{max}: (a) from Martin [17]; data from Mickley and Fairbanks [16]; calculated lines from Eq. (18); see Example 2 for line ①, Prob. 2 for line ②; (b) from Martin [17]; dashed line calculated from Eq. (16); see Example 3.

data of other workers. Note that h drops sharply for Geldart **C** solids, that is, when $d_p < 20$ μm.

Fast Fluidized and Solid Circulation Systems

Most of the studies of heat transfer to exchangers in high-velocity systems were done at ambient conditions and with fine particles. We consider some typical results.

Figure 9(a) shows the findings of Guigon et al. [20] in their large experimental unit, Fig. 9(b) displays the results of Furchi et al. [21] in their small-diameter high-velocity unit, and Fig. 10, prepared by Wu et al. [22], correlates h with the density of the suspension flowing past the heat exchange surface. Note that one should also be able to estimate the effect of G_s/G_g on h in the lean phase from Fig. 9(b). These figures all indicate that h decreases as the fraction of suspended solids in the lean-phase mixture is lowered.

Radiant Heat Transfer

With radiometer probes, Ozkaynak et al. [27] and Mathur and Saxena [28] directly measured the radiant heat transfer coefficient h_r between hot beds of coarse particles and small vertical surfaces immersed therein. Figure 11(a) shows that the radiation contribution to heat transfer rises rapidly with temperature. Figure 11(b) shows that radiant heat transfer is hindered appreciably when emulsion packets bathe the receiving surface.

Lindsay et al. [29] measured emissivities in the freeboard of a fluidized bed combustor, with the results shown in Fig. 12(a). The overall heat transfer coefficient was also measured. Figure 12(b) displays the theoretical breakdown of the overall heat transfer coefficient and shows that radiation is the major contributor, almost $100\ \mathrm{W/m^2 \cdot K}$, at this high temperature.

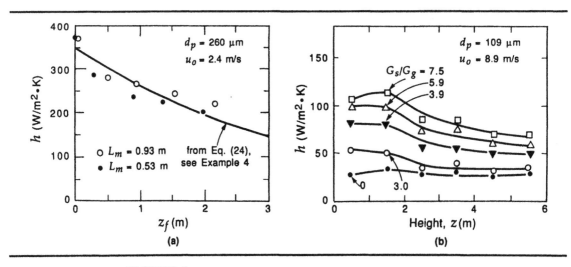

FIGURE 9
h in fast circulating fluidized beds: (a) horizontal tube banks, $d_{ti} = 50$ mm in a 1.19 m $\times$ 0.79 m bed of sand; data from Guigon et al. [20]; (b) on the wall of a $d_t = 0.072$-m tube of glass beads; data from Furchi et al [21].

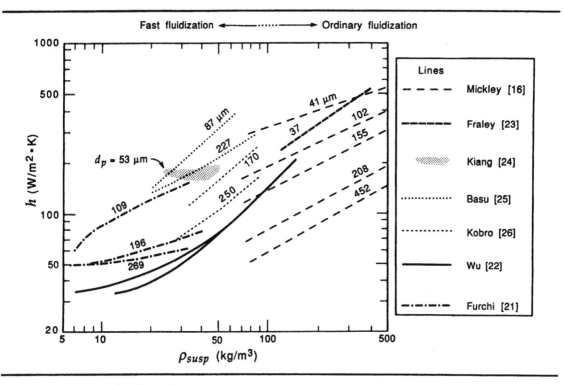

FIGURE 10
Comparison of h in fast fluidized beds with h in ordinary fluidized beds; adapted from Wu et al. [22].

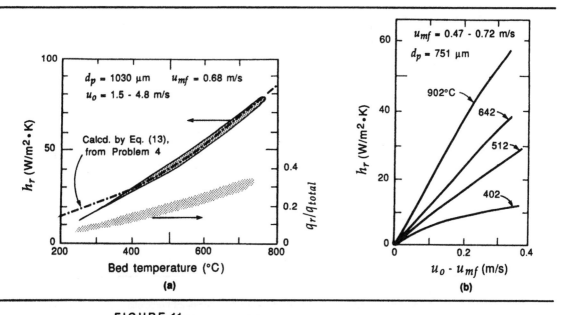

FIGURE 11
Radiant heat transfer from sand beds to vertical surfaces: (a) data from Ozkaynak et al. [27]; (b) adapted from Mathur and Saxena [28].

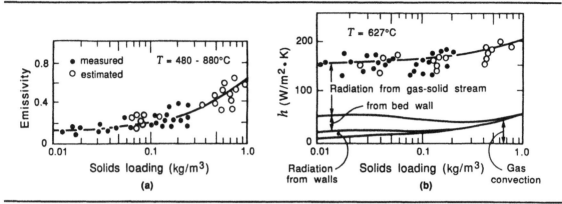

FIGURE 12
Radiant heat transfer to tubes in the freeboard; from Lindsay et al. [29].

Basu and Konuche [25] made similar measurements but in a $0.2 \times 0.2\,\text{m}^2$ circulating bed at 650–885°C, reporting emissitivities of about 0.86 at suspension densities of 14–$40\,\text{kg/m}^3$.

Heat Transfer Properties of Incipiently Fluidized Beds

To understand the mechanism of heat transfer between surface and fluidized bed, we first need information on the heat transfer behavior of incipiently fluidized beds. To this end Fig. 13(a), from Abrahamsen and Geldart [30], shows

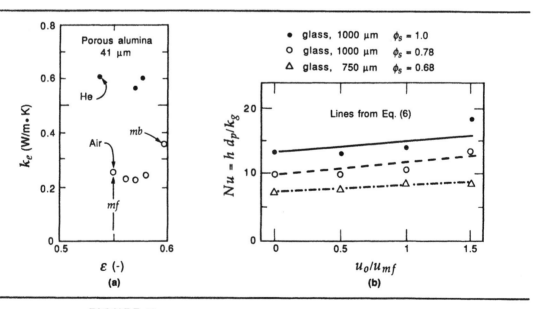

FIGURE 13
Heat transfer in fixed beds and in fluidized beds at incipient fluidization and minimum bubbling conditions: (a) effective thermal conductivity of fine alumina powder, from Abrahamsen and Geldart [30]; (b) h between flat surfaces and stationary beds of large particles, data from Floris and Glicksman [31].

the measured effective thermal conductivity of fine particles under $u_{mf} \sim u_{mb}$ conditions.

Floris and Glicksman [31] measured h between a flat surface and a stationary bed of coarse solid as shown in Fig. 13(b), and Botterill [2] and Colakyan and Levenspiel [32] measured h for moving beds flowing over a horizontal tube.

Heat Transfer at a Distributor Plate

Zhang and Ouyang [33] studied this phenomenon with different materials and particle sizes. They noted rapid fluctuations in surface temperature of the distributor and found h values in the same range as for vessel walls. This suggests that a similar mechanism controls the rate of heat transfer in the two cases.

Theoretical Studies

We start by considering heat transfer in fixed and incipiently fluidized beds, and then extend this analysis to bubbling fluidized beds, to the freeboard region, and to fast fluidized beds. Since the thermal characteristics of materials are essential to this development, Table 1 lists these properties for various frequently used gases and solids.

Fixed and Incipiently Fluidized Beds

Fixed Bed with Stagnant Gas. If heat flows in parallel paths through the gas and the solid, as shown in Fig. 14(a), then the effective thermal conductivity of the fixed bed is given by

$$\mathbf{k}_e^\circ = \varepsilon_{mf}\mathbf{k}_g + (1 - \varepsilon_{mf})\mathbf{k}_s \tag{2}$$

However, to account for the actual geometry and the small contact region between adjacent particles as shown in Fig. 14(b), Kunii and Smith [35] developed the following modification to the parallel path model:

$$\mathbf{k}_e^\circ = \varepsilon_{mf}\mathbf{k}_g + (1 - \varepsilon_{mf})\mathbf{k}_s\left[\frac{1}{\phi_b(\mathbf{k}_s/\mathbf{k}_g) + 2/3}\right] \tag{3}$$

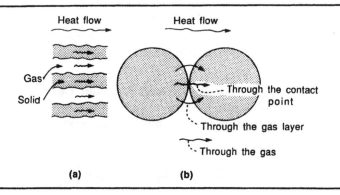

FIGURE 14
Effective thermal conductivity of a fixed bed: (a) unrealistic parallel path model leads to Eq. (2); (b) more realistic model leads to Eq. (3).

TABLE 1 Thermal Properties of Common Solids and Gases at 20°C

Solids	ρ_s (kg/m^3)	C_{ps} $(J/kg \cdot K)$	k_s $(W/m \cdot K)$
Alumina	4070	910	33
Aluminum	2707	896	204
Alundum	3965	778	10
Brass	8550	400	104
Carborundum	3180	640	18
Coal	1350	1260	0.26
Coal ash	2400	1320	3.10
Coke	1800	740	0.50
Copper	8954	383	386
Dolomite	2750	880	1.3
Fire clay	2350	840	0.65
Glass	2700	765	0.90
Gravel	2500	840	0.37
Iron	7897	452	73
Lead	11,373	130	35
Lead glass	3000	655	0.93
Limestone	2500	920	1.7
Marble	2600	810	2.8
Polymer	550	1285	0.27
Quartz	2643	754	6.23
Sand	2600	840	1.9
Silica-alumina catalyst	1250	1060	0.36
SiC	3200	837	18
Steel	7800	480	45

From Xavier and Davidson [3].

Here, the ° refers to stagnant gas conditions, and $\phi_b = d_{eqv}/d_p$ represents the equivalent thickness of gas film around the contact points between particles, which aids in the transport of heat from particle to particle. Since ϕ_b depends on the bed voidage and since we are interested in using Eq. (3) in our fluidized bed development, Fig. 15 gives the values of ϕ_b for the loosest packing of a normal fixed bed, which is at about $\varepsilon = 0.476$.

For most gas-solid systems, $k_s \gg k_g$; thus, the last part of the last term in Eq. (3) is smaller than unity. This means that the thermal conductivity of a fixed bed is lower than for the parallel path model of Eq. (2).

Wall Region with Stagnant Gas. Consider the wall region to extend a half-particle diameter out from the surface of the heat exchange surface. Then similar to Eq. (3), the thermal conductivity in this layer can be represented by

$$k_{ew}^\circ = \varepsilon_w k_g + (1 - \varepsilon_w)k_s \left[\frac{1}{\phi_w(k_s/k_g) + 1/3} \right] \qquad (4)$$

where ε_w is the mean void fraction of this wall layer.

Figure 15 also shows the calculated values for ϕ_w. Note that the thickness of the equivalent gas layer is greater for particle-wall contact than for particle-

Gases	ρ_g (kg/m^3)	C_{pg} $(J/kg \cdot K)$	$\mathbf{k}_g$ $(W/m \cdot K)$	$\mu \times 10^5$ $(kg/m \cdot s)$
Air	1.205	1005	0.026	1.80
Argon	1.675	524	0.017	2.20
Carbon dioxide	1.842	880	0.017	1.45
Freon-12	4.390	670	0.010	1.22
Helium	0.165	5200	0.159	1.85
Hydrogen	0.084	14,444	0.190	0.87
Methane	0.716	2185	0.034	1.09
Nitrogen	1.182	1041	0.026	1.75
Steam (at 100°C)	0.588	2063	0.0251	1.25

particle contact; in addition, because $\varepsilon_w > \varepsilon_{mf}$, these two factors indicate that the wall layer presents a greater resistance to heat transfer than an equivalent layer in the main body of the bed.

We may now define a heat transfer coefficient for this wall region of

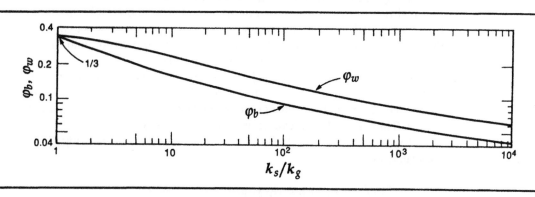

FIGURE 15
Ratio of effective thickness of gas film around a contact point to particle diameter: ϕ_b for contact between adjacent particles, ϕ_w for contact between particle and surface; adapted from Kunii and Smith [35] and Suzuki and Kunii [36].

thickness $d_p/2$, containing stagnant gas, as follows:

$$h_w^\circ = \frac{k_{ew}^\circ}{d_p/2} = \frac{2k_{ew}^\circ}{d_p} \tag{5}$$

Wall Region with Flowing Gas. Figure 13(b) shows that heat transfer in fixed beds is enhanced by gas flow through the bed. This can be attributed to the lateral mixing of gas in the void spaces at the surface with adjacent voids. Yagi and Kunii [36] studied this phenomenon and came up with the following two-term expression:

$$\mathrm{Nu} = \frac{h_w d_p}{k_g} = \left(\begin{array}{c}\text{transfer for}\\\text{no gas flow}\end{array}\right) + \left(\begin{array}{c}\text{extra transfer because}\\\text{of gas flow}\end{array}\right)$$

$$= \frac{h_w^\circ d_p}{k_g} + \alpha_w \Pr \mathrm{Re}_p, \qquad \mathrm{Re}_p = \frac{d_p u_o \rho_g}{\mu} < 2000 \tag{6a}$$

Rearranging this expression gives

$$h_w = h_w^\circ + \alpha_w (C_{pg}\rho_g u_o) = \frac{2k_{ew}^\circ}{d_p} + \alpha_w (C_{pg}\rho_g u_o) \tag{6b}$$

The lines on Fig. 13(b) are drawn for $\alpha_w = 0.05$, and the fit to the data shows that this is a reasonable value for α_w to use in Eq. (6).

Bubbling Beds—Heat Transfer to Emulsion Packets

In a bubbling fluidized bed the rising bubbles sweep past the heat exchange surface, thereby washing away the particles resting there and bringing fresh bed particles into direct contact with the surface. Figure 7(b) indicates that the contact time of these packets of emulsion particles with the surface is about 0.2–0.4 s for the conditions of the experiments reported there. More generally this contact time depends on the experimental conditions and the location of the heat exchange surface.

We now consider heat transfer to these packets of particles.

Large Particles for Short Contact Times. Here the particles are replaced before their mean temperature can change appreciably, the temperature gradient takes place only within the row of particles in direct contact with the exchanger surface, and we can ignore the thermal diffusion into the rest of the emulsion packet. Figure 16(a) shows this situation.

Glicksman and Decker [12] calculated the "heating" time constant of particles resting on a surface. They found that the temperature of particles larger than 1 mm did not change appreciably for a residence time as long as $\tau = 1$ s. Thus this extreme can be used for these large particles.

Experimental data on these large particle systems, shown in Fig. 5, can be correlated by

$$\frac{hd_p}{k_g} \Big/ (1-\delta) = 5.0 + 0.05 \Pr \mathrm{Re}_p \tag{7}$$

Note the similarity in form with the expression for fixed beds, Eq. (6).

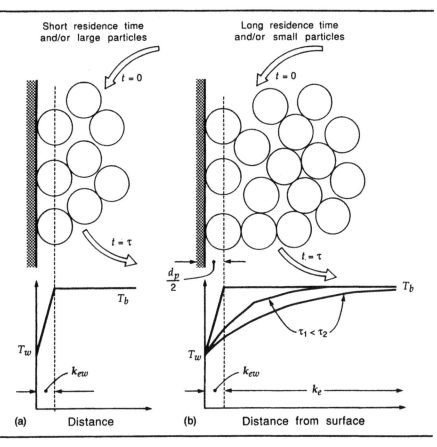

Short residence time
and/or large particles

Long residence time
and/or small particles

FIGURE 16

Models for heat transfer from an emulsion packet to a heat exchanger surface: (a) for large d_p and short contact time τ, the surface temperature is only felt by the first layer of particles; (b) for small d_p and long τ, the surface temperature is felt many layers into the packet.

Small Particles for Long Contact Times. Here the particles near the surface closely approach the surface temperature, the thermal transient is felt many layers from the surface, and, hence, thermal diffusion into the emulsion packet becomes the controlling resistance. This extreme is illustrated in Fig. 16(b).

Botterill and Williams [37] solved the unsteady state heat conduction problem for the first layer of particles at the surface and found that $d_p = 200$-μm particles approach the temperature of the surface in as little as 10 ms. In another estimate of this extreme, Glicksman and Decker [12] suggested that the temperature of particles contacting a surface changes substantially for particles smaller than 500 μm for a contact time of about 1 s.

To account for thermal diffusion through an emulsion packet, Mickley and Fairbanks [16] proposed the "renewal model." Here the instantaneous heat transfer coefficient at the surface of the packet, after the packet has rested on the surface for time t, is

$$h_{\substack{\text{packet at} \\ \text{time } t}} = \left(\frac{k_e^\circ \rho_s (1 - \varepsilon_{\mathrm{mf}}) C_{ps}}{\pi t} \right)^{1/2} \tag{8}$$

Note that this expression ignores the extra resistance right at the exchanger surface where the thermal conductivity and voidage differ from the corresponding values in the main body of the packet.

When all packets of emulsion contact the surface for the same length of time τ, the time-averaged heat transfer coefficient becomes

$$h_{\text{packet}} = \frac{1}{\tau} \int_0^\tau h_{\substack{\text{packet at} \\ \text{time } t}} \, dt = 1.13 \left[\frac{k_e^\circ \rho_s (1 - \varepsilon_{\text{mf}}) C_{ps}}{\tau} \right]^{1/2} \tag{9}$$

Since the fraction of time that the surface is bathed by bubbles is equal to the volume fraction of bubbles δ_w in the vicinity of the surface, we can show that the bubble frequency n_w at the surface is related to the emulsion contact time by the simple expression

$$\tau n_w = 1 - \delta_w \tag{10}$$

Combining Eqs. (9) and (10) gives

$$h_{\text{packet}} = 1.13 \left[\frac{k_e^\circ \rho_s (1 - \varepsilon_{\text{mf}}) C_{ps} n_w}{1 - \delta_w} \right]^{1/2} \tag{11}$$

Bubbling Beds—h at a Heat Exchanger Surface

We are ready to develop the general expression for the heat transfer coefficient between a fluidized bed and the exchanger surface. This expression should account for the fact that part of the time the surface is bathed by gas and part of the time by emulsion packets:

$$h = (h_{\text{bubble at surface}})\delta_w + (h_{\text{emulsion at surface}})(1 - \delta_w) \tag{12}$$

Now, *when a bubble is present* at the surface, there are two contributions to heat transfer: radiation and convection. With emissivities of bed solids and wall given by e_s and e_w, the radiation coefficient becomes

$$h_r = \frac{(5.67 \times 10^{-8})(T_s^4 - T_w^4)}{(1/e_s + 1/e_w - 1)(T_s - T_w)}, \qquad [\text{W/m}^2 \cdot \text{K}] \tag{13}$$

The gas convection contribution when a bubble contacts the surface is normally very small compared to the other contributions to heat transfer. However, for fast fluidized beds and in the freeboard above a dense bed, the fraction of solids is not small, so the convection term can be important. Thus we may write

$$h_{\substack{\text{bubble} \\ \text{present}}} = h_{\text{radiation}} + h_{\text{gas convection}} = h_r + h_g \tag{14}$$

When the emulsion packet is present on the surface, we have heat transfer in series—at the wall region of thickness $d_p/2$ followed by transfer at the emulsion packet. In addition, through the wall region we have both convection and radiation. These three terms sum to

$$h_{\substack{\text{emulsion} \\ \text{present}}} = \frac{1}{1/h_{\text{at wall}} + 1/h_{\text{packet}}} = \frac{1}{1/(h_r + h_w) + 1/h_{\text{packet}}} \tag{15}$$

Putting Eq. (6) into Eq. (15), and Eqs. (14) and (15) into Eq. (12) gives the general expression for heat transfer at a surface:

$$h = [\delta_w(h_r + h_g)]_{\substack{\text{bubble at} \\ \text{surface}}} + \left[\cfrac{1 - \delta_w}{\cfrac{1}{h_r + 2k_{ew}^\circ/d_p + \alpha_w C_{pg}\rho_g u_o} + \cfrac{1}{h_{\text{packet}}}}\right]_{\substack{\text{emulsion at} \\ \text{surface}}} \tag{16}$$

where h_{packet} is given by Eq. (11), h_r by Eq. (13), and k_{ew}° by Eq. (4).

For the Extreme of Fine Particles and High Temperature. Here radiation between emulsion packet and the surface can be ignored because the particles at the surface very quickly approach the surface temperature. Also, gas flow through the emulsion is negligible (small Re_p). Finally, since the wall temperature reaches many particle layers into the emulsion packet resting on the surface, the additional resistance of the first surface layer can be neglected. With these three simplifications, Eq. (16) reduces to

$$h = \delta_w h_r + (1 - \delta_w) h_{\text{packet}}$$

or

$$h = \delta_w h_r + 1.13[k_e^\circ \rho_s (1 - \varepsilon_{mf}) C_{ps} n_w (1 - \delta_w)]^{1/2} \tag{17}$$

For the Extreme of Fine Particles and Low Temperature. Here we ignore radiation, so Eq. (17) reduces to

$$h = 1.13[k_e^\circ \rho_s (1 - \varepsilon_{mf}) C_{ps} n_w (1 - \delta_w)]^{1/2} \tag{18}$$

Line 1 in Fig. 2 shows that this equation can account for the maximum of h at some intermediate gas velocity. In addition, line 1 in Fig. 8(a) shows that this equation can follow the variations in h when different gases are used.

For the Extreme of Large Particles. Heat transfer through the emulsion packet can be ignored because the temperature change occurs only in the first layer at the surface. In bubbling beds h_g can also be ignored. For this situation Eq. (16) reduces to

$$h = \delta_w h_r + (1 - \delta_w)\left(h_r + \frac{2k_{ew}^\circ}{d_p} + 0.05 C_{pg}\rho_g u_o\right)$$

or

$$h = h_r + (1 - \delta_w)\left(\frac{2k_{ew}^\circ}{d_p} + 0.05 C_{pg}\rho_g u_o\right) \tag{19}$$

Equation (16) can account for the effect of particle size on the heat transfer coefficient; see Fig. 8(b) and Example 3.

Alternative Theoretical Approaches. Many models have been proposed to explain the mechanism of heat transfer in fluidized beds [12, 13, 16–18, 34, 38–45]. Some are much too complicated to use for design calculations, and none are general enough to account for all the factors considered in Eq. (16).

h between Moving Beds and Heat Exchange Walls

For gently descending emulsion solids, the residence time of the emulsion in contact with the exchanger wall is very long, the temperature boundary layer extends many particle layers into the bed, no bubbles are present, and radiation can be neglected. In this situation Eq. (16) reduces to

$$h = \frac{1}{d_p/2k_{ew}^\circ + 1/h_{\text{packet}}} \tag{20}$$

where k_{ew}° is given by Eq. (4) and h_{packet} by Eq. (11).

Freeboard Region, Fast Fluidization, and Circulating Solid Systems

Chapters 7 and 8 mention that a thin layer of fine particles flows along the container walls. Also, when horizontal tubes are present in the freeboard, clusters of particles hit these tubes now and then. This behavior results in fairly high heat transfer rates, as shown in Figs. 6, 9, and 10.

Since the gas velocity is high in these systems, the gas-phase heat transfer coefficient may have to be considered (see discussion above Eq. (15)). Also, the exchanger surfaces are bathed by the lean phase most of the time; thus $\delta \cong 1$. With these conditions, Eq. (16) reduces to

$$h = h_r + h_g + (1 - \delta_w)h_{\text{packet}} \tag{21}$$

and with Eq. (11),

$$h = h_r + h_g + 1.13[k_e^\circ \rho_s (1 - \varepsilon_{mf}) C_{ps} n_w (1 - \delta_w)]^{1/2} \tag{22}$$

Here we assume that the rate at which clumps of emulsion solids hit the tubes is related to the upward flux of solids at that level in the bed, or

$$(1 - \delta_w)n_w \propto (\text{upward flux of solids, } G_{su}) \tag{23}$$

Recall from Chap. 7 that $G_{su} \cong \exp(-az_f)$; combining Eqs. (21)–(23) then gives

$$\frac{h - (h_r + h_g)}{h_{z_f=0} - (h_r + h_g)} \cong e^{-az_f/2} \tag{24}$$

Equation (24) tells us how h should change with height in the freeboard of a fluidized bed or in a fast fluidized bed.

Values of a can be estimated from Fig. 7.12 for the freeboard of turbulent beds and from Fig. 8.10 for fast fluidized beds. In Fig. 9 the values of h calculated with Eq. (24) for fast fluidized beds are compared with experimental data (see Example 4).

The mean heat transfer coefficient in a freeboard-exchanger region of height H_f is obtained from the expression

$$\bar{h} = \frac{1}{H_f} \int_0^{H_f} h \, dz \tag{25}$$

Putting Eq. (24) in Eq. (25) and integrating gives

$$\bar{h} = (h_r + h_g) + \frac{2(h_{z_f=0} - h_r - h_g)}{aH_f}(1 - e^{-aH_f/2}) \tag{26}$$

EXAMPLE 1

h on a Horizontal Tube Bank

Calculate h for a horizontal tube bank exchanger immersed in a fluidized bed of fine particles. Compare your results with the experimental data reported in Fig. 2.

Data

Solids: FCC catalyst, $d_p = 57\ \mu m$, $\rho_s = 940\ kg/m^3$, $C_{ps} = 828\ J/kg\cdot K$, $k_s = 0.20\ W/m\cdot K$

Gas: $k_g = 0.035\ W/m\cdot K$

Heat transfer surface: 90 horizontal tubes, 25.4 mm OD, triangular pitch on 50.8-mm centers

Bed: $u_{mf} = 6\ mm/s$, $\varepsilon_{mf} = 0.476$ (estimate)

The bubble frequency at various gas flow rates is estimated as follows:

u_0 (m/s)	0.05	0.1	0.2	0.35
n_w (s^{-1})	2	3.1	3.4	3.5

SOLUTION

We give the calculations for $u_0 = 0.2\ m/s$. Since this is a fine particle bed, we use Eq. (18) to calculate h. But first we evaluate δ and k_{ew}°.

Find δ. To evaluate δ we need u_b, which should be available from Chap. 6. But for the tube-filled beds of fine particles, no correlations are given there, so we extrapolate from related systems and say that bubbles are roughly 1–1.5 times the hydraulic diameter of the tube-filled bed and that their rise velocity is given by Eq. (6.8).

For the hydraulic diameter, Eq. (6.13) gives

$$d_{te} = \frac{4(0.0254\ m)(length)}{2(length)} = 0.0508\ m$$

Hence,

$$d_b = \left(\frac{1+1.5}{2}\right)(0.0508) = 0.0635\ m$$

Then with Eq. (6.7),

$$u_{br} = 0.711(9.8 \times 0.0635)^{1/2} = 0.56\ m/s$$

Since $u_b/u_e = (0.56)(0.476)/(0.006) = 44.5$, this certainly is a vigorously bubbling bed. Thus from Eq. (6.29),

$$\delta = \frac{u_0}{u_b} = \frac{0.2}{0.2 - 0.006 + 0.56} = 0.265$$

Find k_e°. To evaluate k_e°, we first need ϕ_b. For $k_s/k_g = 0.20/0.035 = 5.7$, Fig. 15 gives $\phi_b = 0.19$. Then Eq. (3) gives

$$k_e^\circ = 0.476(0.035) + \frac{(1-0.476)(0.20)}{(0.19)(5.7) + 2/3} = 0.0766\ W/m\cdot K$$

Inserting all into Eq. (18) and approximating the voidage at the tube walls by the bed voidage, or $\delta_w = \delta$, gives

$$h = 1.13[(0.0766)(940)(1 - 0.476)(828)(3.4)(1 - 0.265)]^{1/2} = \underline{315 \; W/m \cdot K}$$

This value is shown as point A in Fig. 2. Similar calculations at other velocities give

$$h = 271 \; W/m^2 \cdot K \qquad \text{at } u_o = \underline{0.05 \; m/s}$$
$$h = 290 \; W/m^2 \cdot K \qquad \text{at } u_o = \underline{3.5 \; m/s}$$

Line 1 on Fig. 2 shows how these equations compare with the reported data.

EXAMPLE 2

Effect of Gas Properties on h

Estimate the heat transfer coefficient h at close to room temperature conditions on a vertical exchanger surface immersed in a small, gently fluidized laboratory-sized fluidized bed of glass beads. Then show how the thermal conductivity of the gas affects the h value, and compare your results with the reported values displayed in Fig. 8(a).

Data

Solids: $d_p = 80 \; \mu m$, $\rho_s = 2550 \; kg/m^3$, $C_{ps} = 756 \; J/kg \cdot K$, $k_s = 1.21 \; W/m \cdot K$
Gas: $k_g = 0.005, \; 0.02, \; \text{and} \; 0.2 \; W/m \cdot K$ for different gases

SOLUTION

Start by considering a gas for which $k_g = 0.02 \; W/m \cdot K$. Then since this is a fine particle bed at not too high a temperature, we calculate h with Eq. (18). To evaluate h with this equation, we need values for δ, n_w, and k_e°. Since neither the bubble frequency nor quantities needed to evaluate δ (see Example 1) are given, we estimate them. First, for a gently bubbling fluidized bed take

$$\delta = 0.1 - 0.3 \quad \text{or, averaging,} \quad \delta = 0.2$$

From Fig. 5.12, at about 30 cm above the distributor,

$$n_w = 3 \; s^{-1}$$

Next k_e° for the clump of emulsion resting on the surface is found from Eq. (3). For the voidage of the clump take $\varepsilon_{mf} = 0.476$ and find ϕ_b from Fig. 15. With $k_s / k_g = 1.21/0.02 = 60.5$, we get $\phi_b = 0.10$. Putting these values in Eq. (3) gives

$$k_e^\circ = 0.476(0.02) + \frac{(1 - 0.476)(1.21)}{(0.10)(60.5) + 2/3} = 0.104 \; W/m \cdot K$$

Substituting all of the values into Eq. (18) gives

$$h = 1.13[(0.104)(2550)(1 - 0.476)(756)(3)(1 - 0.2)]^{1/2} = \underline{567 \; W/m^2 \cdot K}$$

This value is plotted as point A in Fig. 8(a). Using other values for k_g, say 0.005 and 0.2 W/m·K, allows us to draw line 1.

EXAMPLE 3

Effect of Particle Size on h

Calculate how h_{max} changes with particle size in bubbling fluidized beds. Compare the results of these calculations with the experimental findings reported in Fig. 8(b).

Data

Solids: $\rho_s = 2700 \; kg/m^3$, $\varepsilon_{mf} = 0.476$
          $C_{ps} = 755 \; J/kg \cdot K$, $k_s = 1.2 \; W/m \cdot K$
Gas: $k_g = 0.028 \; W/m \cdot K$

Assume the temperature is not too high, in which case $h_r \cong 0$, and ignore transfer when bubble gas contacts the surface, or $h_g \cong 0$. To see how the predicted h_{max} changes with particle size, start with $d_p = 10$ mm, for which Wunder and Mersmann [19] report $h_{max} = 250$ W/m²·K, and assume that $u_o \propto d_p^{1/2}$. Also assume that the intensity of bubbling is roughly constant, or $n_w = 5$ and $\delta_w = \delta = 0.1$, for all particle sizes.

SOLUTION

Since we plan to calculate h_{max} for the whole range of particle sizes from very small to large, we would be wise to use the general expression for h, Eq. (16). Let us evaluate its various terms.

$$\frac{k_s}{k_g} = \frac{1.2}{0.028} = 42.9$$

From Fig. 15 we then get

$$\phi_b = 0.11 \quad \text{and} \quad \phi_w = 0.17$$

From Eq. (3) the effective conductivity of an emulsion packet is

$$k_e^\circ = (0.476)(0.028) + \frac{(1 - 0.476)(1.2)}{(0.11)(42.9) + 2/3} = 0.130 \text{ W/m·K}$$

Substituting into Eq. (11) gives the heat transfer coefficient for the packet of emulsion:

$$h_{packet} = 1.13 \left[\frac{(0.13)(2700)(1 - 0.476)(755)(5.0)}{1 - 0.1} \right]^{1/2} = 992.5 \text{ W/m}^2\text{·K}$$

From Eq. (4),

$$k_{ew}^\circ = (0.476)(0.028) + \frac{(1 - 0.476)(1.2)}{(0.17)(42.9) + 1/3} = 0.0958 \text{ W/m·K}$$

Substituting all the above quantities into Eq. (16) gives

$$h = \frac{1 - 0.1}{\dfrac{1}{2(0.0958)/d_p + \alpha_w C_{pg} \rho_g u_o} + \dfrac{1}{992.5}}$$

Using the one known value, $h = 250$ at $d_p = 10^{-2}$ m, allows us to evaluate the unknown term in the last equation. This gives

$$\alpha_w C_{pg} \rho_g u_o = 366.6$$

We are now ready to calculate h for other particle sizes. For example, for $d_p = 2$ mm $= 2 \times 10^{-3}$ m,

$$h = \frac{0.9}{\dfrac{1}{2(0.0958)/(2 \times 10^{-3}) + 366.6[(2 \times 10^{-3})/10^{-2}]^{1/2}} + \dfrac{1}{992.5}} = \underline{\underline{185 \text{ W/m}^2\text{·K}}}$$

This is shown as point A in Fig. 8(b).

Similar calculations give the dashed curve in Fig. 8(b). Note that Eq. (16) correctly predicts the minimum in h_{max} at about $d_p = 2$ mm. On the contrary for very small particles ($d_p < 20$ μm), it does not predict the observed sharp fall off in h_{max}. However, this is where the system is entering the cohesive regime—Geldart **C** solids—with its very poor fluidization and low h values.

EXAMPLE 4

Freeboard

Heat Exchange

Estimate the point heat transfer coefficient at various levels z_f above the top of the dense bubbling fluidized bed and the mean heat transfer coefficient in the $H_f = 4$-m-high freeboard above a large-diameter, large particle, high-temperature fluidized bed. At the gas velocity to be used, $u_o = 2.4$ m/s, a considerable amount of solid is ejected into the freeboard. Compare your calculated point coefficient with the data reported in Fig. 9(a).

Data. At the bottom of the freeboard region,

$$h = 350 \text{ W/m}^2 \cdot \text{K} \quad \text{at } z_f = 0$$

In the equivalent gas stream, but free of solids,

$$h = 20 \text{ W/m}^2 \cdot \text{K}$$

SOLUTION

In a large-diameter fluidized bed, the decay coefficient that characterizes the decrease in density in the freeboard, from Fig. 7.12, is

$$au_o = 1.5 \text{ s}^{-1} \quad \text{or} \quad a = \frac{1.5}{2.4} = 0.625 \text{ m}^{-1}$$

With Eq. (24) the point coefficient at height z_f is

$$\frac{h - 20}{350 - 20} = e^{-(0.833/2)z_f} \tag{i}$$

This gives

$$h = 350 \text{ W/m}^2 \cdot \text{K} \quad \text{at } z_f = 0$$

and

$$h = 114 \text{ W/m}^2 \cdot \text{K} \quad \text{at } z_f = 4 \text{ m}$$

Equation (i) is plotted in Fig. 9(a) and compares favorably with the reported data there. The mean heat transfer coefficient for the 4-m-high freeboard exchanger is then obtained from Eq. (26). This gives

$$\bar{h} = 20 + \frac{2(350 - 20)}{(0.625)(4)} (1 - e^{[0.625/2]4}) = \underline{208 \text{ W/m}^2 \cdot \text{K}}$$

PROBLEMS

1. Calculate h on a horizontal tube bundle with wall temperature $T = 80°C$ immersed in a $110°C$ bed of 82-μm sand and fluidized by air at various velocities. Compare your results with the data and line 2 of Fig. 2.

   Data

Solids:	$\rho_s = 2700 \text{ kg/m}^3$, $C_{ps} = 756 \text{ J/kg} \cdot \text{K}$, $k_s = 1.2 \text{ W/m} \cdot \text{K}$
Air:	at $110°C$, $\mathbf{k}_g = 0.033 \text{ W/m} \cdot \text{K}$
	at $180°C$, $\mathbf{k}_g = 0.038 \text{ W/m} \cdot \text{K}$
Emissivities:	$e_w = 0.8$, $e_s = 0.9$, $T_w = 80°C$

u_o (m/s)	0.05	0.2	0.4
δ (–)		0.059	0.214	0.38
n_w (s^{-1}) at 110°C	0.50	1.4	2.4
n_w (s^{-1}) at 180°C	0.83	1.9	3.0

2. Repeat Prob. 1 with the bed at 180°C and compare your results with the data and with line 3 of Fig. 2.

3. A cold heat exchanger tube (120°C) is immersed in a hot fluidized bed (600°C) of fine particles. Estimate h at a point on this tube where the bubble frequency is estimated to be $n_w = 2\,s^{-1}$.

Data

Solids: $\rho_s = 2700\ \text{kg/m}^3$, $C_{ps} = 756\ \text{J/kg·K}$, $\mathbf{k}_s = 1.2\ \text{W/m·K}$
Gas: $\mathbf{k}_g = 0.063\ \text{W/m·K}$, $u_o = 0.4\ \text{m/s}$, $u_b = 1.2\ \text{m/s}$
Emissivities: $e_w = 0.8$, $e_s = 0.9$

4. Calculate the radiant heat transfer coefficient at a heat exchanger surface in a bed of coarse particles. Compare your calculation with the data reported in Fig. 11(a).

Data

Solids: $d_p = 1030\ \mu\text{m}$, $\rho_s = 2700\ \text{kg/m}^3$, $C_{ps} = 756\ \text{J/kg·K}$
Temperature of bed: 300°C, 550°C, 800°C
Temperature of wall: 150°C
Emissivities: $e_s = 0.9$, $e_w = 0.85$

5. Estimate h on a heat exchanger tube located 0.7 m above the surface of a large, dense, vigorously bubbling bed of fine particles.

Data

In the dense bed: $h = 400\ \text{W/m}^2\text{·K}$
$u_o = 0.6\ \text{m/s}$, which is many multiples of u_{mf}
In the freeboard: $au_o = 2s^{-1}$
Far up in the freeboard where no solids are present:

$$h_r + h_g = 40\ \text{W/m}^2\text{·K}$$

REFERENCES

1. N.I. Gelperin and V.G. Einstein, in *Fluidization*, J.F. Davidson and D. Harrison, eds., p. 471, Academic Press, New York, 1971.

2. J.S.M. Botterill, *Fluid-Bed Heat Transfer*, Academic Press, New York, 1975; A.E. Denloye and J.S.M. Botterill, *Powder Technol.*, **19**, 197 (1977).

3. A.M. Xavier and J.F. Davidson, in *Fluidization*, 2nd ed., J.F. Davidson et al., eds., p. 437, Academic Press, New York, 1984.

4. A.P. Baskakov, in *Fluidization*, 2nd ed., J.F. Davidson et al., eds., p. 465, Academic Press, New York, 1984.

5. H.J. Bock and O. Molerus, in *Fluidization*, J.R. Grace and J.M. Matsen, eds., p. 217, Plenum, New York, 1980; *German Chem. Eng.*, **6**, 57 (1983).

6. H. Seko, S. Tone, and T. Otake, in *Fluidization IV*, D. Kunii and R. Toei, eds., p. 331, Engineering Foundation, New York, 1983.

7. H.W. Piepers, P. Wiewiorski, and K. Rietema, in

Fluidization IV, D. Kunii and R. Toei, eds., p. 339, Engineering Foundation, New York, 1983.

8. C. Beeby and O.E. Potter, *AIChE J.*, **30**, 977 (1984).

9. G.M. Carlomagno, R. Festa, and L. Massimilla, *AIChE Symp. Ser.*, **79(222)**, 66 (1983).

10. H.J. Bock and J. Schweinzer, *German Chem. Eng.* **1**, 16 (1986).

11. F.W. Staub and G.S. Canada, in *Fluidization*, J.F. Davidson and D.L. Keairns, eds., p. 339, Cambridge Univ. Press, New York, 1978.

12. L.R. Glicksman and N. Decker, "Heat Transfer in Fluidized Beds of Large Particles," report from Mech. Eng. Dept., M.I.T., Cambridge MA, 1983.

13. A.P. Baskakov et al., *Powder Technol.*, **8**, 273 (1973); in *Fluidization and Its Applications*, p. 293, Cepadues, Toulouse, 1974.

14. A. Kortleven et al., XVI ICHMT Symp., Dubrovnik, paper no. 3-3, 1984.

15. N.S. Grewal et al., *Chem. Eng. Comm.*, **39**, 43 (1985); **60**, 311 (1987).

16. H.S. Mickley and C.A. Fairbanks, *AIChE J.*, **1**, 374 (1955); H.S. Mickley, D.F. Fairbanks, and R.D. Hawthorn, *Chem. Eng. Prog. Symp. Ser.*, **57(32)**, 51 (1961).

17. H. Martin, Chem. Eng. Comm., **13**, 1 (1981); XVI ICHMT Symp., Dubrovnik, paper no. 2-5, 1984; *Chem. Eng. Process.*, **18**, 157, 199 (1984).

18. E. Wicke and F. Fetting, *Chem.-Ing.-Techn.*, **26**, 30 (1954).

19. R. Wunder and A. Mersmann, *Chem.-Ing.-Techn.*, **51**, 241 (1979).

20. P. Guigon et al., in *Proc. Second Int. Conf. Circulating Fluidized Beds*, Compiegne, p. 65, 1988.

21. J.C.L. Furchi et al., in *Proc. Second Int. Conf. Circulating Fluidized Beds*, Compiegne, p. 69, 1988.

22. R.L. Wu et al., *AIChE J.*, **33**, 1888 (1987).

23. L.D. Fraley et al., paper no. 83-HT-92, ASME Meeting, Seattle, 1983.

24. K.D. Kiang et al., in *Fluidization Technology*, vol. 2, D.L. Keairns, ed., p. 471, Hemisphere, Washington, D.C., 1975.

25. P. Basu and F. Konuche, in *Proc. Second Int. Conf. Circulating Fluidized Beds*, Compiegne, p. 73, 1988.

26. H. Kobro and C. Brereton, in *Circulating Fluidized Bed Technology*, P. Basu, ed., p. 263, Pergamon, New York, 1985.

27. T.F. Ozkaynak, J.C. Chen, and T.R. Frankenfield, in *Fluidization IV*, D. Kunii and R. Toei, eds., p. 371, Engineering Foundation, New York, 1983.

28. A. Mathur and S.C. Saxena, *AIChE J.*, **33**, 1124 (1987).

29. J.J. Lindsay, W. Morton, and D.C. Newey, in *Fluidization V*, K. Østergaard and A. Sørensen, eds., p. 385, Engineering Foundation, New York, 1986.

30. A.R. Abrahamsen and D. Geldart, *Powder Technol.* **26**, 57 (1980).

31. F. Floris and L.R. Glicksman, XVI ICHMT Symp., Dubrovnik, paper no. 2-2, 1984.

32. M. Colakyan and O. Levenspiel, *AIChE Symp. Ser.*, **80(241)**, 156 (1984).

33. G.T. Zhang and F. Ouyang, *I.E.C. Proc. Design Dev.*, **24**, 430 (1985).

34. A.M. Xavier and J.F. Davidson, in *Fluidization*, J.F. Davidson and D.L. Keairns, eds., p. 333, Cambridge Univ. Press, 1978.

35. D. Kunii and J.M. Smith, *AIChE J.*, **6**, 71 (1960).

36. S. Yagi and D. Kunii, *AIChE J.*, **6**, 97 (1960); M. Suzuki and D. Kunii, *J. Facul. Eng.*, Univ. of Tokyo, (**B**) **xxx(1)**, 1 (1969).

37. J.S.M. Botterill and J.R. Williams, *Trans. Inst. Chem. Eng.*, **41**, 217 (1963); J.S.M. Botterill et al., in *Proc. Int. Symp. on Fluidization*, A.A.H. Drinkenburg, ed., p. 442, Netherlands Univ. Press, Amsterdam, 1967.

38. W.R.A. Goosens and L. Hellinckx, in *Fluidization and Its Applications*, p. 303, Cepadues, Toulouse, 1974.

39. N.M. Čatipovič et al., in *Fluidization*, J.R. Grace and J.M. Matsen, eds., p. 225, Plenum, New York, 1980.

40. A.M. Xavier et al., in *Fluidization*, J.R. Grace and J.M. Matsen, eds., p. 209, Plenum, New York, 1980.

41. O. Levenspiel and J.S. Walton, *Chem. Eng. Prog. Symp. Ser.*, **50(9)**, 1 (1954).

42. K. Yoshida et al., *Chem. Eng. Sci.*, **29**, 77 (1974).

43. H.J. Bock and O. Molerus, *German Chem. Eng.*, **6**, 57 (1983); H.J. Bock et al., *German Chem. Eng.*, **6**, 301 (1983).

44. R. Chandran and J.C. Chen, *AIChE J.*, **31**, 244 (1985).

45. Y. Filtris et al., *Chem. Eng. Comm.*, **72**, 189 (1988).

14

The RTD and Size Distribution of Solids in Fluidized Beds

— Particles of Unchanging Size
— Particles of Changing Size

In the continuous treatment of solids, fresh particles are fed to a bed and are removed by an overflow pipe or are entrained by the gases. In the bed they may be transformed into a different material (e.g., fresh catalyst into deactivated form, ZnS to ZnO), grow (e.g., by deposition from a sprayed evaporating solution, by chemical deposition of silicon from silane gas), or shrink (e.g., by breakup and attrition of friable solid, by sublimation, by combustion of carbon, by chlorination of ilmenite). Individual particles of the same size also have different lengths of stay in the bed. In addition, elutriation, growth, shrinkage, and reaction act differently for the different-sized particles in the bed. All these effects must be accounted for if we wish to control and predict the behavior of a bed that processes solid.

In this chapter we first treat particles of unchanging size. We then treat the more general case of growing and shrinking particles. Except for a few simple cases, numerical methods are needed to find the residence time and size distribution in the bed; however, the procedure is straightforward and follows the same strategy throughout. Our development is limited to steady state operations.

Particles of Unchanging Size

Feed of One Size, Single and Multistage Beds

For a given bed weight and flow rate of solids of unchanging density, as shown in Fig. 1, a material balance gives

$$F_0 = F_1 + F_2 \tag{1}$$

and the mean residence time of solids in a single fluidized bed becomes

$$\bar{t} = \frac{W}{F_0} = \frac{W}{F_1 + F_2} \tag{2}$$

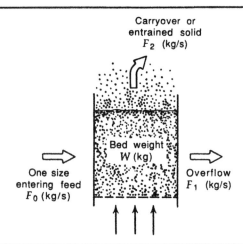

FIGURE 1
Flows in a single-stage fluidized bed with a single-size feed.

Now, with complete mixing of solids in the bed, the equation for backmix flow holds, and the residence time distribution (RTD) of solids in the bed becomes (see [1–5])

$$\mathbf{E}(t) = \frac{1}{\bar{t}}\,e^{-t/\bar{t}} \tag{3}$$

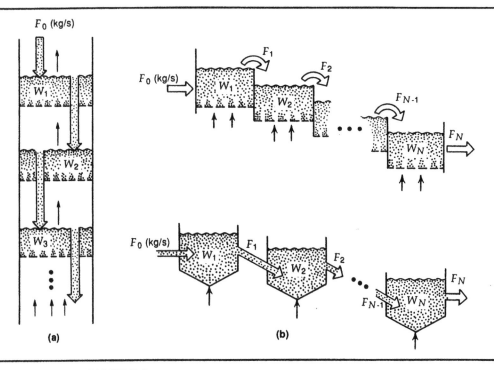

FIGURE 2
Multistage fluidized bed operations, counter- and crossflow operations.

where $\mathbf{E}(t)\,dt$ is the fraction of solids staying in the bed for a time between t and $t + dt$. Entrainment does not change these equations.

The wide RTD of a single bed gives a nonuniform solid product and is inefficient for high conversion of solids. In particular, where the rigorous control of residence time is required for each particle (e.g., in the activation of charcoal), the single fluidized bed should not be used. However, the RTD can be narrowed and greatly improved by multistage operations, such as shown in Fig. 2. For N equal-sized beds in series, with complete mixing and no entrainment from the beds, we have, for each bed,

$$\bar{t}_i = \frac{W_i}{F_0} \tag{4}$$

and for the beds as a whole, the RTD of solids is

$$\mathbf{E}(t) = \frac{1}{(N-1)!\,\bar{t}_i} \left(\frac{t}{\bar{t}_i}\right)^{N-1} \exp\left(-\frac{t}{\bar{t}_i}\right) \tag{5}$$

The RTD curves for solids in single and multistage beds are shown in Fig. 3. Here the line corresponding to $N = 1$ represents Eq. (3). With larger values of N, the RTD of solids approaches that of plug flow.

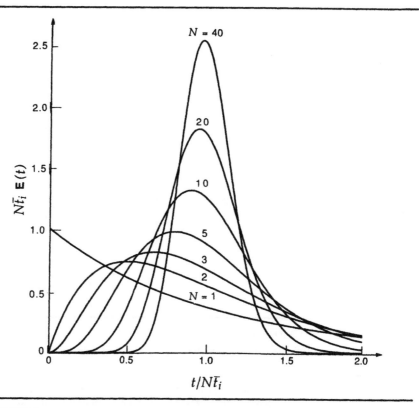

FIGURE 3

Exit age distribution (RTD) for solids in ideal multistaged fluidized beds, from Eq. (5).

Feed of Wide Size Distribution

When particles of wide size distribution $\mathbf{p}_0(R)$* are fed continuously into a bed as shown in Fig. 4, fine particles are likely to be entrained by the gas stream while the remainder of the solids is discharged through an overflow pipe.

Normally, knowing the bed weight W, the feed properties F_0, $\mathbf{p}_0(R)$, and elutriation constant $\kappa(R)$ for all sizes of solids, we want to find the composition $\mathbf{p}_b(R)$ and what leaves the vessel: F_1, $\mathbf{p}_1(R)$, F_2, $\mathbf{p}_2(R)$. The function $\mathbf{p}(R)$, introduced in Chap. 3, describes the size distribution of solids, and $\kappa(R)$ is found by the correlations of Chap. 7 and Eq. (7.8).

We first assume backmix flow of solids in the bed, in which case the size distribution in the overflow stream is that of the bed, or

$$\mathbf{p}_1(R) = \mathbf{p}_b(R) \tag{6}$$

From Chap. 7, the elutriation constant is a function of particle size and, for any

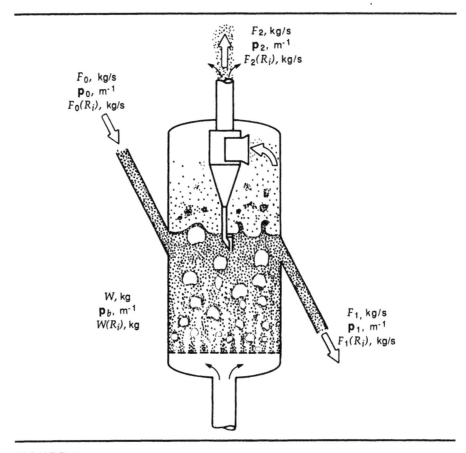

FIGURE 4
The fluidized bed operating with a wide size distribution of solids.

*For convenience in the derivations of this chapter, we use the radius R rather than the diameter d_p to characterize particle sizes. The illustrative examples use d_p. Note that

$$\mathbf{p}(R)\, dR = \mathbf{p}(d_p)\, d(d_p)$$

size R in the bed, is

$$\kappa(R) = \frac{F_2 \mathbf{p}_2(R)}{W \mathbf{p}_b(R)} \tag{7}$$

Now an overall mass balance for all of the flowing solid streams gives

$$F_0 = F_1 + F_2 \tag{8}$$

and for any size interval between R and $R + dR$ a mass balance gives

$$F_0 \mathbf{p}_0(R) \, dR = F_1 \mathbf{p}_1(R) \, dR + F_2 \mathbf{p}_2(R) \, dR \tag{9}$$

For the solids as a whole, the mass average residence time is

$$\bar{t} = \frac{W}{F_0} = \frac{W}{F_1 + F_2} \tag{10}$$

Since entrainment preferentially removes the small particles from the bed, the fines on the average spend a shorter time in the bed. Thus, the average length of stay of particles varies with size, and for any particular size it is

$$\bar{t}(R) = \frac{\text{weight of a particular size in the bed}}{\text{flow rate of that size into the bed}}$$

$$= \frac{W \mathbf{p}_b(R)}{F_0 \mathbf{p}_0(R)} = \frac{W \mathbf{p}_b(R)}{F_1 \mathbf{p}_1(R) + F_2 \mathbf{p}_2(R)} \tag{11}$$

Combining Eq. (11) with Eqs. (6) and (7) gives

$$\bar{t}(R) = \frac{1}{F_1/W + \kappa(R)} \tag{12}$$

This expression shows that the fines have a smaller $\bar{t}(R)$ than the coarse, the reason being elutriation [larger $\kappa(R)$ values for the fines]. In the extreme, for fine solids where $\kappa(R) \gg F_1/W$,

$$\bar{t}_{\text{fines}} = \frac{1}{\kappa(R)} \tag{13}$$

At the other extreme, for the coarse fractions that are not entrained, $\kappa(R) = 0$, in which case

$$\bar{t}_{\text{coarse}} = \frac{W}{F_1} \tag{14}$$

Now, using these $\bar{t}(R)$ values and recalling that each size is in backmix flow in the bed, we obtain the RTD for each size of solid in the bed:

$$\mathbf{E}(t, R) = \frac{1}{\bar{t}(R)} \, e^{-t/\bar{t}(R)} \tag{15}$$

To find the composition of the outflow streams, combine Eqs. (6), (7), and (9) and rearrange to give

$$F_1 \mathbf{p}_1(R) = \frac{F_0 \mathbf{p}_0(R)}{1 + (W/F_1)\kappa(R)} \tag{16}$$

Noting that the area under any probability distribution curve is unity,

$$\int_{0 \text{ or } R_m}^{R_M \text{ or } \infty} \mathbf{p}(R)\, dR = 1\,, \qquad \begin{array}{l} R_M = \text{largest size in feed} \\ R_m = \text{smallest size in feed} \end{array}$$

we find, on integrating and rearranging Eq. (16), that

$$\frac{F_1}{F_0} = 1 - \frac{F_2}{F_0} = \int_{R_m}^{R_M} \frac{\mathbf{p}_0(R)\, dR}{1 + (W/F_1)\kappa(R)} \tag{17}$$

The flow rate and composition of the two outflow streams for a given feed, bed weight, and $\kappa(R)$ are found as follows:

1. Guess and adjust F_1 until Eq. (17) is satisfied. Since F_1 is the only unknown under the integral, the procedure is straightforward.
2. Substitute the value found for F_1 in Eq. (16) to find $\mathbf{p}_1(R)$ for the different R values.
3. Substitute the value of $\mathbf{p}_1(R)$ in Eq. (7) to obtain $\mathbf{p}_2(R)$.

Simple Extensions

If we have a cyclone with a collection efficiency $\eta(R)$ for particles of size R, then we simply replace $\kappa(R)$ by $\kappa(R)[1 - \eta(R)]$ in the above equations and elsewhere in this chapter.

If some degree of segregation of sizes occurs in the bed, then the upper portion of the bed should have more fines, whereas the lower should have more coarse. This should also be reflected in the composition of the overflow stream. We can account for this by replacing Eq. (6) by

$$\mathbf{p}_1(R) = \psi(R)\mathbf{p}_b(R) \tag{18}$$

where

$$\begin{array}{l} \psi_{\text{fines}} > 1 \\ \psi_{\text{coarse}} < 1 \end{array} \qquad \text{for discharge from the top of the bed}$$

and where

$$\begin{array}{l} \psi_{\text{fines}} < 1 \\ \psi_{\text{coarse}} > 1 \end{array} \qquad \text{for discharge from the bottom of the bed}$$

If the solid changes in density, but not in size, during its stay in the bed (e.g., chemical change or volatilization) to β times the feed density, irrespective of size, then we simply replace F_0 by βF_0 in the above equations.

With all these extensions Eqs. (1)–(17) must be appropriately modified. This is easily done. For example, Eq. (12) becomes

$$\bar{t}(R) = \frac{1}{F_1 \psi(R)/W + \kappa(R)[1 - \eta(R)]} \tag{19}$$

and Eq. (17) becomes

$$\frac{F_1}{\beta F_0} = 1 - \frac{F_2}{\beta F_0} = \int_{R_m}^{R_M} \frac{\mathbf{p}_0(R)\, dR}{1 + \kappa(R)[1 - \eta(R)][W/(F_1 \psi(R))]} \tag{20}$$

EXAMPLE 1

Flow with

Elutriation

A feed consisting of two sizes of solids is to be treated in a pilot-plant fluidized bed reactor. The solids remain unchanged in size and density, and elutriation occurs. Find the composition and flow rates of the entrained and overflow streams as well as the mean residence time in the bed for the two sizes of solids.

Data

$$\text{Feed:} \quad F_0 = 2.7, \ F_{0,\text{fines}} = 0.9, \ F_{0,\text{coarse}} = 1.8 \ \text{kg/min}$$
$$\text{Bed weight:} \quad W = 17 \ \text{kg,}$$
$$\text{Elutriation:} \quad \kappa_{\text{fines}} = 0.8, \ \kappa_{\text{coarse}} = 0.0125/\text{min}$$

SOLUTION

This problem illustrates the use of the discrete equivalent of the continuous size distribution equations developed here (see Chap. 3 for more on discrete distributions). By following the suggested procedure, we have for Eq. (17),

$$F_1 = \sum_{i=1}^{2} \frac{F_0(R_i)}{1 + (W/F_1)\kappa(R_i)} = \frac{0.9}{1 + (17/F_1)0.8} + \frac{1.8}{1 + (17/F_1)0.0125}$$

By trial and error, we find that the value of F_1 that satisfies this equation is

$$F_1 = 1.7 \ \text{kg/min}$$

Putting this value in Eq. (16) gives, for the fines,

$$F_{1,\text{fines}} = \frac{F_{0,\text{fines}}}{1 + (W/F_1)\kappa_{\text{fines}}} = \frac{0.9}{1 + (17/1.7)(0.8)} = \underline{\underline{0.1 \ \text{kg/min}}}$$

and, similarly,

$$F_{1,\text{coarse}} = \underline{\underline{1.6 \ \text{kg/min}}}$$

Substituting in Eq. (9) gives

$$F_{0,\text{fines}} = F_{1,\text{fines}} + F_{2,\text{fines}}$$

or

$$F_{2,\text{fines}} = 0.9 - 0.1 = \underline{\underline{0.8 \ \text{kg/min}}}$$

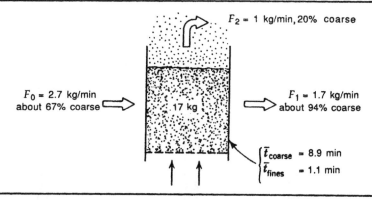

FIGURE E1
Display of the answer to Example 1.

and, similarly,

$$F_{2,\text{coarse}} = \ = 0.2 \text{ kg/min}$$

The mean residence times of the two sizes of solid are then found by substituting into Eq. (12):

$$\bar{t}_{\text{fines}} = \frac{1}{1.7/17 + 0.8} = \underline{1.1 \text{ min}}$$

$$\bar{t}_{\text{coarse}} = \frac{1}{1.7/17 + 0.0125} = \underline{8.9 \text{ min}}$$

Figure E1 shows the various quantities found for this bed.

EXAMPLE 2

Flow with Elutriation and Change in Density of Solids

Solids of wide size distribution are fed continuously to a reactor. During their stay, the density of the solids increases by 20%, but the particle size does not change. For a given feed and bed size, determine the flow rate and size distribution of the two outflow streams from the vessel and the mean residence time for each size of solid in the bed. Entrained solids are not returned to the reactor; hence $\eta(d_p) = 0$.

Data

Diameter of reactor:	$d_t = 4$ m,
Void fraction of static bed:	$\varepsilon_m = 0.40$,
Density of solid in the bed:	$\rho_s = 2500 \text{ kg/m}^3$
Height of static bed:	$L_m = 1.2$ m
Feed rate:	$F_0 = 3000 \text{ kg/hr}$

TABLE E2   Size Distribution and Elutriation Data for Example 2

Size, $d_p \times 100$ (mm) (given)	$p_0(d_p)$ (mm^{-1}) (given)	$\kappa(d_p) \times 10^4$ (s^{-1}) (given)	$\bar{t}(d_p)$ (hr) (calculated)
3	0	—	—
4	0.3	10	0.28
5	0.8	9.75	0.28
6	1.3	9.5	0.29
7	1.9	8.75	0.32
8	2.6	7.5	0.37
9	3.5	6.0	0.46
10	4.4	4.38	0.70
11	5.7	2.62	1.0
12	6.7	1.20	1.5
13	7.5	0.325	3.0
14	7.8	0	9.0
16	7.5	0	9.0
18	6.3	0	9.0
20	5.0	0	9.0
22	3.6	0	9.0
24	2.4	0	9.0
26	1.3	0	9.0
28	0.5	0	9.0
30	0	0	9.0

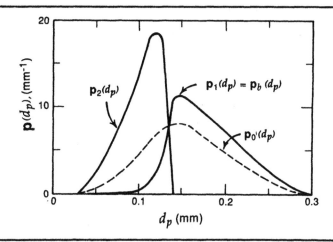

FIGURE E2
Display of the answer to Example 2.

The size distribution of feed solids $p_0(d_p)$ and the elutriation constant $\kappa(d_p)$ are given in Table E2. Take $\psi(d_p) = 1$.

SOLUTION
Use particle diameter rather than radius as the size characteristic. Then the weight of solids in the bed is

$$W = A_t L_m (1 - \varepsilon_m)\rho_s = \frac{\pi}{4} (4)^2 (1.2)(1 - 0.4)(2500) = 22{,}600 \text{ kg}$$

Substituting known values, Eq. (20) becomes

$$\frac{F_1}{1.2(3000)} = \int_0^{0.3 \text{ mm}} \frac{p_0(d_p)\, d(d_p)}{1 + (22{,}600/F_1)\kappa(d_p)}$$

Guessing $F_1 = 2700$, 2600, and 2500 kg/hr, evaluating the integral for each guess, and graphically interpolating gives the flow rates of leaving streams to be

$$F_1 = \underline{\underline{2550 \text{ kg/hr}}} \qquad \text{and} \qquad F_2 = \underline{\underline{1050 \text{ kg/hr}}}$$

With βF_0 in place of F_0, substituting in Eqs. (16), (7), and (11) for various selected diameters gives $p_1(d_p) = p_b(d_p)$, $p_2(d_p)$, and $\bar{t}(d_p)$ at these values. These results are presented in Fig. E2 and Table E2.

Particles of Changing Size

Consider the class of operations where solid particles shrink and lose mass because of reaction, incineration, attrition, or sublimation, or else grow larger and increase in mass because of deposition of material from the fluid, and where the density of deposited materials is that of the particles themselves.

Rate Expressions for Growing and Shrinking Particles

The following are examples of rate expressions used to describe the change in particle size. For linear particle growth by deposition or condensation of material,

$$\mathscr{R}(R) = \frac{dR}{dt} = k \tag{21}$$

For linear particle shrinkage by dissolution, sublimation, reaction, or slow attrition,

$$\mathscr{R}(R) = \frac{dR}{dt} = -k \tag{22}$$

In these expressions k is independent of particle size, but does depend on bed conditions. These kinetics are relatively easy to handle and closely represent most growth and shrinkage phenomena.

Other simple expressions, sometimes applicable to shrinkage by attrition and abrasion, are

$$\mathscr{R}(R) = \frac{dR}{dt} = -k'R \tag{23}$$

or

$$\mathscr{R}(R) = \frac{dR}{dt} = -k'(R - R_{min}) \tag{24}$$

or for growth to a maximum size

$$\mathscr{R}(R) = \frac{dR}{dt} = k'(R_{max} - R) \tag{25}$$

Whatever the form of $\mathscr{R}(R)$, it is used in the material balance expressions to find the flows and size distribution of streams entering and leaving fluidized beds.

General Performance Equations

Here we present the general material balance equations and their various final integrated forms. The derivations of these equations are given elsewhere [6].

Assume steady state operations, an entering stream of solids F_0 (kg/s), an outflow stream F_1 (kg/s), an elutriating outflow stream F_2 (kg/s), spherical particles of constant density ρ_s, backmix flow in the bed (hence $\mathbf{p}_b(R) = \mathbf{p}_1(R)$), and particle growth or shrinkage given by a general rate expression $\mathscr{R}(R)$. Then a mass balance on particles of size between R and $R + dR$ gives

$$\begin{pmatrix} \text{solids} \\ \text{entering in} \\ \text{feed} \end{pmatrix} - \begin{pmatrix} \text{solids} \\ \text{leaving in} \\ \text{overflow} \end{pmatrix} - \begin{pmatrix} \text{solids} \\ \text{leaving in} \\ \text{carryover} \end{pmatrix} + \left[\begin{pmatrix} \text{solids growing into} \\ \text{the interval from} \\ \text{a smaller size} \end{pmatrix} \right.$$

$$\left. - \begin{pmatrix} \text{solids growing out} \\ \text{of the interval to} \\ \text{a larger size} \end{pmatrix} \right] + \begin{pmatrix} \text{solid generation} \\ \text{due to growth} \\ \text{within interval} \end{pmatrix} = 0$$

or, in symbols,

$$F_0\mathbf{p_0}(R) - F_1\mathbf{p}_1(R) - W\kappa(R)\mathbf{p}_1(R) - W\frac{d[\mathscr{R}(R) \cdot \mathbf{p}_1(R)]}{dR} + \frac{3W}{R}\mathbf{p}_1(R)\mathscr{R}(R) = 0, \quad \text{[kg/s]} \tag{26}$$

An overall mass balance represents the rate of generation or disappearance of all solids in the bed and is given by

$$F_1 + F_2 - F_0 = \int_{\text{all } R} \frac{3W\mathbf{p}_b(R)\mathcal{R}(R)\,dR}{R} \begin{cases} >0, & \text{for growth} \\ <0, & \text{for shrinkage} \end{cases} \quad (27)$$

For a feed of wide size distribution, integration of Eq. (26) relates the flows and size distributions of the flowing streams. Thus, for growing particles the flows F_0 and F_1 are related by

$$\frac{W}{F_0} = \int_{R=R_m}^{R_t \to \infty} \frac{R^3}{|\mathcal{R}(R)|} I(R, R_m) \int_{R_i=R_m}^{R_i=R} \frac{\mathbf{p}_0(R_i)_i\,dR_i}{R_i^3 I(R_i, R_m)}\,dR \quad (28)$$

and, for shrinking particles, by

$$\frac{W}{F_0} = \int_{R_t \to \infty}^{R_M} \frac{R^3}{|\mathcal{R}(R)|} I(R, R_M) \int_{R_i=R}^{R_i=R_M} \frac{\mathbf{p}_0(R_i)\,dR_i}{R_i^3 I(R_i, R_M)}\,dR \quad (29)$$

For growing particles, the size distributions of streams 0 and 1 are related by

$$\mathbf{p}_1(R) = \frac{F_0 R^3}{W|\mathcal{R}(R)|} I(R, R_m) \int_{R_i=R_m}^{R_i=R} \frac{\mathbf{p}_0(R_i)\,dR_i}{R_i^3 I(R_i, R_m)} \quad (30)$$

and, for shrinking particles, by

$$\mathbf{p}_1(R) = \frac{F_0 R^3}{W|\mathcal{R}(R)|} I(R, R_M) \int_{R_i=R}^{R_i=R_M} \frac{\mathbf{p}_0(R_i)\,dR_i}{R_i^3 I(R_i, R_M)} \quad (31)$$

Finally, the elutriating stream 2 is related to the bed or exiting stream 1 by Eqs. (7) and (27).

In the above expressions R_m represents the smallest feed size for particle growth, R_M represents the largest feed size for particle shrinkage, and

$$I(R, R_i) = \exp\left[-\int_{R_i}^{R} \frac{F_1/W + \kappa(R)}{\mathcal{R}(R)}\,dR \right] \quad (32)$$

These equations were developed for spherical particles. For cylindrical particles, such as fibers, simply replace the 3 in Eqs. (26) and (27) by 2, for flat plate particles replace 3 by 1. When $\mathbf{p}_0(R)$, $\kappa(R)$, and $\eta(R)$ are given as a smooth function of particle size, these equations can then be used to give the size distribution and other bed properties.

Overturf and Kayihan [7] warn that taking the discrete analog (size slices) of Eqs. (27)–(32) may lead to very large errors unless many size slices are taken, in some cases as many as 3000. We give the numerical procedure recommended by Overturf and Kayihan [7], which discretizes directly the basic differential form of the material balances, Eqs. (26) and (27).

Calculation Procedure for the General Situation

To write the governing differential equation in discrete form, consider a size interval ΔR_i with mean size R_i. For these solids we write

$$F(R_i) = \text{flow rate (kg/s or m}^3/\text{s)}$$

$$W(R_i) = \text{amount in the bed (kg or m}^3)$$

$$\mathcal{R}(R_i) = \text{growth rate (m/s)}$$

$$\kappa(R_i) = \text{elutriation rate (s}^{-1})$$

General Shrinkage Kinetics. Rearranging Eqs. (26) and (27) in discrete form and noting that $\mathcal{R}(R_i) < 0$ gives

$$\frac{W(R_i)}{W} = \frac{F_1(R_i)}{F_1} = \frac{F_0(R_i) - W \dfrac{F_1(R_{i+1})}{F_1} \dfrac{\mathcal{R}(R_{i+1})}{\Delta R_i}}{F_1 + W\kappa(R_i) - W \dfrac{\mathcal{R}(R_i)}{\Delta R_i} - 3W \dfrac{\mathcal{R}(R_i)}{R_i}} \tag{33}$$

$$\sum_i \frac{F_1(R_i)}{F_1} = 1 \tag{34}$$

$$F_1 + F_2 - F_0 = 3W \sum_i \frac{F_1(R_i)\mathcal{R}(R_i)}{F_1 R_i} \tag{35}$$

and

$$F_2(R_i) = W(R_i)\kappa(R_i) = W \frac{F_1(R_i)}{F_1} \kappa(R_i) \tag{36}$$

The recommended calculation procedure for shrinking particles is as follows:

1. Choose reasonable size intervals ΔR_i, either all equal or based on the screen sizes being used. Then find the mean radius in each interval: $R_1, R_2, R_3, \ldots, R_i, \ldots, R_n$, where R_n is that interval containing the largest size of feed solid.

2. Guess F_0, F_1, or W, whichever is unknown.

3. Evaluate the composition of the unknown flow stream from Eq. (33). Be sure to put $F_1(R_{n+1}) = 0$, start with size interval R_n and work down to R_{n-1}, R_{n-2}, and so on.

4. See if Eq. (34) is satisfied. If it is, continue to step 5. If not, go back to step 2 and guess again.

5. Evaluate the flow rate of the entire carryover stream F_2 from Eq. (35).

6. Evaluate the composition of the carryover stream $F_2(R_i)$ from Eq. (36). Now F_0, F_1, F_2, and the three size distributions are known.

General Growth Kinetics. Again, rearranging Eqs. (7), (26), and (27) in discrete form and noting that $\mathcal{R}(R_i) > 0$ gives

$$\frac{W(R_i)}{W} = \frac{F_1(R_i)}{F_1} = \frac{F_0(R_i) + W \dfrac{F_1(R_{i-1})}{F_1} \dfrac{\mathcal{R}(R_{i-1})}{\Delta R_i}}{F_1 + W\kappa(R_i) + W \dfrac{\mathcal{R}(R_i)}{\Delta R_i} + 3W \dfrac{\mathcal{R}(R_i)}{R_i}} \tag{37}$$

as well as Eqs. (34)–(36), which remain unchanged.

The calculation procedure for these growing particles is as follows:

1. Choose reasonable size intervals ΔR_i, either all equal or based on the screen sizes being used. Then find the mean radius in each interval: R_1, R_2, R_3, ..., R_i, ..., where R_1 is the interval containing the smallest size of feed solid.

2. Guess F_0, F_1, or W, whichever is unknown.

3. Evaluate the composition of the unknown flow stream from Eq. (37). Start with size interval R_1 and work upward to R_2, R_3, ..., continuing until the $F(R_i)$ value or the $W(R_i)$ value drops to zero. Note that $F_1(R_0) = 0$.

4. See if Eq. (34) is satisfied. If it is, continue to step 5. If not, go back to step 2 and guess again.

5. Evaluate the flow rate of carryover stream, F_2 kg/s, from Eq. (35).

6. Evaluate the composition of carryover stream $F_2(R_i)$ from Eq. (36).

Examples 4 and 5 illustrate the use of these numerical procedures.

In certain special cases one can bypass this or any numerical procedure and evaluate the integrals of Eqs. (27)–(32) directly. We consider some of these cases, all for spherical particles. Additional cases are treated elsewhere [8].

Linear Shrinkage, Single-Size Feed, No Elutriation

For the special case of linear shrinkage, single-size feed, and no elutriation, shown in Fig. 5(a),

$$\mathcal{R}(R) = dR/dt = -k$$
$$\text{feed size} = R_M$$
$$\kappa(R) = 0 \text{ for all } R \text{ values}$$

Introducing these quantities in Eqs. (27)–(32) gives the size distribution of

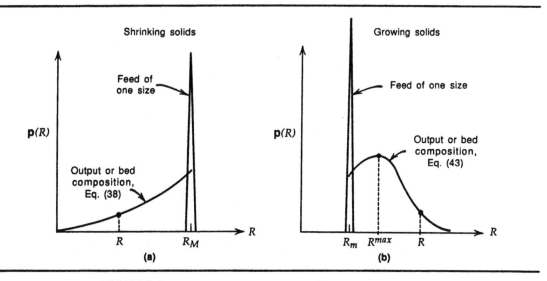

FIGURE 5
Size distribution of product stream from a single-size feed to a fluidized bed.

solids in the vessel and exit stream as

$$\mathbf{p}_b(R) = \mathbf{p}_1(R) = \frac{F_0}{Wk} \frac{R^3}{R_M^3} e^{-F_1(R_M-R)/Wk} \tag{38}$$

the exit flow rate of solids in terms of the feed as

$$\frac{F_1}{F_0} = 1 - 3y + 6y^2 - 6y^3(1 - e^{-1/y}) \tag{39}$$

where

$$y = \frac{\bar{t}}{\tau} = \left(\frac{W/F_1}{\text{time for complete reaction of a particle}}\right) = \frac{Wk}{F_1 R_M} \tag{40}$$

and the surface mean particle size as

$$\bar{R}_s = \frac{3WkR_M}{R_0(F_0 - F_1)} \tag{41}$$

In the extreme where the particles shrink to zero and no solids leave the reactor,

$$\frac{F_0 R_M}{Wk} = 4 \quad \text{and} \quad \frac{\bar{R}_s}{R_M} = \frac{3}{4} \tag{42}$$

Linear Growth, Single-Size Feed, No Elutriation

For the special case of linear growth, single-size feed, and no elutriation, shown in Fig. 5(b),

$$\mathscr{R}(R) = dR/dt = k$$
$$\text{feed size} = R_m$$
$$\kappa(R) = 0 \text{ for all } R \text{ values}$$

Here the general expressions simplify as follows: the size distribution of bed solids and of exit stream becomes

$$\mathbf{p}_1(R) = \frac{F_0}{Wk} \frac{R^3}{R_m^3} e^{-F_1(R-R_m)/Wk} \tag{43}$$

The exit flow rate of solids is related to the feed rate by

$$\frac{F_1}{F_0} = 1 + 3y + 6y^2 + 6y^3, \quad \text{where } y = \frac{Wk}{F_1 R_m} \tag{44}$$

The maximum of the size distribution curve is

$$R^{\max} = 3yR_m \tag{45}$$

and the surface average mean size of the solids in the bed and in the exit stream is

$$\bar{R}_s = \frac{3yR_m F_1}{F_1 - F_0} \tag{46}$$

Growth from Seed. In the special case where the entering particles are of negligible size or where no solids enter the bed and where seeding occurs within the bed, the seed rate

$$n_i = \frac{F_0}{(4/3)\pi R_m^3 \rho_s} = \frac{\text{number of seed particles introduced}}{\text{time}}$$

is finite, even though F_0 and R_m both approach zero.

Here the size distribution of bed and exit stream is given by

$$\mathbf{p}(R) = \frac{1}{2R^{\max}}\left(\frac{3R}{R^{\max}}\right)^3 \exp\left(-\frac{3R}{R^{\max}}\right) \tag{47}$$

The exit flow rate is related to the seed rate by

$$F_1^4 = 8\pi n_i \rho_s W^3 k^3 \tag{48}$$

and the mean size of the bed and of the leaving solids is

$$R^{\max} = \bar{R}_s = \frac{3Wk}{F_1} \tag{49}$$

Comments and Limitations

The development of this chapter is limited to processes where particles grow continuously, shrink to zero continuously, or only reach their limiting size in infinite time. The kinetic forms of Eqs. (21)–(25) all satisfy this requirement.

The agglomeration of particles or the fracturing of particles into two or more smaller particles has to be treated differently. Even with this restriction, attrition can be treated if the tiny abraded fragments worn off the particles are borne away as part of the fluidizing gas and are not considered to be part of the population of solids. Chen and Saxena [9] show how to extend the present treatment to situations where these fines are considered to be part of the solid population.

Processes that lead to a limiting nonzero particle size in a finite time also require a modification of the above analysis, and Chen and Saxena [10] show how to do this. As an example of such a process, consider a two-component particle that consists of a core of unchanging solids surrounded by a layer that shrinks by linear kinetics. In this situation the vessel would contain a significant fraction of solids consisting only of core material. The use of these population models in conjunction with various fluidized bed reactor models to represent coal combustors has been proposed by various authors [11–13].

EXAMPLE 3

Single-Size Feed of Shrinking Particles

Particles of size $d_p = 1$ mm are fed continuously to a fluidized bed where they sublime, shrink, and finally disappear. All entrained fines are returned to the bed with an efficient cyclone, and no solids leave the bed. Find the bed weight for a feed rate of 10 kg/min.

Data. Particle shrinkage is described by

$$-\frac{d(d_p)}{dt} = k = 0.1 \text{ mm/min}$$

SOLUTION

In terms of particle radius, the linear shrinkage rate is

$$-\mathcal{R}(R) = k = 0.05\ \text{mm/min}$$

Putting this value in Eq. (42) then gives

$$W = \frac{F_0 R_M}{4k} = \frac{(10\ \text{kg/min})(0.5\ \text{mm})}{4(0.05\ \text{mm/min})} = 25\ \text{kg}$$

EXAMPLE 4

Wide Size Distribution of Shrinking Particles

Solids of known wide size distribution are fed continuously to a fluidized bed reactor, where they shrink as they react and gasify. There is no overflow stream, but elutriation of fines from the bed is significant. What are the bed weight and size distribution of bed solids for a feed rate of 1 kg/s?

Data. The rate of particle shrinkage is

$$-\mathcal{R}(d_p) = -\frac{d(d_p)}{dt} = k = 1.58 \times 10^{-5}\ \text{mm/s}$$

The continuous size distribution for the feed solids is given in Example 4 of [14]. Here we take size intervals $\Delta(d_{pi}) = 0.1$ mm, and in column 3 of Table E4 we give the fraction of feed solids in each of these size intervals.

SOLUTION

We solve the problem in terms of particle diameter. Then since $F_1 = 0$ and $F_1(d_{pi})/F_1 = W(d_{pi})/W$, Eq. (33) reduces to

$$W(d_{pi}) = \frac{F_0(d_{pi}) - W(d_{p,i+1})\mathcal{R}(d_{p,i+1})/\Delta(d_{pi})}{\kappa(d_{pi}) - \mathcal{R}(d_{pi})/\Delta(d_{pi}) - 3\mathcal{R}(d_{pi})/d_{pi}}$$

With $\mathcal{R}(d_{pi}) = -1.58 \times 10^{-5}$ mm/s and $\Delta(d_{pi}) = 0.1$ mm, this expression reduces to the final working equation

$$W(d_{pi}) = \frac{F_0(d_{pi}) + (1.58 \times 10^{-4})W(d_{p,i+1})}{\kappa(d_{pi}) + (1.58 \times 10^{-4}) + (4.74 \times 10^{-5})/d_{pi}} \tag{i}$$

TABLE E4   Data and Intermediate Calculations for Example 4

1 Size Interval Number, i	2 Mean Size d_{pi} (mm) (given)	3 Feed Rate $F_0(d_{pi}) \times 100$ (kg/s) (given)	4 Elutriation $\kappa(d_{pi}) \times 10^5$ (s^{-1}) (given)	5 Bed Solids $W(d_{pi})$ (kg) (calculated)	6 Bed Fraction $[W(d_{pi})/W] \times 100$ (—) (calculated)
11	1.05	0	0	0	0
10	0.95	0.5	0	38.2	0.5
9	0.85	3.5	0	374.5	5.2
8	0.75	8.8	0	957.9	13.3
7	0.65	13.5	0	1374.1	19.1
6	0.55	17.0	0	1475.9	20.5
5	0.45	18.2	0	1308.0	18.2
4	0.35	17.0	0	945.5	13.1
3	0.25	13.5	2.0	527.5	7.3
2	0.15	7.3	12.5	181.5	2.5
1	0.05	0.7	62.5	24.0	0.3

Total mass of bed $= \Sigma W(d_{pi}) = 7207.1$ kg.

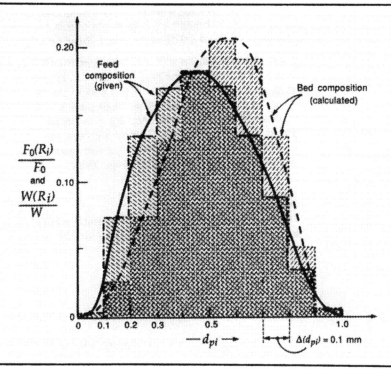

FIGURE E4
Display of the answer to Example 4.

Now, as outlined in the text, start with the largest size interval that contains solids. So, for $i = 10$, Eq. (i) becomes

$$W(d_{p,10}) = \frac{0.005 + 0}{0 + 1.58 \times 10^{-4} + (4.74 \times 10^{-5})/0.95} = 24.04 \text{ kg}$$

Then, for $i = 9$, Eq. (i) becomes

$$W(d_{p,9}) = \frac{0.035 + (1.58 \times 10^{-4})(24.04)}{0 + 1.58 \times 10^{-4} + (4.74 \times 10^{-5})/0.85} = 181.50 \text{ kg}$$

Proceeding in a similar way down to $i = 1$ gives the values in column 5 of Table E4, from which we get column 6. Thus, the total mass in the bed is

$$W = \sum_i W(d_{pi}) = \underline{7207 \text{ kg}}$$

Figure E4 shows the size distribution of feed and bed solids. Even though the particles all shrink, the bed mean is larger than the feed mean. Also note that since F_1 is known, the solution to this problem is direct, with no trial and error needed.

EXAMPLE 5

Elutriation and Attrition of Catalyst

In the fluid catalytic reactor of Fig. 4, the size of catalyst particles decreases by attrition caused by the violent bubbling action and agitation in the bed. The catalyst also slowly deactivates, so to maintain the catalyst activity at some prescribed level solids from the bed are drawn off by an overflow pipe at $F_1 = 36 \text{ kg/hr} = 0.01 \text{ kg/s}$. Because of the high gas velocity, significant entrainment of catalyst occurs; however, the cyclone is quite efficient in returning the coarser of these solids to the bed. The fines are lost.

In order to keep the bed at $W = 40{,}000$ kg despite attrition, entrainment, and overflow, what should be the feed rate of make-up catalyst? Also calculate the size distribution of bed solids and the carryover rate F_2.

Data. Experiments in attrition of radioactive particles show that particles of initial size $d_p = 0.11$ mm shrink to $d_p = 0.085$ mm after 20.8 days in the bed and approach $d_{p,min} = 0.01$ mm after a very long stay in the bed. Assume $\psi = 1$.

The continuous variations with particle size of feed composition, elutriation constant, and cyclone efficiency are given in Example 5 of [14]. Here we take a size interval of $\Delta(d_{pi}) = 2 \times 10^{-2}$ mm and evaluate the feed fraction and elutriation constant with cyclone efficiency in each size interval. These values are listed in columns 3 and 4 of Table E5, respectively.

SOLUTION

The existence of a minimum size to which a particle shrinks suggests that Eq. (24) is a reasonable representation of the shrinking process. Separating variables and integrating gives

$$\ln \frac{R_1 - R_{min}}{R_2 - R_{min}} = k't$$

With known values substituted, we obtain

$$\ln \frac{0.11 - 0.01}{0.085 - 0.01} = k'(20.8 \text{ days})(24 \text{ hr/day})(3600 \text{ s/hr})$$

Hence

$$k' = 1.6 \times 10^{-7} \text{ s}^{-1}$$

and the rate of size change is

$$\mathcal{R}(d_p) = \frac{d(d_p)}{dt} = (-1.6 \times 10^{-7})(d_p - 0.01) \quad \text{(mm/s)} \tag{i}$$

Next apply Eq. (33) in terms of particle diameter and include the effect of the cyclone. Thus

$$\frac{F_1(d_{pi})}{F_1} = \frac{\dfrac{F_0(d_{pi})}{F_0} F_0 - W \dfrac{\mathcal{R}(d_{p,i+1})F_1(d_{p,i+1})}{F_1 \, \Delta d_{p,i}}}{F_1 + W\kappa(d_{pi})[1 - \eta(d_{pi})] + W \dfrac{\mathcal{R}(d_{pi})}{\Delta(d_{pi})} - 3W \dfrac{\mathcal{R}(d_{pi})}{d_{pi}}}$$

TABLE E5   Data and Intermediate Calculations for Example 5

1	2	3
Size Interval Number, i	Mean Size d_{pi}, (mm) (given)	Feed Composition $[F_0(d_{pi})/F_0] \times 100$ (given)
9	0.17	0
8	0.15	0.95
7	0.13	2.45
6	0.11	5.2
5	0.09	10.1
4	0.07	23.2
3	0.05	35.65
2	0.03	20.0
1	0.01	2.45

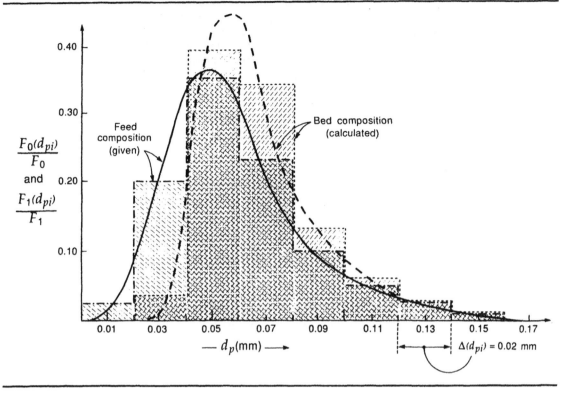

FIGURE E5
Display of the answer to Example 5.

Putting in known values gives the working equation

$$\frac{F_1(d_{pi})}{F_1} = \frac{\dfrac{F_0(d_{pi})}{F_0} F_0 - (2 \times 10^6)\mathcal{R}(d_{p,i+1})F_1(d_{p,i+1})}{F_1 + W\kappa(d_{pi})[1 - \eta(d_{pi})] - (2 \times 10^6)\mathcal{R}(d_{pi}) - 120{,}000\,\dfrac{\mathcal{R}(d_{pi})}{d_{pi}}} \qquad \text{(ii)}$$

4 *Elutriation and* *Cyclone Efficiency* $\kappa(d_{pi})[1 - \eta(d_{pi})] \times 10^6$ *(given)*	*5* $-\mathcal{R}(d_{pi}) \times 10^9,\ (mm/s)$ *from Eq. 1* *(calculated)*	*6* *Bed Composition* $[F_1(d_{pi})F_1] \times 100$ *from Eq. (ii)* *(calculated)*
0	25.6	0
0	22.4	0.7
0	19.2	2.4
0	16.0	6.1
0	12.8	13.65
0	9.6	34.0
0.625	6.4	39.65
10.225	3.2	3.5
159.25	0.4	~0

Now guess F_0, start with $i = 8$, and evaluate successively the value of $F_1(d_{pi})/F_1$ from $i = 8$ to $i = 1$, and see if $\Sigma\, F_1(d_{pi})/F_1 = 1$. Keep adjusting the guessed value of F_0 until Eq. (34) is satisfied.

Following this procedure gives $F_0 = 0.0519$ kg/s, the size distribution of bed and outflow solids, as shown in column 6 of Table E5. This is graphed in Fig. E5.

Repeating this procedure with successively smaller size slices gives the following results:

$$\Delta(d_{pi}) = 20\ \mu\text{m} \cdots \qquad F_0 = 0.0519\ \text{kg/s}$$
$$\Delta(d_{pi}) = 10\ \mu\text{m} \cdots \qquad F_0 = 0.0485\ \text{kg/s}$$
$$\Delta(d_{pi}) = 5\ \mu\text{m} \cdots \qquad F_0 = 0.0476\ \text{kg/s}$$

Extrapolating down to $\Delta(d_{pi}) \rightarrow 0$ gives the required feed rate to the FCC reactor of

$$F_0 = 0.047\ \text{kg/s} = \underline{\underline{169\ \text{kg/hr}}}$$

PROBLEMS

1. A fluidized bed that contains 10 kg of solids is to process 1 kg/min of feed solid of unchanging size consisting of 30% of 50-μm radius particles, 40% of 100-μm radius particles, 30% of 200-μm radius particles. The gas velocity used is high, some solids are blown out of the bed, and the elutriation velocity constant under these operating conditions is

$$\kappa(R) = (500\ \mu\text{m}^2/\text{min})R^{-2}$$

   From this information determine
   (a) the outflow and carryover rates of solids.
   (b) the size distribution of these two streams.
   (c) the mean residence time of the three sizes of solids in the bed.

2. Calculate the mean residence time of the different sizes of zinc blende particles present in a fluidized roaster operated at steady state. Assume $\psi = 1$.

   Data. $W = 40,000$ kg, $F_1 = 4000$ kg/hr

Size, d_p (μm)	60	80	100	120	130	140
$\kappa(d_p) \times 10^3$ (s^{-1})	3.8	3.0	1.7	0.48	0.14	0

3. Consider a feed of one size R_M of solids to a fluidized bed, steady state operations, unchanging density, linear shrinkage, no carryover or entrainment, but with an overflow stream, and 90% consumption of feed solids, or $F_1 = 0.1F_0$.
   (a) What will be the percent consumed if the size of the feed particles is doubled?
   (b) What will be the percent consumed if the reactor size (mass of solids) is halved?
   (c) What must be the size of the reactor for the particles to be completely consumed?

4. At steady state, fresh solids of uniform size R_i enter a bed; the radius of these particles increases linearly with length of stay in the bed, and the outflow rate is eight times the inflow rate. What is the effect on the outflow rate of the following individual changes:
   (a) the inflow rate of solids is quadrupled.
   (b) the radius of inflow solids is quadrupled.
   (c) the bed weight is quadrupled.

5. Suppose that the operation of the previous problem was run with no introduction of feed solids but with self-seeding in the bed. Find the effect of the following single changes on the outflow rate of condensed solids and on the size distribution of this outflow stream:
   (a) the seed rate in the bed is quadrupled.
   (b) the growth rate of particles is quadrupled.
   (c) the bed weight is quadrupled.

6. Consider a modification of the process of Example 3, in which particles are discharged through an overflow pipe at a rate $F_1 = 5$ kg/min. In order to keep the bed weight unchanged at 25 kg, calculate the necessary rate of fresh feed F_0. Also find the size distribution of particles leaving the bed. Is the amount of solids sublimed higher or lower than in Example 3?

7. In Example 4 determine the size distribution of solids in the entrained gas stream and plot this on a graph with the size distribution of bed and feed solids. Also determine the fraction of feed solid converted into gaseous product.

8. As a modification of the process of Example 4, bed particles are to be discharged through an overflow pipe at a rate $F_1 = 0.2$ kg/s. Calculate the necessary feed rate F_0 that would keep the bed weight unchanged at 7120 kg, all other conditions remaining fixed.

9. In the reactor of Example 5 the cyclone collector is disconnected. To keep the bed weight unchanged, how much must the feed rate be raised?

10. Solids of size $d_p = 1000$ μm enter a fluidized bed reactor at a rate of 200 kg/hr and react away to form gaseous product by linear shrinkage at a rate of 500 μm diameter/hr, and no solids leave the reactor. What quantity of solids must the reactor hold? (Solve by the numerical procedure outlined in this chapter, taking five size intervals and check your answer with the exact answer obtained from the integrated forms presented in this chapter.)

REFERENCES

1. S. Yagi and D. Kunii, *Kagaku Kikai (Chem. Eng. Japan)*, **16**, 283 (1952); *J. Chem. Soc. Japan (Ind. Eng. Section)*, **56**, 131, 133 (1953).
2. P. V. Danckwerts, *Chem. Eng. Sci.*, **2**, 1 (1953).
3. Y. Kamiya, *Kagaku Kogaku (Chem. Eng. Japan)*, **19**, 412 (1955).
4. S. Yagi and D. Kunii, *Fifth Symp. (Int.) on Combustion*, 231, Van Nostrand Reinhold, New York, 1955; *Kagaku Kogaku (Chem. Eng. Japan)*, **19**, 500 (1955).
5. S. Yagi and D. Kunii, *Chem. Eng. Sci.*, **16**, 364, 372, 380 (1961).

6. O. Levenspiel, D. Kunii, and T. Fitzgerald, *Powder Technol.*, **2**, 87 (1968/69).

7. B.W. Overturf and F. Kayihan, *Powder Technol.*, **23**, 143 (1979).

8. O. Levenspiel, *Chemical Reactor Omnibook*, Chap. 54, OSU Book Stores, Corvallis, OR, 1989.

9. T.P. Chen and S.C. Saxena, *Powder Technol.*, **18**, 279 (1977).

10. T.P. Chen and S.C. Saxena, *Powder Technol.*, **15**, 283 (1976).

11. T.P. Chen and S.C. Saxena, *AIChE Symp. Ser.*, **24(176)**, 149 (1978).

12. P. Stecconi, *Powder Technol.*, **32**, 35 (1982).

13. B.W. Overturf and G.V. Reklaitis, *AIChE J.*, **29**, 813 (1983).

14. D. Kunii and O. Levenspiel, *Fluidization Engineering*, Chap. 11, Wiley, New York, 1969.

CHAPTER

15

Circulation Systems

The discovery of how to maintain a stable and continuous circulation of solids in a gas-solid system has led to the development of various processes using such a scheme. Chapter 2 shows that these processes are usually large-scale operations, especially in the petroleum industry, with its giant FCC and similar processes. It is fair to say that the key to the commercial success of these processes, large and small, rests primarily on the proper design of their systems of circulating solids.

The basic problem then is to choose the type of circulation system, the size of its piping, the gas flow rate, and so forth—in effect, to design an assembly that is stable and has a circulation rate that meets the demands of the particular process. This chapter gives the principles and relevant data to design such systems.

Circuits for the Circulation of Solids

Classification of Circulation Loops

The development of fluid catalytic cracking (FCC) represented the first use of solid circulation systems, and it was composed of a rather complex arrangement of two circulation loops. The Synthol process was developed, using a much simpler single-loop system. Both processes used very fine Geldart **A** solids. Lurgi's sand cracker then extended the single-loop system to coarse Geldart **BD** solids. The single-loop system has since formed the heart of various industrial processes from the calcination of inorganic solids to the combustion of coal. Chapter 2 mentions some of these applications.

Figures 1–3 illustrate the three main types of circulating systems. Figure 1 shows two examples of the single-loop, one-gas system. Figure 2 shows two examples of the single-loop, two-gas stream system, designed to keep the two gas streams apart. Since downflowing solids usually entrain gas, special care is needed in these systems to keep these gases from mixing. Figure 3 shows two examples of two-loop, two-gas systems. One of these processes uses very fine Geldart **A** solids, and the other uses large Geldart **B** solids. Obtaining a positive

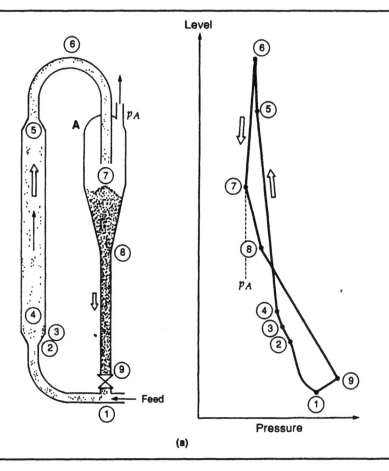

FIGURE 1
Single-loop solids circulation systems involving only one gas stream: (a) Synthol process;

gas seal between the two gas streams is much easier than for the one-loop system because comparatively high standpipes are used.

Figures 1–3 also show the static pressure at all points in these processes. The driving force for the circulation of solids in these loops is governed by the difference in static head at the base of the two arms of the loops, and a quantitative treatment of this phenomenon will be given later.

The figures show that different forms of gas-solid contacting and flow occur in the various sections of any circulation systems—for example:

1. For upward transport of solids: pneumatic transfer line or fluidized bed
2. Upper receiver of solids: fluidized bed or moving bed
3. For downflow of solids (vertical or inclined): standpipe with moving bed or aerated or fluidized solids
4. Device for the control of solid flow: slide valve, L valve, J valve, etc.

From flow considerations the standpipe is the heart of a circulation system because it allows solids to flow from a vessel at low pressure to one at higher pressure, and its design requires careful consideration.

As one may suspect, the dual-loop system of Fig. 3 is complex, and thus

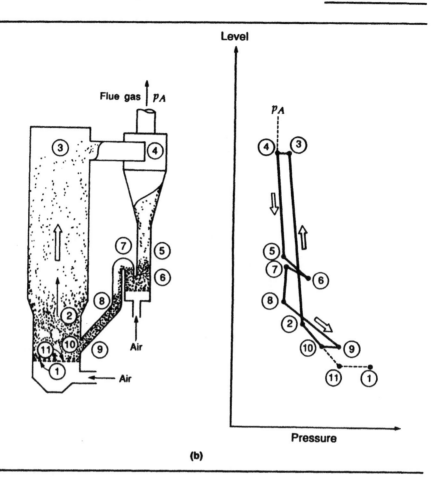

(b)

(b) circulating fluidized bed coal combustor.

difficult and expensive to design. Considerable effort has gone into developing equally effective but simpler circulation systems [1–5].

Pressure Balance in a Circulation Loop

Consider section 1–2 of the circulation loop 1–2–3–4–5–6–7–8–9–1 of Fig. 1(a). Denote the height of point 1 by h_1, the mean bulk density of the gas-solid mixture by $\bar{\rho}_{12}$, and the frictional loss including acceleration loss by $\Delta p_{12} > 0$. Then the mechanical energy balance becomes

$$p_2 - p_1 + \frac{\bar{\rho}_{12} g (h_2 - h_1)}{g_c} + \Delta p_{12} = 0 \tag{1}$$

Similarly, we may write

$$p_3 - p_2 + \frac{\bar{\rho}_{23} g (h_3 - h_2)}{g_c} + \Delta p_{23} = 0 \tag{2}$$

$$\cdots$$

$$p_9 - p_8 + \frac{\bar{\rho}_{89} g (h_9 - h_8)}{g_c} + \Delta p_{89} = 0 \tag{3}$$

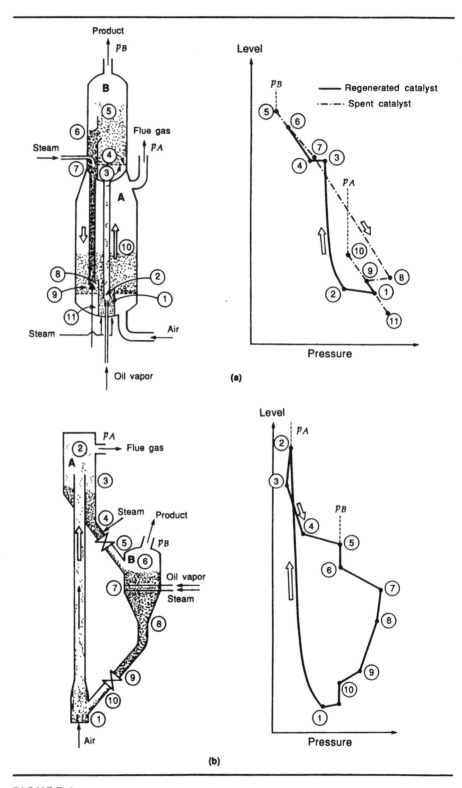

FIGURE 2
Single-loop solids circulation systems involving two gas streams: (a) Exxon's Orthoflow FCC unit; (b) sand cracker.

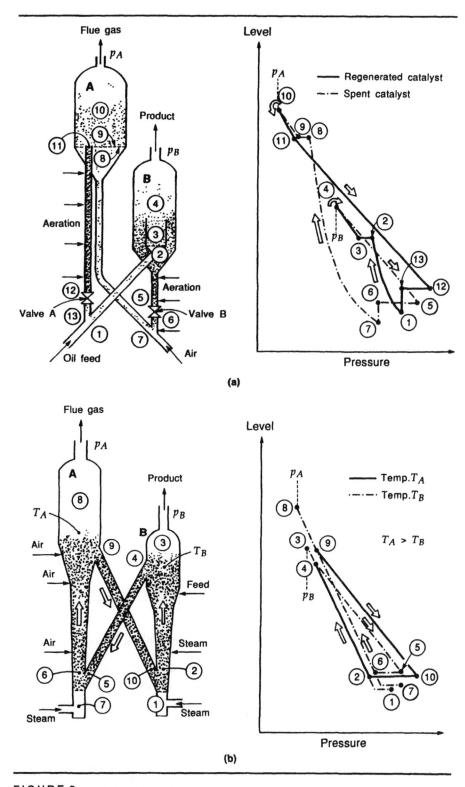

FIGURE 3

Double-loop solids circulations systems involving two gas streams: (a) Exxon's FCC unit; (b) KK and Pyrox processes.

With Δp_v representing the frictional loss at the slide value, we have, across the valve,

$$p_1 - p_9 + \Delta p_v = 0 , \qquad \text{with } h_9 \cong h_1 \tag{4}$$

Summing the above equations gives, for the whole loop,

$$\sum_{i=1}^{8} \left[\frac{\bar{\rho}_{i,i+1} g(h_{i+1} - h_i)}{g_c} + \Delta p_{i,i+1} \right] + \Delta p_v = 0 \tag{5}$$

This equation shows that the solids will settle down to a circulation rate at which the sum of the static head terms just balances the sum of all the frictional resistance terms, including solids acceleration losses, bends, constrictions, and the valve between points 9 and 1.

Equation (5) can be extended to dual circulations systems, such as shown in Fig. 3, provided one sums around all sections of the dual loop. Here one usually encounters the additional restriction in that the pressures p_A and p_B in the two process vessels are fixed by the process requirements. Where $p_B > p_A$ is required, vessel A should be higher than vessel B, as illustrated in Fig. 3(a). The circuit of Fig. 2(a) illustrates the opposite situation.

As an example of the calculation procedure for a dual circulation system, consider Fig. 3(a). A pressure balance at the oil feed under vessel A gives

$$p_B + \sum_{i=1}^{3} \frac{\bar{\rho}_{i,i+1} g(h_{i+1} - h_i)}{g_c} + \sum_{i=1}^{3} \Delta p_{i,i+1}$$
$$= p_A + \sum_{i=10}^{11} \frac{\bar{\rho}_{i,i+1} g(h_i - h_{i+1})}{g_c} - \sum_{i=10}^{11} \Delta p_{i,i+1} - \Delta p_{vA} \tag{6}$$

where $\Delta p_{vA} = \Delta p_{12,13}$, $p_{13} \cong p_1$, $h_{13} \cong h_1$, and all Δp terms are positive. Thus, combining terms gives

$$p_A - p_B = \sum_{i=1,10}^{3,11} \left[\frac{\bar{\rho}_{i,i+1} g(h_{i+1} - h_i)}{g_c} + \Delta p_{i,i+1} \right] + \Delta p_{vA} \tag{7}$$

In a similar fashion a pressure balance at the air feed point under vessel B gives

$$p_B - p_A = \sum_{i=4, \text{ and } 7}^{9} \left[\frac{\bar{\rho}_{i,i+1} g(h_{i+1} - h_i)}{g_c} + \Delta p_{i,i+1} \right] + \Delta p_{vB} \tag{8}$$

where $\Delta p_{vB} = \Delta p_{56}$, $p_6 \cong p_7$, and $h_6 \cong h_7$.

In general, one can control the operation of the circulation system by proper adjustment of valves A and B. Thus, to keep the solid circulation rate constant when the operating pressure difference $p_B - p_A$ increases, either increase the resistance of valve B or decrease the resistance of valve A.

The circulation rate of solids can be controlled in various ways. One common method is by use of slide valves. For precise control of the flow, the frictional loss of these valves should be comparable to the sum of the frictional losses in the rest of the circuit. J valves and L valves can also be used.

A different approach to the control of the solid circulation rate involves adjusting the density of the gas-solid mixture in different parts of the circuit. This is done by varying the aeration rates into the flowing gas-solid mixture. For example, increasing the aeration rate in either or both of the riser sections of the dual circuit of Fig. 3(b) increases the solid circulation rate in that system. Whatever means are adopted, it is important to know the relationship between the control setting and the corresponding circulation rate.

Finding Required Circulation Rates

Solid circulation loops have been developed for a variety of processes; however, their most extensive and important applications have been in the field of solid-catalyzed gas-phase reactions, where they are used in one of the following two situations. The first is where the catalyst deactivates rapidly and requires frequent regeneration. Here a steady state solid circulation with catalyst continually being regenerated and returned to the reactor has obvious advantages over a batch system. A second use is where much heat must be brought into or removed from the reactor. Gases have a very small heat capacity relative to heats of reaction, whereas solids have a relatively large heat capacity. Hence, a continuous circulation of solids between two vessels, say reactor and heat exchanger, can effectively transport heat from one vessel to the other and control the temperatures in the units.

Where catalyst decay is very rapid, this factor will determine the rate of circulation; otherwise, the heat removal criterion will control it. Both criteria must be met by proper temperature control, heat input, catalyst addition rate, and so on. In catalytic reactions the catalyst itself is used as the heat-transporting solid. For other types of gas-phase reactions, any convenient solid may be used as the heat carrier.

Since the capacity of the reactor directly depends on the solids circulation rate, it is essential that the circulation system be well designed. We now show the type of calculation involved in finding required circulation rates when catalyst deactivation is limiting, and when heat removal or addition is limiting.

Circulation Rate for Deactivating Catalysts

Consider the simplest of situations—a catalytic reactor and a catalyst regenerator, both fluidized beds, with solid catalyst circulating continuously at a uniform rate F_s (kg/s) from one to the other. In the reactor the catalyst affects the decomposition of reactant A present in the gas stream and is itself slowly deactivated in the process. This "worn-out" or partly deactivated catalyst is continually regenerated and returned to the reactor to repeat its job.

The relative effectiveness of the catalyst is measured by its activity, defined as

$$\mathbf{a} = \frac{\text{rate of reaction of A using catalyst in a given condition}}{\text{rate of reaction with fresh catalyst}}$$

$$= \frac{r_A}{r_{A0}} \tag{9}$$

The simplest realistic expression that describes the rate of deactivation of catalyst assumes it to be directly proportional to the present activity of the catalyst; thus

$$-\frac{d\mathbf{a}}{dt} = K_a \mathbf{a} \tag{10}$$

where K_a is the rate constant for this first-order deactivation.

Next consider the variation in residence time of individual particles in the reactor. Because of their somewhat haphazard motion and their relatively long stay in the bed, we may take the solids to be in backmix flow (solids completely and uniformly mixed). Thus the stream leaving the reactor is representative of the vessel's contents, and its residence time distribution or probability density function is

$$\mathbf{E}(t) = \frac{1}{\bar{t}} e^{-t/\bar{t}} \tag{14.3) or (11}$$

where $\bar{t} = W/F_s$ is the mean residence time of solids in the reactor and $\mathbf{E}(t)dt$ is the fraction of exit stream with age between t and $t + dt$.

With these assumptions about flow and kinetics and an additional one stating that the catalyst activity is completely restored on regeneration (or $\mathbf{a} = 1$ for catalyst returning to the reactor), we have the situation shown in Fig. 4, and we can then determine the average activity $\bar{\mathbf{a}}$ of catalyst in the reactor in two steps. First, the activity of each catalyst particle decreases from unity with length of stay in the reactor. This is found by integration of Eq. (10), which gives

$$\mathbf{a} = e^{-K_a t} \tag{12}$$

Next, the exit stream contains particles of all ages with their corresponding activities. Thus the mean activity of leaving catalyst is

$$\bar{\mathbf{a}} = \sum_{\substack{\text{particles of all ages, or} \\ \text{the whole exit stream}}} \left(\begin{matrix} \text{activity of a particle} \\ \text{of age between } t \\ \text{and } t + dt \end{matrix} \right) \left(\begin{matrix} \text{fraction of particles} \\ \text{in the exit stream} \\ \text{in this age interval} \end{matrix} \right) \tag{13}$$

In symbols, using Eqs. (11) and (12), we have

$$\bar{\mathbf{a}} = \int_0^\infty \mathbf{a}\mathbf{E}(t)\,dt = \int_0^\infty e^{-K_a t}\left(\frac{1}{\bar{t}}\right)e^{-t/\bar{t}}\,dt \tag{14}$$

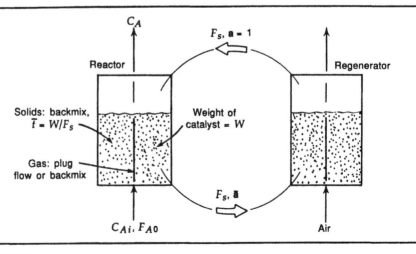

FIGURE 4
A catalytic reactor-regenerator in which deactivating catalyst is completely regenerated.

Performing the integration, we get

$$\bar{a} = \frac{1}{1 + K_a \bar{t}} = \frac{1}{1 + K_a W/F_s} \tag{15}$$

Rearrangement gives the required circulation rate of solids as

$$F_s = \frac{K_a W \bar{a}}{1 - \bar{a}} \tag{16}$$

This expression shows that an increase in bed size or an increase in deactivation rate of catalyst requires a corresponding increase in circulation rate if the activity level in the reactor is to remain unchanged. Thus for the first-order catalytic reaction (see Chap. 12),

$$A \rightarrow R, \quad -\frac{1}{V_s}\frac{dN_A}{dt} = K_r C_A, \quad K_r = [\text{m}^3 \text{ gas}/\text{m}^3 \text{ solid·s}] \tag{12.1}$$

simply replace K_r in the reactor equations of Chap. 12 by $K_r \bar{a}$.

Numerous other factors may have to be added to this simple treatment. For example:

- The catalyst may not be in backmix flow.
- The kinetics of deactivation and of reaction may be more complicated.
- The catalyst returning to the reactor may not be completely regenerated because of insufficient time in the regenerator. For this situation see Chap. 17.
- With long use and repeated regeneration, the catalyst may lose some activity, which cannot be restored. This is called *permanent deactivation*.
- Fresh catalyst may be introduced into the circulating system, and some may be removed to control the changes in size distribution caused by attrition, to make up for stack losses and cyclone inefficiency, and to counter the continual decrease in activity caused by permanent deactivation.

A treatment of most of these extensions may be found elsewhere [6].

Circulation Rate for a Required Heat Removal Rate

Consider a reactor-cooler system, both fluidized beds, an exothermic gas phase reaction (either homogeneous or catalytic) occurring in the reactor, and circulating solids taking up part of this heat and giving it up in the cooler. Solids are in backmix flow, and, irrespective of the gas flow pattern, the temperature of the solids is the same as the temperature of the gas leaving that unit. Experimental evidence (see Chap. 11) shows that this is a very good assumption. Finally, in these energy calculations it is more convenient to use mass than molar units for enthalpies, heats of reaction (− for exothermic, + for endothermic), and flow rates of gases and solids. Figure 5 shows the situation with the pertinent symbols and temperatures of the various streams.

The problem is to find the solid circulation rate necessary to keep the reactor at a desired temperature, given the condition of the incoming streams. This problem is solved by making the overall and the individual energy balances for the two units and combining the resulting equations. Thus the overall energy

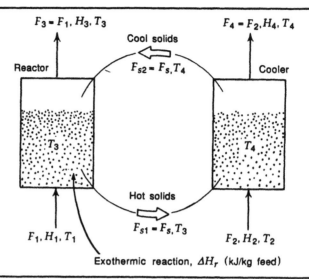

FIGURE 5
A heat-limiting reactor-regenerator showing nomenclature used.

balance (follow Fig. 5) gives

$$\begin{pmatrix} \text{heat released} \\ \text{by reaction} \end{pmatrix} = \begin{pmatrix} \text{heat gained by} \\ \text{reactant gas} \end{pmatrix} + \begin{pmatrix} \text{heat gained} \\ \text{by coolant} \end{pmatrix} \qquad (17a)$$

Since only heat, not matter, is being transferred from unit to unit, $F_1 = F_3$, $F_2 = F_4$, and $F_{s1} = F_{s2} = F_s$; thus letting H be the enthalpy, Eq. (17a) becomes

$$(-\Delta H_r)F_1 = F_1(H_3 - H_1) + F_2(H_4 - H_2) \qquad (17b)$$

For the reactor, the energy balance is

$$\begin{pmatrix} \text{heat released} \\ \text{by reaction} \end{pmatrix} = \begin{pmatrix} \text{heat gained by} \\ \text{reactant gas} \end{pmatrix} + \begin{pmatrix} \text{heat gained} \\ \text{by solid} \end{pmatrix} \qquad (18a)$$

or,

$$(-\Delta H_r)F_1 = F_1(H_3 - H_1) + F_s C_{ps}(T_3 - T_4) \qquad (18b)$$

Similarly, for the cooler

$$(\text{heat lost by solids}) = (\text{heat gained by coolant}) \qquad (19a)$$

or

$$F_s C_{ps}(T_3 - T_4) = F_2(H_4 - H_2) \qquad (19b)$$

These three energy balances represent but two independent equations, so any of the three balances can be obtained from the other two.

The overall energy balance gives the outlet coolant enthalpy (hence temperature). Then either of the individual energy balances gives the necessary circulation rate. Thus

$$F_s = \frac{F_1(-\Delta H_r + H_1 - H_3)}{C_{ps}(T_3 - T_4)} = \frac{F_2(H_4 - H_2)}{C_{ps}(T_3 - T_4)} \qquad (20)$$

For endothermic reactions ΔH_r is positive, and the cooler is replaced by a heater, but the derivation and equations remain unchanged.

The type of treatment just shown may take on many different forms and may be extended in various ways to include the following additional factors:

- There may be transfer of matter from one fluid stream to the other by deposition or absorption on the solid in one unit and by removal in the other. In this case $F_1 \neq F_3$, $F_2 \neq F_4$, and $F_{s1} \neq F_{s2}$.
- Solids may enter and leave the circulation system.
- Reaction may occur in both units.
- Heat losses may be considered.

Example 2 introduces all these factors. However, the procedure is always the same: determine the energy balances, and from them extract the required temperatures and solid circulation rate.

EXAMPLE 1

Circulation Rate When Deactivation Controls

A catalyst with a 1-s half-life of activity is to be used in a cracker-regenerator system treating 960 tons/day of feed oil. Satisfactory conversion of the feed can be obtained as long as the mean activity of the catalyst in the reactor is no lower than 1% that of fresh catalyst. If the reactor contains 50 tons of catalyst that is completely regenerated, find the catalyst circulation rate needed to maintain this catalyst activity in the bed.

SOLUTION

Since no information is given about the decay rate of catalyst, assume first-order kinetics. Then Eq. (12) gives

$$0.5 = e^{-K_a(1 \text{ s})}$$

from which the rate constant is

$$K_a = 0.6932 \text{ s}^{-1}$$

The required circulation rate is then given by Eq. (16):

$$F_s = \frac{(0.6932 \text{ s}^{-1})(50 \text{ tons})(0.01)}{1 - 0.01} = 0.35 \text{ ton/s} = \underline{\underline{21 \text{ tons/min}}}$$

or

$$\frac{F_s}{F} = \frac{21 \times 60 \times 24}{960} = \underline{\underline{31.5 \text{ tons of solid circulated/ton feed oil}}}$$

EXAMPLE 2

Circulation Rate When Heat Duty Controls

Hydrocarbon oil, fed at 260°C, is to be cracked at 500°C in the fluidized bed of a reactor-regenerator system. During the endothermic reaction ($\Delta H_{r1} = 1260 \text{ kJ/kg}$ feed), carbon is deposited on the catalyst (7% by weight of feed) and is removed entirely* by burning with excess air in the regenerator ($\Delta H_{r2} = -33,900 \text{ kJ/kg}$ carbon, and combustion to CO_2 is complete*). The hot catalyst then returns to the reactor to supply the heat needed for the cracking reaction. Find the necessary solid circulation rate and air injection rate.

*These assumptions are made to simplify the example. In actual practice, flue gases usually contain a CO–CO_2 mixture, even in the presence of excess oxygen.

Additional Data. To ensure complete combustion, use 5–10% excess air.

Enthalpy of feed oil at 260°C = 703 kJ/kg (basis 20°C).
Enthalpy of cracked product at 500°C = 1419 kJ/kg (basis 20°C).
Air for regenerator enters at 20°C.
Specific heat of entering air: $\quad C_{pa} = 1.09$ kJ/kg·°C
$\qquad\qquad\qquad$ flue gases: $\quad C_{pf} = 1.05$ kJ/kg·°C
$\qquad\qquad\qquad$ solids: $\quad C_{ps} = 1.01$ kJ/kg·°C
$\qquad\qquad\qquad$ vaporized feed: $\quad C_{pv} = 3.01$ kJ/kg·°C

SOLUTION

The nomenclature is shown in Fig. E2. Because mass is moving from one gas stream to the other via deposited and burned carbon, $F_3 < F_1$ and $F_4 > F_2$. Thus the energy balance for the cracker is

$$\text{(heat released by reaction)} = \text{(heat gained by gas)} + \text{(heat gained by solids)}$$

Ignoring the small enthalpy contribution by the carbon deposit, put $F_{s1} = F_{s2} = F_s$. This results in no error in calculated T_4 and only a slight error in calculated F_s. Then the energy balance becomes

$$(-\Delta H_{r1})F_1 = (F_3 H_3 - F_1 H_1) + F_s C_{ps}(T_3 - T_4)$$

and with values inserted we get

$$-1260 F_1 = 0.93 F_1(1419) - 703 F_1 + F_s(1.01)(500 - T_4) \tag{i}$$

Similarly, for the regenerator,

$$(-\Delta H_{r2})(0.07 F_1) = F_4 H_4 - F_2 H_2 + F_s C_{ps}(T_4 - T_3)$$

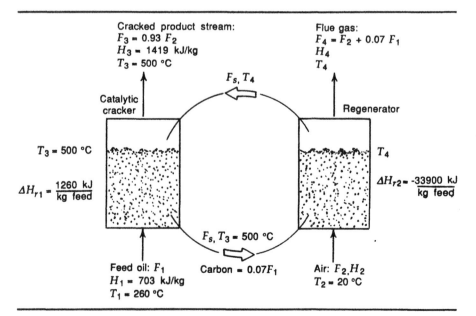

FIGURE E2
Nomenclature and values of various streams.

or

$$(33900)(0.07F_1) = (F_2 + 0.07F_1)(1.05)(T_4 - 20) - F_2(1.09)(20 - 20)$$
$$+ F_s(1.01)(T_4 - 500) \tag{ii}$$

Combining Eqs. (i) and (ii) gives

$$\frac{F_s}{F_1} = \frac{1863}{T_4 - 500} \quad \text{and} \quad \frac{F_2}{F_1} = \frac{479}{T_4 - 20} - 0.07$$

Hence, the flow rates of air and solid depend on the selected flue gas temperature. At this point one restriction previously noted must be met: the amount of air introduced must be 5–10% greater than the minimum needed for complete combustion of all the carbon to CO_2. In general, for a mass fraction of carbon β_c this becomes

$$\frac{F_2}{F_1} \geq \beta_c \left(\frac{22.4 \text{ N-m}^3}{12 \text{ kg}} \right) \left(\frac{1.293 \text{ kg/N-m}^3}{0.21} \right) = 11.49 \beta_c$$

For $\beta_c = 0.07$,

$$\left(\frac{F_2}{F_1} \right)_{min} = 0.80$$

Selecting various flue gas temperatures, we find the corresponding air and solid circulation rates.

T_4 (C°)	520	540	560	580	600	620	640	660
F_s/F_1	93.2	46.6	31.1	23.3	18.6	15.5	13.3	11.6
F_2/F_1	0.958	0.921	0.887	0.855	0.826	0.798	0.763	0.726
Excess air (%)	19.1	14.5	10.3	6.3	2.6	—	—	—

This table shows the range of satisfactory operating conditions to be

Solid circulation rate: $\frac{F_s}{F_1} = \underline{31\text{–}23}$

Air to feed oil ratio: $\frac{F_2}{F_1} = \underline{0.89\text{–}0.86}$

Regenerator temperature: $\underline{560\text{–}580°C}$

These values are consistent with practical data from commercial FCC units.

Flow of Gas-Solid Mixtures in Downcomers

As mentioned earlier, the downcomer is the crucial element in a pneumatic circulation system for solids, its function being to transfer solids from an upper zone or vessel at low pressure to a lower zone of higher pressure while providing a gas seal between vessels. In this operation, gas tries to flow up the pipe to the low-pressure region but is hindered by the downflowing solids. With proper design, the gas may remain approximately stationary.

For gas to not short-circuit up the pipe, one should have dense downflowing solids—in other words, either a moving bed or a fluidized bed at nearly minimum fluidizing conditions. Small pressure gradients in the pipe will lead to moving bed flow; larger gradients sufficient to counter the weight of solids in the pipe will result in aerated flow. Gradients larger than this cannot be maintained. In these situations one must use longer downcomers or provide a restriction at the bottom of the pipe and, perhaps, at the top. In any case, for proper control of the pressure gradient and the flow of solids in downcomers, such restrictions are usually called for.

Figure 6(a) shows three configurations for the top of a downcomer. The one used depends on how solids approach the downcomer. For the bottom of the downcomer, Fig. 6(b) shows the different types of flow restrictions used to control the pressure drop and the flow rate of solids.

Any combination of top and bottom can be used. For example, in Figs. 1–3 we have

Top of downcomer	Bottom of downcomer	Figures
Downflow from fluidized bed	Slide or cone valve	1(a), 2(a), 3(a)
Downflow from fluidized bed	Into a fluidized bed	3(b)
Overflow from fluidized bed	Into a fluidized bed	1(b)
Downflow from moving bed	Slide valve	2(b)

As shown in Fig. 3(b), inclined standpipes are also used in practice.

Before considering flow in the downcomer proper, we first consider the discharge of solids from a reservoir into the downcomer and the discharge of solids from a vertical pipe into a receiving vessel.

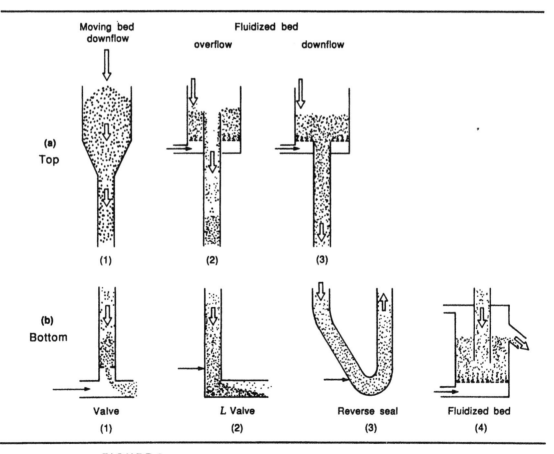

FIGURE 6
Configuration of the ends of downcomers.

Downward Discharge from a Vertical Pipe

When solids descend without aeration in a long vertical pipe having a restriction such as an orifice of diameter d_{or} at its lower end, pistonlike flow with slightly lower velocity at the walls occurs everywhere except near the discharge point. There the solids funnel out, leaving an annulus of stagnant material. The height of this exit region is approximately $(d_t/2) \tan \theta_f$, where θ_f is the angle of internal friction. This angle and the angle of repose θ_r are illustrated in Fig. 7, and their values are given in Table 1 for a variety of substances.

The discharge rate of solids from orifices has been studied by several investigators, and Zenz and Othmer [8] surveyed their findings. Rausch [9] used tubes from 7.6 to 20.3 cm ID, orifices from 0.059 to 5.08 cm ID, 10 different kinds of particles from 0.127 to 12.7 mm with bulk densities from 0.73 to 6.74 g/cm³, and correlated his findings and the experimental data of several investigators as well as, in any consistent set of units, such as SI or c.g.s., by the expression

$$\frac{F_s(\tan \theta_r)^{1/2}}{C_w C_0 g^{1/2} \bar{\rho} d_p^{2.5}} = 0.161 \left(\frac{d_{or}}{d_p}\right)^{2.75} \tag{21}$$

where F_s is the mass discharge rate of solids, and $\bar{\rho}$ is the bulk density of the gas-solid mixture. For a flat orifice plate where $d_{or}/d_p > 10$, $C_w C_0$ can be taken as unity. When $\theta_0 = 60°$ in Fig. 7, $C_w C_0 \cong 1$ at $d_{or}/d_p > 20$. Note that Eq. (21) was developed for moving beds whose upper surface is kept at atmospheric pressure.

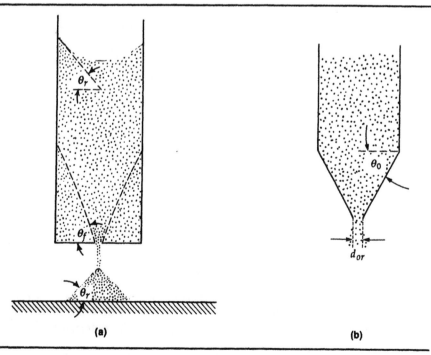

(a)   (b)

FIGURE 7
Downward discharge of particles without aeration through an orifice.

TABLE 1   Repose Angle for Solid Particles (abridged from [7])

Solids	Mean Bulk Density (kg/m^3)	Repose Angle θ_r (deg)	Angle of Internal Friction θ_f (deg)
Calcium oxide, powdered	433	43	—
Catalyst, fluid cracking, 60 μm	513	32	79
TCC beads, 3 mm	729	35	72
Coke, pulverized	400	34	—
Dolomite, pulverized	738	41	—
Glass beads, 290 μm	1470	26	—
5.4 mm	1360	32	—
Iron powder, 130 μm–3.6 mm	2200–2400	40–42	—
Lead shot, 1.3–6.4 mm	6600–6800	23–33	—
Limestone, pulverized	1360	47	—
Portland cement	1630	39	—
Sand, 480 μm	1460	37	64
Steel balls, 8.9–13 mm	4800–5000	33–37	—
Wheat	770	23	55

For coarse solids such as TCC catalyst, cereal grains, or crushed coal ($d_p > 127$ μm), it has been found that the rate of discharge is given by Eq. (21) and is not influenced by the length of the pipe as long as it is over three to four pipe diameters.

With fine cracking catalyst and iron ore, 100% smaller than 100 μm, Judd et al. [10, 11] measured the pressure drop across an orifice at the bottom of a standpipe ($d_t = 25$ and 50.8 mm, $d_{or}/d_t = 0.25$–0.77). Their results fitted the following equation, which is based on a momentum balance across the orifice:

$$G_s = \frac{F_s}{A_t} = 0.7\left[\left(\frac{1}{1-(d_{or}/d_t)^4}\right)2\rho_s(1-\varepsilon)\Delta p_{or}\right]^{1/2}$$ (22)

where Δp_{or} is the pressure drop across an orifice.

Provided that the gas and solid issue from the orifice at the same velocity, a material balance gives

$$\frac{G_s/\rho_s}{1-\varepsilon} = \frac{G_g/\rho_g}{\varepsilon}$$ (23)

The particle discharge rate from hoppers such as shown in Fig. 7 can be increased by introducing aeration gas near the orifice opening. Altiner and Davidson [12, 13] studied this phenomenon theoretically and experimentally.

Moving Bed Downflow

Coarse particles are usually made to move down vertical pipes rather slowly and in moving bed flow. In this type of flow the movement of gas relative to the solids, not relative to the pipe walls, determines the pressure gradient in the pipe.

Consider a fixed bed of solids of voidage ε_m moving downward at a constant velocity u_s (downward positive). Let

> u_g be the upflow velocity of gas
> $\Delta u = u_g + u_s$ be the relative velocity of gas with respect to the solids
> $u_o = \varepsilon_m u_g$ be the superficial rise velocity of gas up the downcomer pipe

These velocities are related by the equation

$$\Delta u = u_g + u_s = \frac{u_o}{\varepsilon_m} + u_s \tag{24}$$

The frictional pressure drop between any two levels in the downcomer is obtained by a slight modification of Eq. (3.6), as follows:

$$\frac{\Delta p_{fr}}{L} g_c = 150 \frac{(1 - \varepsilon_m)^2}{\varepsilon_m^2} \frac{\mu(\Delta u)}{(\phi_s d_p)^2} + 1.75 \frac{1 - \varepsilon_m}{\varepsilon_m} \frac{\rho_g(\Delta u)^2}{\phi_s d_p} \tag{25}$$

Equation (25) should apply as long as the particles do not fluidize—that is, whenever $\Delta u / \varepsilon_{mf} < u_{mf}$. The adequacy of Eqs. (24) and (25) were confirmed by Yoon and Kunii [14] in their experiments with moving beds of glass beads, $\bar{d}_p = 130$–$1130\ \mu m$, in $d_t = 41$- and 70-mm downcomer pipes.

Now, when $u_g > u_s$, gas is able to work its way up the pipe against the downflowing solids. Conversely, when $u_g < u_s$, gas is dragged downward against the pressure gradient. Between these two situations, when $u_g \cong u_s$, the gas can make no headway against the downflowing solids and stays in place.

In moving beds of fine particles the relative velocity Δu becomes very small. Thus the solids carry with them most of their accompanying void gas, and we have $-u_o = \varepsilon_m u_s$.

Fluidized Downflow

Suppose the gas flow rate is controlled so as to maintain incipient fluidization in the downflowing solids. With increasing gas flow, bubbles form and move upward relative to the downflowing solids. Letting u_b be the mean bubble rise velocity in a stationary bed, and upward flow rate of gas is reasonably approximated as follows:

For Geldart **A** solids for which gas and solids descend at close to the same velocity,

$$u_o \cong \delta(u_b - u_s) - (1 - \delta)\varepsilon_{mf} u_s \tag{26}$$

For Geldart **B** solids for which gas and solids flow at significantly different velocities,

$$u_o \cong \delta(u_b - u_s) + (1 - \delta)(u_{mf} - \varepsilon_{mf} u_s) \tag{27}$$

When gas bubbles are present in fine particle (Geldart **A**) downcomers and when u_s is large enough to give $u_s \cong u_b$, bubbles remain stationary in the bed. In this situation Eq. (26) shows that gas flows downward in the pipe. Also, these stationary bubbles likely sweep up and coalesce with smaller bubbles to form large slugs.. To prevent this, one should try to control the bubble size so that u_b is very different from u_s.

In fluidized downflow the pressure drop is approximated by the hydrostatic expression

$$\frac{\Delta p_{fr}}{L} g_c = \bar{\rho} g = \rho_s (1 - \varepsilon_f) g \tag{28}$$

Fluidized Downflow in Tall Downcomers

For tall downcomers, say 30 m, the pressure at the bottom can be two to three times as high as at the top. This can lead to serious difficulties for smooth operations. Ideally one would want minimum fluidizing conditions throughout the pipe, wherein the emulsion flows like a liquid, without bubbles. Design should try to closely approach this condition.

With Geldart **B** solids, bubbles form whenever gas flow exceeds u_{mf} relative to the descending solids. Once formed, these bubbles coalesce and grow into slugs that sometimes bridge and block the downflow of solids. One way of preventing this is to design the downcomer with cross-sectional area inversely proportional to the gas pressure—in other words, larger on top and smaller at the bottom.

In contrast to this behavior, Geldart **A** solids flow downward much more smoothly and stably than do Geldart **B** solids. However, here one encounters a different type of problem. Consider a gas-solid mixture entering the top of a tall downcomer. Since these particles are very small, gas flows downward with the solids, and only a little slower than the solids. In flowing down the pipe, the pressure rises, reducing the voidage of the mixture to less than ε_{mf}. Thus fluidized bed flow is converted to moving bed flow, with the bed becoming more compact as it descends. To prevent this from happening, one needs aeration in the lower portions of tall downcomers.

Where is aeration needed, and how much is needed in the downcomer to prevent the transition from fluidized to moving bed flow? Consider two levels in a downcomer pipe and the same descending velocity for gas and solid, a reasonable approximation for the very fine Geldart **A** solids that are moving downward at many multiples of u_{mf}. Then from Eq. (23) the mass flux ratio of gas to solid at the upper level 1 in the downcomer is

$$\frac{G_{g1}}{G_s} = \frac{\rho_{g1}}{\rho_s} \left(\frac{\varepsilon_1}{1 - \varepsilon_1} \right) \tag{29}$$

Lower in the downcomer, at level 2, the voidage of the mixture decreases to ε_2 because the gas density has increased from ρ_{g1} to ρ_{g2}. Taking ratios of Eq. (29) with the ideal gas law for close to isothermal conditions gives

$$\frac{\rho_{g2}}{\rho_{g1}} = \frac{\varepsilon_1(1 - \varepsilon_2)}{\varepsilon_2(1 - \varepsilon_1)} = \frac{\text{pressure at level 2}}{\text{pressure at level 1}} \tag{30}$$

The amount of aeration gas needed to bring the voidage at level 2 up from ε_2 to ε_1 is then found from Eq. (29) to be

$$\frac{G_{g,\text{aeration}}}{G_s} = \left(\frac{G_{g2}}{G_s} \right)_{\text{after}} - \left(\frac{G_{g2}}{G_s} \right)_{\text{before}} = \frac{\rho_{g2}}{\rho_s} \left(\frac{\varepsilon_1}{1 - \varepsilon_1} - \frac{\varepsilon_2}{1 - \varepsilon_2} \right)$$

$$= \left(\frac{\varepsilon_1}{1 - \varepsilon_1} \right) \frac{\rho_{g1}}{\rho_s} \left(\frac{\rho_{g2}}{\rho_{g1}} - 1 \right) \qquad . \tag{31}$$

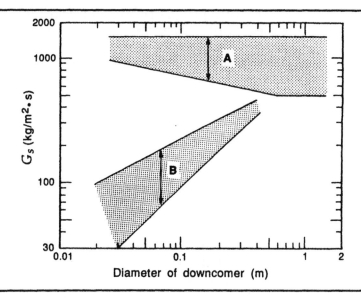

FIGURE 8
Range of solids mass flux: (a) vertical, for Geldart **A** particles (FCC, fluid coking); from Matsen [15,17]; (b) inclined, for coarser particles (Geldart **B**); best estimate.

This equation can be used to determine where and how much aeration is needed to keep the bed voidage between prescribed limits ε_1 and ε_2; see Matsen [15], Dries [16], and Example 3.

For fine cracking catalyst in fluidized downflow, a typical value for the mass flux of solids in downcomers is about 1 ton/m^2·s (see Fig. 8).

Comments on Downcomers

The flow patterns that may exist or coexist in a downcomer are influenced by the end conditions on the downcomer, such as flow restrictions, types of flow, and pressure difference above and below the pipe. In addition, one can have more than one flow behavior in the pipe for given end conditions, depending on the past history of flow in the pipe. This is a complex subject in which our state of knowledge is still very incomplete. Here we only briefly examine this subject. We refer the interested reader to pertinent references that reflect today's knowledge on the subject.

Consider the downcomer of Fig. 9 that connects two vessels.

1. Its purpose is to transfer solids from vessel to vessel against the normal pressure gradient.

2. The largest pressure gradient that can be had in the downcomer is that corresponding to dense fluidized flow. Moving bed flow has a somewhat smaller pressure gradient, and lean-phase flow can only occur with a negligible pressure gradient. So, if the pressure gradient in the pipe is not to be negligible, one should maintain a dense gas-solids mixture. Stay away from lean-phase flow.

3. If, as a result of an upset in operations, the downcomer switches to lean-phase flow, gas would rush up the pipe, making it difficult if not impossible to return to dense-phase operations. Consequently, one has to guard against and make provisions to counter this type of upset.

4. If the pressure gradient in fluidized flow is insufficient to satisfy the required pressure difference $p_2 - p_1$, one has to design a longer downcomer.

5. If the upper restriction in Fig. 9 is dominant, solids may run out of the pipe faster than they enter; thus one may have lean-phase flow in the upper section of the pipe. This situation may also occur when the solids enter the pipe from a nonaerated hopper. Aerating the lower portion of the hopper would be helpful in such cases. Fluidized feed to a downcomer does not pose such problems.

6. If the restriction at the lower end of the downcomer dominates and is of the right size, one should be able to maintain fluidized flow throughout the downcomer.

7. The following references concern various aspects of downcomers.

- Pressure gradient in downcomers, at orifices, friction loss [10, 11, 15, 17–19, 30]
- Transition and coexistence of flow regimes [10, 15, 16, 19–23]
- Slug formation and bridging [15–17, 23]
- Sealing of downcomers, stripping of adsorbed gases [24, 25]
- Unstable operations [6, 20–22, 26–29]
- Aeration of downcomers [12, 13, 15–17, 19, 27, 37]
- Erosion of walls [25]
- Inclined standpipes [23, 25, 30–32]
- Mass flux [15, 17]
- Comprehensive reviews [21, 30, 33]

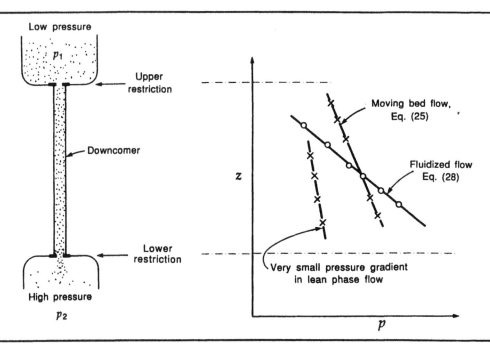

FIGURE 9
Typical pressure gradients in various forms of solid downflow.

EXAMPLE 3

Aeration
of a Fine
Particle
Downcomer

To prevent compaction or bubbling of the fine fluidized solids flowing at $F_s =$ 100 kg/s down a 25-m-long, 340-mm ID downcomer, we plan to operate the unit in a voidage range of $\varepsilon = 0.50 - 0.55$. If solids enter the standpipe at $\varepsilon_1 = 0.55$, determine where aeration should be introduced and at what rates.

Data

$$p_1 = 1.2 \text{ bar} = 120 \text{ kPa}, \ \rho_s = 1000 \text{ kg/m}^3, \ \rho_{g1}/\rho_s = 10^{-3}$$

SOLUTION

Without aeration the flowing gas is compressed as it moves down the pipe, and the pressure at the lower level 2, at which the voidage of the downflowing mixture becomes $\varepsilon_2 = 0.50$, is given by Eq. (30):

$$\frac{\rho_{g2}}{\rho_{g1}} = \frac{\varepsilon_1(1 - \varepsilon_2)}{\varepsilon_2(1 - \varepsilon_1)} = \frac{0.55(1 - 0.5)}{0.5(1 - 0.55)} = 1.222 = \frac{p_2}{p_1}$$

Thus

$$p_2 - p_1 = 1.222p_1 - p_1 = 0.222(120) = 26.7 \text{ kPa}$$

Since flow is fluidized in the pipe, the static head expression of Eq. (28) applies. Thus noting that the mean voidage in this pipe is $\bar{\varepsilon} = (0.50 + 0.55)/2 = 0.525$, we rearrange Eq. (28) to get

$$\Delta h_{12} = \frac{\Delta p_{12} g_c}{\rho_s(1 - \bar{\varepsilon})g} = \frac{(26,700)(1)}{(1000)(1 - 0.525)(9.8)} = 5.74 \text{ m}$$

To return the voidage to $\varepsilon = 0.55$, the amount of aeration gas needed is found from Eq. (31) to be

$$\frac{G_{g,\text{aeration}}}{G_s} = \frac{0.55}{1 - 0.55} \frac{1}{1000} (1.222 - 1) = 2.716 \times 10^{-4}$$

The flux of solids in the downcomer is

$$G_s = \frac{F_s}{A_t} = \frac{100 \text{ kg/s}}{(\pi/4)(0.34)^2 \text{ m}^2} = 1101 \text{ kg/m}^2 \cdot \text{s}$$

Hence the gas aeration rate needed is

$$G_{g,\text{aeration}} = (2.716 \times 10^{-4})(1101) = 0.30 \text{ kg/m}^2 \cdot \text{s}$$

or

$$F_g = G_g A_t = 0.30 \left[\frac{\pi}{4} (0.34)^2 \right] = 0.0272 \text{ kg/s}$$

Since the first aeration point is only 5.7 m below the top of the 25-m pipe, we may expect that additional aeration points may be required lower down the pipe. Repeating the above calculations, we find that three aeration points are needed. Summarizing these calculations gives

Aeration point	Location below top of downcomer (m)	Amount of aeration gas needed (kg/s)
1	5.74	0.0272
2	12.74	0.0366
3	21.29	0.0447

EXAMPLE 4

Circulation in Side-by-Side Beds

Determine the diameter of downcomers needed in a circulation system similar to Fig. 3(a), if the required circulation rate of solid is to be 600 kg/s. Also calculate the height of downcomers Δh_A and Δh_B necessary to provide the driving force for this circulation rate.

Data

The mean size of solids is 60 μm.
Referring to Fig. 3(a),

$$p_A = 120 \text{ kPa}, \qquad p_B = 180 \text{ kPa}$$

Bed heights are

$$L_{f,A} = h_{10} - h_9 = 8 \text{ m}, \qquad L_{f,B} = h_4 - h_3 = 8 \text{ m}$$

The bulk densities in kg/m^3 are

$$\rho_{1,2} = 100, \quad \rho_{3,4} = 400, \quad \rho_{4,5} = 550, \quad \rho_{6,7} = \rho_{7,8} = 200$$
$$\rho_{9,10} = \rho_{10,11} = 400, \qquad \rho_{11,12} = 550, \quad \rho_{13,1} = \rho_{1,2} = 100$$

The pressure drops across the distributors in the regenerator and reactor, respectively, are

$$\Delta p_{8,9} = \Delta p_{d,A} = 7 \text{ kPa}, \qquad \Delta p_{2,3} = \Delta p_{d,B} = 7 \text{ kPa}$$

The friction loss and Δp needed to accelerate the solids through the transfer lines are

$$\Delta p'_{1,2} = (9 + 4) \text{ kPa}, \qquad \Delta p'_{7,8} = (15 + 3) \text{ kPa}$$

The friction losses across the reactor's stripper-downcomer and the regenerator's downcomer are

$$\Delta p_{4,5} = 20 \text{ kPa}, \qquad \Delta p_{11,12} = 4 \text{ kPa}$$

The pressure drops assigned for the two control valves are

$$\Delta p_{12,13} = \Delta p_{v,A} = 5 \text{ kPa}, \qquad \Delta p_{5,6} = \Delta p_{v,B} = 15 \text{ kPa}$$

Take the heights of the risers to be

$$h_2 - h_{13} = 15 \text{ m}, \qquad h_8 - h_6 = 30 \text{ m}$$

Finally we make the following reasonable approximations:

$$p_{13} = p_1, \quad p_6 = p_7, \quad \text{and} \quad h_{13} = h_1, \quad h_6 = h_7$$

SOLUTION

Diameter of downcomer. From Fig. 8, select $G_s = 900 \text{ kg/m}^2 \cdot \text{s}$. Then the diameter of downcomers needed is

$$A_t = \frac{\pi}{4} d_t^2 = \frac{F_s}{G_s} = \frac{600}{900}; \quad \text{thus, } d_t = \underline{\underline{0.92 \text{ m}}}$$

Height of downcomer A Equations (7) and (8) apply directly. With Eq. (7) we have

$$(p_B - p_A)g_c = \rho_{1,2}g(h_1 - h_2) + \rho_{3,4}g(h_3 - h_4) + \rho_{10,11}g(h_{10} - h_{11})$$
$$+ \rho_{11,12}g(h_{11} - h_{12}) + \rho_{13,1}g(h_{13} - h_1)$$
$$- (\Delta p_{1,2} + \Delta p_{d,B} + \Delta p_{11,12} + \Delta p_{v,A})g_c$$

Inserting values gives

$$(180 - 120)10^3 = 100(9.8)(-15) + 400(9.8)(-8) + 400(9.8)(7)$$
$$+ 550(9.8)\,\Delta h_1 + 100(9.8)0 - [(9 + 4) + 7 + 4 + 5]10^3$$

From which

$$\Delta h_A = \underline{\underline{20\,m}}$$

Height of downcomer B Similarly, with Eq. (8) we have

$$(p_B - p_A)g_c = \rho_{4,5}g(h_5 - h_4) + \rho_{6,7}g(h_7 - h_6) + \rho_{7,8}g(h_8 - h_7)$$
$$+ \rho_{9,10}g(h_{10} - h_9) + (\Delta p_{4,5} + \Delta p_{v,B} + \Delta p_{7,8} + \Delta p_{d,A})g_c$$

and, with values inserted,

$$(180 - 120)10^3 = 550(9.8)(-\Delta h_B) + 200(9.8)(0) + 200(9.8)(30)$$
$$+ 400(9.8)(8) + [20 + 15 + (15 + 3) + 7]10^3$$

from which the height of downcomer should be

$$\Delta h_B = \underline{\underline{16.7\,m}}$$

Note: By choosing different riser heights, we can raise and lower reactor A with respect to B and thereby change the pressure difference $p_A - p_B$.

EXAMPLE 5

**Steam Seal
of a Coarse
Particle
Downcomer**

Coarse 1-mm solids descend in moving bed flow at $u_s = 0.15\,m/s$ through a 0.8-m ID downcomer 15 m long that connects two vessels. The pressure is 300 kPa in the lower vessel and 240 kPa in the upper, so gas percolates through the solids from the lower to the upper vessel. This is unacceptable.

To counter this flow, steam will be introduced into the downcomer 10 m from the top in such a way that the pressure at the injection point is equal to the pressure in the lower chamber.

(a) Determine the upward leak rate of gas in the absence of steam sealing.

(b) Sketch a pressure versus height profile in the downcomer with and without steam sealing.

(c) Determine the necessary steam feed rate for steam sealing. Note that some of the injected steam will flow downward, some upward.

Data

$$\phi_s = 0.8, \quad \varepsilon_m = 0.45, \quad \mu = 4 \times 10^{-5}\,kg/m{\cdot}s$$
$$\rho_{g,lower} = 2\,kg/m^3, \quad \rho_{g,upper} = 1.6\,kg/m^3, \quad \text{or} \quad \bar{\rho}_g = 1.8\,kg/m^3$$

SOLUTION

(a) *Without steam seal.* Equation (25) allows us to find the velocity of upflowing gas relative to the downflowing solids. Inserting values gives

$$\frac{60,000}{15}\,(1) = 150\,\frac{(1 - 0.45)^2}{(0.45)^2}\,\frac{(4 \times 10^{-5})\,\Delta u}{(0.8 \times 10^{-3})^2} + 1.75\,\frac{1 - 0.45}{0.45}\,\frac{(1.8)(\Delta u)^2}{0.8 \times 10^{-3}} \qquad (i)$$

from which

$$\Delta u = u_g - u_s = 0.2620\,m/s$$

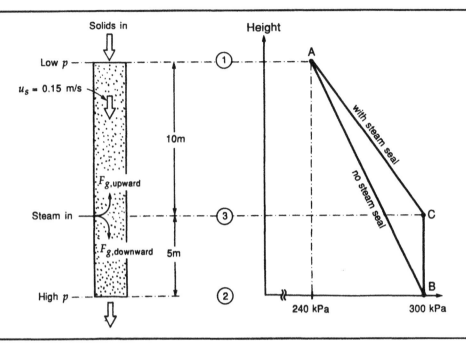

FIGURE E5
Pressure profile in a steam-sealed downcomer.

and

$$u_o = u_g \varepsilon_m = (\Delta u + u_s)\varepsilon_m = (0.2620 - 0.15)(0.45) = 0.0504 \text{ m/s}$$

Thus the flow rate of gas up the tube is

$$F_g = \bar{\rho}_g u_o A_t = 1.8(0.0504)\,\frac{\pi}{4}\,(0.8)^2 = \underline{0.0456 \text{ kg/s}} \quad \text{(or } \sim 25 \text{ L/s)}$$

(b) *p versus h graph* In Fig. E5 line AB represents no steam seal, just straight downflow of solids, and ACB represents the steam seal solution.

(c) *With steam seal* Referring to Fig. E5, we see that the gas at locations 2 and 3 is at the same pressure, so there is no movement of gas relative to the solids in this zone. Hence, between 2 and 3 the gas moves downward with the solids. Gas also moves up the bed from 3 to 1. We now calculate these two flow rates.

For section 1 to 3—upflow From Eq. (25) we obtain quantities identical to Eq. (i), except that the 15 m is replaced by 10 m. Following the same procedure, we get

$$\Delta u = 0.3791 \text{ m/s}$$
$$u_{o,\text{upward}} = 0.1031 \text{ m/s}$$
$$F_{g,\text{upward}} = 0.0933 \text{ kg/s} \quad \text{(or } \sim 52 \text{ L/s)}$$

For section 3 to 2

$$u_{g,\text{downward}} = 0.15 \text{ m/s}$$

$$\therefore u_{o,\text{downward}} = 0.15(0.45) = 0.0675 \text{ m/s}$$

$$\therefore F_{g,\text{downward}} = \rho_g u_o A_t = (2)(0.0675)\left(\frac{\pi}{4}\,0.8^2\right) = 0.0679 \text{ kg/s}$$

Summing, the total flow rate of steam is

$$F_{g,\text{total}} = F_{g,\text{up}} + F_{g,\text{down}} = 0.0933 + 0.0679 = \underline{0.1612 \text{ kg/s}} \quad \text{(or} \sim 86 \text{ L/s)}$$

Note: Steam should be sent into the moving bed through a distributor ringing the pipe wall to prevent the formation of bubbles.

Flow in Pneumatic Transport Lines

Vertical Upflow of Solids

As mentioned in Chap. 3, the pneumatic transport regime involves very dilute gas-solid mixtures (<1% solids) or a solids-to-gas mass ratio of less than about 10 to 20 (see Fig. 3.14). This flow regime is sometimes incorporated into the upflow section of solids circulation systems (see Figs. 1–3).

The general characteristics of this flow regime are shown in Fig. 10. Consider a constant feed rate of solids G_s to the bottom of this vertical transport line, represented by line CDE. Start at a high gas velocity (point C), and let the gas flow rate be gradually reduced. Two things happen: the frictional resistance decreases while the voidage decreases, which increases the static head. From point C to point D the decrease in frictional resistance dominates, so the total pressure drop decreases. Past point D a further lowering in gas flow rate causes a rapid rise in solid inventory and static head, and this dominates, causing a rise in total pressure. In approaching point E the bulk density of the mixture becomes too great to keep the particles apart, and they collapse into a fluidized mass in the transfer line. This phenomenon is called *choking*, and the superficial gas velocity at point E is called the *choking velocity* u_{ch} (see Zenz and Othmer [8]).

Choking can lead to immediate shutdown in conventional pneumatic conveying systems, whose blowers are not designed to deliver a high enough pressure to handle fluidized transport of solids. Actually engineers are more concerned with finding the limits of safe operations rather than the choking phenomenon itself. These are the stable conditions wherein a slight change in

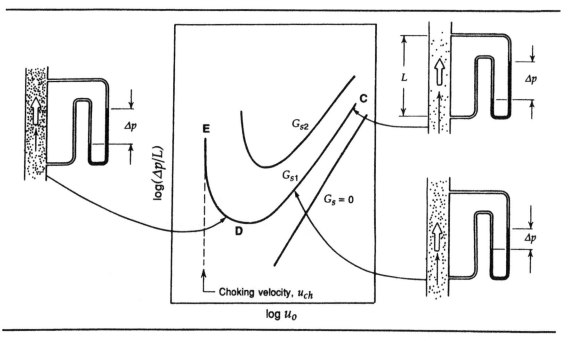

FIGURE 10
Behavior of lean upflow mixtures; adapted from Zenz and Othmer [8].

operating conditions or solid flux will not cause the system to rapidly fall into choking behavior. In Fig. 10 this means operating on the right arm of the curves, or section CD rather than DE, and not too close to point D.

Following are some pertinent references on the design of pneumatic transport lines [32, 34–36, 38–41].

Horizontal Flow

Horizontal flow is possible only with suspended or fluidized solids. If suspended solids are fed without additional aeration, say from a fluidized bed to a horizontal pipe, the behavior shown in Fig. 11 observed. Near the pipe entrance the solids retain their fluidized state with uniform density, and if the flow velocity is high enough, this state is maintained along the length of the pipe (see Fig. 11(a)). At lower velocities particles begin to settle in the pipe and form dunes (see Fig. 11(b)). This is called *saltation*. As solids progressively settle in the pipe, the dunes flow, sometimes almost filling the pipe. Small ripples seem to travel along the top of the thick solid layer, and the lower portion appears to be practically stationary (see Fig. 11(c)). Alternatively, depending on the solid-gas ratio, intermittent flow of solids (see Fig. 11(d)) may occur in place of dune formation. Qualitatively similar behavior is observed in inclined transport lines.

Figure 12 illustrates the phenomenon of saltation. Point C on the G_{s1} curve represents a high enough gas velocity so that all the particles are conveyed in suspension without sedimentation or saltation. Next, keeping the feed rate of solids unchanged at G_{s1}, slowly reduce the gas velocity. The frictional loss decreases, and at point D particles begin to settle to the bottom of the pipe. This reduces the effective flow channel and causes a sharp rise in pressure drop, up to point E. Point D is called the *saltation velocity* $u_{o,cs}$. At velocities below point E the flow channel is further restricted and the pressure drop rises steadily.

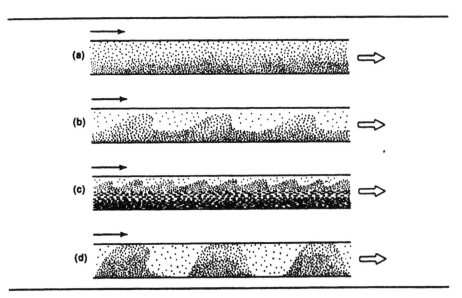

FIGURE 11
Observed flow patterns of horizontal flowing gas-solid mixtures; from Wen and Simons [43].

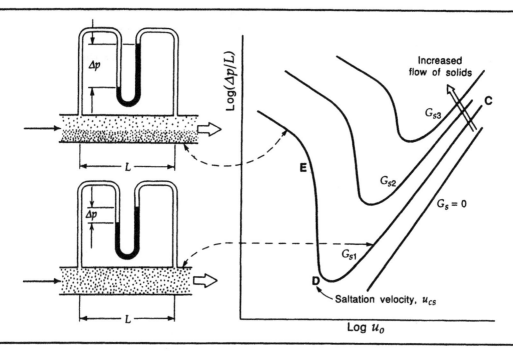

FIGURE 12
Behavior of lean horizontal flowing gas-solid mixtures; adapted from Zenz and Othmer [8].

Jones and Leung [44] compared eight published correlations, using the accumulated data of many workers in the field, and recommended the correlation of Thomas [45] as the most accurate in predicting the saltation velocity. Naturally, one should operate well above this velocity.

Safe Gas Velocity for Pneumatic Transport

When a pneumatic transport line is designed for a solid circulation system, choking and saltation should definitely be avoided. From wide experience, conventional pneumatic conveying technology has come up with suggested safe gas velocities for horizontal and upward vertical lean-phase transport of solids (see Table 2), and they are larger than $u_{o,ch}$ and $u_{o,cs}$. Also, as may be seen, faster gas velocities are usually needed for horizontal flow.

Warnings: In some physical and chemical processes, fine solids become sticky and agglomerate. Also note that the thickness of an agglomeration or a settled layer usually increases with time. Thus, avoid horizontal transfer lines wherever possible and replace them with inclined pipes or goosenecks (upside-down U tubes).

Pressure Drop in Pneumatic Transport

The pressure difference between two points of a pneumatic transport pipe is found by the ordinary Bernoulli equation modified to account for the flow of the gas-solid mixture rather than just one phase. Consider the pipe inclined upward

TABLE 2   Safe Values for Pneumatic Transport (abridged from [7])

Material	Approx. Size (μm)	Av. Bulk Density (kg/m^3)	Min. Safe Air Velocity (m/s)		Max. Safe Density for Flow (kg/m^3)	
			↔	↕	↔	↕
Alumina	100% < 105	930	7.6	1.5	96	480
Ash	90% < 150	720	4.6	1.5	160	480
Bauxite	100% < 105	1440	7.6	1.5	130	640
Bentonite	95% < 76	770–1040	4.6	1.5	160	480
Cement	95% < 88	1040–1440	7.6	1.5	160	960
Coal	100% < 380, 75% < 76	560	4.6	1.5	110	320
	100% < 6.4 mm	720	12.2	9.2	16	24
	100% < 12.7 mm	720	15.3	12.2	12	16
Magnesite	90% < 76	1600	9.2	3.1	160	480
Phosphate rock	90% < 152	1280	9.2	3.1	110	320
Silica	95% < 105	800–960	6.1	1.5	80	320
Soda ash (dense)	50% < 177	1040	12.2	3.1	48	160
Soda ash (light)	66% < 150	560	9.2	3.1	80	240
Sodium sulphate	100% < 500, 55% < 105	1280–1440	12.2	3.1	80	240
Uranium dioxide	100% < 152, 50% < 76	3520	18.3	6.1	160	960
Wheat	4.8 mm	750	12.2	9.2	24	32

at an angle θ from the horizontal, and let solids be introduced at point 1 as shown in Fig. 13. The gas flows at high velocity, so the kinetic energy of the accelerating solid may be significant and may have to be accounted for. However, since the volume fraction of solids is small in pneumatic conveying, the change of gas velocity is small and can be ignored. Under these conditions

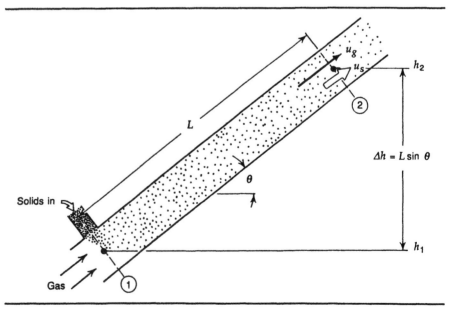

FIGURE 13
Sketch used to develop the mechanical energy balance for a flowing gas-solid mixture.

the pressure difference is made up of three terms that account for the static head, the kinetic energy of solids, and the frictional resistance Δp_f of the mixture with the pipe wall, or

$$p_1 - p_2 = \frac{\bar{\rho}gL \sin \theta}{g_c} + \frac{u_s G_s}{g_c} + \Delta p_f \tag{32}$$

Assuming plug flow for gas and for solid, the mean density of the mixture is

$$\bar{\rho} = \rho_s(1 - \varepsilon_g) + \rho_g \varepsilon_g = \frac{G_s}{u_s} + \frac{G_g}{u_g} = \rho_g \varepsilon_g \left(1 + \frac{G_s}{G_g} \frac{u_o}{u_s}\right) \tag{33}$$

where

$$G_g = u_o \rho_g = u_g \rho_g \varepsilon_g$$
$$G_s = u_s \rho_s(1 - \varepsilon_g)$$

When $G_s/G_g \gg 1$ and $\varepsilon_g \cong 1$, we have

$$\bar{\rho} \cong \rho_g \frac{G_s}{G_g} \frac{u_o}{u_s} \tag{34}$$

In fully developed flow the acceleration term in Eq. (32) can be dropped.

The friction loss term in Eq. (32) has been estimated in a number of ways. The simplest uses a friction factor f'_s with the mean density of the flowing mixture given by Eq. (34), or

$$\Delta p_f = \frac{2f'_s \bar{\rho} u_o^2 L}{g_c d_t} \tag{35}$$

This expression is often used in practice, with values of f'_s given in Table 3.

An alternative approach considers the frictional loss to consist of two components: one due to fluid, the other, to solid. Thus

$$\Delta p_f = \Delta p_{f,g} + \Delta p_{f,s}$$
$$= \frac{2f_g \rho_g u_g^2 L}{g_c d_t} + \frac{2f_s G_s u_s L}{g_c d_t} \tag{36}$$

TABLE 3   Values of Coefficient f'_s versus Air Velocities (abridged from [7])

Solids	Size	Pipe Dia. (mm) (Horizontal or Vertical)	Air Velocity (m/s)			
			10	15	20	30
Coal	0–1 mm	25.4(H)	—	0.0014	0.0011	0.0011
Limestone	Various sizes up to 3.2 mm	51(H)	0.017	0.0040	0.0033	0.0033
Salt	76–252 μm	44.4(H)	0.0065	—	—	—
		44.4(V)	0.018	0.016	—	—
Sand	0.8–1.4 mm	44.4(H)	0.005	0.0045	—	—

where the gas friction factor f_g is a function of the tube Reynolds number, $Re_t = d_t u_o \rho_g / \mu$, as follows:

$$f_g = 0.0791\, Re_t^{-0.25} \qquad \text{for } 3 \times 10^3 < Re_t < 10^5 \tag{37}$$

and

$$f_g = 0.0008 + 0.0552\, Re_t^{-0.237} \qquad \text{for } 10^5 < Re_t < 10^8 \tag{38}$$

For f_s near atmospheric conditions, Leung and Wiles [36] recommend the following equation, which they say is particularly applicable to commercial-scale risers:

$$f_s = \frac{0.05}{u_s} \qquad (u_s \text{ in m/s}) \tag{39}$$

For high-pressure systems Knowlton and Bachovchin [35] correlated their findings up to 40 bar by the equation

$$f_s = 0.0252 \left(\frac{G_s}{\rho_g u_g} \right)^{0.0415} \left(\frac{u_s}{u_g} \right)^{-0.859} - 0.03 \tag{40}$$

Pressure Drop in Bends

An additional pressure drop is caused by a bend: the smaller the ratio of bend radius r_b to pipe diameter d_t, the larger the pressure drop Δp_b. According to [7], this loss can be estimated from the expression

$$\Delta p_b = 2 f_b \bar{\rho} u_o^2 \tag{41}$$

where f_b = 0.375, 0.188, and 0.125 for r_b/d_t = 2, 4, 6 or more, respectively. A small bend radius gives a high-pressure drop and may also cause severe erosion of the pipe wall and attrition of solids. Because of all these undesirable consequences, long radius bends should be used in the transport lines of a circulation system.

Despite these warnings, right-angle exits are sometimes designed for the top of riser reactors operating in the fast fluidized regime. Here some particles stop at the top of the riser to form a "cushion" of particles. Erosion of pipe and particles is said to be reduced dramatically compared to an ordinary bend. This design aims to solve the erosion problem by a sacrifice of pressure drop.

Practical Considerations

Mechanical Valves for the Control of Solids Flow. Valves located in the lower section of moving and fluidized beds or in the downflow section of transportation lines are used to control the flow of gas-solid mixtures. *Disk valves* are commonly used when solids are in the aerated state, whereas *slide valves* are used for both moving bed and aerated flow. Since these valves do not give a tight seal, clean purge gas is commonly used to remove solids where the clearance is small. For longer-lasting, better control of the flow, these valves are sometimes used in pairs. *Cone valves* are used for flow control of fine solids drawn from a fluidized bed into a pneumatic transport system, as shown in Fig. 2(a), and for the regulation of flow from a moving bed. Because of their conical shape, cone valves can maintain their effectiveness even after considerable wear; however, they should not be used where a tight seal is needed.

When these valves are used in downcomers, as shown in Figs. 1–3, solids just above the valve are compacted to form moving beds, which then may unduly restrict the flow of solids. For smooth and adjustable control of solids flow, adequate aeration should be incorporated into the design of valves.

Aeration Devices for the Control of Solids Flow. In place of mechanical valves one can control the flow of solids by adjusting the flow of aeration gas in the devices shown in Fig. 14. Parts (a), (b), and (c) show, respectively, L, J, and W valves. Knowlton and Hirsan [46, 47] investigated the flow characteristics of L and J valves and correlated the mass flux of solids, gas velocity, and pressure drop across the valve for a variety of solids.

Inventory. The amount of solids in a fluidized bed can easily be found by measuring the difference in pressure at two levels in the bed. To prevent clogging of the pressure taps by fine solids, some gas is bled into the bed at low velocity through the taps, as shown in Fig. 15.

Transfer of Solids between Side-by-Side Fluidized Beds. Jin et al. [4] and Kuramoto et al. [5] measured the flow rate of solids across openings between fluidized beds. Figure 16 shows these rates calculated from their reported data.

Dense Upflow of Solids. For Geldart **B** or **D** particles, pneumatic transport requires very high gas velocities and considerable pumping power. Because of these high velocities, this type of flow often leads to serious erosion problems at key points in the transport system. To bypass these problems, some processes have adopted upflow fluidized solids in their circulation systems; for example, see Figs. 2.16(b), 2.21(a), and Fig. 3(b).

In these operations the upflow rate of solids is controlled by adjusting the amount of aeration gas fed to the bottom of the vertical transport line, which should have ready access to fresh fluidized solids and have a higher pressure than the top.

According to Sugioka et al. [1], G_s up the vertical transport line varies

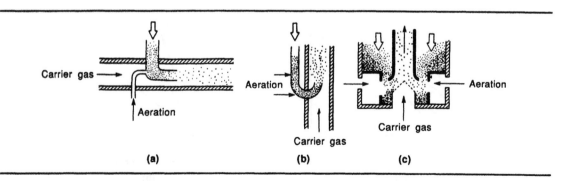

FIGURE 14
Control of flow rate of solids by aeration gas; from Zenz and Othmer [8]: (a) the L valve; (b) the J valve; (c) the W valve.

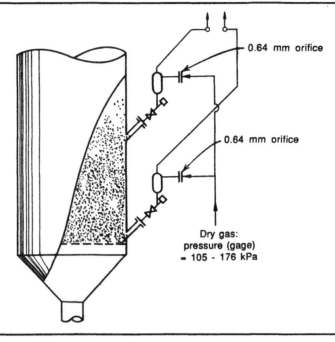

FIGURE 15
Differential pressure level instrument; adapted from Kraft et al. [48].

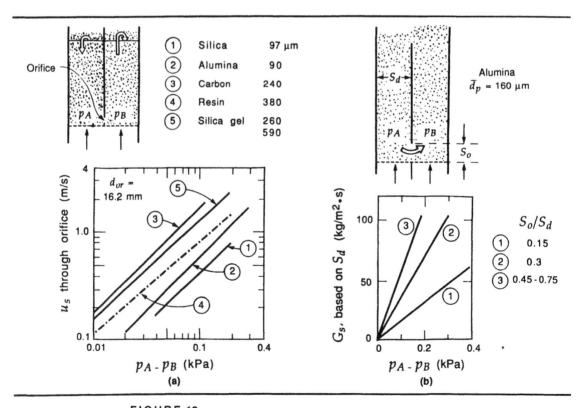

FIGURE 16
Solids mass flux across an orifice and an opening: (a) from Jin et al. [4]; (b) from Kuramoto et al. [5]. S_d is the cross sectional area of downcomer.

linearly with pressure difference at the same level at the bottom junction of the downflow and upflow lines (see points 5 and 6, and points 2 and 10 of Fig. 3(b)). A sample of their results are as follows:

	Pressure difference = 10 kPa		
Size of iron ore (μm)	50	100	200
Mass flux, G_s (kg/m^2·s)	34	45	70

The flow characteristics and driving force for solids circulation have been investigated in simple circuits by several workers [2–5].

Cyclones. In large systems with rapid solid circulation, the entrainment of fines by gas is a serious problem. To avoid rapid loss of solids, cyclones and electrostatic precipitators should have a very high collection efficiency, even after possible deformation and deterioration due to use. For this reason two- or three-stage cyclone collectors are often used for systems of fine particles. Figure 17(a) illustrates a three-stage cyclone. The level of the inlet to the first cyclone is determined by the TDH (see Chaps. 3 and 7), and diplegs play an important role as pressure seals against the considerable back pressure generated. The diameter of the diplegs should be made progressively smaller to reflect the decreasing separation rate of fines in the three stages.

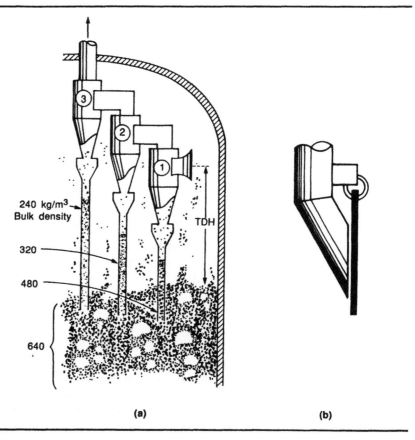

(a)   (b)

FIGURE 17
Three-stage cyclone and trickle valve of Ducon Co.

During start-up, when the diplegs are not primed with solids, gas and solid may short-circuit through diplegs, resulting in serious loss of solids. Many devices have been used to prevent this, one of which, a flap-type device known as a *trickle valve* (Ducon Company), is shown in Fig. 17(b). It can be used under all conditions except high-temperature corrosive atmospheres.

Stable Multistage Operations. Proper design of overflow pipes and downcomers is important because the stable downflow of solids through them is easily disturbed by slugging, gas bypassing, and backward flow of solids. In multistage operations this may result in one stage being completely clogged with solids while the next stage is practically bare. To be quite safe from these disruptive and unbalancing phenomena, devices such as the trickle valve are most effective.

Static Electricity. Many other factors enter into design. One is the problem of static electricity, generated when solids are transported in pipelines, and the accompanying hazards of explosion of combustible powders. These effects are more serious with dry gas and solids of low electrical conductivity. Proper grounding will minimize these effects.

Horizontal Flow and U Tubes. Another problem is the settling of solids on the bottom of U bends and horizontal pipes when transporting high-bulk-density mixtures. The number and length of such sections should be minimized, and then those that are used should be equipped with adequate aeration. This factor should be considered in an early stage of design.

Start-Up. Start-up of a large bed of solids can be difficult and dangerous because the large excess of pressure needed to lift the solids, and the large initial slug of the many tons of solid, can rip out bed internals and cause serious structural damage to the unit. Thus, solids are always removed before shutdown and are progressively introduced during start-up. Moreover, design should incorporate provisions to avoid any accidental collapse of a bed full of solids.

Reflections. Designing successful circulation systems is not easy. It requires a knowledge and careful analysis of all pertinent factors, but above all it requires good judgment to know which factors are important and which can be ignored. Much work is involved; however, probably more than in any other aspect of design, success and failure here will determine the success or failure of the process as a whole.

PROBLEMS

1. Given the fluid catalytic reactor of Fig. 3(a) with a bed inventory of $W = 80$ tons, find the rate of solid circulation that would keep the mean activity of the catalyst in the bed at 10% of the fresh catalyst. Assume the catalyst is fed into the bed completely regenerated. Determine the ratio of solid circulation to the feed hydrocarbon if 600 tons/day of feed are treated by the reactor. According to experiment, the activity of the catalyst drops to one-half that of fresh catalyst after a 1-min stay in the reactor.

2. In the circulation system of Example 2 the deposit of carbon is reduced to $\beta_c = 5\%$ by using a different catalyst and different reaction conditions. To satisfy the heat balance, we must vaporize and preheat the feed before introducing it into the reactor. Determine the relation between the circulation ratio F_s/F_1 and both T_1 and F_2/F_1 when the fresh feed is vaporized and preheated to 420°C. Use the numerical data of Example 2 plus the specific heat of feed vapor of 3.01 kJ/kg·K and latent heat of vaporization of feed oil of 335 kJ/kg.

3. Figure P3 shows a proposed process using two fluidized beds to recover volatile organic matter from wet solid particles. Fresh solid is fed to the evaporator, where its moisture and volatile matter are immediately vaporized. The remaining solids are circulated to the burner, where a small fixed fraction of the solids is burned by the fluidizing air. Part of this combustion heat is then returned to the evaporator by the hot circulating solids, and the spent solids are discharged from the burner. Find the necessary circulation rate of the solids and the temperature of fluidizing gas entering the burner that would keep the evaporator at 700°C and the burner at 800°C.

Data. All feeds except the burner gas enter at room temperature, 20°C. Specific heat:

Combustion gas:	1.00 kJ/kg·K
Carrier gas:	1.00 kJ/kg·K
Volatile free solids:	0.96 kJ/kg·K

On the basis of 1 kg of fresh feed (water-free basis):

Nonvolatile solids:	0.75 kg
Entering combustion air:	0.70 kg
Entering carrier gas:	0.06 kg
Volatile matter:	0.25 kg
Nonvolatile solid burned:	0.03 kg

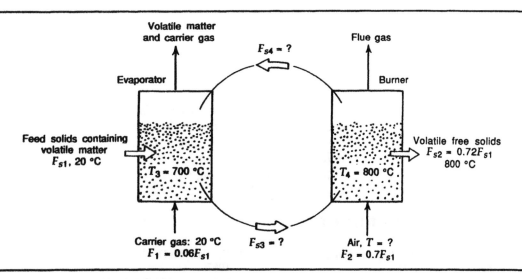

FIGURE P3
Variables of the proposed vaporization process.

Heat needed to bring moisture and volatile matter from 20 to 700°C
(latent heat included):   2193 kJ/kg
Heat of combustion at 800°C: 32350 kJ/kg
Estimated heat loss in each unit: 2% of combustion heat

4. In Example 5 we used a considerable amount of steam (86 L/s) to seal the pipe, and some of it flowed down the pipe, which seems wasteful and unnecessary. With steam introduced at the same location in the pipe, what is the minimum amount of steam needed, and what is the pressure, to completely seal the pipe from the upflow of gas from the lower to the upper vessel?

5. Again referring to the downcomer of Example 5, where along the pipe should we introduce steam, and at what pressure, to minimize the use of steam? How much steam would be needed for this operation?

6. Find the resulting pressure drop when solid particles are transported pneumatically by a gas through a vertical pipe 0.2 m and 20 m long. At the bottom end of the pipe the solids are accelerated.

Data

Solids: $\bar{d}_p = 200\ \mu m$, $\rho_s = 2000\ kg/m^3$, $u_t = 1.3\ m/s$
Gas: $\rho_g = 1.0\ kg/m^3$, $\mu = 2 \times 10^{-5}\ kg/m \cdot s$, $u_o = 20\ m/s$

Take $G_s/G_g = 10$, $f'_s = 0.0033$.

7. In a process similar to Fig. 2(a), determine the pressure drop needed in the two cone valves to keep a steady circulation of solid.

Data

$p_A = 160\ kPa$, $p_B = 120\ kPa$, $\bar{d}_p = 60\ \mu m$

Height of fluidized bed:

$L_{fA} = h_{10} - h_1 = 9\ m$, $h_9 - h_1 = 1\ m$, $L_{fB} = h_5 - h_4 = 8\ m$

Densities (kg/m^3):

$\rho_{10,1} = 400$, $\rho_{2,3} = 60$, $\rho_{4,5} = 400$, $\rho_{5,6} = 500$, $\rho_{7,8} = 600$

Across the distributor in the reactor: $\Delta p_{3,4} = 6\ kPa$
Friction and acceleration loss in the transport line: $\Delta p_{2,3} = 7\ kPa$
Friction loss by moving bed flow in the standpipe: $\Delta p_{7,8} = 50\ kPa$
Take $h_3 - h_2 = 20\ m$, $h_7 - h_8 = 19\ m$,
$\rho_{4,5}(h_5 - h_4) = \rho_{6,7}(h_6 - h_7)$.

REFERENCES

1. R. Sugioka et al., *Kagaku Kogaku Ronbunshu*, **11**, 93 (1985); **14**, 444 (1988).

2. H.A. Masson, *AIChE J.*, **32**, 177 (1986).

3. D. Fox, Ph.D. thesis, Tech. Univ. Compiegne, 1987.

4. Y. Jin et al., *Fluidization '85, Science and Technol-*

ogy, M. Kwauk et al., eds., p. 172, Kunming, Science Press, Beijing, 1985.

5. M. Kuramoto et al., *Powder Technol.*, **44**, 77 (1985); **47**, 141 (1986); *Kagaku Kogaku Ronbunshu*, **12**, 241 (1986).

6. O. Levenspiel, *Chemical Reactor Omnibook*, Chap. 33, OSU Book Stores, Corvallis, OR, 1989.

7. Engineers Equipment Users Association, *Pneumatic Handling of Powdered Materials*, Constable and Company, London, 1963.

8. F.A. Zenz and D.F. Othmer, *Fluidization*, Van Nostrand Reinhold, New York, 1956.

9. J.M. Rausch, Ph.D. thesis, Princeton Univ., 1948.

10. M.R. Judd and P.N. Rowe, in *Fluidization*, J.F. Davidson and D.L. Keairns, eds., p. 110, Cambridge Univ. Press, New York, 1978.

11. M.R. Judd and P.D. Dixon, *AIChE Symp. Ser.*, **74(176)**, 38 (1978).

12. H.K. Altiner and J.F. Davidson, in *Fluidization*, J.R. Grace and J.M. Matsen, eds., p. 461, Plenum, New York, 1980.

13. H.K. Altiner, *AIChE Symp. Ser.*, **79(222)**, 55 (1983).

14. S.M. Yoon and D. Kunii, *Ind. Eng. Chem. Process Des. Dev.*, **9**, 559 (1970).

15. J.M. Matsen, Professional Develop. Seminar, 2nd World Congress of Chem. Eng., Montreal, Oct., 1981.

16. H.W.A. Dries, in *Fluidization*, J.R. Grace and J.M. Matsen, eds., p. 493, Plenum, New York, 1980.

17. J.M. Matsen, in *Fluidization Technology*, vol. 2, D.L. Keairns, ed., p. 135, Hemisphere, Washington, D.C., 1976.

18. T. Hikita et al., in *Fluidization IV*, D. Kunii and R. Toei, eds., p. 219, Engineering Foundation, New York, 1983.

19. J.M. Matsen, *Powder Technol.*, **7**, 93 (1973).

20. L.S. Leung and P.J. Jones, in *Fluidization*, J.F. Davidson and D.L. Keairns, eds., p. 116, Cambridge Univ. Press, New York, 1978.

21. L.S. Leung, in *Fluidization*, J.R. Grace and J.M. Matsen, eds., p. 25, Plenum, New York, 1980.

22. P.J. Jones, C.S. Teo, and L.S. Leung, in *Fluidization*, J.R. Grace and J.M. Matsen, eds., p. 469, Plenum, New York, 1980.

23. R.A. Sauer et al., *AIChE Symp. Ser.*, **80(234)**, 1 (1984).

24. T.M. Knowlton et al., *AIChE Symp. Ser.*, **77(205)**, 184 (1981).

25. J.R. Bernard et al., *AIChE J.*, **32**, 273 (1986).

26. L.S. Leung, in *Fluidization Technology*, vol. 2, D.L. Keairns, ed., p. 125, Hemisphere, Washington, D.C., 1976.

27. J.C. Ginestra, S. Rangchari, and R. Jackson, in *Fluidization*, J.R. Grace and J.M. Matsen, eds., p. 477, Plenum, New York, 1980.

28. S. Rangachari and R. Jackson, *Powder Technol.*, **31**, 185 (1982).

29. P.J. Jones and L.S. Leung, in *Fluidization*, 2nd ed., J.F. Davidson et al., eds., p. 293, Academic Press, New York, 1985.

30. Y.O. Chong et al., *AIChE Symp. Ser.*, **80(234)**, 149 (1984).

31. T.Y. Chen et al., *Powder Technol.*, **22**, 89 (1979); *AIChE Symp. Ser.*, **74(176)**, 75 (1978); *AIChE J.*, **26**, 24, 31 (1980).

32. R.M. Enick and G.E. Klinzing, *Chem. Eng. Comm.*, **49**, 127 (1986).

33. T.M. Knowlton, in *Gas Fluidization Technology*, D. Geldart, ed., p. 341, Wiley, New York, 1986.

34. W.C. Yang, *AIChE J.*, **20**, 605 (1974); **21**, 1013 (1975).

35. T.M. Knowlton and M. Bachovchin, in *Fluidization Technology*, vol. 2, D.L. Keairns, ed., p. 253, Hemisphere, Washington, D.C., 1976.

36. L.S. Leung and R.J. Wiles, *Ind. Eng. Chem. Process Des. Dev.*, **15**, 552 (1976).

37. D.D. Do et al., in *Proc. Particle Technology*, H. Brauer and O. Molerus, eds., D23, Nuremberg, 1977.

38. J.M. Matsen, *Powder Technol.*, **32**, 21 (1982).

39. W.C. Yang, *Powder Technol.*, **35**, 143 (1983); *AIChE J.*, **30**, 1025 (1984).

40. J. Yerushalmi and A. Avidan, in *Fluidization*, 2nd ed., J.F. Davidson et al., eds., p. 226, Academic Press, New York, 1985.

41. R.A. Smith and G.E. Klinzing, *AIChE J.*, **32**, 313 (1986).

42. D.J. Wildman et al., *AIChE Symp. Ser.*, **83(255)**, 58 (1987).

43. C.Y. Wen and H.P. Simons, *AIChE J.*, **5**, 263 (1959).

44. P.J. Jones and L.S. Leung, *Ind. Eng. Chem. Process Des. Dev.*, **17**, 571 (1978).

45. D.G. Thomas, *AIChE J.*, **7**, 423 (1961); **8**, 373 (1962).

46. T.M. Knowlton and I. Hirsan, *9th Synthetic Pipeline Gas Symp.*, Chicago, Oct.–Nov., 1977; *Hydrocarbon Process.*, **57**, 149 (March 1978); *Proc. Int. Powder and Bulk Solid Handling and Processing Conf.*, Philadelphia, p. 46, May 1979.

47. T.M. Knowlton and I. Hirsan, *5th Int. Conf. Pneumatic Transport of Solids in Pipes*, E3, London, April 1980.

48. W.W. Kraft et al., in *Fluidization*, D.F. Othmer, ed., Van Nostrand Reinhold, New York, 1956.

16

Design for
Physical
Operations

— Heat Transfer
— Mass Transfer
— Drying of Solids

In this chapter the material previously presented is applied to the problems of design of fluidized beds for the physical operations of heat transfer, mass transfer, and drying.

Information Needed for Design

For fluidized bed operations the following information is needed:

1. The rate constant for single particles in the environment expected in the fluidized bed
2. The tendency of solids to agglomerate, break, or erode
3. The tendency of solids to coat the wall surfaces of the freeboard or exit duct
4. The effective bubble diameter expected in the bed
5. The possibility of an explosion in the exiting gas stream

Bench-scale experiments may suffice to yield information for items 1 to 3. For item 1 the kinetics of heat and mass transfer between gas and solid, as reported in Chap. 11, show that the approach to equilibrium is very rapid. So if hot entering gas contacts cold solids, the gas in effect reaches the temperature of the solids before moving more than 2 to 3 cm into the bed, and it leaves at the bed temperature. This means that the overall heating of these solids is determined by the heat capacity of the entering hot gas, not by the kinetics of the process. As we shall see, this kind of equilibrium approach is the normal situation for heat transfer and absorption, particularly with fine solids. Design for this class of phenomena is taken up in this chapter.

Where fine solids flow into and out of a system, large static charges may build up in the bed and cause spark discharges. When inflammable materials are treated, explosions in the freeboard and gas exit system are an ever-present danger; the engineer in charge of design should be aware of this hazard and should build into the unit the necessary safety features to counter it.

In some operations, agglomerates form and accumulate on the distributors to disrupt the proper distribution of gas in the bed. If this is found in the bench-scale experiments, special care should be given to the selection of a suitable distributor and to the method of discharge of agglomerates during steady state operations.

Heat Transfer

Batch Operations

Are the Bed Solids and the Exit Gas at the Same Temperature? First we show that it is reasonable to assume that gas and solids are at the same temperature everywhere in the bed. Consider a batch of solids originally at temperature T_{p0} fluidized by gas entering at temperature T_{gi}, as shown in Fig. 1. Because of their rapid movement and their relatively large heat capacity, the temperature of solids can reasonably be taken to be independent of location in the bed. Then assuming plug flow for the gas stream and a quasi–steady state condition for the gas temperature profile with height, the change of gas temperature with height is found from the heat balance

$$\begin{pmatrix} \text{heat given} \\ \text{up by gas} \end{pmatrix} = \begin{pmatrix} \text{heat gained} \\ \text{by solids} \end{pmatrix} \tag{1}$$

or, in symbols, using the whole bed heat transfer coefficient (see Chap. 11),

$$-\rho_g C_{pg} u_o dT_g = \frac{6(1 - \varepsilon_f)}{\phi_s d_p} h_{\text{bed}}(T_g - T_p) \, dz \tag{2}$$

Integration gives the exit temperature of gas T_{ge} in a bed of height $z = L_f$ as

$$\frac{T_{ge} - T_p}{T_{gi} - T_p} = \exp\left[-\frac{\text{Nu}_{\text{bed}}}{\text{Pr Re}_p} \frac{6(1 - \varepsilon_f)}{\phi_s} \frac{L_f}{d_p} \right] \tag{3}$$

where for spherical solids the shape factor $\phi_s = 1$.

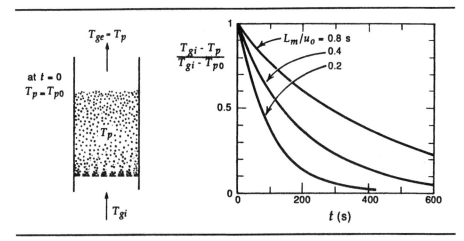

FIGURE 1

Heat exchange between bed and fluidizing gas in a batch operation. The calculated lines are from Eq. (9), with $\rho_g C_{pg}/\rho_s C_{ps} = 10^{-3}$, $\varepsilon_m = 0.5$.

If we take a 95% temperature approach as the measure of an adequate approach to equilibrium, or $(T_{ge} - T_p)/(T_{gi} - T_p) = 0.05$, then from Eq. (3) the criterion for equilibrium conditions becomes

$$\frac{\text{Nu}_{bed}}{\text{Pr Re}_p} \frac{6(1 - \varepsilon_f)}{\phi_s} \frac{L_f}{d_p} > 3 \tag{4}$$

As an illustration of the order of magnitude of these quantities, the data in Example 11.3 and Fig. 11.6 for $d_p = 360$-μm particles give

$$\begin{aligned}
\text{Pr} &= 0.69, & \phi_s &= 0.81 \\
\text{Nu}_{bed} &= 0.4\text{--}1.0, & \varepsilon_{mf} &= 0.45 \\
L_f &> 4.6\text{--}5.5 \text{ mm}, & \text{Re}_p &= 10\text{--}30
\end{aligned}$$

This L_f value shows that as long as the bed is more than 6 mm high the temperature of the exiting gas will approach to within 5% of the temperature of the bed solids. Thus, for ordinary fluidizing conditions we can reasonably take the exiting gas to be at the temperature of the solids, or

$$T_{ge} \cong T_p \tag{5}$$

Changing Bed Temperature with Time. Applying Eq. (1) to the whole bed during a short time interval dt gives

$$\rho_g C_{pg} u_o (T_{gi} - T_p)\, dt = \rho_s C_{ps} (1 - \varepsilon_f) L_f\, dT_p \tag{6}$$

Solving this heat balance with the initial condition

$$T_p = T_{p0} \qquad \text{at } t = 0 \tag{7}$$

gives

$$\frac{T_{gi} - T_p}{T_{gi} - T_{p0}} = \exp\left[-\frac{\rho_g C_{pg}}{\rho_s C_{ps}} \frac{u_o}{(1 - \varepsilon_f) L_f} t \right] \tag{8}$$

For normal fluidizing conditions

$$\frac{\rho_g C_{pg}}{\rho_s C_{ps}} \approx 10^{-3}, \qquad \text{with } (1 - \varepsilon_f) L_f = (1 - \varepsilon_m) L_m, \quad \varepsilon_m \approx 0.5$$

in which case

$$\frac{T_{gi} - T_p}{T_{gi} - T_{p0}} \approx \exp\left[-2 \times 10^{-3} \frac{t}{L_m / u_o} \right] \tag{9}$$

This expression shows that the time needed to heat the solids to a given temperature is proportional to the static bed height and inversely proportional to the gas velocity. Lines calculated by Eq. (9) are shown in Fig. 1.

Summary. In most cases the following simplifications can be made in the analysis of the heating and cooling of a batch of solids.

- At any instant the bed solids are all at the same temperature, and the gas leaves the bed at the temperature of the solids.

- The temperature of the solids changes exponentially with time with rate constant proportional to the gas velocity and inversely proportional to bed height, hence the amount of solids present.

Continuous Operations

Consider the steady state heating of a continuous stream F_0 (kg/s) of cold solids by hot gas in a single fluidized bed as shown in Fig. 2. If we neglect heat loss to the surroundings, reasonable for commercial-scale beds, the energy balance of Eq. (1) can be applied about the fluidized bed. Then, from the simplification found reasonable from the batch analysis—in effect that the exiting gas, bed solids, and exiting solids are all at the same temperature T_1—the energy balance becomes

$$A_t \rho_g u_o C_{pg}(T_{gi} - T_1) = F_0 C_{ps}(T_1 - T_{pi}) \tag{10}$$

Rearranging and introducing the bed weight

$$W = A_t L_m (1 - \varepsilon_m)\rho_s \tag{11}$$

gives

$$T_1 = \frac{T_{pi} + \phi T_{gi}}{1 + \phi} \tag{12}$$

where

$$\phi = \frac{A_t \rho_g u_o C_{pg}}{F_0 C_{ps}} = -\frac{\Delta T_p}{\Delta T_g} \tag{13}$$

The efficiencies of heat utilization of gas and solid in this system are then

$$\eta_g = \left(\frac{\begin{array}{c}\text{temperature drop}\\ \text{of gas}\end{array}}{\begin{array}{c}\text{maximum possible}\\ \text{temperature drop}\end{array}}\right) = \frac{T_{gi} - T_1}{T_{gi} - T_{pi}} = \frac{1}{1 + \phi} \tag{14}$$

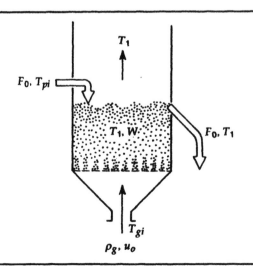

FIGURE 2
Continuous heat exchange between solids and fluidizing gas.

and

$$\eta_s = \left(\frac{\begin{array}{c}\text{temperature rise}\\\text{of solids}\end{array}}{\begin{array}{c}\text{maximum possible}\\\text{temperature rise}\end{array}} \right) = \frac{T_1 - T_{pi}}{T_{gi} - T_{pi}} = \frac{\phi}{1 + \phi} = 1 - \eta_g \qquad (15)$$

As an example of the significance of these equations, suppose that solids at 0°C are to be heated to 90°C by gas entering at 100°C. Equations (14) and (15) show that the efficiency of heat utilization is 90% for the solids but only 10% for the gas. This means that relatively much gas has to be used for this process. These efficiency equations are equally applicable to the cooling of solids.

No matter what the exit temperature of solids may be, the efficiency of single-stage operations is always low, and this prompts the desire for multistage contacting with its improved efficiencies of heat utilization. Figure 3 shows three practical alternatives for multistage contacting—two countercurrent and one crosscurrent schemes. We evaluate the efficiencies of these schemes.

For *countercurrent* contacting in N fluidized beds of Fig. 3(a), an energy

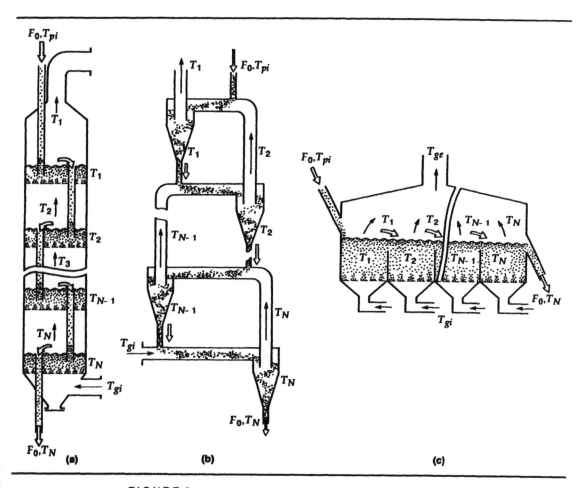

FIGURE 3
Heat exchange in multistage fluidized beds: (a) and (b) countercurrent contacting; (c) crosscurrent contacting.

balance that neglects heat losses to the surroundings gives

bed 1 $\qquad T_1 - T_{pi} = \phi(T_2 - T_1)$
bed 2 $\qquad T_2 - T_1 = \phi(T_3 - T_2)$

$\qquad\qquad\qquad \cdot \qquad\qquad \cdot$
$\qquad\qquad\qquad \cdot \qquad\qquad \cdot$ $\qquad\qquad\qquad\qquad\qquad$ (16)
$\qquad\qquad\qquad \cdot \qquad\qquad \cdot$

bed $N \qquad T_N - T_{N-1} = \phi(T_{gi} - T_N)$

Rearranging and eliminating the intermediate temperatures gives, for the heat recovery efficiencies,

$$\eta_g = \frac{T_{gi} - T_1}{T_{gi} - T_{pi}} = \frac{\sum\limits_{n=0}^{N-1} \phi^n}{\sum\limits_{n=0}^{N} \phi^n} \qquad\qquad (17)$$

and

$$\eta_s = \frac{T_N - T_{pi}}{T_{gi} - T_{pi}} = \frac{\sum\limits_{n=1}^{N} \phi^n}{\sum\limits_{n=0}^{N} \phi^n} = \phi\eta_g \qquad\qquad (18)$$

Adjusting the flow rates so that $\phi = 1$ gives identical efficiencies for gas and solids

$$\eta_g = \eta_s = \frac{N}{N+1} \qquad\qquad (19)$$

Also, for many stages plug flow is approached, and this gives, as expected,

$$\eta_g = \eta_s \cong 1 \qquad\qquad (20)$$

The above equations also apply to the suspension heat exchanger system with its N cyclones shown in Fig. 3(b).

For *crosscurrent* contacting of Fig. 3(c), an energy balance gives

bed 1 $\qquad T_1 - T_{pi} = \phi'(T_{gi} - T_1)$
bed 2 $\qquad T_2 - T_1 = \phi'(T_{gi} - T_2)$

$\qquad\qquad\qquad \cdot \qquad\qquad \cdot$
$\qquad\qquad\qquad \cdot \qquad\qquad \cdot$ $\qquad\qquad\qquad\qquad\qquad$ (21)
$\qquad\qquad\qquad \cdot \qquad\qquad \cdot$

bed $N \qquad T_N - T_{N-1} = \phi'(T_{gi} - T_N)$

where $\phi' = \phi/N$ is based on the flow through each of the N beds. For equal-sized beds,

$$T_{ge} = \frac{T_1 + T_2 + \cdots + T_N}{N} \qquad\qquad (22)$$

Hence, solving Eq. (21) for the heat recovery efficiency, we have

$$
\eta_g = \frac{T_{gi} - T_{ge}}{T_{gi} - T_{pi}} = \frac{1}{N\phi'} \left[1 - \frac{1}{(1 + \phi')^N} \right]
\tag{22}
$$

and

$$
\eta_s = \frac{T_N - T_{pi}}{T_{gi} - T_{pi}} = 1 - \frac{1}{(1 + \phi')^N} = N\phi'\eta_g
\tag{23}
$$

Comparing the countercurrent exchanger of Fig. 3(a) with the crosscurrent exchanger of Fig. 3(c), we find that each has certain advantages. Thus, for a fixed number of stages, countercurrent contacting is always more efficient; however, it is not a simple matter to maintain a stable downflow of solids and to avoid an imbalance among beds. Thus, it requires careful design to ensure smooth operations. Crosscurrent contacting of Fig. 3(c) is simpler to design and operate, and, where a high value of η_g is not needed, it is preferred.

The fluidized designs of Figs. 3(a) and (c) need large pumping power, which in large-scale operations may be costly. In contrast to these fluidized operations, the suspension exchanger of Fig. 3(b) can operate at a much smaller pumping power and still maintain the advantage of countercurrent contacting. These factors led to the choice of this design in the commercial calcination system for fine limestone shown in Fig. 2.22(b).

Heat Loss to Surroundings

In small heat exchangers, the heat loss to the surroundings should be accounted for. For single-stage operations, letting this loss be q_l in Fig. 2, Eq. (10) should be replaced by

$$
A_t \rho_g u_o C_{pg}(T_{gi} - T_1) = F_0 C_{ps}(T_1 - T_{pi}) + q_l
\tag{24}
$$

With β as the ratio of the heat loss to the surroundings to the total heat given up by the hot gas stream, Eq. (24) becomes

$$
A_t \rho_g u_o C_{pg}(T_{gi} - T_1)(1 - \beta) = F_0 C_{ps}(T_1 - T_{pi})
\tag{25}
$$

Thus, for the heating of solids where heat loss is to be considered, simply replace ϕ in Eqs. (12) to (23) by $\hat{\phi}$, where

$$
\hat{\phi} = \phi_{\text{with heat loss}} = (1 - \beta)\phi_{\text{without heat loss}}
\tag{26}
$$

Similarly, for the cooling of solids, use $\hat{\phi}$, defined as

$$
\hat{\phi} = \phi_{\text{with heat loss}} = (1 + \beta)\phi_{\text{without heat loss}}
\tag{27}
$$

where β is the ratio of heat lost to the surroundings to the heat gained by the gas stream. In multistage operations, if β is approximately constant for all stages, simply replace ϕ by the appropriate $\hat{\phi}$ in all equations.

EXAMPLE 1

**Single-Stage
Limestone
Calciner**

Determine the fuel consumption and thermal efficiency of a fluidized limestone calciner operating at 1000°C and unaccompanied by any heat recovery provisions (see Fig. E1). The thermal efficiency is defined as

$$\eta = \frac{\text{heat necessary for decomposition}}{\text{heat generated by the combustion of fuel oil}}$$

Data. The strongly endothermic calcination reaction is

$$CaCO_3 \longrightarrow CaO + CO_2, \qquad \Delta H_r = 1795 \text{ kJ/kg } CaCO_3$$

	$CaCO_3$	CaO	CO_2	Air	Combustion Gas
Molecular weight (kg/mol)	0.100	0.056	0.044	0.029	~0.029
Specific heat (kJ/kg·K)	1.13	0.88	1.13	1.00	1.17

Combustion at 10% excess air requires 15 kg air/kg fuel oil.
Net combustion heat of fuel oil 41,800 kJ/kg.
Temperature of feed and surroundings 20°C.

SOLUTION   An energy balance about the calciner gives

$$\left(\begin{array}{c}\text{heat liberated by}\\ \text{combustion of fuel}\end{array}\right) = \left(\begin{array}{c}\text{heat in exit}\\ \text{combustion gas}\end{array}\right) + \left(\text{heat to } CO_2\right) + \left(\text{heat to solids}\right)$$

$$+ \left(\begin{array}{c}\text{heat to}\\ \text{decompose } CaCO_3\end{array}\right)$$

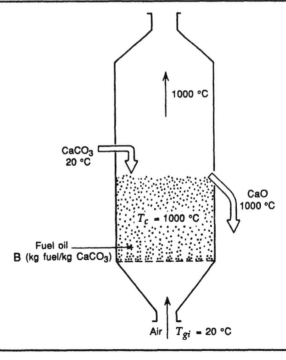

FIGURE E1
A single-bed fluidized calciner.

Let B (kg fuel/kg $CaCO_3$) be the fuel consumption; then the energy balance based on 1 kg $CaCO_3$ becomes

$$41,800B = (15 + 1)B(1.17)(1000 - 20) + (0.44)(1.13)(1000 - 20)$$
$$+ (0.56)(0.88)(1000 - 20) + 1795$$

from which

$$B = 0.118 \text{ kg fuel/kg } CaCO_3 = \underline{0.211 \text{ kg fuel/kg CaO}}$$

The heat requirement for calcination is then

$$(0.211)(41,800) = \underline{8820 \text{ kJ/kg CaO}}$$

and the thermal efficiency is

$$\eta = \frac{1795}{(0.118)(41,800)} = 0.364 = \underline{36\%}$$

When coarse limestone is calcined in the usual shaft kiln, the heat consumption is approximately 4200 to 4600 kJ/kg CaO; hence, the thermal efficiency $\eta = 69$ to 75%. This indicates that the single-stage fluidized calciner of this problem has a considerably lower efficiency and, consequently, cannot compete economically with the shaft kiln. To save fuel and raise the thermal efficiency, we must recover heat from the combustion gas and from the spent solids by multistaging. This is shown in the next example.

EXAMPLE 2

Multistage Limestone Calciner

Determine the fuel consumption and thermal efficiency of the multistage fluidized calciner of Fig. E2. For a feed rate of 400 tons $CaCO_3$/day determine the diameter of the stages if the gas velocity is to be maintained at $u_o = 0.8$ m/s throughout. See Example 1 for additional data.

SOLUTION

(a) *Calcining bed.* An energy balance about the calcining bed gives

$$\left(\begin{array}{c} \text{heat liberated by} \\ \text{combustion of fuel} \end{array} \right) + \left(\begin{array}{c} \text{heat in entering} \\ \text{preheated solids} \end{array} \right) + \left(\text{heat in entering preheated air} \right)$$

$$= \left(\begin{array}{c} \text{heat in leaving} \\ \text{combustion gas} \end{array} \right) + \left(\begin{array}{c} \text{heat in leaving} \\ CO_2 \end{array} \right) + \left(\begin{array}{c} \text{heat in leaving} \\ \text{solids} \end{array} \right)$$

$$+ \left(\text{heat needed to decompose } CaCO_3 \right)$$

Based on 1 kg of $CaCO_3$ and referring to the nomenclature of Fig. E2, this energy balance becomes

$$41,800 + (1.13)(T_3 - 20) + (15)B(1.00)(T_r - 20)$$
$$= (15 + 1)(1.17)(1000 - 20) + (0.44)(1.13)(1000 - 20)$$
$$+ (0.56)(0.88)(1000 - 20) + 1795$$

from which we find

$$T_3 + 13.34(T_r - 20) + 20,800B = 2470 \tag{i}$$

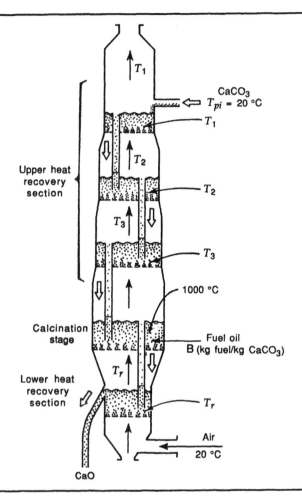

FIGURE E2
A multibed fluidized calciner.

From Eq. (13), we also have

$$\phi = \frac{\text{heat capacity of combustion gas and of } CO_2 \text{ formed}}{\text{heat capacity of solids}}$$

$$= \frac{(15+1)(1.17)B + (0.44)(1.13)}{1.13} = 16.6B + 0.44 \tag{ii}$$

(b) *Upper heat recovery section.* Equation (17) gives

$$T_3 = \frac{T_{pi} + (\phi + \phi^2 + \phi^3)T_{gi}}{1 + \phi + \phi^2 + \phi^3} \tag{iii}$$

where $T_{pi} = 20°C$, $T_{gi} = 1000°C$.

(c) *Lower heat recovery section.* An energy balance similar to that for the upper section gives, analogous to Eqs. (ii) and (iii),

$$\phi^+ = 30.6B \tag{iv}$$

and

$$T_r = \frac{1000 + 20\phi^+}{1 + \phi^+} \tag{v}$$

(d) *Evaluation of temperatures and energy requirements.* Substituting Eqs. (ii)–(v) in Eq. (i) and solving for B by trial and error or graphical fit gives

$$B = 0.0631 \text{ kg fuel oil/kg CaCO}_3$$
$$\phi = 1.488, \qquad \phi^+ = 1.934$$

Therefore, the temperatures of the various stages are

$$T_1 = \frac{20 + (1.488)(696)}{1 + 1.488} = 424°\text{C}$$

$$T_2 = \frac{20 + (1.488 + 1.488^2)879}{1 + 1.488 + 1.488^2} = 696°\text{C}$$

$$T_3 = \frac{20 + (1.488 + 1.488^2 + 1.488^3)1000}{1 + 1.488 + 1.488^2 + 1.488^3} = 879°\text{C}$$

$$T_r = \frac{1000 + 20 \times 1.934}{1 + 1.934} = 354°\text{C}$$

Also, the thermal efficiency is

$$\eta = \frac{1795}{(0.0631)(41,800)} = 0.68 = \underline{68\%}$$

and the heat requirement is

$$\frac{(0.0631)(41,800)}{0.56} = \underline{4728 \text{ kJ/kg CaO}}$$

(e) *Evaluation of bed dimensions.* In the lower heat recovery section the volumetric flow rate of gas is

$$\frac{\left(\dfrac{400,000 \text{ kg CaCO}_3}{24 \times 3600 \text{ s}}\right)\left(0.0631 \dfrac{\text{kg fuel}}{\text{kg CaCO}_3}\right)\left(15 \dfrac{\text{kg air}}{\text{kg fuel}}\right)}{1.293\left(\dfrac{273}{273 + 354}\right) \dfrac{\text{kg air}}{\text{m}^3}} = 7.78 \text{ m}^3/\text{s}$$

Thus the diameter of the lower bed is

$$\frac{\pi}{4} d_t^2 = \frac{7.78 \text{ m}^3/\text{s}}{0.8 \text{ m/s}} \qquad \text{or} \qquad d_t = \underline{3.52 \text{ m}}$$

Similar calculations for the calcination section give

Flow of combustion gas: $16.9 \text{ m}^3/\text{s}$
Flow of CO_2 formed: $4.8 \text{ m}^3/\text{s}$
Diameter of section: $d_t = \underline{5.88 \text{ m}}$

Similarly, for the three fluidized beds of the upper recovery section, we find

Bed number	1	2	3	
d_t (m)	4.35	5.13	5.59	‖

Comments about Examples 1 and 2. First, it was assumed that heat loss to the surroundings could be neglected. In a commercial plant, this heat loss is usually a few percent. Knowing the size of the process unit and its temperature one can calculate this loss. From this one can determine the value of β and thus replace ϕ by $\hat{\phi}$ in the foregoing equations.

Second, it was also assumed that fuel was completely burned in the calcination stage. Proper design of an efficient burning system requires additional information. In fact, a part of the combustion gas produced by the thermal cracking of fuel oil may leave the calcination stage without mixing with air. Hence, further combustion may take place in the freeboard above this stage thereby slightly raising the temperature. For this reason the distributor for the bed directly above the calcination stage should be made of heat-resistive material.

For solids with a wide size distribution, elutriation occurs, the various stages will have different size distributions, and fines will accumulate in the upper stages and will be entrained from the bed. For complete calcination of feed solids, these fines should be recovered and returned to the calciner.

It is not easy to maintain a stable downflow of solids and to avoid an imbalance among the beds in a setup like that in Fig. E2. For the processing of fine solids, this difficulty can be bypassed by use of suspension heaters, such as shown in Figs. 3(b) or 2.22(b). The calculation procedure for these units is the same as in the previous example.

Mass Transfer

Batch Operations

As shown in Fig. 4, we fluidize a batch of solids with an inert carrier gas containing dilute A of mass concentration C_{Ai} (kg/m^3), which is adsorbed isothermally by the solids. Let Q be the weight fraction on a dry basis of adsorbed vapor (adsorbate) on the solids, and let C_A^* be the vapor concentration of A in equilibrium with solids having a moisture fraction Q. Figure 5 shows the equilibrium relationships found by experiment for various adsorbent systems.

Concentration of Adsorbate on Particles. Calculations show that the difference in concentration of diffusing vapor between the center and surface of

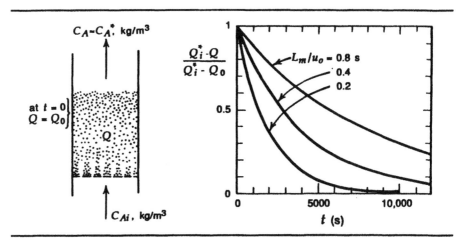

FIGURE 4
Adsorption of vapor by a batch of solids. The calculated lines are from Eq. (32) for the adsorption at 25°C of water vapor from air by activated alumina.

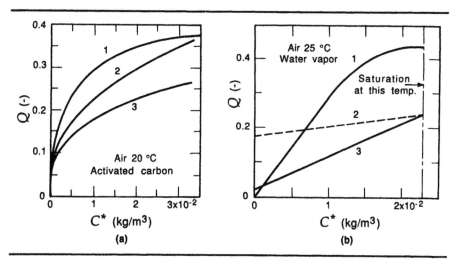

FIGURE 5
Examples of equilibrium isotherms between adsorbate on the solid and in air, Q (kg adsorbed/kg solid): (a) line 1, benzene; 2, ethanol; 3, acetone; (b) line 1, silica gel; 2, molecular sieve; 3, activated alumina.

adsorbent particles is negligibly small except for an extremely fast change in vapor composition with time in the bed. Also, for normal fluidizing conditions such as $d_p < 1$ mm and $L_m > 0.2$ to 0.3 m, the exiting gas can be taken to be in equilibrium with the solids in the bed, or

$$C_{Ae} = C_A^* \qquad (28)$$

Changing Adsorbate with Time. A material balance about the whole bed gives

$$\left(\begin{array}{c} \text{vapor lost} \\ \text{by gas} \end{array} \right) = \left(\begin{array}{c} \text{vapor adsorbed} \\ \text{by solids} \end{array} \right)$$

From Eq. (28) this expression becomes

$$A_t u_o (C_{Ai} - C_A^*)\, dt = A_t L_m (1 - \varepsilon_m) \rho_s\, dQ\,, \qquad \text{[kg]} \qquad (29)$$

Separating and integrating for an initial moisture content on the solids Q_0 gives

$$\int_{Q_0}^{Q} \frac{dQ}{C_{Ai} - C_A^*} = \frac{u_o t}{\rho_s L_m (1 - \varepsilon_m)} \qquad (30)$$

Given the relationship between C_A^* and Q for the system at hand, we can then find how Q changes with time. For example, for the adsorption of water vapor from air by activated alumina, as shown in Fig. 5, we have approximately

$$C_A^* = b_1 + b_2 Q \qquad (31)$$

Substituting Eq. (31) in Eq. (30) and solving gives

$$\frac{Q_i^* - Q}{Q_i^* - Q_0} = \exp\left[- \frac{b_2 u_o t}{\rho_s (1 - \varepsilon_m) L_m} \right] \qquad (32)$$

where Q_i^* is the vapor fraction of the solid that would be in equilibrium with the incoming gas, or

$$C_{Ai} = b_1 + b_2 Q_i^* \tag{33}$$

With $b_2 = 0.1 \, \text{kg/m}^3$, $\rho_s = 2000 \, \text{kg/m}^3$, $\varepsilon_m = 0.5$ for alumina, Eq. (32) can be written as

$$\frac{Q_i^* - Q}{Q_i^* - Q_0} = \exp\left[-\frac{1 \times 10^{-4} t}{L_m/u_o}\right] \tag{34}$$

Analogous to the heat transfer equation derived earlier, Eq. (34) shows that the moisture fraction in the solids rises exponentially with time with rate proportional to gas velocity and inversely proportional to bed height. This result is presented in Fig. 4 and shows that the approach to equilibrium for the solids is rather slow under normal fluidizing conditions. Also, since the adsorption of vapor liberates energy, heat exchange may be necessary if isothermal operations are required.

For desorption of volatile matter from porous adsorbed solids, a similar analysis gives

$$-\int_{Q=Q_0}^{Q} \frac{dQ}{C_A^* - C_{Ai}} = \frac{u_o t}{\rho_s(1 - \varepsilon_m)L_m} \tag{35}$$

With the linear equilibrium relationship of Eq. (33), this integrates to

$$\frac{Q - Q_i^*}{Q_0 - Q_i^*} = \exp\left[-\frac{b_2 u_o t}{\rho_s(1 - \varepsilon_m)L_m}\right] \tag{36}$$

Again, Q_0 and Q_i^* are the weight fractions of volatiles initially in the solid and in equilibrium with the entering gas, respectively.

Continuous Operations

This section considers only reversible mass transfer. Irreversible transfer or adsorption followed by reaction with solid can be treated by the methods in Chap. 18.

A rigorous treatment of continuous operations is quite complicated. However, a great simplification can be made with the fairly good assumption, suggested from batch operations, that the bed solids are in equilibrium with the leaving gas. Thus, for beds that are not too shallow, say $L_m/d_p > 100$ to 200, assume an equilibrium operation and take a mass balance. Referring to Fig. 6, a mass balance for the transferring vapor gives

$$A_t u_o(C_{Ai} - C_A^*) = F_0(Q - Q_i) \qquad \text{[kg moisture/s]} \tag{37}$$

where each gas-solid system has its own particular equilibrium relationship, such as given by Eq. (33). In general, Eqs. (33) and (37) can then be solved simultaneously and directly without integration.

For example, for water vapor on activated alumina, equilibrium can reasonably be approximated by a linear relationship as shown in Fig. 5. Then, Eq. (33) becomes, in turn, for the bed solids and exit gas

$$C_A^* = b_1 + b_2 Q \tag{38}$$

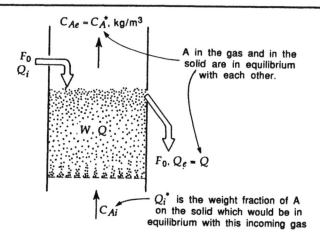

FIGURE 6
Continuous mass transfer operations between solids and gas in a single fluidized bed.

For gas in equilibrium with entering solids,

$$C_{Ai}^* = b_1 + b_2 Q_i \tag{39}$$

and for solids in equilibrium with entering gas,

$$C_{Ai} = b_1 + b_2 Q_i^* \tag{40}$$

Rearranging Eq. (37) allows the following efficiency measures to be defined:

$$\eta_g' = \left(\frac{\text{amount of solute}}{\text{maximum that could}} \right) = \frac{C_{Ai} - C_{Ae}}{C_{Ai} - C_{Ai}^*} = \frac{1}{1 + \phi'} \tag{41}$$

and

$$\eta_s' = \left(\frac{\text{amount of solute}}{\text{maximum that could}} \right) = \frac{Q - Q_i}{Q_i^* - Q_i} = \frac{\phi'}{1 + \phi'} = 1 - \eta_g' \tag{42}$$

where

$$\phi' = \frac{A_t u_o b_2}{F_0} \tag{43}$$

These equations show that a 99% efficiency in removing solute from the gas is possible only at the expense of a 1% efficiency in the use of solids as an adsorbent. This requires using large amounts of solids and illustrates the general finding that single-stage contacting is accompanied by low efficiencies. This can be remedied by multistaging, as shown in Fig. 7. For such operations efficiency

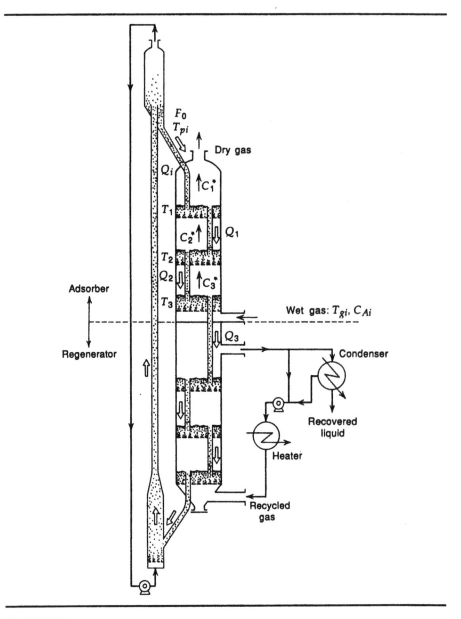

FIGURE 7
A schematic of an adsorber-regenerator tower.

equations analogous to those for heat transfer, Eqs. (17) and (18), may be derived, and Example 3 illustrates the calculation procedure.

Up to this point heat effects have been ignored, and mass transfer is assumed to take place isothermally. In practice, however, the heat of condensation on the solid may be significant, in which case it should be included when estimating the bed temperature. Thus, heat exchange may be needed if strict isothermal conditions are required in the multistage unit. Since internal exchangers would increase the cost of operations and may introduce additional attrition problems, adiabatic operations with interstage cooling are often preferred.

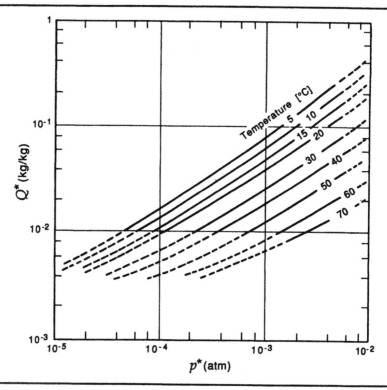

FIGURE 8
Water content of silica gel in equilibrium with moist air; adapted from Ermenc [1].

The heats of sorption (in kJ/kg) for various adsorbed materials are as follows:

For water in silica gel:	122
For water in activated alumina:	59
For benzene in activated charcoal:	84–138
For benzene in silica gel:	46

Figure 8 illustrates the variation of equilibrium with temperature for a specific system.

EXAMPLE 3

Multistage Adsorber

The three-stage adsorption tower of Fig. 7 is to be used to reduce the moisture content of air from 0.01 to 0.0001 atm. For isothermal operations at 20°C and close to 1 atm, determine the required circulation rate of silica gel and its leaving moisture content.

Data

Air rate (dry, standard basis): $10\ m^3/s$
Entering silica gel is dry: $Q = 0$

Figure E3 shows the equilibrium curve for this system.

SOLUTION We solve this problem graphically by a modified McCabe-Thiele method. First, a material balance from the top of the adsorber, represented by Q_j

and C_1^*, to any level j in the adsorber gives

$$Q_j - Q_i = \frac{A_t u_o}{F_0}(C_j - C_1^*)$$

Now at low moisture fractions in the air we have

$$C \cong \frac{pM}{RT} = \frac{(p \text{ atm})(0.018 \text{ kg/mol})}{(82.06 \times 10^{-6} \text{ m}^3 \cdot \text{atm/mol} \cdot \text{K})(293 \text{ K})} = 0.749p \text{ kg/m}^3$$

Inserting this in the material balance and noting that

$$A_t u_o = 10\left(\frac{273 + 20}{273}\right) \text{m}^3/\text{s}, \qquad Q_i = 0 \text{ kg/kg}, \qquad p_1^* = 0.0001 \text{ atm}$$

gives

$$Q_j = \frac{8.04}{F_0}(p_j - 0.0001)$$

which represents a family of operating lines passing through point A in Fig. E3 with slope $8.04/F_0$.

By trial and error find which line of this family will raise the partial pressure of moisture to 0.01 atm in three steps. The solution is shown as line AB on Fig. E3, and its slope is found from the graph to be 10.2. Therefore the flow rate of solids is

$$F_0 = \frac{8.04}{10.2} = 0.788 \text{ kg/s}$$

and from Fig. E3 the moisture content of the leaving solids is

$$Q_3 = 0.101 \text{ kg H}_2\text{O/kg dry solids}$$

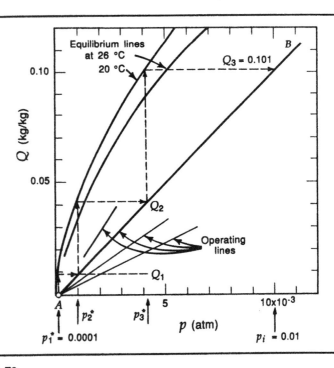

FIGURE E3
Graphical solution of Example 3.

Comment. This solution assumes that each stage acts as a theoretical contacting stage. For fluidized adsorption with fine porous solids, this approximation is satisfactory for reasonable bed heights, uniform gas distribution, and stable operations.

   For adiabatic and other nonisothermal operating schemes, we must first determine the temperatures of the various stages and then use the appropriate equilibrium curves for the individual stages. The design procedure is outlined in unit operations texts and by Ermenc [1].

Drying of Solids

Batch Operations

The analysis of drying in fluidized beds is more complicated than heat transfer or isothermal mass transfer alone. Different mechanisms can control, and different drying regimes, commonly called the *constant-rate* and the *falling-rate* regimes, may be observed successively in a single run.

   Constant Rate Drying or Heat Transfer Limiting. Consider a batch of solids with moisture fraction Q (kg moisture/kg dry solid) being dried by passing hot air at temperature T_{gi} up through the bed as shown in Fig. 9. If the solids are small, very porous, and sufficiently wet to contain free moisture, then they will dry at a constant rate. In this period the approach to equilibrium is rapid for both heat and mass transfer, so the bed and the leaving gas will remain close to the adiabatic saturation temperature of the entering gas stream. Very porous solids such as silica gel and activated charcoal exhibit this type of drying behavior.

   The change in moisture content of solids with time is found from an energy accounting

$$\left(\begin{array}{c}\text{heat lost by} \\ \text{entering gas}\end{array}\right) = \left(\begin{array}{c}\text{heat transferred to solids} \\ \text{to vaporize the liquid}\end{array}\right)$$

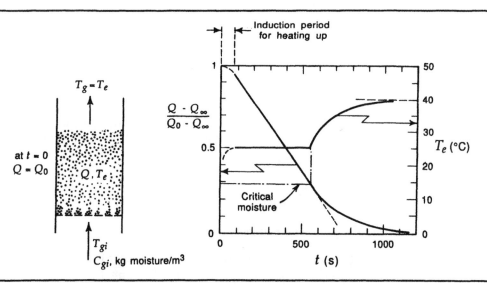

FIGURE 9
Batch drying of solids showing the constant-rate and falling-rate regimes.

If $\mathscr{L}$ is the latent heat of vaporization, this becomes

$$A_t \rho_g u_o C_{pg}(T_{gi} - T_e)\, dt = -A_t \rho_s L_m(1 - \varepsilon_m)\mathscr{L}\, dQ \tag{44}$$

When the moisture fraction in the particles is greater than some critical value, the drying rate stays constant, and integration of Eq. (44) gives

$$Q_0 - Q = \frac{\rho_g}{\rho_s}\, \frac{C_{pg}(T_{gi} - T_e)}{\mathscr{L}}\, \frac{t}{(1 - \varepsilon_m)L_m/u_o}\,, \qquad Q_0 > Q > Q_{cr} \tag{45}$$

where Q_{cr} is the critical moisture fraction below which the drying rate begins to fall because of diffusional effects. For fine porous solids, such as cracking catalyst or activated alumina, Q_{cr} is very close to Q_i^*, the moisture fraction in equilibrium with the entering gas stream. For these solids the drying rate will be essentially constant throughout the drying operation.

This drying regime is somewhat similar to the equilibrium processes discussed earlier, in that the carrying capacity of the gas limits the process. Thus, the rate of change in moisture fraction of the particles varies proportionately with gas velocity and inversely with bed height.

Falling-Rate Drying, or Diffusion in Solids Limiting. For resins and other materials where the volatile components are bound within or strongly wet the particle, the diffusion of moisture to the surface may be slow enough to control the overall drying operation. Let us see how the moisture fraction changes with time for this regime.

The diffusion of moisture in a spherical particle of diameter $d_p = 2R$ is governed by Fourier's law of conduction, or

$$\frac{\partial Q_m}{\partial t} = \mathscr{D}_m\left(\frac{\partial^2 Q_m}{\partial r^2} + \frac{2}{r}\frac{\partial Q_m}{\partial r}\right) \tag{46}$$

where Q_m is the moisture fraction at any position r, and $\mathscr{D}_m$ is the diffusivity of moisture within the particle. The boundary conditions representing this process are

$$
\begin{aligned}
t &= 0 \quad \text{any } r, \quad Q_m = Q_0 \\[4pt]
t &= t, \quad
\begin{cases}
r = 0, & \dfrac{\partial Q_m}{\partial r} = 0 \\[8pt]
r = R, & Q_m = Q^*
\end{cases}
\end{aligned}
\tag{47}
$$

where Q^* is the moisture fraction at the surface of the particle that would be in equilibrium with the gas bathing the particle (see Fig. 10). Further, the average moisture content in a particle is

$$Q = \int_0^R \frac{4\pi^2 Q_m\, dr}{(4/3)\pi R^3} \tag{48}$$

Solving the preceding equations gives [2]

$$\frac{Q - Q^*}{Q_0 - Q^*} = \frac{Q_f}{Q_{f0}} = \frac{6}{\pi^2}\sum_{n=1}^{\infty}\frac{1}{n^2}\exp\left[-(n\pi)^2\frac{\mathscr{D}_m t}{R^2}\right] \tag{49}$$

where Q_f is the free moisture fraction of the solids.

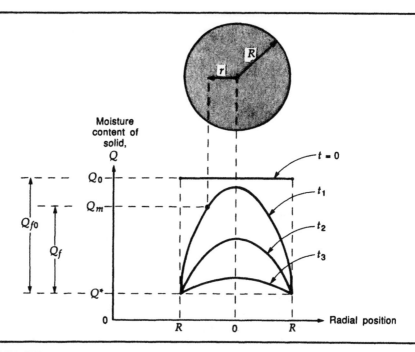

FIGURE 10
Distribution of moisture in a particle drying in the falling-rate regime.

When diffusion controls, the moisture fraction decreases in a complex way but close to exponentially with time. Also, the rate of drying should closely approximate that of a single isolated particle bathed in the entering gas stream.

Intermediate Case and Critical Moisture Content. Figure 11 illustrates the magnitude of the two rate processes if they occur singly and shows how the primary resistance can shift from one to the other in a particular drying run. For example, in a deep bed, or $L_m/u_o = 0.2$ s, the equilibrium process (line 4) is slower than the diffusion process (line 1); hence, it controls the overall rate of drying. On the other hand, for a shallower bed, or $L_m/u_o = 0.02$ s, the diffusion process (line 1) is slower than the equilibrium process (line 2), and it controls. For intermediate conditions where $L_m/u_o = 0.1$ s, the drying may start as equilibrium-controlled (line 3); however, at about $t = 400$ s it becomes diffusion-influenced, and later diffusion-controlled.

Figure 11 shows that the critical moisture content at which the controlling mechanism shifts is roughly given by

$$\frac{Q_{cr} - Q^*}{Q_0 - Q^*} \cong 0.1 \qquad \text{for } \frac{L_m}{u_o} = 0.1 \text{ s}$$

In focusing attention on a single drying particle, we find that the critical moisture content is reached when the vapor pressure at the surface drops below the vapor pressure of the pure liquid at the drying conditions of the inlet gas stream. Moreover, a theoretical analysis for constant drying conditions indicates that Q_{cr} is small for high $\mathscr{D}_m$, small d_p, and slow drying.

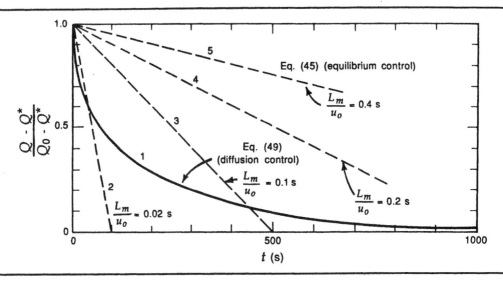

FIGURE 11
Estimation of the limiting step in the fluidized drying of a batch of solids. Data used for these curves: in Eq. (45), $\rho_g/\rho_s = 5 \times 10^{-4}$, $Q - Q^* = 0.5$, $C_{pg}(T_{gi} - T_e)/\mathscr{L} = 0.1$; in Eq. (49), $\mathscr{D}_m = 10^{-10}$ m²/s, $d_p = 1$ mm.

Continuous Operations

At any instant during batch operations the solids are all at the same stage of drying; consequently, there is little interaction between the particles as far as drying is concerned. The situation is quite different in continuous operations because here we have a small fraction of more or less wet particles interspersed in a bed of almost dry solids. The temperatures of the dry and wet solids may differ, and the interaction between particles may be important. These factors suggest that an analysis of the drying may differ significantly for the two cases.

We focus on a single wet particle of temperature T_p surrounded by gas of vapor concentration C_e. Since the probability of its being bathed by bubble gas is very small, we assume that it is present in the emulsion all the time and that its rate of drying is determined by its interaction with the emulsion and its neighboring almost dry particles, both of which are at temperature T_e.

The drying rate of this typical particle of size d_p can be represented by

$$\left(\begin{array}{c} \text{decrease of free} \\ \text{moisture in a} \\ \text{particle} \end{array} \right) = \left(\begin{array}{c} \text{mass transferred} \\ \text{from the wet particle} \\ \text{to the neighboring} \\ \text{emulsion fluid} \end{array} \right) = \left(\begin{array}{c} \text{heat transferred from} \\ \text{neighboring solids} \\ \hline \text{latent heat of} \\ \text{vaporization} \end{array} \right)$$

In terms of the free moisture fraction, $Q_f = Q - Q^*$, this equation becomes

$$-\left(\frac{\pi}{6} d_p^3 \right) \rho_s \, dQ_f = \pi d_p^2 k_d (C_p^* - C_e) \, dt = \frac{\pi d_p^2 h_p (T_e - T_p) \, dt}{\mathscr{L}} \tag{50}$$

where

$$C_p^* = \left(\begin{array}{c} \text{concentration of vapor} \\ \text{in equilibrium with the} \\ \text{moisture content at the} \\ \text{surface of the solid} \end{array} \right) = f(Q_f, T_p) \tag{51}$$

k_d = mass transfer coefficient between the wet solid surface and the neighboring void space

h_p = heat transfer coefficient between the particle at T_p and the emulsion at T_e

The temperature of the emulsion T_e is found by an energy balance:

$$\left(\begin{array}{c}\text{heat transferred}\\\text{to the emulsion}\end{array}\right) = \left(\begin{array}{c}\text{heat from the}\\\text{entering gas } T_{gi}\end{array}\right) + \left(\begin{array}{c}\text{heat from}\\\text{exchanger tubes}\end{array}\right)$$

$$= \left(\begin{array}{c}\text{heat needed to raise the temperature}\\\text{and vaporize the liquid in solids}\end{array}\right)$$

$$+ \left(\begin{array}{c}\text{heat to raise}\\\text{temperature of solids}\end{array}\right) + \left(\begin{array}{c}\text{heat}\\\text{loss}\end{array}\right) \qquad (52)$$

which becomes

$$A_t u_o \rho_g C_{pg}(T_{gi} - T_e) + q_h = F_0(Q_{fi} - \bar{Q}_f)[\mathscr{L} + C_{pl}(T_e - T_{pi})]$$
$$+ F_0 C_{ps}(T_e - T_{pi}) + q_1 \qquad (53)$$

For very wet feed solids, nearly dry leaving solids, no heat loss, and no heat input by heat exchange, Eq. (53) simplifies to

$$T_e = T_{gi} - \frac{F_0 Q_{fi} \mathscr{L}}{A_t u_o \rho_g C_{pg}} \qquad (54)$$

Equations (53) and (54) show how the bed temperature can be adjusted or controlled. Thus, at given fluidizing conditions, the bed temperature can be raised by reducing the feed rate of solids F_0 or by adding heat q_h to the bed. In addition, Eq. (50) shows that a rise in bed temperature increases the drying rate of the wet particles within the bed, thereby decreasing the necessary drying time.

With this simple model for the interaction of drying particles, we follow the drying history of a particle in the bed and find the behavior of the bed as a whole during continuous operations. First, the change in moisture fraction with time of an individual particle in the bed is found from Eq. (50):

$$-dQ_f = \frac{6h_p(T_e - T_p)\,dt}{\rho_s d_p \mathscr{L}} \qquad (55)$$

For constant-rate drying of individual particles, these remain at a constant temperature close to the wet-bulb temperature of the emulsion. Thus, the drying progression of an individual particle is obtained by integrating Eq. (55) for constant ΔT and for $Q_f = Q_{fi}$ at $t = 0$:

$$\frac{Q_f}{Q_{fi}} = 1 - \frac{t}{\tau} \qquad (56)$$

where

$$\tau = \frac{\rho_s d_p Q_{fi} \mathscr{L}}{6h_p(T_e - T_p)} = \left(\begin{array}{c}\text{time needed to completely}\\\text{dry a feed particle}\end{array}\right) \qquad (57)$$

Consider the bed of Fig. 12. The distribution of residence times of

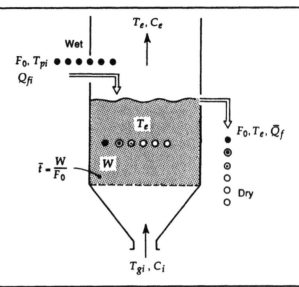

FIGURE 12
A model for the continuous fluidized drying of solids.

particles is

$$\mathbf{E}(t) = \frac{1}{\bar{t}} e^{-t/\bar{t}} \tag{14.3) or (58}$$

where

$$\bar{t} = \frac{W}{F_0} \tag{14.2) or (59}$$

Also, the average moisture fraction of leaving solids $\bar{Q}_f$ is defined by

$$\bar{Q}_f = \int_0^\tau Q_f \mathbf{E}(t)\, dt \tag{60}$$

Combining Eqs. (56), (58), and (60) and integrating from $t = 0$ to $t = \tau$, since the moisture content is zero for all lengths of stay greater than τ, gives

$$\frac{\bar{Q}_f}{Q_{fi}} = 1 - \frac{1 - e^{-\tau/\bar{t}}}{\tau/\bar{t}} \tag{61}$$

Where the diffusion of moisture is rate-limiting, the temperature of a drying particle will rise with time as it dries, and this can be approximated by

$$T_e - T_p \propto e^{-B't} \tag{62}$$

Introducing Eq. (62) in Eq. (55) gives for the progressive drying of an individual particle

$$\frac{Q_f}{Q_{fi}} = e^{-B't} \tag{63}$$

where the constant B' is a complex function of the bed and drying conditions.

For the bed as a whole, the mean moisture fraction of the leaving solids is found by substituting Eq. (63) in Eq. (60)* and integrating. This gives

$$\frac{\bar{Q}_f}{Q_{fi}} = \frac{1}{1 + B'\bar{t}} \tag{64}$$

Wide Size Distribution of Solids. If the feed contains a distribution of particle sizes including small solids that are entrained by the gas stream, various additional factors must be considered. First, when the entrained fines are not returned to the bed, we find, using the treatment in Chap. 14, that the mean residence time for the different sizes of solids in the bed is

$$\bar{t}(d_p) = \frac{1}{F_1/W + \kappa(d_p)} \tag{14.12) or (65}$$

where F_1 is the overflow rate of solids, as shown in Fig. 14.1.

Next, we need to know how the various drying factors vary with particle size. Thus, we need to know τ or B' versus d_p, and κ versus d_p and u_o. Approximately, we may take $\tau \propto d_p$ and $B' \propto d_p^{-2}$. For $\kappa(d_p)$, see Chap. 7. For given particle size and fluidizing conditions, τ and B' depend on the bed temperature T_e, which suggests that the effect of τ and B' can be found by operating the bed at different temperature levels.

Remarks. Although the average moisture content of the leaving solids can be very low, the stream from a single-stage dryer inevitably contains a small fraction of very wet particles. For processes where only the average moisture is important, single-stage operations may be quite satisfactory. However, for processes where solids are needed at nearly the same moisture content, such as in the manufacture of synthetic fibers, these few wet particles can be harmful. For this requirement multistaging should be used to improve the residence time distribution of the solids. Since it is the residence time distribution of the solids, not gas, that is of particular concern, either of the multistaging schemes of Fig. 3 can be used. However, the ease of construction, simple regulation of the solid stream, and lower power consumption recommend the crossflow scheme of Fig. 3(c) for these drying operations.

When considerable heat is needed for the drying, one may use internal heaters. These reduce the size needed for the bed, cyclones, and solid collection systems, and these reductions can sometimes more than offset the additional cost of the heaters themselves. Example 5 treats this situation.

Fluidized dryers can be used to recover valuable solvents from granular or powdery materials by stripping with superheated steam or with superheated vapor of the solvent itself, as illustrated in Example 5.

Care should be paid to the explosion hazard in fluidized drying. For combustible materials, rupture disks should be installed and the oxygen content should be carefully controlled to remain outside the explosion limits. Since static charges accumulated by the fines can cause ignition, it is important to ground all internal metal parts of the system, and when combustible vapors are used for stripping, the feed and discharge sections should be carefully and properly sealed [3, 4]. Additional considerations about fluidized bed drying are covered by Vaněček et al. [5].

---

*In this case, $\tau = \infty$.

EXAMPLE 4

Dryer Kinetics and Scale-up

Wet, porous solids are dried from $Q_{fi} = 0.20$ to $\bar{Q}_f = 0.04$ in a steady state, continuous flow, fluidized bench dryer operated under the following conditions:

Dry solids: $\rho_s = 2000 \text{ kg/m}^3$, $C_{ps} = 0.84 \text{ kJ/kg·K}$
             $F_0 = 7.6 \times 10^{-4} \text{ kg/s}$, $T_{si} = 20°C$
Gas: $\rho_g = 1 \text{ kg/m}^3$, $C_{pg} = 1.00 \text{ kJ/kg·K}$
             $u_o = 0.3 \text{ m/s}$, $T_{gi} = 200°C$, gas is dry
Liquid: $\mathscr{L} = 2370 \text{ kJ/kg}$, $C_{pl} = 4.2 \text{ kJ/kg·K}$
Bed: $d_t = 0.1 \text{ m}$, $L_m = 0.1 \text{ m}$, $\varepsilon_m = 0.45$

(a) Calculate the bed temperature, neglecting the heat loss from the dryer to the surroundings.
(b) Estimate the time necessary to completely dry a single feed particle in the bench dryer.
(c) Design a larger fluidized dryer for the same feed and product as above but with a capacity of $F_0 = 3.6$ tons/hr $= 1$ kg/s on a dry solids basis.

SOLUTION

(a) *Bed temperature.* The heat balance of Eq. (53) becomes, for the bench dryer,

$$\frac{\pi}{4}(0.1)^2(0.3)(1)(1.00)(200 - T_e) = (7.5 \times 10^{-4})(0.20 - 0.04)[2370 + 4.2(T_e - 20)]$$
$$+ (7.5 \times 10^{-4})(0.84)(T_e - 20)$$

from which the bed temperature is $T_e = \underline{60°C}$.

(b) *Drying time for a particle.* Since these solids are porous, the drying would most likely be in the constant-rate regime. With no additional information given, we assume that it is. Then from Eq. (61),

$$\frac{0.04}{0.20} = 1 - \frac{1 - e^{-\tau/\bar{t}}}{\tau/\bar{t}}$$

Solving by trial and error, we get

$$\frac{\tau}{\bar{t}} = 0.46 \qquad\qquad\qquad\qquad (i)$$

The weight of solids in the bench dryer is

$$W = \frac{\pi}{4}(0.1)^2(0.1)(1 - 0.45)(2000) = 0.863 \text{ kg}$$

Their mean residence time, from Eq. (59), is

$$\bar{t} = \frac{W}{F_0} = \frac{0.863}{7.5 \times 10^{-4}} = 1150 \text{ s}$$

Hence the time for complete drying of a particle, from Eq. (i), is

$$\tau = (1150)(0.46) = \underline{529 \text{ s}}$$

(c) *Commercial-scale dryer.* For the larger dryer, Eq. (59) gives the bed weight as

$$W = (1150)(1.0) = 1150 \text{ kg} = 1.15 \text{ tons}$$

Operating at the same temperature as the bench dryer, the heat balance of Eq. (53) then gives

$$A_t(0.30)(1)(1.00)(200 - 60) = (1)(0.20 - 0.04)[2370 + 4.2(60 - 20)]$$
$$+ (1)(0.84)(60 - 20)$$

from which

$$A_t = 10.5\,\text{m}^2 \quad \text{or} \quad d_t = 3.65\,\text{m}$$

The flow rate of gas necessary for the operation is

$$A_t u_o \rho_g = (10.5)(0.3)(1) = \underline{\underline{3.14\,\text{kg/s}}}$$

EXAMPLE 5

Solvent
Recovery
from Polymer
Particles

A continuous single-stage dryer is to recover heptane solvent from wet polymer particles, using superheated steam and solvent vapor as the fluidizing gas. Batch experiments at the planned operating conditions show that drying is diffusion-controlled and follows Eq. (63) with a heptane half-life of 140 s. Compare the following designs:

(a) A dryer without an internal heater, hence a freely bubbling bed
(b) A dryer using vertical heating tubes with one-eighth the entering gas flow rate as the free bubbling bed

Figure E5 shows the overall operations.

Data

Dry solids:	$\rho_s = 1600\,\text{kg/m}^3$, $C_{ps} = 1.25\,\text{kJ/kg·K}$
	$F_0 = 1800\,\text{kg/hr} = 0.5\,\text{kg/s}$, $T_{si} = 20°C$
Moisture:	water, to be dried from 1.0 to 0.2 kg/kg dry solids
	heptane, to be dried from 1.1 to 0.1 kg/kg dry solids
Fluidizing gas:	$T_{gi} = 240°C$
Bed:	$T_e = 110°C$, $\varepsilon_m = 0.5$, $\varepsilon_f = 0.75$,
	maximum allowable superficial gas velocity, $u_o = 0.6\,\text{m/s}$
Heater:	vertical tubes $d_i = 0.08\,\text{m}$ in a $l_i = 0.2$-m square array
	$h_w = 400\,\text{W/m}^2\text{·K}$
	saturated steam condenses in the tubes at 238°C

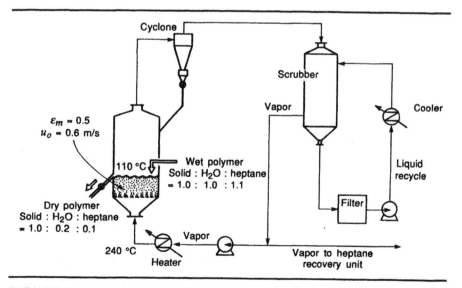

FIGURE E5
A simplified flowsheet of a fluidized dryer that uses superheated steam to recover solvent from wet polymer particles.

Specific heats in kJ/kg·K
    water liquid: 4.18   heptane liquid: 2.05
    water vapor: 1.92   heptane vapor: 1.67
Latent heat of vaporization in kJ/kg
    water: 2260   heptane   326
Density of vapor in kg/m^3 at operating conditions
    water: 0.56   heptane: 3.1

Neglect heat losses to the surroundings.
For the dryer with immersed vertical tubes take $L_f = 2L_m = 1.5\,m$.

SOLUTION

(a) *Dryer without internals.* Figure E5 shows that the composition of the recycle vapor is identical to the vapor forming in the bed. Thus

$$\text{Water/heptane weight ratio:}\qquad \frac{1.0-0.2}{1.1-0.1} = \frac{0.8}{1}$$

$$\text{Water/heptane volume ratio:}\qquad \frac{0.8/18}{1/100} = \frac{0.816}{0.184}$$

Hence, the mean density of the vapor mixture is

$$\bar{\rho}_g = 0.816(0.56) + 0.184(3.1) = 1.027\,kg/m^3$$

and

$$\bar{C}_{pg} = 0.816(0.56)(1.92) + 0.184(3.1)(1.67) = 1.83\,kJ/m^3 \cdot K$$

With u_o as the superficial vapor velocity just above the distributor, the heat balance of Eq. (53) gives

$$(A_t u_o)(1.83)(240 - 110) = (0.5)(1.0 - 0.2)[2260 + 4.18(110 - 20)]$$
$$+ (0.5)(1.1 - 0.1)[326 + 2.05(110 - 20)] + (0.5)(1.25)(110 - 20)$$
$$= 1366\,kW$$

Thus, the heat transferred to the emulsion is 1366 kW, and the volumetric flow rate of recycle gas to the dryer is

$$A_t u_o = 5.742\,m^3/s$$

Next, the rate of formation of vapor in the bed is

$$(0.5)\left[\frac{1.0-0.2}{0.56} + \frac{1.1-0.1}{3.1}\right] = 0.8756\,m^3/s$$

Since the superficial gas velocity at the top of the bed is to be $u_o = 0.6\,m/s$, the superficial velocity just above the distributor should be

$$u_o = (0.6)\,\frac{5.742}{5.742 + 0.8756} = 0.5206\,m/s$$

from which the cross-sectional area and diameter of the bed are

$$A_t = \frac{5.742}{0.5206} = 11.03\,m^2 = \frac{\pi}{4}\,d_t^2 \qquad \text{and} \qquad \underline{d_t = 3.75\,m}$$

To find the bed height needed, first use Eq. (63):

$$0.5 = e^{-B'(140)}; \qquad \text{thus } B' = 0.00495\,s^{-1}$$

Then from Eq. (64) we find the mean residence time of solids:

$$\frac{0.1}{1.1} = \frac{1}{1 + (0.00495)\bar{t}} \quad \text{or} \quad \bar{t} = 2020 \text{ s}$$

From Eq. (59) the bed weight is

$$W = F_0 \bar{t} = (0.5)(2020) = 1010 \text{ kg}$$

Hence the static and the fluidized bed heights are

$$L_m = \frac{W}{A_t(1 - \varepsilon_m)\rho_s} = \frac{1010}{(11.03)(1 - 0.5)(1600)} = 0.115 \text{ m}$$

and

$$L_f = (0.115)\frac{1 - 0.5}{1 - 0.75} = 0.23 \text{ m}$$

(b) *Dryer with internal heaters.* With one-eighth the flow rate of recirculating gas as in part (a), the flow into the bed is

$$A_{bed}u_{o,distributor} = \frac{5.742}{8} = 0.7178 \text{ m}^3/\text{s}$$

But to keep $u_o = 0.6$ m/s at the top of the bed, $u_{o,distributor}$ must be

$$u_{o,distributor} = (0.6)\left[\frac{0.7178}{0.7178 + 0.8756}\right] = 0.2703 \text{ m/s}$$

from which we find the cross-sectional area to be made available to the drying solids:

$$A_{bed} = \frac{0.7178}{0.2703} = 2.6557 \text{ m}^2$$

Next, the energy balance of Eq. (53) allows us to determine the heat to be added to the drying polymer particles by the condensing steam, or

$$(2.6557)(0.2703)(1.83)(240 - 110) + q = 1366 \text{ kW}$$

from which

$$q = 1195 \text{ kW}$$

Now for the heat exchange tubes,

$$q = h_w A_w (T_w - T_e)$$

Thus, the total surface area of heat exchanger tubes needed is

$$A_w = \frac{1,195,000}{(400)(238 - 110)} = 23.34 \text{ m}^2$$

Hence, the total length, the total number of 1.5-m-long tubes, and the total cross-sectional area of these tubes are

$$L_t = \frac{A_w}{\pi d_i} = \frac{23.34}{\pi(0.08)} = 92.88 \text{ m}$$

$$N_t = \frac{92.88}{1.5} = \underline{61.9 \text{ tubes}}$$

$$A_{tubes} = N_t\left(\frac{\pi}{4} d_i^2\right) = 62\left(\frac{\pi}{4}\right)(0.08)^2 = 0.3116 \text{ m}^2$$

Thus the total cross-sectional area of the tube-filled dryer is

$$A_{total} = A_{bed} + A_{tubes} = 2.6557 + 0.3116 = 2.9673 \text{ m}^2$$

and the diameter of vessel is

$$d_{dryer} = \left[\frac{A_{total}}{\pi/4} \right]^{1/2} = \left[\frac{(2.9673)(4)}{\pi} \right]^{1/2} = \underline{\underline{1.944 \text{ m}}}$$

Finally, in a square array of tubes, the center-to-center distance between tubes (the pitch) is

$$l_i = \left(\frac{A_{total}}{N_t} \right)^{1/2} = \left(\frac{2.967}{62} \right)^{1/2} = 0.2188 \cong \underline{\underline{0.22 \text{ m}}}$$

Comparison of designs and comments. The results of the foregoing calculations are as follows:

	Bed Diameter (m)	Recycle Vapor Flow (m^3/s)
Without internal heater	3.75	5.74
With heating tubes	1.94	0.72

Accounting for construction costs and power consumption of the recycle blower, the latter design is usually preferred. However, for sticky and easily agglomerable solids, the former design may be safer for long-term continuous operations. The choice of design should be made after careful observation of bench-scale experiments.

This example shows that drying with superheated vapor from which solvent can easily be recovered is practical. In view of ever-stricter environmental regulations controlling the discharge of toxic solvents into the atmosphere, the type of closed recirculation drying process illustrated here becomes more and more economically attractive.

In any drying process with superheated vapor, the discharge line for the dry solids should be carefully sealed to prevent contact with air. Consecutive sealing with superheated steam and then hot nitrogen can do this.

Finally, note that only the average moisture content is required in this example. Where the same moisture content is required for all the leaving particles, we must try to have an almost uniform residence time for the solids. The easiest way to do this is to position vertical plates in the bed (see Figs. 2.4(b), (c), and (g) and 14.2) to encourage the solids to approach plug flow.

PROBLEMS

1. Design a heat exchanger as shown in Fig. 3(a) (number of stages, temperature, and diameter of each stage) for continuously heating solids from $T_{si} = 20°C$ to $T_N = 820°C$ with gas at $T_{gi} = 1020°C$. Take identical thermal efficiencies for gas and solid, or $\eta_g = \eta_s$.

   Data

   Solids: $\rho_s = 2000 \text{ kg/m}^3$, $F_0 = 3 \text{ kg/s}$
   Gas: $\rho_g = 1 \text{ kg/m}^3$ at 20°C, $u_0 = 0.3 \text{ m/s}$ at all temperatures
   $C_{pg} \cong C_{ps}$
   Bed: $\varepsilon_m = 0.5$, $L_m = 0.3 \text{ m}$

2.   If the solids in Prob. 1 are heated in a four-stage crossflow exchanger, estimate the necessary gas flow rate and its thermal efficiency. Compare the power consumption of the two schemes. The pressure drop across each stage is $\Delta p_d = 3.6$ kPa.

3.   (a) Suggest a bed geometry (diameter and height) and power requirement to cool 10 tons/min of hot solids from 1000 to 100°C by direct contact with cold gas at 0°C in a fluidized bed heat exchanger.
      (b) Suppose the design could be modified to incorporate vertical heat exchanger tubes in the bed to take up 75% of the heat load. How would this affect the power requirement and bed geometry?

Data

Solids: $d_p = 800$ μm, $\rho_s = 1600$ kg/m^3, spherical
            $C_{ps} = 0.8$ kJ/kg·K
Gas: $\rho_g = 0.4$ kg/m^3, $\mu = 5 \times 10^{-5}$ kg/m·s
            $C_{pg} = 1.2$ kJ/kg·K, $k_g = 0.08$ kJ/m·hr·K
Bed: $\varepsilon_{mf} = 0.5$, $\varepsilon_f = 0.6$
            Operable limits: $u_o = 0.8$ to 1.6 m/s
                               $L_f = 0.3$ to 1.0 m

4.   Three hundred tons per day of limestone are to be calcined in a multistage fluidized calciner consisting of one bed for calcination, one bed for the upper heat recovery section, and one bed for the lower heat recovery section. Determine the fuel consumption and the diameters of the fluidized beds. Also find the thermal efficiency and compare it with the results of Example 1. For additional data see Example 1.

5.   (a) What would be the vapor pressure of the leaving air if the absorption tower of Example 3 were operated at 26°C?
      (b) How should the flow rate of solids be adjusted if the tower is to operate at 26°C and the leaving air is to have a moisture content of 0.0001 atm?

6.   Suppose the solids of Example 4 are a resinous material whose drying is diffusion-controlled according to Eq. (63). What then would be the size of unit needed to treat a feed $F_0 = 3.6$ tons/hr from $Q_{fi} = 0.20$ to $\bar{Q}_f = 0.04$?

REFERENCES

1. E.D. Ermenc, *Chem. Eng.*, **68**, 87 (May 29, 1961).
2. J. Crank, *The Mathematics of Diffusion*, Oxford Univ. Press, New York, 1957.
3. A. Klinkenberg and J.J. van der Minne, *Electrostatics in the Petroleum Industry*, American Elsevier, New York, 1958.
4. F.G. Eichel, *Chem. Eng.*, **74**, 153 (March 13, 1967).
5. V. Vaněček, M. Markvart, and R. Drbohlav, *Fluidized Bed Drying*, translated by J. Landau, Leonard Hill, London, 1966.

17 Design of Catalytic Reactors

— Bench-Scale Reactors
— Pilot-Plant Reactors
— Design Decisions
— Deactivating Catalysts

This chapter discusses the main factors in design and scale-up of solid-catalyzed gas-phase reactors. As the examples show, material is drawn from many chapters to solve the design problem; however, Chap. 12 is particularly important because it shows that the overall conversion of reactant gas can be approximately predicted if the effective diameter of bubbles in the bed is known. Therefore, a reasonable estimate of the expected bubble size is essential for design. This chapter also suggests what to look for in the various stages of experimentation and development work in the scale-up from batch to pilot plant to commercial operations.

Bench-Scale Reactors

Preliminary experiments with small fixed beds of catalyst should give the following information:

- A reasonable estimate of the stoichiometry and rate constant K_r for the desired reaction under various conditions
- An indication of possible side reactions
- The various regimes (size of pellets, effect of catalyst formulation, temperature) where physical factors such as pore diffusion intrude to modify the rate and the distribution of products
- An indication of the stability of the catalyst and a preliminary measure of its rate of deactivation

From thermodynamics we should also know the heat of reaction and the equilibrium conversion.

Information from these exploratory experiments should suggest favorable conditions for running the reaction, expected yields for ideal flow patterns, and whether strict temperature control and/or frequent catalyst regeneration are needed. Unfortunately, the feed may be a complex mixture in which many reactions occur, and catalyst decay may be rapid. In such situations reasonable quantitative measures are hard to get —often they are found only in the later steps of experimentation. Nevertheless, we assume, on the basis of the information available, that the decision has been reached to proceed with fluidized bed operations.

Experiments with a bench-scale fluidized bed are then made and should give the following information:

- Data on $1 - X_A$ versus $K_r \tau$ under various bed conditions
- Effective bubble diameter in the bed during the runs
- Possible secondary reactions either in the fluidized bed or in the freeboard, and their heat effects
- The kinetics and the deactivation rate constant of the catalyst, if its activity changes
- Tendency of the catalyst to agglomerate, erode, break, or shatter

Since the most powerful tool for development work is a reliable experiment-tested theory, the first step to scale-up is to experimentally confirm the applicability of the theoretical equations and to determine their parameters. Thus, at this stage we should be able to fit our data with theoretical curves such as in Figs. 12.1–12.3, using measured values for the rate constant, bubble size, and so forth.

We should not use bench-scale conversion data directly for design of large-scale units because the flow pattern is not the same in different-sized units. Since the interchange coefficient is much smaller in large beds, more bypassing of gas and lowered conversion may be expected in these larger units. Ignorance of this fact has caused problems in the past.

Pilot-Plant Reactors

Based on the information gained so far, we should be able to design the main features of the pilot plant, including the arrangement of units to achieve safe steady state operations, leaving the detailed design and construction to the mechanical engineer.

The first purpose of the pilot plant is to compare the observed conversion with that predicted from theory using bench-scale results. Good agreement will heighten confidence in these theoretical predictions for further scale-up.

It is important that the last three listed bench-scale phenomena be carefully examined, for although they may seem to be unimportant in small units with short run times, they may become serious problems in larger units. Therefore, the second purpose of the pilot plant is to discover whether these factors and other harmful phenomena, normally undetected in bench-scale operations, may become serious as the scale of operations is enlarged. These factors usually cannot be predicted from theory, but sometimes can be expected from past experience in the development of similar processes. If none of these problems become evident in the pilot plant, scale-up can often proceed directly to the commercial unit.

A final purpose of the pilot plant is to get confidence in steady state operations, to acquire data on stability, control, and rates of reaction for further development, to obtain an economic estimate for the operations, to produce sufficient product for market studies, and so on.

What size of pilot plant should we build, and how many stages of scale-up should we plan for? This depends on the extent of our knowledge and experience with similar processes and cost and development time pressures.

TABLE 1   Scale-up of the FCC Process

	Small Pilot Plant	Large Pilot Plant	Commercial Scale
Reactor			
diameter (m)	0.05	0.38	4.57
height (m)	6.1	6.1	8.5
Regenerator			
diameter (m)	0.102	0.56	5.95
height (m)	6.1	9.5	11.3
Feed rate (L/hr)	1.33	66	9900
Conversion %			
gasoline (vol)	50.2	49.7	49.5
gas oil (vol)	50.0	50.0	50.0
dry gas (wt)	4.2	5.0	6.2
carbon (wt)	3.2	2.8	2.8

From Carlsmith and Johnson [1].

Table 1 shows how this question was answered in the development of the FCC process. For engineers familiar with these factors, it is not unusual to scale up to the commercial plant directly from bench-scale experimentation; see Katsumata and Dozono [2].

esign ecisions

With bench-scale data available, the following factors can be considered in the design of larger units.

Use of a Circulation System. If the catalyst deactivates rapidly and if regeneration can supply (or remove) the heat needed (or generated) by the reaction, then a circulation system should be considered for the pilot plant, because the development of a safe, stable, steady state circulation system requires much experience. We should also choose a catalyst that is resistant to attrition and breakage.

Selection of Bed Type and Bubble Size. The two commonly used bed types are the freely bubbling bed and the bed with internals. Since circulation systems use flowing solids for heat control, freely bubbling beds can be used there. In systems with no circulation of solids, heat exchange in the reactor is needed and this usually involves bed internals. Thus, circulation systems can have freely bubbling beds or beds with internals only for control of bubble size and its rise velocity, whereas beds with a batch of solids usually have internals for heat exchange and control of bubble size and bubble rise velocity.

Although numerous freely bubbling beds have been operated commercially, little information exists on the effective size and rise velocity of bubbles and on the relationship between bubble properties and conversion in these beds. This is not serious if the heat transport requirement controls (see Chap. 15); however, it becomes a serious problem if the conversion of reactant controls. Thus, if conversion is expected to be the limiting factor, it is recommended that internals be used to achieve rational scale-up and to reduce the time and cost of development. If heat transport is expected to control, then freely bubbling beds can be used.

Two types of internals commonly used are vertical and horizontal bundles of tubes. When a uniform temperature profile is required throughout the bed, a vertical bundle is recommended because horizontal bundles are apt to hinder the vertical movement of catalyst, possibly giving an appreciable vertical temperature profile. When a uniform vertical temperature is required with two or more horizontal heat exchange bundles, they should be carefully designed and controlled to remove or add the necessary heat appropriate for that level of bed. Also, the bundles should be positioned to avoid channeling of solids near the reactor wall.

Horizontal perforated plates are sometimes used to separate the fluidized bed into two or more compartments to hinder gulf streaming of solids and to attain higher conversion and selectivity. This design is particularly useful when the catalyst fouling rate is relatively high and when the catalyst residence time distribution has to be controlled.

Bed Aspect Ratio (Height-to-Diameter Ratio). In a bed with internals the bubble size is close to constant, so for a given weight of solids and volumetric gas flow rate, the aspect ratio has only a small effect on conversion. However, to avoid possible short-circuiting of gas and to ensure good temperature control, very shallow beds should be avoided. As a safe value the minimum aspect ratio should be about unity.

It is desirable to use a catalyst with properties of domain A' in Fig. 3.9, and containing 10–30% of fines less than 44 μm. With such solids the mean bubble size changes with height as indicated by Fig. 6.7, but remains small throughout the bed.

For large freely bubbling beds of large aspect ratio, one may expect severe gulf circulation to develop to give a faster bubble rise velocity, as mentioned in Chap. 6. This could lower the conversion of gaseous reactant appreciably.

A bundle of vertical tubes or other internals reduces the hydraulic diameter of the bed, d_{te}, and reduces the rise velocity u_b of bubbles, resulting in their longer residence time in the bed. For higher conversion, therefore, control of the bubble rise velocity by use of vertical internals is definitely effective; see Example 1.

Freeboard. As explained in Chap. 12, in vigorously bubbling beds catalyst is entrained above the splash zone; thus reaction continues in the freeboard. Hence, to predict the overall conversion at the reactor outlet requires reliable information on solids entrainment.

If no secondary reactions occur in the freeboard, then the gas outlet can be located at the TDH (see Chap. 7). If harmful secondary reactions do occur, they can be countered by reducing the freeboard and using more efficient cyclones or by immediately quenching the gas stream with a heat exchanger or direct injection of diluent gas. The latter procedure, however, can result in a considerable heat loss.

Distributors. Good distributors such as porous sintered metal or ceramic plates are commonly used with bench-scale equipment, but they are not normally used in larger units because of their high cost and poor resistance to the high mechanical and thermal stresses.

The selection of a good distributor may be of first importance for the success of the development as a whole. Chapter 4 considers this question; however, a few additional remarks may be appropriate here:

- Thermal expansion of the distributor should be taken into account in design, otherwise severe stresses may cause problems.
- In a bed with internals to limit bubble size, any conventional type of distributor should operate satisfactorily.
- In a freely bubbling bed a distributor with relatively small openings should be used.
- When a perforated plate is used for a gas feed stream that is entraining fine solids, the orifice openings should be large enough (0.6 to 2.5 cm) to prevent clogging by the solids.
- It is recommended that the type of distributor intended for the larger units be used in the bench-scale reactor even though it may have originally used a fine distributor. A comparison of bench-scale performance with the two distributors may be useful.

Activity Level of Catalyst. In general, catalysts deactivate because of poisoning, aging, or fouling. *Poisoning* is usually a slow, irreversible process from which the catalyst activity cannot be recovered or regenerated by ordinary techniques, such as burning off coke. This permanent lowering of activity is frequently caused by small quantities of impurities in the feed. They may be inorganic or organic salts of metals, such as vanadium, iron, nickel, and copper. *Aging* also is a slow, irreversible process. This permanent deactivation is attributed to heat-induced changes in physical properties of the catalyst, such as sintering. *Fouling* is caused by deposition of reaction products on the catalyst surface. Usually this product is a carbonaceous material, in which case the process is called *coking*.

When deactivation and regeneration occur, we should know their rates and plan in our experimentation to use catalyst activities comparable with those expected in the larger units. Incidentally, the cost of the catalyst strongly influences the activity level at which we can afford to operate. This means that even in bench-scale operations we may want to use a catalyst with far lower activity than that of the fresh catalyst.

High-Temperature Stability and Aging of Catalyst. For highly exothermic and fast reactions the catalyst particles dispersed in rising bubbles or resting in stationary pockets near the inlet of an oxidizing gas stream may jump to the high-temperature stable point and there deactivate rapidly. Careful observation of spent catalyst can show this. We have pointed out the importance of this phenomenon in mass and heat transfer operations (see Chaps. 5 and 11). This situation is also likely to occur when catalyst fouled by carbonaceous material is regenerated by air.

Control of Size Distribution. Time and wear gradually break down the particles and move the size distribution in the bed toward the fines. This is countered by possible inefficiency in the solids recovery system, which causes a loss of fines and a shift of size distribution to the coarse. The direction the distribution moves depends on the friability of the solids and the collection efficiency of the cyclones and electrostatic precipitators. The proper addition of coarse or fine solids and the adjustment of the cyclone efficiency are used to control the size distribution in the bed and to keep the bed lubricated with sufficient fines for good fluidization.

Surface of Heat Exchanger. The cooling or heating surface needed in a reactor can be calculated by the methods of Chap. 13, and, in light of the uncertainties of this complex phenomenon, it is wise to be generous in the estimate. Except for small-diameter beds, it is recommended that immersed tubes be used, in which case the temperature of a hot bed can easily be controlled by adjusting the flow rate of a coolant. However, the temperature of coolant should be sufficiently high to avoid condensation of any reaction product at the heat exchange surface.

Experience shows that hairline cracks are liable to develop in bent tubes of small radius, and since even a small leakage of coolant can seriously harm a catalytic reaction special care should be taken to guard against this.

As mentioned, it is best to construct the heat exchanger tubes to also act as internals to control bubble size and bubble rise velocity. Finally, to allow rapid modification of the exchanger assembly, it is suggested that the assembly be designed to be fitted into the bed from the top of the reactor vessel.

EXAMPLE 1

Reactor Development Program

Ikeda [3] reported on a development program for the production of acrylonitrile by the ammoxidation of propylene with air:

$$C_3H_6 + NH_3 + \tfrac{3}{2}O_2 \longrightarrow CH_2{:}CH{\cdot}CN + 3H_2O , \qquad \Delta H_r = -5.15 \times 10^5 \text{ J}$$

Three successive stages of scale-up were studied, and all the reactors used contained vertical heat exchanger tubes. The conditions in these reactors were as follows:

Reactor	Reactor Diameter, d_t (m)	Equivalent Diameter, d_{te} (m)	Estimated Bubble Size, d_b (m)
Laboratory unit	0.081	0.04	0.05
Pilot-plant unit	0.205	0.12	0.057
Semicommercial unit	3.6	0.70	0.07

The reaction stoichiometry and kinetics were simplified and represented by

$$A \xrightarrow{K_{r1}} R \xrightarrow{K_{r3}} S$$
$$\scriptstyle K_{r2}$$

where A = propylene, R = acrylonitrile, S = HCN, CO, CO_2, etc. From bench-scale experiments the kinetic constants at the reactor conditions were reported to be

$$K_{r1} = 1.3889 \text{ s}^{-1}, \quad K_{r2} = 0.6111 \text{ s}^{-1}, \quad K_{r3} = 0.022 \text{ s}^{-1}$$

Thus

$$K_{r12} = K_{r1} + K_{r2} = 1.3889 + 0.6111 = 2.000 \text{ s}^{-1}$$

(a) For the three reactors, relate the effective rate constant for the reaction of A, K_{f12}, to the gas flow rate represented by u_o.
(b) Relate the selectivity for acrylonitrile to conversion of propylene at different values of K_{f12}.
(c) For all three reactors relate the exit gas composition to bed height.
(d) Calculate the bed height needed to maximize formation of intermediate in the pilot-plant reactor with $u_o = 0.2\,m/s$.

Data

$$d_p = 60\ \mu m. \quad \varepsilon_m = 0.50, \quad \varepsilon_{mf} = 0.55, \quad u_{mf} = 0.006\ m/s$$

$$\mathscr{D}_A = \mathscr{D}_R = \mathscr{D}_s = 2 \times 10^{-5}\ m^2/s, \quad \gamma_b = 0.005$$

SOLUTION

Preliminary calculations. The reaction scheme of this problem is a special case of Eq. (12.23) in which $K_{r34} = K_{r3}$. We illustrate a typical calculation sequence by selecting $u_o = 0.2\,m/s$ for the pilot-plant reactor for which $d_b = 0.057\,m$. Then Eq. (6.7) gives

$$u_{br} = 0.711(9.8 \times 0.057)^{1/2} = 0.5314\ m/s$$

For Geldart **A** solids Eq. (6.11) gives

$$u_b = 1.55[(0.2 - 0.006) + 14.1(0.057 + 0.005)](0.12)^{0.32} + 0.5314$$
$$= 1.372\ m/s$$

From Eq. (6.29),

$$\delta = \frac{0.2}{1.372} = 0.1458$$

and from Eq. (6.20),

$$1 - \varepsilon_f = (1 - 0.1458)(1 - 0.55) = 0.3844$$

Evaluate γ_c from Eq. (6.36) with Fig. 5.8. Thus

$$\gamma_c = (1 - 0.55)\left[\frac{3}{(0.5314)(0.55)/0.006 - 1} + 0.6 \right] = 0.2983$$

and from Eq. (6.35),

$$\gamma_e = \frac{(1 - 0.1458)(1 - 0.55)}{0.1458} - 0.005 - 0.2983 = 2.333$$

Next, we calculate the interchange coefficients from Eqs. (10.27) and (10.34),

$$K_{bc} = \frac{4.5(0.006)}{0.057} + \frac{5.85(2 \times 10^{-5})^{1/2}(9.8)^{1/4}}{(0.057)^{5/4}} = 2.136\ s^{-1}$$

$$K_{ce} = 6.77\left[\frac{(2 \times 10^{-5})(0.55)(0.5314)}{(0.057)^3} \right]^{1/2} = 1.203\ s^{-1}$$

Now to the effective rate constants. Equation (12.32) gives

$$K_{f12} = \left[0.005(2) + \cfrac{1}{\cfrac{1}{2.136} + \cfrac{1}{0.2983(2) + \cfrac{1}{\cfrac{1}{1.203} + \cfrac{1}{2.333(2)}}}} \right] \frac{0.1458}{0.3844}$$

$$= 0.3448\ s^{-1}$$

Substituting all values in Eqs. (12.33)–(12.36) then gives

$$K_{f3} = 0.0207 \text{ s}^{-1}, \qquad K_{fA} = 0.0121 \text{ s}^{-1}$$

Hence

$$K_{fAR} = \frac{1.3889}{2}(0.3448) - 0.0121 = 0.2394 - 0.0121 = 0.2273 \text{ s}^{-1}$$

K_{fA} is very tedious to calculate, but since it is very small compared to the first term in the above equation we may safely ignore it. We do this from now on. Thus use

$$K_{fAR} \cong \frac{K_{r1}}{K_{r12}} K_{f12} = 0.2394 \text{ s}^{-1}$$

Problem Solving Strategy: The effective rate constant for reactant, K_{f12}, is related to u_o for each of the three reactors by Eqs. (12.26) and (12.31). In addition, K_{f12} is a useful parameter for measuring changes in selectivity and conversion; so we use it to tie together all other variables.

(a) *Relationship between u_o and K_{f12}.* Point A on Fig. E1(a) shows the condition calculated above. With other values of u_o and then with bubble sizes corresponding to the other two development reactors, we can draw the three lines on this figure. The experimental points measured by Ikeda in these reactors are also shown on this graph.

(b) *Relate selectivity with conversion in the three reactors.* The conversion of reactant A is given by Eq. (12.26), and the selectivity of acrylonitrile is defined by

$$S_R = \frac{C_R/C_{Ai}}{X_A}$$

With $u_o = 0.2$ m/s in the pilot-plant reactor, take $K_{f12} = 0.3448$ and $X_A = 0.9$. Then from Eq. (12.26),

$$K_{f12}\tau = -\ln(1 - X_A) = -\ln 0.1 = 2.303$$

Substituting in Eq. (12.27) gives

$$\frac{C_R}{C_{Ai}} = \frac{0.2394}{0.0207 - 0.3448}\left[\exp(-2.303) - \exp\left(-\frac{2.303 \times 0.0207}{0.3448}\right)\right]$$
$$= 0.5694$$

Thus,

$$S_R = \frac{0.5694}{0.9} = 0.6327 \text{ mol R formed/mol A reacted}$$

This condition is represented by point B in Fig. E1(b).

Following this procedure and choosing $K_{f12} = 0.3, 0.4,$ and 0.5 gives the lines shown in the figure. Also shown are the experimental values reported by Ikeda.

(c) *Relate the exit composition to $\tau = L_f(1 - \epsilon_f)/u_o$.* Consider the pilot-plant unit in which $u_o = 0.2$ m/s and $K_{f12} = 0.3448$, and take $\tau = 5$ s. Then Eqs. (12.26) and (12.27), with the above calculated values, give

$$X_A = 1 - e^{-(0.3448)(5)} = 0.822$$
$$\frac{C_R}{C_{Ai}} = \frac{0.2394}{0.0207 - 0.3448}[\exp(-0.3448 \times 5) - \exp(-0.0207 \times 5)]$$
$$= 0.5343$$

These values are shown as points C and D in Fig. E1(c).

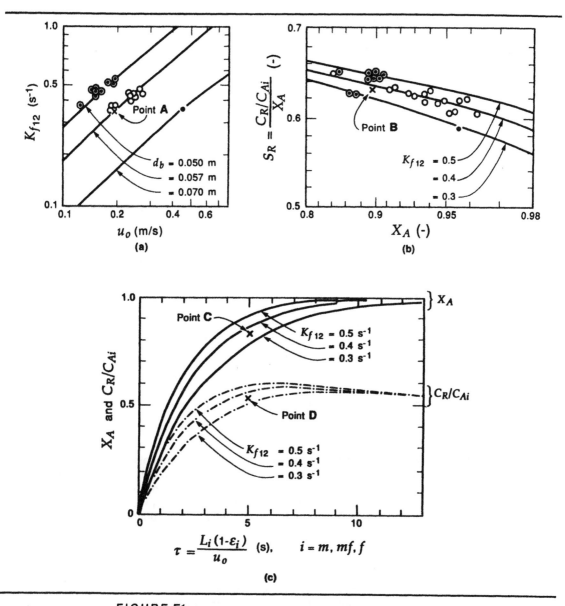

FIGURE E1
Comparison of calculated lines with the experimental data reported by Ikeda [3]: ⊙ laboratory unit; ○ pilot-plant unit; ● semicommercial unit.

Following the same procedure but with $K_{f12} = 0.3$, 0.4, and 0.5 allows us to draw the lines shown in Fig. E1(c). With Fig. E1(a), one can then easily find the performance in any of the three reactors at any feed rate. For example, if one used $u_0 = 0.46$ m/s in the semicommercial unit, Fig. E1(a) gives $K_{f12} = 0.35$. Then Fig. E1(c) gives the conversion and selectivity at any τ value and, hence, for any bed height.

(d) *Calculate the height of bed needed to maximize the production of acrylonitrile in the pilot-plant reactor.* For $u_0 = 0.2$ m/s Eq. (12.37) gives

$$\frac{C_{R,max}}{C_{Ai}} = \frac{0.2394}{0.3448}\left(\frac{0.3448}{0.0207}\right)^{0.0207/(0.0207-0.3448)} = 0.586$$

Equations (12.38) and (12.31) give

$$\tau = \frac{L_f(1 - \varepsilon_f)}{u_o} = \frac{\ln(0.0207/0.3448)}{0.0207 - 0.3448} = 8.6789 \text{ s}^{-1}$$

Thus when fully fluidized,

$$L_f = \frac{0.2}{0.3844}(8.6789) = \underline{4.52 \text{ m}}$$

When settled, the bed height is

$$L_m = \frac{L_f(1 - \varepsilon_f)}{1 - \varepsilon_m} = \frac{4.52(0.3844)}{1 - 0.5} = \underline{\underline{3.47 \text{ m}}}$$

The conversion of propylene at this condition is given by Eq. (12.26). Thus

$$X_A = 1 - e^{-(0.3448)(8.6789)} = \underline{0.950}$$

Remarks. The height of freeboard required is determined by the methods of Chap. 7, and the distributor is designed by the methods of Chap. 4.

To prevent any possible explosion in the feed line, the air should be sent to the bed through a distributor at the bottom of the bed, whereas the propylene-ammonia mixture should be introduced higher up the bed through pipe spargers with downward-pointing orifices, as shown in Fig. 4.4. With this feed arrangement any carbonaceous material deposited on the catalyst is burned when the particles pass through the oxygen-rich zone just above the air distributor, producing in situ regeneration of the fouled catalyst.

EXAMPLE 2

Design of a Commercial Acrylonitrile Reactor

From the information gained from the three experimental reactors of Example 1, design a large commercial reactor to produce 50,000 tons of acrylonitrile per 334-day year. We would like to operate at the same conversion of propylene, $X_A = 95\%$, and with a product selectivity of at least $S_R = 60\%$. Thus, determine the superficial gas velocity, bed diameter and height, and the arrangement of the vertical heat exchanger tubes needed.

Data

Same solids as in Example 1 $\varepsilon_m = 0.5$, $\varepsilon_{mf} = 0.55$
In the reactor $T = 460°C$, pressure = 2.5 bar
Feed gas enters at the bed temperature with composition

$$C_3H_6 : NH_3 : air = 1.0 : 1.1 : 11$$

Heat exchanger: Vertical tubes 0.08 m OD,

$$h_{outside} = 300 \text{ W/m}^2 \cdot \text{K}, \quad h_{inside} = 1800 \text{ W/m}^2 \cdot \text{K}$$

Coolant boiling water at 253.4°C

SOLUTION

Preliminary. For scale-up, choose the bubble size and effective rate constant for reactant to be the same as in the 3.6-m ID reactor. This fixes the effective bed diameter as well. Thus, from Example 1

$$d_b = 0.07 \text{ m}, \quad d_{te} = 0.7 \text{ m}, \quad \text{and} \quad K_{f12} = 0.35 \text{ s}^{-1}$$

Then from Fig. E1(a) we find

$$u_o = \underline{0.46 \, m/s}$$

From Fig. E1(c) the highest concentration of desired product R is obtained at $\tau = L_m(1 - \varepsilon_m)/u_o = 8 \, s$. Thus

$$L_m = \frac{(0.46)(8)}{0.5} = \underline{7.36 \, m}$$

With these bed heights we choose heat exchanger tubes 7 m long.
From Fig. E1(c) we also find

$$\frac{C_R}{C_{Ai}} = 0.58 \quad \text{and} \quad X_A = 0.95$$

in which case

$$S_R = \frac{0.58}{0.95} = 0.61, \text{ or } 61\%$$

This figure matches the value obtained from Fig. E1(b) and meets the requirements of the problem.

Cross-sectional area of reactor. The production rate of acrylonitrile is

$$\frac{50,000 \times 10^3}{(334)(24)(3600)} = 1.733 \, kg/s$$

so the feed rate of propylene is

$$\frac{(1.733)(1/0.053)}{(0.61)(0.95)(1/0.042)} = 2.37 \, kg/s$$

If we ignore the small increase in number of moles on reaction, the volumetric flow rate of gases through the reactor is

$$\left(2.37 \, \frac{kg}{s}\right) \frac{22.4 \, m^3}{42 \, kg} \left(\frac{733}{273}\right) \left(\frac{1 \, bar}{2.5 \, bar}\right) \left(\frac{1 + 1.1 + 11}{1}\right) = 17.8 \, m^3/s$$

Thus, the cross-sectional area of reactor needed for the fluidized bed is

$$A_t = \frac{17.8 \, m^3/s}{0.46 \, m/s} = 38.7 \, m^2$$

Heat exchanger calculation. The rate of heat liberation in the reactor is

$$q = \left(2.373 \, \frac{kg}{s}\right)(0.95)\left(\frac{1 \, kmol}{42 \, kg}\right)\left(5.15 \times 10^8 \, \frac{J}{kmol}\right) = 2.764 \times 10^7 \, W$$

The overall heat transfer coefficient is

$$U = \frac{1}{1/300 + 1/1800} = 257.1 \, W/m^2 \cdot K$$

Hence, the exchanger surface needed to remove this heat is

$$A_w = \frac{q}{U \, \Delta T} = \frac{2.764 \times 10^7}{257.1(460 - 253.4)} = 520.4 \, m^2$$

The number of 7-m-long tubes required is

$$N_t = \frac{520.4}{\pi(0.08)(7)} = \underline{295.8}$$

For a square tube arrangement, the distance between the centers of neighboring tubes, the pitch, is

$$l_i = \left(\frac{A_t}{N_t}\right)^{1/2} = \left(\frac{38.7}{295.8}\right)^{1/2} = \underline{\underline{0.362\ m}}$$

The hydraulic diameter of the bed equivalent to this spacing is determined from Eq. (6.13) to be

$$d_{te} = \frac{4[l_i^2 - (\pi/4)d_i^2]}{\pi d_i} = \frac{4[(0.362)^2 - (\pi/4)(0.08)^2]}{\pi(0.08)} = 2.01\ m \tag{i}$$

This calculated equivalent bed diameter is much larger than $d_{te} = 0.7$ m, which is required to control the rise velocity of bubbles. If we add dummy tubes to the bed to make $d_{te} = 0.7$ m, then, from Eq. (i) rearranged, the pitch becomes

$$l_i = \left(\frac{\pi d_{te} d_i}{4} + \frac{\pi}{4}\,d_i^2\right)^{1/2} = \left[\frac{\pi(0.7)(0.08)}{4} + \frac{\pi}{4}\,(0.08)^2\right]^{1/2} = 0.221\ m$$

Thus, the fraction of bed cross section taken up by tubes is

$$l_i^2 - \frac{\pi}{4}\,d_i^2 = (0.221)^2 - \frac{\pi}{4}\,(0.08)^2 = 0.0438$$

So the reactor diameter, including all its tubes, is

$$\frac{\pi}{4}\,d_t^2 = \frac{38.7}{1 - 0.0438}$$

or

$$d_t = \underline{\underline{7.18\ m}}$$

Remarks. Vertical internals have two functions: they remove the exothermic reaction heat, and they give the required small equivalent bed diameter d_{te}. The above calculations show that the tube spacing for heat removal alone would not meet the hydrodynamic requirement. Thus, we need additional tubes, 2.5 times as many as for heat transfer alone. An alternative to these extra tubes is to use thin vertical I beams. These have less mass per unit surface area than do tubes, which is desirable.

Ikeda et al. used the procedure outlined in this problem for their experimentation, development work, and scale-up to large commercial acrylonitrile reactors, 8 m ID. Chapter 2 discusses this operation, and Fig. 2.10(a) sketches their final design.

Deactivating Catalysts

This subject should be covered more completely in good books on catalytic kinetics. However, because fluidized reactors often use deactivating catalyst, we briefly summarize some of the important points on the kinetics and the design of such systems.

Kinetics with No Catalyst Deactivation. If the catalyst retains its activity during use, the reaction rate constant K_r can be measured without too much difficulty using a batch of solids in an experimental reactor of just about any type.

Kinetics with Slow Catalyst Deactivation. Probably the simplest experimental system that will give the kinetic constants for reaction and deactivation is backmix flow of gas with a batch of solids. Gas with reactant of concentration C_{Ai} is fed to the reactor, and the slowly changing outlet concentration is recorded with time.

We illustrate the method of interpreting data with the simple kinetics considered in Chap. 15, or a first-order reaction and a first-order deactivation, as follows:

$$A \rightarrow R$$

$$-r_A = -\frac{1}{V_s}\frac{dN_A}{dt} = K_r C_A \mathbf{a} \tag{1}$$

$$-\frac{d\mathbf{a}}{dt} = K_a \mathbf{a} \tag{2}$$

where

$$\mathbf{a} = \frac{\text{rate after using the catalyst for time } t}{\text{reaction rate with fresh catalyst}} \tag{3}$$

For a batch of slowly deactivating solids at run time t, and backmix flow of gas, a material balance gives

$$A_t u_0 (C_{Ai} - C_A) = [A_t(1 - \varepsilon_m)L_m]K_r C_A \mathbf{a}$$

or

$$\frac{C_A}{C_{Ai}} = \frac{1}{1 + K_r\mathbf{a}[L_m(1 - \varepsilon_m)/u_0]} = \frac{1}{1 + K_r\mathbf{a}\tau} \tag{4}$$

The variation of activity with run time is found by integrating Eq. (2) to give

$$\mathbf{a} = e^{-K_a t} \tag{5}$$

Combining Eqs. (4) and (5) and rearranging then gives

$$\ln\left(\frac{C_{Ai}}{C_A} - 1\right) = \ln(K_r\tau) - K_a t \tag{6}$$

Making a long-term run at fixed τ, following the changing exit concentration of A with time, plotting the results as in Fig. 1, and evaluating the slope and intercept on this figure give the rate constant for reaction and for deactivation, K_r and K_a.

For other ways of running the experimental reactor and for more involved kinetics, the approach to the evaluation of the kinetic constants is similar to that outlined here but more involved in some cases (see [4]).

Kinetics with Rapid Catalyst Deactivation. When deactivation is rapid, it becomes impractical to use the catalyst only once. It has to be regenerated. It also becomes impractical to try to evaluate the kinetic constants with a batch of solids. In such a case a fluidized reactor setup is recommended. This should be fed continuously with a stream of fresh catalyst, mixed with diluent solids if necessary, and spent solids should be discharged through an overflow tube. Gas should also be in backmix flow either by using a differential reactor (small conversion) or by recycling. The kinetic constants are then found

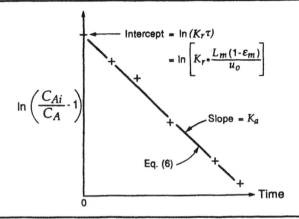

FIGURE 1
A long time run with a batch of slowly deactivating catalyst and backmix flow of gas gives the rate constants for both reaction and deactivation.

by varying the concentration level of reactant in the feed gas or by varying the activity level of catalyst in the bed.

We illustrate the method of interpreting data from this type of setup with the following simple kinetics:

$$\text{For reaction} \qquad -r_A = K_r C_A \mathbf{a} \qquad\qquad (1) \text{ or } (7)$$

$$\text{For deactivation} \qquad -\frac{d\mathbf{a}}{dt} = K_{a1}\mathbf{a} \qquad\qquad (8)$$

$$\text{For regeneration} \qquad \frac{d\mathbf{a}}{dt} = K_{a2}(1 - \mathbf{a}) \qquad\qquad (9)$$

For a *single reactor* with backmix flow of solids and with activities shown in Fig. 2(a), a slight extension of Eq. (15.16) gives, for the solids,

$$K_{a1}\bar{t}_1 = \frac{\bar{\mathbf{a}}_2 - \bar{\mathbf{a}}_1}{\bar{\mathbf{a}}_1} \qquad\qquad (10)$$

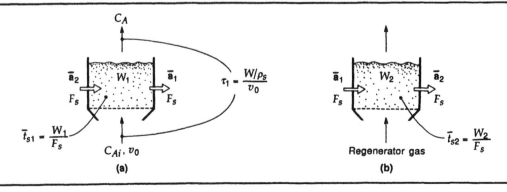

FIGURE 2
(a) A flowing solids reactor with decaying catalyst. (b) A flowing solids catalyst-regenerator.

with the following reactor performance equations, for the gas:

For backmix flow of gas
$$\frac{C_A}{C_{Ai}} = \frac{1}{1 + K_r \bar{a}_1 \tau} \tag{11}$$

For plug flow of gas
$$\frac{C_A}{C_{Ai}} = \exp(-K_r \bar{a}_1 \tau) \tag{12}$$

For the fluidized bed
$$\frac{C_A}{C_{Ai}} = \exp(-K_f \bar{a}_1 \tau) \tag{13}$$

For a *regenerator*, again with backmix flow of solids as shown in Fig. 2(b), we may write

$$K_{a2}\bar{t}_2 = \frac{\bar{a}_2 - \bar{a}_1}{1 - \bar{a}_2} \tag{14}$$

For a *recirculating system* in which solids are only partially regenerated, as shown in Fig. 3, we combine Eqs. (10) and (14) and eliminate $\bar{a}_2$ to get

$$\bar{a}_1 = \frac{K_{a2}\bar{t}_2}{K_{a1}\bar{t}_1 + K_{a1}\bar{t}_1 K_{a2}\bar{t}_2 + K_{a2}\bar{t}_2} \tag{15}$$

We use this expression with the appropriate equation for the gas. With the kinetics known, these equations allow us to find the circulation rate and bed weights needed for reactor performance.

Optimum Size Ratio of Reactor and Regenerator. Reflection shows that for a given gas feed rate, conversion of gaseous reactant, and solid circulation rate, there should be a size ratio that needs the least catalyst. This is found by minimizing $\bar{t}_1 + \bar{t}_2$ for a given value of $\bar{a}_1 \bar{t}_1$ because $\bar{t}_1$ is proportional to the reactor size. Now from Eqs. (10) and (14),

$$\bar{t}_1 + \bar{t}_2 = \frac{\bar{a}_2 - \bar{a}_1}{K_{a1}\bar{a}_1} + \frac{\bar{a}_2 - \bar{a}_1}{K_{a2}(1 - \bar{a}_2)}$$

Minimizing with respect to the only variable, $\bar{a}_2$, gives, after manipulation, the

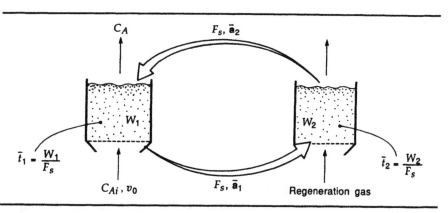

FIGURE 3
A reactor-regenerator system where solids are only partly regenerated.

optimum size ratio

$$\frac{\bar{t}_1}{\bar{t}_2} = \left(\frac{K_{a2}}{K_{a1}}\right)^{1/2} \tag{16}$$

which shows that the slower of the two processes should have the larger unit.

Comments. When the activity change can be described by Eqs. (8) and (9), the system is easily analyzed. When other kinetic expressions must be used, the mathematics may become more involved but it follows the same procedure. One such form is the power equation

$$-\frac{d\mathbf{a}}{dt} = K_r \mathbf{a}^m$$

As pointed out by Szepe and Levenspiel [5], this equation form encompasses many of the previously proposed expressions as special cases, can reasonably represent a variety of deactivation mechanisms, and is certainly easy to treat mathematically compared to other equation forms. For more on other equation forms, rapid deactivation and regeneration, and circulation systems, see Levenspiel [4].

EXAMPLE 3

Reactor-
Regenerator
with Circulating
Catalyst:
Catalytic
Cracking

Determine the optimum sizes of fluidized reactor and regenerator in a solid circulation system of the type in Fig. 3, designed to crack an oil feed to the reactor. By optimum, we mean the design that minimizes the total inventory of catalyst.

Requirements

Feed oil:	$F_1 = 6000 \text{ m}^3/\text{day} = 4800 \text{ tons}/\text{day} = 55.6 \text{ kg}/\text{s}$
Reactor:	conversion of hydrocarbon, $X_A = 63\%$
	superficial gas velocity, $u_o = 0.6 \text{ m}/\text{s}$
Temperature:	of reactor 500°C
	of regenerator 580°C
Solid circulation rate:	from Example 15.2, $F_s/F_1 = 23.3$

Let the regenerator and reactor heights be the same.

Data

Catalyst:	$\rho_s = 1200 \text{ kg}/\text{m}^3$, $\bar{d}_p = 60 \ \mu\text{m}$
Fluidized reactor:	$\varepsilon_m = 0.50$, $\varepsilon_{mf} = 0.55$, $u_{mf} = 0.006 \text{ m}/\text{s}$
	$d_t = 8 \text{ m}$, $d_b = 0.08 \text{ m}$, $\mathscr{D} = 2 \times 10^{-5} \text{ m}^2/\text{s}$

Bench-scale experiments show that the kinetics are reasonably represented by Eqs. (7)–(9) with rate constants

at 500°C: $K_r = 8.6 \text{ s}^{-1}$, $K_{a1} = 0.06 \text{ s}^{-1}$

at 580°C: $K_{a2} = 0.012 \text{ s}^{-1}$

Assume that the upward gulf stream velocity in the large reactor levels off in beds larger than 2 m ID.

SOLUTION

Calculate the parameters for the fluidized reactor. With Eq. (6.7),

$$u_{br} = 0.711(9.8 \times 0.08)^{1/2} = 0.6295 \text{ m}/\text{s}$$

From Eq. (6.11) for Geldart **A** solids,

$$u_b = 1.55[(0.60 - 0.006) + 14.1(0.08 + 0.005)](2)^{0.32} + 0.6295$$
$$= 4.559 \, \text{m/s}$$

From Eqs. (6.29), (6.36), (6.35), and (6.20) we find

$$\delta = \frac{u_o}{u_b} = \frac{0.6}{4.559} = 0.1316$$

$$\gamma_c = (1 - 0.55)\left[\frac{3}{(0.6295)(0.55)/0.006 - 1} + 0.60\right] = 0.2938$$

$$\gamma_e = \frac{(1 - 0.1316)(1 - 0.55)}{0.1316} - (0.005 + 0.2938) = 2.671$$

$$1 - \varepsilon_f = (1 - 0.1316)(1 - 0.55) = 0.3908$$

From Eqs. (10.27) and (10.34) the interchange coefficients are

$$K_{bc} = 4.5\left(\frac{0.006}{0.08}\right) + 5.85\left[\frac{(2 \times 10^{-5})^{1/2}(9.8)^{1/4}}{(0.08)^{5/4}}\right] = 1.425 \, \text{s}^{-1}$$

$$K_{ce} = 6.77\left[\frac{(2 \times 10^{-5})(0.55)(0.6295)}{(0.08)^3}\right]^{1/2} = 0.7873 \, \text{s}^{-1}$$

Find the bed height versus catalyst activity in the reactor. By trial and error we choose $\bar{a}_1$ and calculate the corresponding L_m. Guess $\bar{a}_1 = 0.07$; thus $K_r\bar{a}_1 = (8.6)(0.07) = 0.6 \, \text{s}^{-1}$. Then Eqs. (12.14) and (12.16) give

$$K_f \frac{\delta}{1 - \varepsilon_f} = \left[0.005(0.6) + \cfrac{1}{\cfrac{1}{1.425} + \cfrac{1}{(0.2938)(0.6) + \cfrac{1}{\cfrac{1}{0.7873} + \cfrac{1}{(1.671)(0.6)}}}}\right]\frac{0.1316}{0.3908}$$

$$= 0.1597 \, \text{s}^{-1}$$

and

$$1 - X_A = 1 - 0.63 = \exp(-0.1597\tau)$$

But

$$\tau = \frac{L_m(1 - \varepsilon_m)}{u_o} = \frac{L_m(1 - 0.5)}{0.6} = 6.226 \, \text{s}$$

from which

$$L_m = 7.47 \, \text{m}$$

Similar calculations with different values of $\bar{a}_1$ give the following table of corresponding values of bed heights:

$\bar{a}_1$ (−)	0.0233	0.0465	0.0698	0.0930	0.116	0.140	
$K_r\bar{a}_1$ (s^{-1})	0.2	0.4	0.6	0.8	1.0	1.2	(i)
L_m (m)	11.83	8.65	7.47	6.80	6.36	6.02	

Find the optimum size ratio for various $\bar{a}_1$. Start by taking $L_m = 8 \, \text{m}$. Then the corresponding bed weight is

$$W_1 = \frac{\pi}{4}(8)^2(100)(1 - 0.5)(8) = 241,280 \, \text{kg}$$

The mean residence time of solids in the reactor is

$$\bar{t}_1 = \frac{W_1}{F_s} = \frac{W_1}{23.3F_1} = \frac{241,280}{(23.3)(55.6)} = 186.2 \text{ s}$$

At the optimum, Eq. (16) gives

$$\bar{t}_2 = \bar{t}_1\left(\frac{K_{a1}}{K_{a2}}\right)^{1/2} = 186.2\left(\frac{0.06}{0.012}\right)^{1/2} = 416.4 \text{ s}$$

and from Eq. (15),

$$\bar{a}_1 = \frac{(0.012)(416.4)}{(0.06)(186.2) + (0.06)(186.2)(0.012)(416.4) + (0.012)(416.4)} = 0.0694$$

and thus

$$K_r\bar{a}_1 = 8.6(0.0694) = 0.597 \text{ s}^{-1}$$

Similar calculations with different values of L_m give

L_m (m)	5	6	7	8	10	12	
$\bar{a}_1$ (−)	0.0978	0.0860	0.0769	0.0694	0.0581	0.050	(ii)

Final design values. For the reactor alone, condition (i) must be satisfied; for the optimum size ratio, condition (ii) must be satisfied. The final design must satisfy both conditions. Thus, for the reactor we have

$$L_m = 7.3 \text{ m}, \quad \bar{a}_1 = 0.0744, \quad \bar{a}_1 K_r = 0.64 \text{ s}^{-1}, \quad d_t = \underline{8 \text{ m}}$$

$$W_1 = \frac{\pi}{4}(8)^2(1200)(1 - 0.5)(7.3) = 220,200 \text{ kg} = \underline{220 \text{ tons}}$$

$$\bar{t}_1 = \frac{220,200}{(23.3)(55.6)} = 170 \text{ s}$$

For the regenerator, Eqs. (10) and (16) give

$$\bar{a}_2 = (1 + K_{a1}\bar{t}_1)\bar{a}_1 = (1 + 0.06 \times 170)(0.0744) = 0.833$$

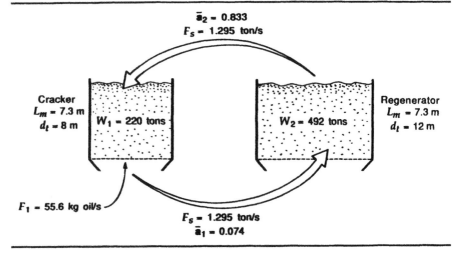

FIGURE E3
Final design of an FCC reactor-regenerator system.

$$\bar{t}_2 = 170\left(\frac{0.06}{0.012}\right)^{1/2} = 380.1 \text{ s}$$

$$W_2 = W_1\left(\frac{\bar{t}_2}{\bar{t}_1}\right) = (220 \text{ tons})\left(\frac{380.1}{170}\right) = \underline{492 \text{ tons}}$$

Choosing the same static bed height for the reactor and regenerator, as suggested, we have

$$d_{t2} = (8 \text{ m})\left(\frac{492}{220}\right)^{1/2} = \underline{12 \text{ m}}$$

The solid circulation rate is

$$F_s = 23.3F_1 = 23.3(55.6) = 1295 \text{ kg/s} = \underline{4662 \text{ tons/hr}}$$

Figure E3 shows these final values.

Comments. If the circulation rate of solids can be raised, then the height of fluidized solids in the reactor, and hence the inventory of solids, can be reduced. Roughly, the reactor size and circulation rate vary inversely proportionally if all other factors remain unchanged.

PROBLEMS

1. The company's research group has developed a more active catalyst for the acrylonitrile synthesis reaction of Examples 1 and 2, having rate constants $K_{r1} = 2 \text{ s}^{-1}$, $K_{r2} = 1 \text{ s}^{-1}$, $K_{r3} = 0.025 \text{ s}^{-1}$. For 95% conversion of propylene determine the bed diameter and static bed height L_m that will maximize the selectivity S_R of acrylonitrile using this catalyst. Then determine this selectivity. For additional information, see Examples 1 and 2 and let the arrangement of the bed internals be the same as in Example 2.

2. After completing our calculations for the FCC unit of Example 3, but before construction has started, we are told that we need to make some design changes because the unit will now be processing a lower grade of oil, which has $K_{a1} = 0.1 \text{ s}^{-1}$; everything else is unchanged. How should we modify our design?

3. The FCC unit of Example 3 has been built and is operating satisfactorily when we are told that we must process a lower grade of oil, for which $K_{a1} = 0.1 \text{ s}^{-1}$, all other rate constants remaining unchanged. With this new feed can we still maintain 63% conversion of hydrocarbon; if so, how would we need to modify our present operations? *Note:* Bed heights can be increased up to $L_m = 10 \text{ m}$, and the solid circulation rate can be increased by up to 15%.

4. Design a fluid catalytic reactor for the commercial production of monovinyl acetate. At 200°C the reaction is exothermic and proceeds as follows:

$$C_2H_2 + CH_3 \cdot COOH \rightarrow CH_2 : CH \cdot OCOCH_3, \qquad \Delta H_r = -117 \text{ kJ}$$

Also determine the addition rate of fresh catalyst.

Requirements

Production rate of monovinyl acetate:	100 tons/day
Ratio of C_2H_2 to $CH_3 \cdot COOH$ in the inlet gas:	4:1
Required conversion of $CH_3 \cdot COOH$:	72%
Theoretical maximum conversion:	80%
$u_o = 0.3$ m/s:	

Data

The reactor is to operate at 200°C and 1 atm.
The reaction is approximated by first-order kinetics with respect to the limiting component, $CH_3 \cdot COOH$, with the following rate constants:

with fresh catalyst, $K_r = 0.1$ s^{-1}
with actual catalyst in the bed, $\bar{a} = 0.5$

Deactivation is slow, with rate given by Eq. (2), where $K_a = 0.2$ day^{-1}. Properties of the solid, the bed, and the flowing gas

$$\bar{d}_p = 450 \ \mu m, \quad u_{mf} = 0.1 \ m/s, \quad \varepsilon_m = 0.55, \quad \varepsilon_{mf} = 0.6$$

$$\rho_s = 1500 \ kg/m^3, \quad \rho_g = 0.80 \ kg/m^3, \quad \mathscr{D} = 2 \times 10^{-5} \ m^2/s$$

Assume the same data as in Examples 1 and 2 for the design of the heat exchanger. In addition, assume a negligible density change of the flowing gas, that all entrained particles are returned to the bed, and $\gamma_b = 0.005$. For these Geldart **B** solids, assume $d_b \cong \frac{1}{2}d_{te}$, and, from Fig. 5.8, $f_w = 0.30$.

REFERENCES

1. L.E. Carlsmith and F.B. Johnson, *Ind. Eng. Chem.*, **37**, 451 (1945).
2. T. Katsumata and T. Dozono, *AIChE Symp. Ser.*, **83(255)**, 86 (1987).
3. Y. Ikeda, *Kagaku Kogaku*, **34**, 1013 (1970).
4. O. Levenspiel, *Chemical Reactor Omnibook*, Chaps. 31–33, OSU Book Stores, Corvallis, OR, 1989.
5. S. Szepe and O. Levenspiel, *Proc. 4th European Symp. on Chemical Reaction Engineering*, Brussels, 1968, Pergamon, New York, p. 265, 1971.

The Design of Noncatalytic Gas-Solid Reactors

— Kinetic Models for the Conversion of Solids

— Conversion of Solids of Unchanging Size

— Conversion of Shrinking and Growing Particles

— Conversion of Gas and Solids

— Miscellaneous Extensions

This chapter deals with the chemical transformation of solid particles that react with fluidizing gas. These numerous reactions, industrially very important, are represented as follows:

$$A(gas) + bB(solid) \rightarrow \begin{array}{l} \text{gaseous product} \hfill (1) \\ \text{solid product} \hfill (2) \\ \text{gaseous and solid product} \hfill (3) \end{array}$$

During reaction the particles may grow, shrink, or remain unchanged in size. Examples of reactions where particles remain essentially unchanged in size include the roasting of sulfide ores, the reduction of iron ore, the nitrogenation of calcium carbide, the calcination of limestone, and the activation of charcoal. Solids shrink in the combustion or gasification of carbonaceous materials and in the chlorination of the oxides of titanium, zirconium, and uranium. The thermal cracking of crude oil in a bed of carbon and the production of pure silicon in a bed of seed particles are examples of particle growth during reaction.

For operations with a continuous feed of solids, the exit stream consists of particles of different ages and degrees of conversion. The average conversion of this stream thus depends on two factors:

• The rate of reaction of single particles in the reactor environment
• The residence time distribution of the solids in the reactor

We want to combine these factors to predict reactor behavior. We also apply the methods of this chapter to the reaction of solids that do not need the action of a gaseous reactant. Throughout, A refers to the gaseous reactant, B to the solid reactant.

**Kinetic
Models
for the
Conversion
of Solids**

The conversion of solids can follow one of two extremes of behavior. At one extreme the diffusion of gaseous reactant into a particle is rapid enough compared to chemical reaction so that solid reactant B is consumed more or less uniformly throughout the particle. This is the *uniform-reaction model* (Fig. 1).

At the other extreme, diffusion into the reactant particle is so slow that the reaction zone is restricted to a thin front that advances from the outer surface into the particle. This model is called the *unreacted-* or *shrinking-core model* (Fig. 2).

Real situations lie between these extremes; however, because these extremes are easy to treat, we like to use them whenever possible to represent the real situation. Naturally, the first and most important consideration is to select the model that most closely represents reality, and only then should we proceed to the detailed mechanism and evaluation of the rate constants.

Uniform-Reaction Model for Porous Solids of Unchanging Size

As shown in Fig. 1, gaseous reactant A is present evenly, or close to evenly, throughout the particle and reacts with solid reactant B everywhere. Although further analysis in terms of a detailed mechanism may lead to a more complicated reaction rate expression, we may take

$$\left(\begin{array}{c} \text{rate of consumption} \\ \text{of B} \end{array} \right) \propto \left(\begin{array}{c} \text{concentration of A} \\ \text{bathing the particle} \end{array} \right) \left(\begin{array}{c} \text{amount of reactant} \\ \text{B left unreacted} \end{array} \right)$$

as a first approximation. In terms of the fraction of B converted, X_B, and for a uniform concentration of gaseous reactant, C_A, the rate expression becomes

$$\frac{dX_B}{dt} = k_r C_A (1 - X_B) \qquad (4)$$

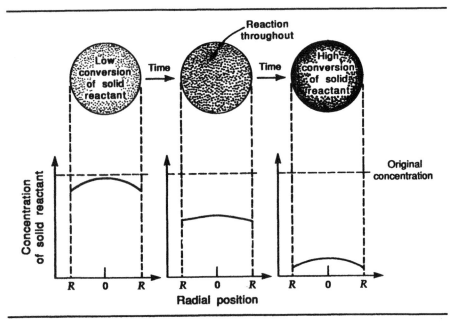

FIGURE 1
Uniform-reaction model. Here reaction proceeds throughout the particle.

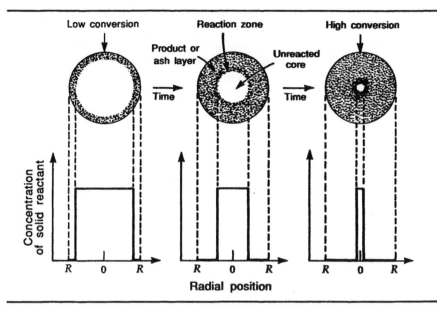

FIGURE 2
Shrinking-core model. Here reaction proceeds at a narrow front that advances into the particle. Reactant is completely converted as the front passes by.

Since C_A is a constant, integration gives the progress of conversion with time, or

$$1 - X_B = \exp(-k_r C_A t) \tag{5}$$

where k_r (m^3 gas/mol A·s) is the rate coefficient based on unit volume of solid.

Shrinking-Core Model for Solids of Unchanging Size

Figure 2 shows that the reaction front advances from the outer surface into the particle leaving behind a layer of completely converted and inert material called the *ash* or *product layer*. At the same time, the core of unreacted solid shrinks and finally disappears. For a reacting particle at some intermediate stage of conversion, the following steps can occur in series:

Step 1. Gaseous A diffuses through the film surrounding the particle to its surface.

Step 2. Gaseous A penetrates and diffuses through the blanket of product solid to reach the reaction front.

Step 3. Gaseous A reacts with reactant B in the narrow reaction zone.

Step 4. Gaseous reaction products diffuse through the product layer from the reaction zone to the surface of the particle.

Step 5. Gaseous reaction products diffuse into the main gas stream.

A reaction need not involve all these steps. For example, if no gaseous product forms, only the first three steps occur.

We present the kinetic expressions and integrated conversion equations when one of the above resistances controls. The detailed derivations of these expressions are given elsewhere [1, 2] and are based on the assumptions that reaction is irreversible, as given by Eq. (1), that particles are spherical, and that the thickness of reaction zone is small compared with the dimensions of the particle. This last assumption allows us to use the shrinking-core model. However, we must be careful to check this assumption, for although this model may reasonably represent large particles it may well be that the uniform-reaction model better represents the state of affairs for small particles in the same environment.

Finally, the distinguishing feature of the conversion equations that follow is that they are expressed in terms of a characteristic time τ, the time required to completely convert an unreacted particle into product. This differs from what usually is encountered in other areas of kinetics.

Chemical Reaction Controls. Here the rate of conversion of solid is proportional to the area of reaction front. Thus, for an unreacted core of radius r_c in a particle of radius R, the rate of reaction of A can reasonably be represented by

$$-\frac{1}{4\pi r_c^2}\frac{dN_A}{dt} = -\frac{1}{4\pi r_c^2 b}\frac{dN_B}{dt} = k_c C_A \tag{6}$$

where k_c (m^3 gas/m^2 solid·s) is a rate constant for the chemical reaction, but is proportional to the volume fraction of B in the solid.

The progress of reaction in a single particle, in terms of the core size or conversion, is

$$\frac{t}{\tau} = 1 - \frac{r_c}{R} = 1 - (1 - X_B)^{1/3} \tag{7}$$

where the relation between conversion and radius of shrinking core is

$$\left(\frac{r_c}{R}\right)^3 = 1 - X_B \tag{8}$$

In Eq. (7) the time for complete conversion is

$$\tau = \frac{\rho_B R}{b k_c C_A} = \frac{\rho_B d_p}{2 b k_c C_A} \tag{9}$$

where τ is independent of the volume fraction of B in the solid, and ρ_B (mol/m^3) is the molar density of B in the solid; see Kimura et al. [3].

Diffusion through Gas Film Controls. This mechanism can only control in the early stages of conversion when no product layer is present. As soon as a product layer forms, its resistance dominates; consequently, for engineering applications the resistance to diffusion through the gas film surrounding the particle can safely be ignored whenever a product layer remains on the particle.

Diffusion through the Product Layer Control. The direct application of Fick's law for diffusion of reactant A through the ash blanket gives the progress of reaction with time as

$$\frac{t}{\tau} = 1 - 3\left(\frac{r_c}{R}\right)^2 + 2\left(\frac{r_c}{R}\right)^3 \tag{10a}$$

where τ is the time for complete conversion of a fresh particle and is given by

$$\tau = \frac{\rho_B R^2}{6b\mathcal{D}_s C_A} = \frac{\rho_B d_p^2}{24b\mathcal{D}_s C_A} \qquad (10b)$$

where $\mathcal{D}_s$ is the effective diffusivity of gaseous reactant through the product blanket, and ρ_B is the molar density of B in the unreacted solid.

Combination of Resistances. When the resistances of the chemical reaction and diffusion steps are comparable, we can approximately represent the overall progress of the reactions by Eqs. (7) and (9), where the reaction rate constant k_c is replaced by $\bar{k}$, defined by

$$\frac{1}{\bar{k}} \cong \frac{1}{k_c} + \frac{d_p}{12\mathcal{D}_s} , \qquad [\text{m}^2 \text{ solid·s/m}^3 \text{ gas}] \qquad (11)$$

For completely porous spherical particles we can define a Thiele-type modulus

$$M_T = R\sqrt{\frac{K_r}{\mathcal{D}_s}} \quad \text{with} \quad K_r = \frac{(3 \text{ to } \frac{1}{3})k_c}{R} = \frac{K_r \times \rho_B}{b} , \qquad [\text{m}^3 \text{ gas/m}^3 \text{ solid·s}] \qquad (12)$$

to represent the relative rates of diffusion of A into the particle and the reaction of A in the particle. Then, as shown by Kimura et al. [4],

When $M_T \leq 1$,   gaseous reactant A can easily penetrate the particle and is close to evenly distributed therein, thus the uniform-conversion model of Eq. (5) applies.

When $M_T > 20$,   the progression of reaction of the particle follows the shrinking-core model with diffusion through the product layer controlling, as given by Eq. (10).

Intermediate Models for Particles of Unchanging Size

Numerous models have been developed to account for particle behavior between the extremes of the uniform-conversion and the shrinking-core models. These intermediate models fall into two classes: those for porous particles and those for particles that start as nonporous but then become porous on reaction.

For porous particles we have the well-known porous-pellet model of Ishida and Wen [5] and the grain model of Sohn and Szekely [6]. The conversion-versus-time curves for these models are closely similar to the curves for the shrinking-core model—high rate of conversion at the beginning, slowing progressively as conversion rises.

For particles that start as nonporous, we have the crackling-core model of Park and Levenspiel [7] shown in Fig. 3. By action of reactant gas the pellet transforms progressively from the outside in, by crackling and fissuring, to form an easily penetrated (no diffusional resistance) porous structure consisting of grain material that then reacts away to the final product according to the shrinking-core model. Of special interest is that this model can account for the sometimes observed S-shaped conversion-versus-time curves—thus, slow conversion at the start, then fast, and finally slow.

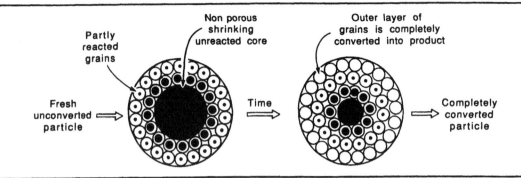

FIGURE 3
Crackling-core model. Here a nonporous particle becomes porous as the front advances into the particle. The grains formed behind the front then slowly react.

Szekely and Themelis [8] summarized various findings on the reduction of very dense iron ore by stating that a shrinking core of ore is observed, but that behind it the solid is only partly converted. This general observation is consistent with the crackling-core model. Levenspiel [9, Chap. 55] considers many of the intermediate regime models as well as the types of experiments needed to discriminate among them.

Finally, in any particular reacting system the controlling resistance not only may shift from diffusion to reaction, but it may even change from the shrinking-core model to the uniform-reaction model as particle size and temperature of operations are changed.

Models for Shrinking Particles

When a flaking ash or no ash forms, as in the burning of coal in air, the particle shrinks and finally disappears (see Fig. 4). We visualize the following steps occurring in succession:

Step 1. Gaseous A diffuses through the gas film to the surface of the particle.

Step 2. Gaseous A either reacts at the surface of the particle or penetrates a short distance into the fine pores of the solid before reacting. In any case, reaction occurs in a narrow zone at the exterior surface of the particle.

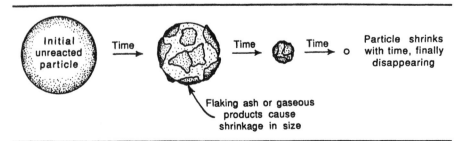

FIGURE 4
Particles shrink by formation of either gaseous product or flaky solid, or by attrition.

Step 3. Gaseous reaction products diffuse through the gas film into the main body of gas.

Here the rate of conversion of a particle depends on the amount of exterior surface area exposed to the gas; consequently, the rate of reaction for a spherical particle of size R is

$$\frac{1}{4\pi R^2}\frac{dN_A}{dt} = -\frac{1}{4\pi R^2 b}\frac{dN_B}{dt} = -\frac{\rho_B}{b}\frac{dR}{dt} = k_c C_A \tag{13}$$

The progressive shrinkage of a particle from initial size R_i to size R at time t is found by integrating Eq. (13):

$$\frac{t}{\tau} = 1 - \frac{R}{R_i} = 1 - (1 - X_B)^{1/3} \tag{14}$$

where the time for complete disappearance of the particle is

$$\tau = \frac{\rho_B R_i}{b k_c C_A} = \frac{\rho_B d_{pi}}{2b k_c C_A} \tag{15}$$

For faster reactions where the resistance to diffusion and reaction are comparable in magnitude we may define an approximate overall reaction rate coefficient $\bar{k}$ as follows:

$$\frac{1}{\bar{k}} \cong \frac{1}{k_c} + \frac{1}{\bar{k}_d}, \qquad [\text{m}^2\,\text{solid}\cdot\text{s}/\text{m}^3\,\text{gas}] \tag{16}$$

where $\bar{k}_d$ is the mean value of the mass transfer coefficient between the shrinking particle and the gas stream.

Finding the Right Model

The controlling resistance to reaction depends primarily on the structure and porosity of the solid. In addition, for any particular solid the resistances of the various steps can differ widely with operating conditions; thus it is important to know which step controls in a particular environment. Normally we expect the following behavior.

Temperature effect. Generally, when a reaction step is rate-controlling, the temperature dependence is very strong; when mass transfer controls, the temperature dependence is minor. Hence, a rise in temperature causes the controlling resistance to shift from reaction to mass transfer.

Particle size effect. Small particles follow the uniform-reaction model, whereas large particles follow the shrinking-core model, with ash diffusion controlling at high temperature but reaction controlling at low temperature.

The controlling mechanism and the value of the rate constant can be estimated by following the conversion with time for different-sized particles. The equations already derived show that

τ independent of d_p for the uniform-reaction model

$\tau \propto d_p$ for the shrinking-core model with reaction controlling

$\tau \propto d_p^2$ for the shrinking-core model with diffusion through the product blanket controlling

In extrapolating to new and untried operating conditions, we must know when to be prepared for a change in controlling step and when we may reasonably expect the rate-controlling step not to change. In any case, instead of extrapolating to new and untried conditions, it is recommended that data be taken whenever possible at the conditions to be used.

Models for the Reaction of Solids Alone

Fluidized beds are well suited for reactions where solids are transformed into particles without the action of gaseous reactant because here the particles can be rapidly heated or cooled; in effect, their temperature environment can be controlled easily. These reactions include thermal decompositions such as the calcination of limestone, the clinkering of pellets for cement, and the carbonization of coal and oil shale.

With large particles such as used in the calcination of coarse limestone,

$$CaCO_3(s) \rightarrow CaO(s) + CO_2 \uparrow$$

the highly endothermic reaction proceeds according to the shrinking-core model, wherein the driving force is not the diffusion of reactant gas through the product layer but the conduction of heat through the product layer to the heat-absorbing decomposition front.

For fine particles the controlling resistance may shift from heat conduction to the decomposition reaction. Also, since these fine particles are likely to be entrained from the fluidized bed, we should consider the possibility of the reverse reaction occurring in the transfer line where the particles find themselves bathed in a CO_2-rich environment at lower temperature.

EXAMPLE 1

Kinetics of Zinc Blende Roasting

Spherical particles of pure zinc blende of size $d_p = 2$ mm are roasted in an 8% oxygen stream at 900°C. The stoichiometry of the reaction is

$$O_2 + \tfrac{2}{3} ZnS = \tfrac{2}{3} SO_2 + \tfrac{2}{3} ZnO$$

a. Assuming that the reaction proceeds by the shrinking-core model, calculate the time needed for complete conversion of the solids, and find the rate-controlling step.
b. Repeat the calculation for particles of size $d_p = 0.1$ mm.

Data

$$\rho_s = 4130 \text{ kg/m}^3, \quad \mathscr{D}_s = 8 \times 10^{-6} \text{ m}^2/s$$
$$k_c = 0.02 \text{ m/s}, \quad \text{Pressure} = 1 \text{ bar} = 10^5 \text{ Pa}$$

SOLUTION

The concentration of gaseous reactant, oxygen, is

$$C_A = \frac{p_A}{RT} = \frac{0.08 \times 10^5}{(8.314)(1173)} = 0.820 \text{ mol/m}^3$$

Since the molecular weight of ZnS is 0.09745 kg/mol, the molar density of pure solid reactant is

$$\rho_B = \frac{4130 \text{ kg/m}^3}{0.09745 \text{ kg/mol}} = 42,381 \text{ mol/m}^3$$

and from the stoichiometry $b = 2/3$.

a. *For 2-mm particles*, Eq. (11) gives

$$\bar{k} = \frac{1}{1/0.02 + (2 \times 10^{-3})/(12)(8 \times 10^{-6})} = \frac{1}{50 + 20.83} = 0.0141 \text{ m/s}$$

Thus the resistance to diffusion through the product layer is nearly as important as the reaction step. Also from Eq. (9),

$$\tau = \frac{(42,381)(2 \times 10^{-3})}{2(2/3)(0.0141)(0.820)} = 5498 \text{ s} = \underline{1 \text{ hr } 31.6 \text{ min}}$$

b. *For 0.1-mm particles*, Eq. (11) gives

$$\bar{k} = \frac{1}{1/0.02 + (1 \times 10^{-4})/(12)(8 \times 10^{-6})} = \frac{1}{50 + 1.04} = 0.0196 \text{ m/s}$$

This result shows that the resistance to diffusion is negligible and that reaction rate controls for these smaller particles.
The time for complete reaction is then given by Eq. (9) as

$$\tau = \frac{(42,381)(1 \times 10^{-4})}{2(2/3)(0.0196)(0.820)} = 198 \text{ s} = \underline{3.3 \text{ min}}$$

EXAMPLE 2

Kinetics of Carbon Burning

Calculate the time needed to burn to completion particles of graphite ($d_{pi} = 1$ mm, $\rho_s = 2200 \text{ kg/m}^3$, $\bar{k} = k_c = 0.2$ m/s) in the environment of Example 1.

SOLUTION

From the data, the density of graphite is

$$\rho_B = \frac{2200 \text{ kg/m}^3}{0.012 \text{ kg/mol}} = 183,000 \text{ mol/m}^3$$

and for the stoichiometry $C + O_2 = CO_2$ we have $b = 1$. Thus, from Eq. (15) the time required for complete combustion of a particle is

$$\tau = \frac{(183,000)(10^{-3})}{2(1)(0.2)(0.820)} = 558 \text{ s} = \underline{9.3 \text{ min}}$$

Conversion of Solids of Unchanging Size

Single-Size Particles and Single Fluidized Bed

Consider the reactor of Fig. 5, with a constant feed rate of both solids and gas, the solids being of uniform size and in backmix flow in the reactor. This model represents a fluidized reactor without carryover of solids.

The conversion X_B of an individual particle of solid depends on its length of stay in the bed. For the appropriate controlling resistance this conversion is given by Eq. (5), (7), or (10). However, the individual particles have different lengths of stay in the bed. For this reason the conversion level varies from particle to particle, and, on accounting for this, the mean conversion of the exit stream of solids $\bar{X}_B$ is

$$\begin{pmatrix} \text{fraction of B} \\ \text{unconverted in} \\ \text{the leaving solids} \end{pmatrix} = \sum_{\substack{\text{particles} \\ \text{of all ages}}} \begin{pmatrix} \text{fraction of B} \\ \text{unconverted in} \\ \text{particles staying} \\ \text{in the reactor} \\ \text{for time between} \\ t \text{ and } t + dt \end{pmatrix} \begin{pmatrix} \text{fraction of exit} \\ \text{stream that stays} \\ \text{this length of time} \\ \text{in the reactor} \end{pmatrix} \quad (17a)$$

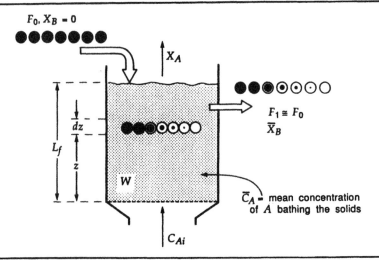

FIGURE 5
Variables for a fluidized reactor treating particles of uniform size. (Shrinking-core behavior shown here.)

In symbols,

$$1 - \bar{X}_B = \int_{t=0}^{\infty} (1 - X_B)_{\text{particle}} E(t) \, dt \tag{17b}$$

where the exit age distribution for the solids in a single fluidized bed is

$$E(t) = \frac{1}{\bar{t}} e^{-t/\bar{t}} \tag{14.3 or 18}$$

and where the mean residence time of these solids is

$$\bar{t} = \frac{W}{F_1} \tag{14.2 or 19}$$

We now give the conversion expressions for a single-fluidized bed for the various controlling resistances. For the *uniform-reaction model* and first-order reaction with respect to the reactant gas, substitution of Eqs. (5) and (18) in Eq. (17b) gives

$$1 - \bar{X}_B = \int_{t=0}^{\infty} \exp(-k_r C_A t) \frac{e^{-t/\bar{t}}}{\bar{t}} \, dt = \frac{1}{1 + k_r C_A \bar{t}} \tag{20}$$

For *shrinking-core reaction-controlling kinetics*, substitution of Eqs. (7) and (18) in Eq. (17b) gives

$$1 - \bar{X}_B = \int_0^{\tau} \left(1 - \frac{t}{\tau}\right)^3 \frac{e^{-t/\bar{t}}}{\bar{t}} \, dt \tag{21}$$

The range of integration is from 0 to τ rather than from 0 to ∞, because a particle that stays in the bed longer than time τ does not contribute to $1 - \bar{X}_B$. Integrating this expression gives

$$1 - \bar{X}_B = 1 - 3\left(\frac{\bar{t}}{\tau}\right) + 6\left(\frac{\bar{t}}{\tau}\right)^2 - 6\left(\frac{\bar{t}}{\tau}\right)^3(1 - \bar{e}^{\tau/\bar{t}}) \tag{22}$$

or, in equivalent expanded form, useful for $\bar{t}/\tau > 1$,

$$1 - \bar{X}_B = \frac{1}{4}\left(\frac{\tau}{\bar{t}}\right) - \frac{1}{20}\left(\frac{\tau}{\bar{t}}\right)^2 + \frac{1}{120}\left(\frac{\tau}{\bar{t}}\right)^3 - \cdots \tag{23}$$

or for $\bar{t}/\tau > 5$,

$$1 - \bar{X}_B \cong \frac{1}{4}\frac{\tau}{\bar{t}} \tag{24}$$

In all these expressions τ is given by Eq. (9).

For *shrinking-core diffusion in product-layer-controlling kinetics*, replacing Eq. (10) in Eq. (17b) and integrating gives an expression that, on expansion, reduces to [2, 10]

$$1 - X_B = \frac{1}{5}\left(\frac{\tau}{\bar{t}}\right) - \frac{19}{420}\left(\frac{\tau}{\bar{t}}\right)^2 + \frac{41}{4620}\left(\frac{\tau}{\bar{t}}\right)^3 - 0.00149\left(\frac{\tau}{\bar{t}}\right)^4 + \cdots \tag{25}$$

Again, for high conversions, or where $\bar{t}/\tau > 5$,

$$1 - \bar{X}_B \cong \frac{1}{5}\left(\frac{\tau}{\bar{t}}\right) \tag{26}$$

In these expressions τ is given by Eq. (10b).

Single-Size Particles and Multiple Fluidized Beds

In single-stage contacting a significant portion of the feed solids stays in the vessel for a very short time; consequently, a very large reactor is needed to achieve high conversions. Multistaging for the solids, either by countercurrent or crosscurrent flow, reduces this bypassing, gives a distribution of residence times approaching plug flow, and reduces the size of reactor needed, particularly at high conversion of solids.

If we assume that the time for complete conversion of a particle is independent of its location—that is, it depends solely on its *total* length of stay in the multistage system—we can then apply Eq. (17b) with its appropriate residence time distribution. For an N-stage equal-sized system this is

$$E(t) = \frac{1}{(N-1)!\,\bar{t}_i}\left(\frac{t}{\bar{t}_i}\right)^{N-1} e^{-t/\bar{t}_i} \tag{14.5}$$

where for each stage

$$\bar{t}_i = \frac{W_{\text{per bed}}}{F_0} \tag{14.4}$$

Thus, the mean conversion of solids leaving an N-stage system, from Eq. (17b), is

$$1 - \bar{X}_B = \int_{t=0}^{\infty} (1 - X_B)_{\text{particle}} \frac{1}{(N-1)!\,\bar{t}_i}\left(\frac{t}{\bar{t}_i}\right)^{N-1} e^{-t/\bar{t}_i}\, dt \tag{27}$$

which can be solved for the appropriate kinetics.

For *uniform-reaction kinetics* in an N-stage bed, putting Eq. (5) in Eq. (27) and integrating gives

$$1 - \bar{X}_B = \frac{1}{(1 + k_r C_A \bar{t}_i)^N} \tag{28}$$

For *shrinking-core reaction-controlling kinetics*, putting Eq. (7) in Eq. (27) gives

$$1 - \bar{X}_B = \sum_{m=0}^{m=N-1} \frac{(N - m + 2)!}{(N - m - 1)! m!} \left(\frac{\tau}{\bar{t}_i}\right)^{m-3} e^{-\tau/\bar{t}_i}$$

$$+ \sum_{m=0}^{m=3} \frac{(N + m - 1)3!}{(N - 1)! \, m! \, (3 - m)!} \left(-\frac{\bar{t}_i}{\tau}\right)^m \tag{29}$$

and for large values of $\bar{t}_i/\tau$, or high conversion, this equation reduces to

$$N = 2, \quad 1 - \bar{X}_B = \frac{1}{20} \left(\frac{\tau}{\bar{t}_i}\right)^2 - \frac{1}{60} \left(\frac{\tau}{\bar{t}_i}\right)^3 + \frac{1}{280} \left(\frac{\tau}{\bar{t}_i}\right)^4 - \cdots \tag{30}$$

$$N = 3, \quad 1 - \bar{X}_B = \frac{1}{120} \left(\frac{\tau}{\bar{t}_i}\right)^3 - \frac{1}{280} \left(\frac{\tau}{\bar{t}_i}\right)^4 + \cdots \tag{31}$$

To illustrate the size reduction achieved by multistaging, compare the size requirements for one- and two-stage beds for 99% conversion of solids where reaction at the shrinking core controls. At the high conversion needed, Eqs. (24) and (30) combined give

$$0.01 = \frac{1}{4} \left(\frac{\tau}{\bar{t}'}\right)_{\text{single stage}} = \frac{1}{20} \left(\frac{\tau}{\bar{t}''}\right)^2_{\text{two stages}}$$

Solving for the same τ and same feed rate gives

$$\left(\frac{\text{total bed weight for two beds}}{\text{bed weight for single bed}}\right) = \frac{2\bar{t}''_i}{\bar{t}'_i} = \frac{2 \times 5^{1/2}}{25} = 0.18$$

Thus, using two stages reduces the total bed weight to 18% that of a single bed. This represents a significant improvement.

For *shrinking-core product-layer diffusion-controlling kinetics*, integrating Eq. (27) for $N > 1$ is awkward. However, the size ratio for single- and multistage operations for this case is reasonably approximated by the corresponding ratio for reaction-controlling kinetics.

In practical operations the reaction conditions such as temperature and gas composition may vary among stages. If so, then a stepwise calculation of conversion in each stage must be made. This type of calculation is presented by Tone et al. [10].

Size Distribution of Particles in a Single Bed

When the feed consists of a size distribution of solids, the mean conversion of the exit stream must account for both size and length of stay of particles in the bed. In words,

$$\begin{pmatrix} \text{fraction of} \\ \text{all solids} \\ \text{unconverted} \end{pmatrix} = \sum_{\text{all sizes}} \begin{pmatrix} \text{fraction} \\ \text{unconverted} \\ \text{of size } R \end{pmatrix} \begin{pmatrix} \text{fraction of} \\ \text{that size} \\ \text{in the feed} \end{pmatrix} \tag{32a}$$

where the first term on the right-hand side is given by Eq. (20). In symbols,

$$1 - \bar{\bar{X}}_{\mathrm{B}} = \int \left\{ \int [1 - X_{\mathrm{B}}(R, t)_{\mathrm{particle}}] \mathbf{E}(R, t)\, dt \right\} \mathbf{p}_0(R)\, dR \tag{32b}$$

In this expression, noting that the solids are in backmix flow,

$$\mathbf{E}(R, t) = \frac{1}{\bar{t}(R)}\, e^{-t/\bar{t}(R)} \tag{14.5}$$

where each particle size has its own particular mean residence time in the bed:

$$\bar{t}(R) = \frac{1}{F_1/W + \kappa(R)[1 - \eta(R)]} \tag{14.19}$$

For the *uniform-reaction model* the fraction of B unconverted is independent of particle size and is given by Eq. (20). Thus, inserting Eq. (20) into Eq. (32b) gives

$$1 - \bar{\bar{X}}_{\mathrm{B}} = \int_0^{R_M} \frac{\mathbf{p}_0(R)\, dR}{1 + k_r C_A \bar{t}(R)} \tag{33}$$

or, in discrete form, where $F_0(R_i)/F_0$ is the fraction of feed in size interval R_i,

$$1 - \bar{\bar{X}}_{\mathrm{B}} = \sum_{i=1}^{M} \frac{1}{1 + k_r C_A \bar{t}(R_i)} \frac{F_0(R_i)}{F_0} \tag{34}$$

In Eqs. (33) and (34) $\bar{t}(R)$ is given by Eq. (14.19).

For *shrinking-core reaction-controlling kinetics*, putting Eq. (22) in Eq. (32b) with $\bar{t}(R)/\tau(R) = y$ gives

$$1 - \bar{\bar{X}}_{\mathrm{B}} = \int_0^{R_M} [1 - 3y + 6y^2 - 6y^3(1 - e^{-1/y})] \mathbf{p}_0(R)\, dR \tag{35}$$

or, in discrete form with $\bar{t}(R_i)/\tau(R_i) = y_i$,

$$1 - \bar{\bar{X}}_{\mathrm{B}} = \sum_{i=1}^{M} [1 - 3y_i + 6y_i^2 - 6y_i^3(1 - e^{-1/y_i})] \frac{F_0(R_i)}{F_0} \tag{36}$$

Similarly, for *shrinking-core product-layer diffusion-controlling kinetics*, substituting Eq. (25) in Eq. (32b) gives, in discrete form,

$$1 - \bar{\bar{X}}_{\mathrm{B}} = \sum_{i=1}^{M} \left[\frac{1}{5} y_i - \frac{19}{420} y_i^2 + \frac{41}{4620} y_i^3 - \cdots \right] \frac{F_0(R_i)}{F_0} \tag{37}$$

Remarks

Again, as with physical operations, we have a choice of crosscurrent or countercurrent flow (see Fig. 16.3). If the conversion of solid is of primary interest, the crossflow arrangement is preferred because of its relative simplicity.

An alternative to multistage fluidized beds for obtaining high conversions is a fluidized bed followed by a transfer line or moving bed reactor. The fluidized bed with its positive temperature control and efficient heat removal carries the reaction most of the way to completion, say 80–90%; the second unit then carries it the rest of the way with only minor problems of heat removal.

EXAMPLE 3

Roasting
Kinetics
from Flowing
Solids Data

Zinc blende particles of size $d_p = 110 \ \mu m$ are fed continuously to an experimental roaster that is fluidized with excess air (constant oxygen concentration) and kept at 900°C. The bed weight is kept constant, and the conversion to the oxide is determined for different feed rates of solids. Determine which model, the uniform-reaction or either of the shrinking-core, best fits the reported data. The stoichiometry of the reaction is

$$O_2 + \tfrac{2}{3} ZnS = \tfrac{2}{3} SO_2 + \tfrac{2}{3} ZnO$$

Reported Data		Calculated from Models		
		Uniform-Reaction,	Shrinking-Core Reaction Control,	Shrinking-Core Diffusion Control,
$\bar{t}$ (min)	$\bar{X}_{B,obs}$	$\bar{X}_B$ from Eq. (20)	$\bar{X}_B$ from Eq. (23)	$\bar{X}_B$ from Eq. (25)
3	0.840	0.840	0.840	0.840
10	0.940	0.946	0.947	0.945
30	0.985	0.981	0.982	0.981
50	0.990	0.989	0.989	0.988

SOLUTION

For the first run ($t = 3$ min and $\bar{X}_{B,obs} = 0.840$), find the parameters of the three models. Then calculate the expected conversion for the other values of $\bar{t}$.

For the uniform-reaction model, Eq. (20) gives

$$1 - 0.84 = \frac{1}{1 + k_r C_A (3)}$$

Hence

$$k_r C_A = 1.75 \ \text{min}^{-1}$$

Using the value 1.75 for the rate group, conversions for the other three runs are calculated from Eq. (20) and tabulated as shown.

For reaction control with a shrinking core, Eq. (23) gives for the first run,

$$1 - 0.84 = \frac{1}{4} \left(\frac{\tau}{3} \right) - \frac{1}{20} \left(\frac{\tau}{3} \right)^2 + \frac{1}{120} \left(\frac{\tau}{3} \right)^3 - \cdots$$

from which we find $\tau = 2.22$ min. With this value for τ, we then calculate the conversion for the other runs and tabulate the results.

Similarly, with diffusion in the product layer controlling, we find $\tau = 2.922$ min with Eq. (25), from which we find the values shown in the last column of the table.

Comment. Note that at high conversion of solid, or $\bar{X}_B > 0.8$, all three of these very different models give practically the same conversion predictions! This is a useful finding, which suggests that reliable predictions are possible for a given feed even when one is uncertain about the proper model to use. It also indicates that it is unnecessary to use any of the more complicated intermediate models in design.

EXAMPLE 4

Scale-up of a
Reactor with
Flowing Solids

Solids of uniform size are reacted with gas in a steady flow, bench-scale, fluidized reactor with the following results:

$$W = 1 \ \text{kg}, \qquad \bar{X}_B = 0.85 \quad \text{for } d_p = 200 \ \mu m$$

$$F_1 = 0.01 \ \text{kg/min}, \quad \bar{X}_B = 0.64 \quad \text{for } d_p = 600 \ \mu m$$

The stoichiometry of the reaction is

$$A(gas) + B(solid) = R(gas) + S(solid)$$

a. Select a model to represent this reaction and
b. Design a commercial unit for 98% conversion of 4 tons/hr of solid feed of size $d_p = 600\ \mu$m.

Data

Preliminary experiments show that the rate is quite temperature-sensitive, which suggests that a chemical step rather than mass transport controls. Also take an unchanging particle size and

$$\rho_s = 2500\ \text{kg/m}^3\ , \quad \varepsilon_m = 0.40\ , \quad L_m = 0.5 d_t$$

SOLUTION

a. Chemical reaction control has two possibilities: the uniform-reaction and the shrinking-core models. For the uniform-reaction model, Eq. (20) shows that $\bar{X}_B$ at given $\bar{t}$ is independent of particle size. This does not agree with experiment.

For the shrinking-core model the data replaced in Eq. (23) give

$$\tau = \begin{cases} 69\ \text{min} & \text{for } d_p = 200\ \mu\text{m} \\ 210\ \text{min} & \text{for } d_p = 600\ \mu\text{m} \end{cases} \quad \text{or} \quad \tau \propto d_p$$

But this is close to what is predicted by Eq. (9). Hence, the shrinking-core model with reaction control fits the data and is the one to use in design.

b. Let us first consider a *single-stage fluidized roaster*. Then Eq. (24) gives

$$1 - \bar{X}_B = 1 - 0.98 = \frac{1}{4}\left(\frac{210}{\bar{t}}\right)$$

from which the mean residence time of solids in the reactor is

$$\bar{t} = 2630\ \text{min} = 43.8\ \text{hr}$$

From Eq. (14.4), the weight of bed needed is then

$$W = F_1 \bar{t} = (4)(43.8) = \underline{175\ \text{tons}}$$

With d_t as the diameter of the bed, we have

$$175 \times 10^3\ \text{kg} = \frac{\pi}{4}\ d_t^2 (0.5 d_t)(2500\ \text{kg/m}^3)(1 - 0.4)$$

from which the required bed dimensions are

$$d_t \cong \underline{6.7\ \text{m}} \quad \text{and} \quad L_m \cong \frac{d_t}{2} = \underline{3.35\ \text{m}}$$

which is unacceptably large.

Next try a *two-stage fluidized roaster*. From Eq. (30) we have

$$1 - \bar{X}_B = 1 - 0.98 = \frac{1}{20}\left(\frac{210}{\bar{t}_i}\right)^2$$

from which the mean residence time in each stage is

$$\bar{t}_i = 332\ \text{min} = 5.53\ \text{hr}$$

Thus the weight of solids and the dimensions for each stage are

$$W = (4)(5.53) = \underline{22.1\ \text{tons}}$$
$$d_t \cong \underline{3.4\ \text{m}} \quad \text{and} \quad L_m \cong \underline{1.7\ \text{m}}$$

These results show that this operation can be accomplished in a single bed of 175 tons or in two beds of 22.1 tons each.

Conversion of Shrinking and Growing Particles

For particles that shrink on reaction according to the kinetics of Eqs. (13)–(16), the decrease in volume of a particle (hence its conversion) follows the behavior of the core of the shrinking-core—reaction-control model, the only difference being that the mass of the particle decreases to zero as conversion rises. Therefore, the flow rate of solids leaving a flow reactor reflects directly the conversion level of solids in the exit stream (see Fig. 6).

For a single-size feed of d_{p0}, Eqs. (9) and (22) are modified accordingly to give

$$1 - \bar{X}_B = \frac{F_1}{F_0} = 1 - 3y + 6y^2 - 6y^3(1 - e^{-1/y}) \tag{38}$$

where

$$y = \frac{\bar{t}}{\tau} = \frac{W/F_1}{\rho_B d_{p0}/2bk_cC_A} \tag{39}$$

In the extreme where all particles shrink to zero and no solids leave the reactor, these expressions reduce to

$$\frac{F_0 d_{p0} \rho_B}{Wbk_cC_A} = 8 \tag{40}$$

Extensions to a size distribution of feed solids, to other kinetics, and to other particle shapes as well as to growing particles can be found in Levenspiel [9, Chap. 54].

Conversion of Both Gas and Solids

Until now we have assumed that the reacting solids are bathed by gas of the same mean composition, no matter what changes are made in the operating conditions. Often this approximation is reasonable, such as when reaction is slow and the concentration of gaseous reactant does not change much in passing through the bed. In this case the conversion equations given so far can

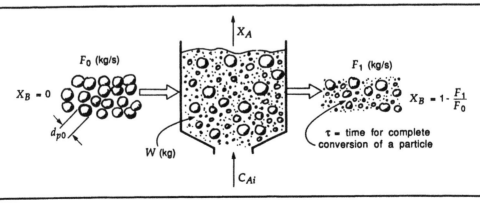

FIGURE 6
Variables for a fluidized reactor treating shrinking particles.

reasonably be expected to apply. This assumption also applies when solids are transformed into product without the action of gaseous reactant, such as in the calcination of limestone.

In the general case, however, the mean gas-phase driving force in the bed is a variable that changes with operating conditions. For example, if the feed rate of solids is lowered, then the concentration of gaseous reactant in the bed rises. Thus, the conversion of solid and the concentration of gaseous reactant leaving and within the bed are interdependent, and a proper analysis of the bed behavior requires accounting for both these changes. We deal with this interaction by a three-step calculation that is applicable to solids of constant size and of changing size.

Step 1: Conversion of gas. Write expressions for the conversion of gaseous reactant A in terms of a mean first-order reaction rate constant $\bar{K}_r$, and for the mean concentration of A encountered by the bed solids. To do this use one of the flow models in Chap. 12, either the model for fine Geldart **A** particles, the model for large Geldart **D** particles, or the intermediate regime model for particles between these extremes.

Step 2: Conversion of solids. Write an expression for the conversion of B in the particles that are bathed by gas of mean composition $\bar{C}_A$. This would be one of the models developed in this chapter for particles of constant size or of changing size.

Step 3: Overall material balance. Then relate the conversion of gaseous reactant A with that of solid reactant B.

Various combinations of these kinetic models can be encountered. For example, for solids of unchanging size we have four possible extremes, designated as **W**, **X**, **Y**, and **Z** of Table 1. The following are situations where one or another of these extremes may be expected to apply.

Combination **W**: Fine particles reacting in a bed of fine solids (example: activation of charcoal).

Combination **X**: Coarse particles reacting in a bed of fine solids (examples: roasting of sulfide ore; reduction of iron ore).

Combination **Y**: Fine particles reacting in a bed of large solids (examples: fine limestone powder reacting with SO_2 gas in a large-particle coal ash bed; gasification of fine char in a bed of agglomerated ash).

Combination **Z**: Reaction in a large-particle fluidized bed (examples: combustion of oil shale; capture of SO_2 by large limestone particles in a large particle bed).

TABLE 1 Combination of Models for Gas-Solid Reactors

		For Solids	
		Uniform-Reaction Model	Shrinking-Core Model, Reaction or Diffusion Control
For Gas	K-L Fast Bubble Model	**W**	**X**
	Slow Bubble Model	**Y**	**Z**

As examples of the calculation procedure we present the equations for two cases: fine particle and coarse particle beds.

Performance Calculations for the Large Particle Bed (Case Z)

Figure 5 illustrates the situation for the large particle bed. Here is the three-step procedure.

Step 1. Assume a first-order reaction for the disappearance of A:

$$-\frac{1}{V_{\text{solid}}}\frac{dN_A}{dt} = \bar{K}_r C_A \tag{41}$$

where $\bar{K}_r$ is the mean value of the rate constant for the reaction of A with all the solids in the bed, some fresh, others highly converted. This constant depends, in general, on the extent of conversion of the solids in the bed. With this kinetic form we can use the equations of Chap. 12, even though we are not dealing with catalytic reactions.

Thus, the large particle conversion expressions, Eqs. (12.53)–(12.57), apply here, and for the conversion of gaseous reactant A leaving the reactor we can write

$$X_A = 1 - \exp\left[-\bar{K}_r\,\frac{L_{\text{mf}}(1-\varepsilon_{\text{mf}})u_{\text{mf}}(1-\delta)}{u_o^2}\right] \tag{12.57 or 42}$$

In addition, we need to know the mean concentration of A which bathes the solids. Since the gas passes in a combination of bypass and plug flow through the bed in this flow regime we can write, with Eq. (42),

$$\bar{C}_A = \frac{1}{L_f}\int_0^{L_f} C_A(z)\,dz = \frac{1}{L_f}\int_0^{L_f} C_{Ai}\exp\left[-\bar{K}_r\,\frac{L_{\text{mf}}(1-\varepsilon_{\text{mf}})u_{\text{mf}}(1-\delta)}{u_o^2}\right]dz$$

$$= \frac{C_{Ai}X_A u_o^2}{\bar{K}_r L_{\text{mf}}(1-\varepsilon_{\text{mf}})u_{\text{mf}}(1-\delta)} = \frac{C_{Ai}X_A}{\bar{K}_r\tau}\,\frac{u_o}{u_{\text{mf}}(1-\delta)} \tag{43}$$

Step 2. For the shrinking-core model with diffusion through the ash layer controlling (for example) and one size of particle, we write

$$\bar{X}_B = 1 - \frac{1}{5}\frac{\tau}{\bar{t}} + \frac{19}{420}\left(\frac{\tau}{\bar{t}}\right)^2 - \cdots \tag{25}$$

where, in the $\bar{C}_A$ environment,

$$\tau = \frac{\rho_B d_P^2}{24b\mathscr{D}_s\bar{C}_A} \tag{10b}$$

Step 3. Here:

$$\begin{pmatrix}\text{disappearance}\\\text{of B from solids}\\\text{(mol/s)}\end{pmatrix} = b\begin{pmatrix}\text{disappearance}\\\text{of A from gas}\\\text{(mol/s)}\end{pmatrix} \tag{44a}$$

With M_B as the molecular weight of B and F_0 (kg/s) as the feed rate of B to the reactor, Eq. (44a) becomes

$$\left(\frac{F_0}{M_B}\right)\bar{X}_B = b(A_t u_o C_{Ai} X_A) \tag{44b}$$

Equations (41)–(44) are also applicable to feed particles of wide size distribution provided that we use Eq. (37) in place of Eq. (25).

Performance Calculations for the Fine Particle Bed (Case W)

Here the K-L model for gas and the uniform-reaction model for solids are likely to apply. Hence, the three-step procedure is as follows.

Step 1. Using the mean value of the first-order reaction rate constant $\bar{K}_r$ (see the previous case considered), we have

$$X_A = 1 - \exp\left[-K_f \frac{\delta L_f}{u_o}\right] \tag{12.16 or 45}$$

where

$$K_f = \left[\gamma_b \bar{K}_r + \cfrac{1}{\cfrac{1}{K_{bc}} + \cfrac{1}{\gamma_c \bar{K}_r + \cfrac{1}{\cfrac{1}{K_{ce}} + \cfrac{1}{\gamma_e \bar{K}_r}}}}\right] \tag{12.14 or 46}$$

In addition, the mean concentration of A bathing the solids was found by Kimura [11], after a tedious derivation, to be given by the simple plug flow expression

$$\bar{C}_A = \frac{C_{Ai} X_A u_o}{\bar{K}_r L_{mf}(1 - \varepsilon_{mf})} = \frac{C_{Ai} X_A}{\bar{K}_r \tau} \tag{47}$$

This type of expression results because gaseous reactant reacts away by a first-order reaction.

Step 2. Applying the uniform-reaction model, we have for the conversion of solids:

$$\bar{X}_B = 1 - \frac{1}{1 + k_r \bar{C}_A W/F_0} \tag{20}$$

For the shrinking-core models, $\bar{X}_B$ is given by Eqs. (22)–(26). However, as shown in Example 3, Eqs. (23) and (25) can reasonably be approximated by Eq. (20) whenever $\bar{X}_B > 0.80$. In design calculations for solids of wide size distribution this is a very useful simplification because it allows us to bypass the tedious calculations with Eqs. (35)–(37).

Step 3. Here Eq. (44) applies.

For Cases **X** and **Y** we just recombine the equations used in Cases **W** and **Z** to correspond to the situation at hand. Example 5 deals with Case **X**, Example 6 with Case **Z**.

EXAMPLE 5

Design of a Roaster for Finely Ground Ore

Zinc blende particles are fed continuously into a fluidized reactor where they are roasted at 900°C in the fluidizing air. Cinder is discharged through an overflow tube, and entrained solids are all returned to the bed from the cyclone collector. As a basis for a reasonable design, we examine various operating combinations as follows.

(a) We are aiming for a conversion of solid of 96–98%. To get an idea of the size of reactor needed, we first determine the bed diameter needed for 100% conversion of solid (B) and 66.67% conversion of oxygen (A) for a feed rate of solid of $F_0 = 172.8$ tons/day $= 7.2$ tons/hr $= 2$ kg/s.

(b) Keeping the bed diameter and flow rates of both gas and solid streams unchanged from part (a), determine by the three-step procedure X_A, $\bar{X}_B$, $\bar{C}_A$, and the time τ for complete conversion of a particle in this environment.

(c) Repeat these calculations for different feed rates of solids keeping all other values unchanged. Try

$$F_0 = 2.5 \text{ kg/s} \quad \text{(less solid than stoichiometric)}$$
$$= 3.0 \text{ kg/s} \quad \text{(the stoichiometric equivalent)}$$
$$= 3.5 \text{ kg/s} \quad \text{(more solid than stoichiometric)}$$

then decide on the flow rate of solids to use.

Data

Solid: $\bar{d}_p$ 150 μm, $\bar{\rho}_s = 4130$ kg/m^3
reaction follows the shrinking core model with $k_c = 0.015$ m/s, $\mathcal{D}_s = 8 \times 10^{-6}$ m^2/s

Gas: atmospheric, $u_0 = 0.6$ m/s, $\mathcal{D}_e = \mathcal{D} = 2.3 \times 10^{-4}$ m^2/s

Bed: $L_m = 1$ m, $u_{mf} = 0.025$ m/s, $\varepsilon_m = 0.45$, $\varepsilon_{mf} = 0.5$
$d_b = 0.20$ m (estimated from Chap. 6), $\gamma_b = 0.005$

The stoichiometry and molecular weight of the reacting components are

$$O_2 + \tfrac{2}{3} \underset{(0.09744)}{\text{ZnS}} = \tfrac{2}{3}SO_2 + \tfrac{2}{3} \underset{(0.08139)}{\text{ZnO}}$$

To remove the exothermic heat of reaction, bundles of exchanger tubes are placed in the bed. From their configuration we calculate the equivalent bed diameter to be $d_{te} = 0.4$ m. The presence of the tubes reduces the open area of bed to 85% of the reactor cross section.

SOLUTION

(a) *Extreme Calculation.* The cross-sectional area of bed needed for complete conversion of solid is obtained by a mass balance, as follows:

$$(A_t \text{ m}^2)\left(0.6 \, \frac{\text{m}}{\text{s}}\right)\left(\frac{273}{1173}\right)(0.21 - 0.07)\left(\frac{1 \text{ mol O}_2}{0.0224 \text{ m}^3}\right)$$
$$= \left(\frac{2 \text{ kg/s}}{0.09744 \text{ kg/mol ZnS}}\right)\left(\frac{3 \text{ mol O}_2}{2 \text{ mol ZnS}}\right)$$

from which the open area

$$A_t = 35.28 \text{ m}^2 = 0.85\left(\frac{\pi}{4} \, d_t^2\right)$$

Thus, the diameter of reactor with all its internals is

$$d_t = \underline{\underline{7.27 \text{ m}}}$$

(b) **Follow the Three-Step Procedure.** We use the values of A_t and W just found.

Step 1. Conversion of gas. A check of Fig. 3.9 shows that we are dealing with a bed of Geldart **B** solids. Next, we want to know which model for gas flow to use. So, from Eq. (6.7),

$$u_{br} = 0.711(9.8 \times 0.20)^{0.5} = 0.9954 \text{ m/s}$$

Now, comparing the rise velocity of bubbles relative to the velocity of gas percolating through the emulsion, we obtain

$$\frac{u_{br}}{u_{mf}/\varepsilon_{mf}} = \frac{0.9954}{0.025/0.5} = 19.9$$

This is well in excess of 6–11 to ensure that we have the fast-bubble thin-cloud K-L model of Chap. 12. We now use this.

For Geldart **B** solids apply Eq. (6.12) to get

$$u_b = 1.6[(0.6 - 0.025) + 1.13(0.20)^{0.5}](0.40)^{1.35} + 0.9954 = 1.497 \text{ m/s}$$

(For comparison, Eq. (6.9) gives $u_b = 1.575$ m/s.)

From Eqs. (6.29) and (6.20),

$$\delta \cong \frac{u_o}{u_b} = \frac{0.6}{1.497} = 0.4008$$

$$1 - \varepsilon_f = (1 - 0.4008)(1 - 0.5) = 0.2996$$

From Eqs. (6.36) and (6.35),

$$\gamma_c = (1 - 0.5)\left[\frac{3}{(0.9954)(0.5)/0.025 - 1} + 0.15\right] = 0.1544$$

$$\gamma_e = \frac{(1 - 0.4008)(1 - 0.50)}{0.4008} - 0.005 - 0.1544 = 0.5881$$

From Eqs. (10.27) and (10.34),

$$K_{bc} = \frac{4.5(0.025)}{0.20} + 5.85 \frac{(2.3 \times 10^{-4})^{0.5}(9.8)^{0.25}}{(0.2)^{1.25}} = 1.736 \text{ s}^{-1}$$

$$K_{ce} = (6.77)\left[\frac{(2.3 \times 10^{-4})(0.5)(0.9954)}{(0.2)^3}\right]^{0.5} = 0.8098 \text{ s}^{-1}$$

From Eq. (6.19) and the above calculations,

$$\frac{L_f}{u_b} = \frac{\delta L_f}{u_o} = \frac{\delta L_m(1 - \varepsilon_m)}{(1 - \varepsilon_f)u_o} = \frac{(0.4008)(1)(1 - 0.45)}{(0.2996)(0.6)} = 1.226$$

Now, for the reaction of oxygen, Eq. (12.16) gives

$$1 - X_A = \exp(-1.226K_f) \tag{i}$$

where, from Eq. (12.14),

$$K_f = \left[0.005\bar{K}_r + \cfrac{1}{\cfrac{1}{1.736} + \cfrac{1}{0.1544\bar{K}_r + \cfrac{1}{\cfrac{1}{0.8098} + \cfrac{1}{0.5881\bar{K}_r}}}} \right] \tag{ii}$$

In addition, from Eq. (47),

$$\frac{\bar{C}_A}{C_{Ai}} = \frac{X_A(0.6)}{\bar{K}_r(1)(1 - 0.45)} = \frac{1.091 X_A}{\bar{K}_r}$$

But from the ideal gas law,

$$C_{Ai} = \frac{p_{Ai}}{RT} = \frac{0.21(101325)}{(8.314)(1173)} = 2.182 \text{ mol } O_2/m^3$$

Hence,

$$\bar{C}_A = \frac{2.3804 X_A}{\bar{K}_r} \text{ mol } O_2/m^3 \tag{iii}$$

Step 2. Conversion of solids. Here we need τ and $\bar{t}$. For this first determine the molar density of feed:

$$\rho_B = \frac{4130 \text{ kg}/m^3}{0.09744 \text{ kg}/\text{mol ZnS}} = 42,385 \text{ mol ZnS}/m^3 \text{ solid}$$

Then from Eq. (11) the modified rate constant for the reaction is

$$\frac{1}{\bar{k}} = \frac{1}{0.015} + \frac{150 \times 10^{-6}}{12(8 \times 10^{-6})} = 66.67 + 1.56$$

Comparing terms shows that the reaction step at the shrinking core contributes about 98% of the overall resistance and

$$\bar{k} = 0.01466 \text{ m/s}$$

Inserting these values in Eq. (9) gives

$$\tau = \frac{(42,385)(150 \times 10^{-6})}{2(2/3)(0.01466)\bar{C}_A} = \frac{325.26}{\bar{C}_A}$$

The mean residence time of solids is obtained from Eq. (14.2):

$$\bar{t} = \frac{W}{F_0} = \frac{A_t L_m (1 - \varepsilon_m)\rho_s}{F_0}$$

$$= \frac{35.28(1)(1 - 0.45)(4130)}{2} = \frac{80,140 \text{ kg}}{2 \text{ kg/s}} = 40,070 \text{ s}$$

The conversion of solid B is given by Eq. (22):

$$\bar{X}_B = 3y - 6y^2 + 6y^3(1 - e^{-1/y}) \tag{iv}$$

where

$$y = \frac{\bar{t}}{\tau} = \frac{40,070}{325.26/\bar{C}_A} = 123.19 \bar{C}_A \tag{v}$$

Step 3. Material balance about both streams. From Eq. (44b) and the quantities already calculated, we must satisfy the following condition:

$$\frac{2}{0.09744} \bar{X}_B = \frac{2}{3}(35.28)(0.6)(2.182)X_A$$

or

$$\bar{X}_B = 1.500 X_A \tag{vi}$$

TABLE E5   Results of Example 5

Feed (kg/s)	$\bar{t}$ (s)	$\bar{X}_B/X_A$	$\bar{K}_r$ (s^{-1})	$\bar{C}_A/C_{Ai}$	τ (s)	X_A (−)	$\bar{X}_B$ (−)
2	40,070	1.5	6.50	0.110	1,360	0.661	0.991
2.5	32,060	1.2	23.0	0.0390	3,820	0.808	0.970
3	26,710	1.0	75	0.0132	11,300	0.903	0.903
3.5	22,900	0.86	145	0.0071	21,000	0.942	0.807

At this point we must find the value of $\bar{K}_r$ that will satisfy Eq. (vi). We proceed as follows:

- Guess $\bar{K}_r$.
- Insert $\bar{K}_r$ in Eq. (ii) and evaluate K_f.
- Insert K_f in Eq. (i) and evaluate X_A.
- Insert X_A and $\bar{K}_r$ in Eq. (iii) and evaluate $\bar{C}_A$.
- Insert $\bar{C}_A$ in Eqs. (iv) and (v) and evaluate $\bar{X}_B$.
- See if Eq. (vi) is satisfied. If not, repeat with different $\bar{K}_r$ values until Eq. (vi) is satisfied.

Following this procedure, we find

$$\bar{K}_r = 6.50 \text{ s}^{-1}, \quad \frac{\bar{C}_A}{C_{Ai}} = 0.110$$

$$\tau = 1360 \text{ s}, \quad X_A = 0.661, \quad \bar{X}_B = \underline{0.991}$$

These results show that the solids are very highly converted, which suggests an overdesign. Therefore, we will try higher treatment rates for solids in the same equipment.

(c) **Try Other Feed Rates of Solids.** Following the procedure of part (b) but with higher flow rates of solids gives the values summarized in Table E5. As may be seen, when the feed rate of solids exceeds stoichiometric, the conversion of solids drops rapidly. The optimal feed rate of solids to satisfy the requirement of this process is

$$F_0 = \underline{2.5 \text{ kg/s}}$$

Comment. This example uses combination **X** of models of Table 1 and is the most tedious to solve. With any other combination of models the solution becomes much simpler. The next example considers one of the other combinations.

EXAMPLE 6

Design of a
Roaster for
Coarse Ore

For Example 5 suppose the energy requirement needed to grind the ore to $d_p = 150$ μm is excessive. How would the conversion change if we use a larger feed particle of $d_p = 750$ μm while keeping all flows and bed weight unchanged: $F_0 = 2.5$ kg/s, $u_0 = 0.6$ m/s, $W = 80,140$ kg. However, with this larger particle size the minimum fluidizing velocity becomes $u_{mf} = 0.5$ m/s instead of 0.025 m/s. For additional information see the data and solution of Example 5.

SOLUTION

Selection of Models to Represent the Roaster. Figure 3.9 shows that we are dealing with a bed of large Geldart **D** particles. Next, we have to decide which flow model to use for the gas. We calculate

$$\frac{u_{br}}{u_{mf}/\varepsilon_{mf}} = \frac{0.9954}{0.5/0.5} = 0.9954 , \quad <1$$

Thus, the reactor is operating in the slow bubble regime represented by Eqs. (12.53)–(12.57). Finally, since 150-μm particles react according to the shrinking-core model, the larger particles here should also react according to this model.

This discussion shows that we have combination **Z** of models for the gas and solid. We then carry out the three-step procedure.

Step 1. From Eqs. (6.26) and (6.8),

$$\delta = \frac{u_0 - u_{mf}}{u_b + 2u_{mf}} = \frac{u_0 - u_{mf}}{(u_0 - u_{mf} + u_{br}) + 2u_{mf}}$$

$$= \frac{0.6 - 0.5}{0.6 + 0.5 + 0.9954} = 0.0477$$

From Eqs. (42) and (43),

$$X_A = 1 - \exp\left[-\bar{K}_r \frac{(1)(1 - 0.45)(0.5)(1 - 0.0477)}{(0.6)^2}\right] = 1 - \exp[-0.7275\bar{K}_r] \tag{i}$$

$$\bar{C}_A = \frac{2.182 X_A (0.6)^2}{\bar{K}_r (1)(1-0.45)0.5(1 - 0.0477)} = 3.000 \frac{X_A}{\bar{K}_r} \tag{ii}$$

Step 2. From Eq. (11),

$$\frac{1}{\bar{k}} = \frac{1}{0.015} + \frac{750 \times 10^{-6}}{12(8 \times 10^{-6})} = 66.67 + 7.81$$

Thus, mass transfer provides about 10% of the overall resistance and

$$\bar{k} = 0.0134 \, \text{m/s}$$

From Eqs. (9), (14.2), and (22),

$$\tau = \frac{(42,385)(750 \times 10^{-6})}{2(2/3)(0.0134)\bar{C}_A} = \frac{1779}{\bar{C}_A}$$

$$\bar{t} = \frac{W}{F_0} = \frac{80,140}{2.5} = 32,056 \, \text{s}$$

$$\bar{X}_B = 3y - 6y^2 + 6y^3(1 - e^{-1/y}) \tag{iii}$$

where

$$y = \frac{\bar{t}}{\tau} = \frac{32,056}{1779/\bar{C}_A} = 18.019\bar{C}_A \tag{iv}$$

Step 3. From Table E5, for $F_0 = 2.5 \, \text{kg/s}$,

$$\bar{X}_B = 1.2 X_A \tag{v}$$

By trial and error we find that the following values satisfy Eqs. (i)–(v):

$$\bar{K}_r = 2.375 \, \text{s}^{-1} \qquad \frac{\bar{C}_A}{C_{Al}} = 0.476$$

$$\tau = 1713 \, \text{s} , \quad X_A = 0.823 , \quad \bar{X}_B = \underline{0.987}$$

Comment. Comparison with the fine particle system of Example 5 shows slightly better performance with the much larger particles. The reason for this somewhat surprising result is that the gas does not pass through the bed in nearly segregated flow in bubbles with thin clouds. There is more contact with coarse solids, as shown by the fact that the mean concentration of A bathing the solids, given by $\bar{C}_A / C_{Ai}$, is 0.476 instead of 0.039, or about 12 times as high here as with fine particle systems. Thus, it pays in many ways to get out of the fine particle, fast bubble with thin cloud regime.

Miscellaneous Extensions

There are many additional topics to discuss as well as various extensions and alternatives to the simple analyses presented here. We briefly consider a few of these.

Wide Size Distribution of a Batch of Particles

Consider shrinking-core reaction-controlled kinetics. For a single particle size the rate of conversion of B in a particle can be found by differentiating Eq. (14) to get

$$\frac{dX_B}{dt} = k(1 - X_B)^{2/3} \qquad (48)$$

where

$$k = \frac{3}{\tau} = \frac{6bk_c C_A}{\rho_B d_p} \qquad (49)$$

For a wide size distribution of solids it would be useful to be able to select a representative particle size such that its conversion-time behavior would reasonably approximate that of the whole size distribution. For a batch of solids of log-normal size distribution with standard deviation σ, Kimura et al. [12] found that the expression

$$\frac{dX_B}{dt} = k(1 - X_B)^m \qquad (50)$$

with

$$m = 2/3 \quad \text{for } \sigma = 0, \text{ single size of particles}$$
$$m = 1 \quad \text{for } \sigma = 0.5$$
$$m = 1.4 \quad \text{for } \sigma = 1.0$$

reasonably represented the conversion-time behavior.

Wide Size Distribution of Feed Solids to a Fluidized Bed

For a rectangular or symmetric triangular size distribution of feed solids having a maximum size $d_{p,\max}$ and minimum size $d_{p,\min}$, Murhammer et al. [13] found that the conversion-time behavior for shrinking-core reaction-controlled kinetics is well approximated by a stream of single-size solids of size equal to the mean of these size distributions, with error limits

$$\Delta \bar{X}_B < 0.01 \qquad \text{for } d_{p,\text{max}}/d_{p,\text{min}} < 2$$

$$\Delta \bar{X}_B < 0.04 \qquad \text{for } d_{p,\text{max}}/d_{p,\text{min}} < 5$$

Numerical calculations show that similar error limits hold for shrinking-core product-layer diffusion kinetics.

Fluidized Coal Combustion

As much as 40% of the combustibles in coal can be in the form of volatile materials. As a result, the burning of coal involves gas-solid and gas-gas reactions. In addition, if the sulfur compounds in coal that are released as volatiles are to be captured in the fluidized bed by CaO solids (formed from the decomposition of limestone), then capture should occur in both the reducing atmosphere of the volatiles and in the oxidizing atmosphere of the air-rich regions of the bed. In the simplest terms, we have the gas-gas and gas-solid reactions shown in Fig. 7.

The key to the development of a reasonable model for coal combustion lies in the order of magnitude of the characteristic times for the following three phenomena—devolatilization of coal particles, dispersion of coal particles in the bed, and reaction of the particles—quantities that strongly depend on particle size. We illustrate what we mean with commercial-type coal combustors.

Since the throughput of air is what limits the duty of coal combustors, we want either a very high airflow rate (meters per second) or high-pressure operations. Also, heat exchanger tubes are present in these large units for heat removal (the desired product of this operation). These needs are met with

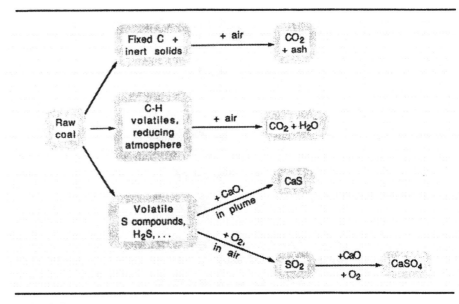

FIGURE 7
Map showing the reaction paths taken by the reacting components in a limestone-containing fluidized coal combustor.

high-velocity, large particle (Geldart **D**) beds operating near u_{mf} in which rise small cloudless bubbles. Note that in this regime of operations the fluidizing air rises in close to plug flow.

We briefly consider two examples of the modeling of such units.

Large Coal Particles Thrown onto a Large-Particle Bed. Here the characteristic times for the three factors mentioned above are roughly

$$t_{mixing} = 10\,s\,,\quad t_{devolatilization} = 100\,s\,,\quad t_{reaction\,of\,solids} = 1000\,s$$

With these different orders of magnitude, the particles first disperse uniformly throughout the oxygen-rich bed and then devolatilize, after which the carbon burns to give a shrinking particle with fine ash or a firm ash particle. Figure 8 represents the situation and shows that the fresh devolatizing particle is surrounded with burning vapors. This behavior is clearly observed.

Large Particle Bed with Fine Coal Powder Blown in from Below at a Discrete Number of Points. Here the characteristic times are approximately

$$t_{devolatilization} = 1\,s\,,\quad t_{mixing} = 10\,s\,,\quad t_{reaction\,of\,solids} = 100\,s$$

In this situation devolatilization occurs first close to the coal injection points, followed by mixing, then reaction. Since the gas rises in close to plug flow, plumes of volatiles will form in the bed, each above a feed point for the powdered coal. Thus the fluidized bed will consist of numerous reducing regions (the rising plumes) in an otherwise oxidizing bed, gas-gas reaction at the boundary between regions, and also afterburning above the bed where the plumes of volatile vapors escape the bed and react with the excess oxygen. The meandering plume model of Park et al. [14], shown in Fig. 9, represents this situation.

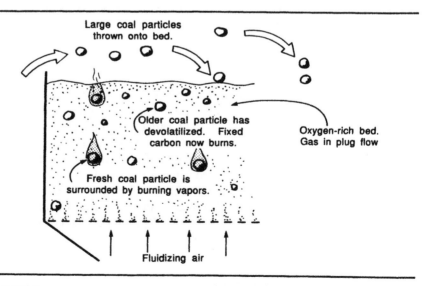

FIGURE 8
Main features of a large-particle fluidized bed combustor that is top-fed with large coal particles.

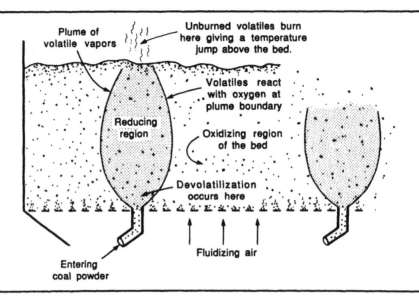

FIGURE 9
Main features of a large-particle fluidized bed combustor that is bottom-fed with fine coal powder.

Many other contacting patterns, such as circulating fast fluidized operations using extremely fine coal and limestone powder, high-pressure operations, dense bubbling beds of fine powder, and so on, can and have been explored in fluidized coal combustion. Each requires the development of an appropriate model, and the choice of assumptions that reasonably reflects that particular physical operation is of paramount importance.

Reactor-Regenerator System for Solids of Changing Size

Consider a circulation system where solids grow in one unit and shrink in the other—for example, feed oil sprayed onto hot carbon particles in a fluidized reactor. The oil cracks, and carbon deposits on the particles, which then grow. In the regenerator, carbon is burned and the particles shrink. The problem is to determine the conditions for stable operation of such a system. This then interrelates the system variables—the size distribution in the two beds, the circulation rate of solids between the beds, the bed sizes, and the required withdrawal or addition rate of solids in the system—and shows how these factors must be adjusted to achieve the requirements of the operation.

Kunii et al. [15, 16] show how to treat this system; and an extension to the case where solids are constantly withdrawn from the system is straightforward, primarily because the size distribution of solids is the same everywhere in each of the fluidized beds.

PROBLEMS

1. The reduction of iron ore pellets of density $\rho_s = 4600 \, kg/m^3$ and size $d_p = 1 \, cm$ by hydrogen can be approximated by the shrinking-core model.

With no water vapor present, the stoichiometry of the reaction is

$$4H_2 + Fe_3O_4 \rightarrow 4H_2O + 3Fe$$

with rate nearly proportional to the concentration of hydrogen in the gas stream. The rate constant k_c defined by Eq. (6) has been measured by Otake et al. [17] to be

$$k_c = 1.93 \times 10^3 \, e^{-12078/T} \quad (m/s)$$

Taking $\mathscr{D}_s = 3 \times 10^{-6} \, m^2/s$ as the average value of the diffusion coefficient for hydrogen penetration of the product blanket, calculate the time needed for complete conversion of a particle from oxide to metal in a stream of hydrogen at atmospheric pressure and 600°C.

2. Small particles of calcium carbide are fed into a 57-mm ID fluidized bed to react with pure fluidizing nitrogen to produce calcium cyanamide according to the reaction

$$CaC_2 + N_2 \rightarrow CaCN_2 + C$$

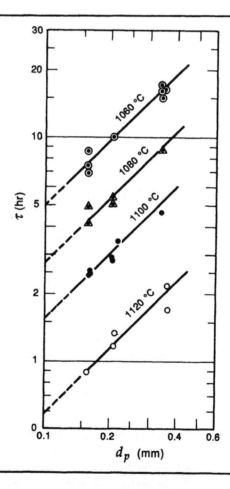

FIGURE P2

Time for complete reaction of a feed particle as a function of particle size and temperature; from [16].

The particles remain unchanged in size, and the time for complete conversion of individual particles is reported in Fig. P2.

(a) Select a model to represent the kinetics of this reaction and determine its rate constants.

(b) Design a commercial fluidized reactor to operate at 1120°C with solids of mean size $\bar{d}_p = 0.3$ mm and for a product flow rate $F_1 = 3.6$ tons/hr of 98% calcium cyanamide; thus $\bar{X}_B = 0.98$.

(c) Repeat part (b) for two equal-sized beds.

(d) Design a reactor system composed of a fluidized bed for the first stage and of a moving bed for the second stage. In the fluidized bed let $\bar{X}_B = 0.8$.

Data. For the fluidized beds take $L_m = 0.5d_t$, $\rho_s = 2290$ kg/m^3, $\varepsilon_m = 0.45$

3.   A fluidized process is planned for removing sulfur from a pyrite ore containing free sulfur. This is planned as a two-stage unit consisting of a vaporizer and roaster (see [18]). In the upper vaporizing stage a hot mixture of sulfur dioxide and nitrogen contacts fresh feed and vaporizes the free sulfur in these particles. In the lower stage the remaining sulfide is roasted in air to supply the hot gases for the upper-stage vaporization. The roasting proceeds according to the reaction

$$FeS + \tfrac{3}{2}O_2 \rightarrow FeO + SO_2$$

This problem is concerned with the design of the upper-stage vaporizer.

Experiments in an experimental vaporizer show that the time for the complete vaporization of free sulfur is $\tau = 15$ s at 400°C for $d_p = 0.22$ mm. Design the vaporizer for a treatment rate of 36 tons/hr and 99.5% removal of free sulfur.

Data. $\rho_s = 2500$ kg/m^3, $\varepsilon_m = 0.45$, $L_m = 0.3d_t$

4.   In a fluidized bed we plan to react to completion 12 tons/hr of solid B of uniform size using a large excess of gas. How large must the bed be?

Data. The stoichiometry is

$$A(gas) + B(solid) \rightarrow R(gas)$$

The time for disappearance of a particle is 1 hr, and runs at different gas velocities suggest that film diffusion is negligible.

$$\rho_s = 5000 \text{ kg/m}^3, \qquad \varepsilon_{mf} = 0.4$$

5.   In Example 6 the flow rate of solids is $F_0 = 2.5$ kg/s, and the corresponding conversion of solids is above 98%. What conversion of solids would we get if the flow rate of solids was increased to 3 kg/s, which would be stoichiometric. Would the conversion here be higher than for the $d_p = 150$-μm solids for the same treatment rate of solids?

REFERENCES

1. O. Levenspiel, *Chemical Reaction Engineering*, 2nd ed., Wiley, New York, 1972.

2. S. Yagi and D. Kunii, *Kagaku Kogaku* (*Chem. Eng. Japan*), **16**, 283 (1952); **19**, 500 (1955); *Proc. 5th Int. Symp. Combustion*, Van Nostrand Reinhold, New York, p. 231, 1955.

3. S. Kimura, S. Tone, and T. Otake, *J. Chem. Eng. Japan*, **15**, 73 (1982).

4. S. Kimura et al., *J. Chem. Eng. Japan*, **14**, 190 (1981).

5. M. Ishida and C.Y. Wen, *AIChE J.*, **14**, 311 (1968).

6. H.Y. Sohn and J. Szekely, *Chem. Eng. Sci.*, **27**, 763 (1972).

7. J.Y. Park and O. Levenspiel, *Chem. Eng. Sci.*, **30**, 1207 (1975).

8. J. Szekely and N. Themelis, *Rate Phenomena in Applied Metallurgy*, Wiley, New York, 1971.

9. O. Levenspiel, *Chemical Reactor Omnibook*, OSU Book Stores, Corvallis, OR, 1989.

10. S. Tone, K. Kawamura, and T. Otake, *Kagaku Kogaku* (*Chem. Eng. Japan*), **31**, 77 (1967).

11. S. Kimura, personal communication, 1977.

12. S. Kimura et al., *J. Chem. Eng. Japan*, **14**, 491 (1981); **16**, 217 (1983).

13. D. Murhammer, D. Davis, and O. Levenspiel, *Chem. Eng. J.*, **32**, 87 (1986).

14. D. Park, O. Levenspiel, and T.J. Fitzgerald, *Fuel*, **60**, 295 (1981); **61**, 578 (1982).

15. O. Levenspiel, D. Kunii, and T. Fitzgerald, *Powder Technol.*, **2**, 87 (1968–69).

16. D. Kunii and O. Levenspiel, *Fluidization Engineering*, Chap. 15, Krieger, Melbourne FL, 1985.

17. T. Otake, S. Tone, and S. Oda, *Kagaku Kogaku* (*Chem. Eng. Japan*), **31**, 71 (1967).

18. D. Kunii and O. Levenspiel, *Symp. on Fluidization*, *Tripartite Chem. Eng. Conf.*, Montreal, 1968.

Author Index

Subject Index

Printed and bound by CPI Group (UK) Ltd, Croydon, CR0 4YY

03/10/2024

01040333-0014